PERGAMON INTERNATIONAL LIBRARY
of Science, Technology, Engineering and Social Studies

The 1000-volume original paperback library in aid of education, industrial training and the enjoyment of leisure
Publisher: Robert Maxwell, M.C.

Histological and Histochemical Methods

Theory and Practice

Other titles of interest

BOOKS

THE MICROSCOPE PAST AND PRESENT
BRADBURY

LIVING EMBRYOS, 3rd edition
COHEN & MASSEY

THE ULTRASTRUCTURE OF THE ANIMAL CELL, 2nd edition
THREADGOLD

JOURNALS

* MICRON

CELLULAR AND MOLECULAR BIOLOGY

* Free specimen copy available on request.

Histological and Histochemical Methods

Theory and Practice

J. A. Kiernan

Department of Anatomy, The University of Western Ontario, London, Ontario, Canada

PERGAMON PRESS

OXFORD · NEW YORK · TORONTO
SYDNEY · PARIS · FRANKFURT

U.K.	Pergamon Press Ltd., Headington Hill Hall, Oxford OX3 0BW, England
U.S.A.	Pergamon Press Inc., Maxwell House, Fairview Park, Elmsford, New York 10523, U.S.A.
CANADA	Pergamon Press Canada Ltd., Suite 104, 150 Consumers Road, Willowdale, Ontario M2J 1P9, Canada
AUSTRALIA	Pergamon Press (Aust.) Pty. Ltd., P.O. Box 544, Potts Point, N.S.W. 2011, Australia
FRANCE	Pergamon Press SARL, 24 rue des Ecoles, 75240 Paris, Cedex 05, France
FEDERAL REPUBLIC OF GERMANY	Pergamon Press GmbH, 6242 Kronberg-Taunus, Hammerweg 6, Federal Republic of Germany

First edition 1981

British Library Cataloguing in Publication Data
Kiernan, J A
Histological and histochemical methods.
—(Pergamon international library).
1. Histology, Pathological—Technique
I. Title
616.07'583 RB43 80.40426

ISBN 0-08-024936-1 Hard cover
ISBN 0-08-024935-3 Flexi cover

Computer set by Page Bros (Norwich) Ltd, England and printed in Great Britain by A. Wheaton & Co., Exeter

Preface

IN THIS text it is my intention to explain the major groups of methods used in the preparation of biological specimens for examination with the optical microscope. The scientific principles involved are illustrated by detailed descriptions of selected practical procedures. The reader should be in a position to understand exactly what he is doing and why he is doing it when carrying out any of the techniques.

It is not feasible to explain everything from first principles, so I have assumed that the reader's education in science includes the subjects of chemistry and biology to standards a little below those of the Advanced Level of the General Certificate of Education in Great Britain or a little above those of first-year university courses in North America. An elementary knowledge of biochemistry will also be helpful, and the interpretation of stained sections will obviously be easier for the reader who knows some histology. The methods discussed in this book are all applicable to the tissues of vertebrate animals. Most of them may also be used for invertebrates and plants.

The number of references to the literature has been kept fairly small. Those cited fall into three categories:

(a) Books and review articles, to which the reader is referred for more information. References of this type are given both for the methods discussed in detail and for others which are mentioned only briefly.

(b) Original papers of recent date when the material is not adequately covered in books or review articles.

(c) Original papers concerned with controversial issues when the interested reader may wish to evaluate the evidence for himself.

Anyone proposing to use a histological or histochemical method in research or for any other important purpose is urged to find out as much as possible about the underlying physics and chemistry. Few of the methods used for staining tissues are fully understood, but sound scientific principles govern both the choice of the technique for a particular purpose and the practical conduct of the procedure. This book is not a collection of recipes to be followed unthinkingly: the student will have to read the "theory" in order to carry out the "practice" intelligently. Although the text is set out in such a way that it may, with advantage, be read from beginning to end, many readers will want to refer only to single chapters. Numerous cross-references are given in order to avoid repetition of explanatory material contained in the various parts of the book. No student taking a course in histological techniques could reasonably be expected to memorize the whole book. Indeed, much of the content is included solely for purposes of reference: much of the material in Chapters 5 and 20 and all the detailed practical instructions, for example, fall into this category. It is not unreasonable that the text be available for consultation by students while sitting written or practical examinations.

The "theoretical" exercises at the end of each chapter should help the student to assess his own progress. These are mostly questions requiring deductive reasoning based on the information presented. The "practical" exercises may serve as a guide to teachers in planning laboratory classes. A few of the suggested experiments involve live animals: in Great Britain these must be carried out by a person holding a Home Office licence with the appropriate certificates. Universities and colleges in most other countries have their own regulations, which should be observed.

Various colleagues have commented on parts of the manuscript. I am especially grateful to Drs R. C. Buck, M. G. Cherian, B. A. Flumerfelt, P. Haase, and J. R. Sparrow. Their advice has resulted in several improvements to the text, but I remain responsible for any errors or failures to explain subjects clearly enough. Comments on the content and style of this book will be welcomed, especially from those readers who are making their first encounters with practical histology and histochemistry. My own understanding of these subjects has been acquired from having to use the methods in research, which has been supported by grants from several organizations. Thanks are due to the British Medical Research Council and the National Fund for Research into Crippling Diseases, and to the J. P. Bickell Foundation (Toronto), the Canadian Medical Research Council, the Multiple Sclerosis Society of Canada, and the Ontario Thoracic Society.

London, Ontario J. A. KIERNAN

Contents

Each chapter is preceded by a detailed table of contents.

Conventions and abbreviations		ix
Chapter	**1.** Introduction to microtechnique	1
	2. Fixation	8
	3. Decalcification	25
	4. Processing and mounting	29
	5. Dyes	38
	6. Histological staining in one or two colours	77
	7. Blood stains	88
	8. Methods for connective tissue	92
	9. Methods for nucleic acids	104
	10. Organic functional groups and protein histochemistry	114
	11. Carbohydrate histochemistry	145
	12. Lipids	169
	13. Methods for inorganic ions	190
	14. Enzyme histochemistry: general considerations	201
	15. Hydrolytic enzymes	206
	16. Oxidoreductases	220
	17. Methods for amines	245
	18. Neurohistological techniques	256
	19. Immunohistochemistry	280
	20. Miscellaneous data	295
Glossary		305
Bibliography		309
Index		325

Conventions and Abbreviations

CONVENTIONS

It is important that the reader be familiar with the conventions listed here before attempting to follow the instructions for any practical procedure.

[ ] (i) Enclose a complex, such as $[Ag(NH_3)_2]^+$ or $[PdCl_4]^{2-}$.
(ii) Indicates "concentration of", in molar terms. Thus, $[Ca^{2+}]^3$ = "the cube of the molar concentration of calcium ions".

Accuracy. Solids should be weighed and liquids measured to an accuracy of ±1%. With quantities less than 10 mg or 1.0 ml, an accuracy of ±10% is acceptable.

Alcohol. Unqualified, this word is used for methanol, ethanol, isopropanol, or industrial methylated spirit (which is treated as 95% v/v). When the use of a specific alcohol is necessary, this is stated. "Absolute" refers to the commercially available "100%" ethanol, which really contains about 1% water and may also be contaminated with traces of benzene. Absolute ethanol is hygroscopic and should be kept in securely capped bottles, which should be only three-quarters filled to allow for evaporation. In an ordinary covered staining tank, ethanol does not remain acceptably "absolute" for more than about 5 days.

When diluting alcohols for any purpose, use distilled or de-ionized water.

Concentrations expressed as percentages. The symbol % is used in various ways:

(i) For solids in solution, % = grams of solid dissolved in 100 ml of the final solution.
(ii) For liquids diluted with other liquids, % = number of millilitres of the principal component present in 100 ml of the mixture, the balance being made up by the diluent (usually water). "70% ethanol" means 70 ml of absolute ethanol made up to 100 ml with water.
(iii) For gases (e.g. formaldehyde), % = grams of the gas contained in 100 ml of solution.
(iv) Where doubt may arise, the symbol v/v, w/v, or w/w is appended to the % sign. For dilution of acids and ammonia, see Chapter 20.

Formalin. This word refers to the commercially obtained solution containing 37–40% (w/v) of formaldehyde in water. The shortened form "formal" is used in the names of mixtures such as formal–saline and formal–calcium. The term "formol" will be found in some books, but this is wrong because the ending -ol suggests, incorrectly, that formaldehyde is an alcohol.

Safety precautions. The precautions necessary in any laboratory, especially the prevention of fire, should be observed at all times. Some reagents used in histology and histochemistry have their special hazards. These are mentioned as they arise in the text.

Concentrated mineral acids (especially sulphuric) must be diluted by adding acid to water (**not** water to acid) slowly with stirring.

Formaldehyde and hydrochloric acid should not be thrown down a sink together: their vapours can react together in the air to form *bis*-chloromethyl ether, a potent carcinogen. Each substance should be flushed down the drain separately, with copious running tap water.

Concentrated nitric acid must not be allowed to come into contact with organic liquids, especially alcohol: the strongly exothermic reaction may result in an explosion.

Salts—water of crystallization. The crystalline forms of salts are shown in instructions for mixing solutions. If the form stated is not available, it will be necessary to calculate the equivalent amount of the alternative material. This is simply done by substitution in the formula

$$\frac{W_1}{M_1} = \frac{W_2}{M_2}$$

in which W = weight, M = molecular weight, and subscripts 1 and 2 refer to the prescribed and the alternative compounds respectively.

For example: 125 mg of cupric sulphate ($CuSO_4$) are prescribed, but only the hydrated salt, $CuSO_4.5H_2O$, is available. Molecular weights are 223.14 and 249.68 respectively. Then

$$\frac{125}{223.14} = \frac{W_2}{249.68}$$

$$W_2 = \frac{125 \times 249.68}{223.14}$$

$$= 139.9$$

It will therefore be necessary to use 139.9 (i.e. 140) mg of $CuSO_4.5H_2O$ in place of 125 mg of the anhydrous salt.

Solutions. If a solvent is not named (e.g. "1% silver nitrate"), it is assumed to be water. See also under **Water** below.

Structural formulae. Aromatic rings are shown as Kekulé formulae, with alternating double bonds. Thus benzene is

Modern practice favours the designation

which indicates the equivalence of all the bonds in the ring. The Kekulé formulae are used in this book because, with them, it is easier to understand structural changes associated with the formation of coloured compounds, such as the formation of quinonoid from aromatic rings.

A few deviations from standard chemical notation (e.g. in formulae for lipids) are explained as they arise.

Temperature. Unless otherwise stated, all procedures are carried out at room temperature, which is assumed to be 15–25°C. The other commonly used temperatures are 37°C and about 60° C. A histological laboratory should have ovens or incubators maintained at these temperatures.

If an oven containing melted paraffin wax is used as a 60°C incubator, make sure that any aqueous or alcoholic solutions put in it are covered. Water or alcohol vapour may otherwise contaminate the wax.

Water. *When "water" is prescribed in practical instructions, it means distilled or deionized water.* When water from the public supply may be used, it is specifically mentioned as "tap water".

ABBREVIATIONS

α, β — (i) Used to indicate the configuration at position C1 in glycosides (see Chapter 11).
(ii) In aliphatic compounds the α-carbon atom is adjacent to the carbon atom bearing the principal functional group (i.e. α is carbon number 2). The use of numbers and Greek letters is shown below for *n*-hexanol:

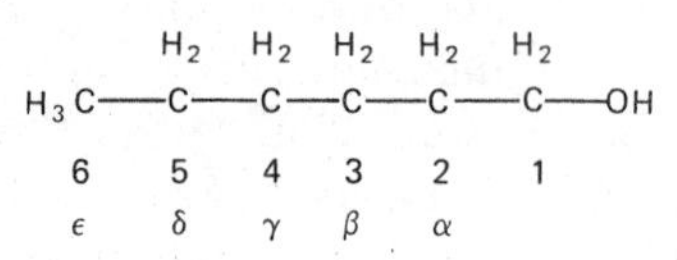

(iii) In glycerol and its derivatives, the middle carbon atom is designated as β and the carbons on either side as α and α'.
(iv) In derivatives of naphthalene, indicate the position of a substituent relative to the site of fusion of the rings:

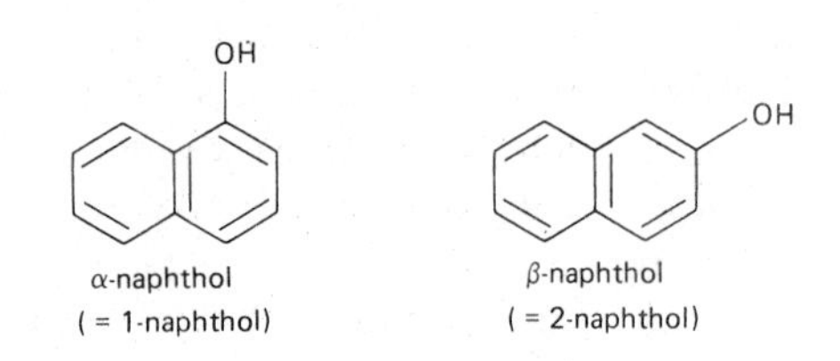

Δ	Symbol used to indicate double bonds in lipids (see Chapter 12).
ε-	Indicates carbon number 6 or a substituent on this atom, as in the case of the amino group at the end of the side-chain of lysine.
µg	microgram (= 10^{-6} g or 10^{-3} mg).
µm	micrometre (= 10^{-6} m or 10^{-3} mm); also sometimes called a "micron".
ACh	Acetylcholine.
AChE	Acetylcholinesterase.
APUD	Amine precursor uptake and decarboxylation.
Ar	An aryl radical (in formulae).
AThCh	Acetylthiocholine.
ATP	Adenosine triphosphate.
ATP-ase	Adenosine triphosphatase.

bis-	Twice (in names of compounds).
B.P.	British Pharmacopoeia; Boiling point.
BuThCh	Butyrylthiocholine.
B.W. 284C51	1,5-*bis*-(4-diallyldimethylammonium-phenyl)-pentan-3-one. (The initials are for Burroughs Wellcome, the manufacturer).
°C	Degrees Celsius (Centigrade).
ChE	Cholinesterase (= pseudocholinesterase).
C.I.	*Colour Index* (see Chapter 5).
cis-	Indicates a geometrical isomer in which two substituents lie on the same side of the molecule.
CNS	Central nervous system.
Con A	Concanavalin A.
CPC	Cetylpyridinium chloride.
cyt.	Cytochrome (with identifying letter, a, b, c, etc.).
D-	Indicates a compound, usually a sugar, of the D-series. The compound itself is not necessarily dextrorotatory.
DAB	3,3′-diaminobenzidine.
dansyl	The 5-(dimethylamino)-1-naphthalenesulphonyl radical.
DFP	Diisopropylfluorophosphate.
DMAB	*p*-dimethylaminobenzaldehyde (also known as Ehrlich's reagent).
DMP	2,2-dimethoxypropane.
DNA	Deoxyribonucleic acid.
DN-ase	Deoxyribonuclease.
DOPA	*β*-3,4,dihydroxyphenylalanine.
DPN	Diphosphopyridine nucleotide (an obsolete name for NAD^+, q.v.).
DPX	A resinous mounting medium. The initials stand for its three components, distrene (a polystyrene), a plasticizer, and xylene.
E_0, E_0'	Symbols for oxidation–reduction potentials (see Chapter 16).
E600	Diethyl-*p*-nitrophenyl phosphate.
E.C.	Enzyme Commission (see Chapter 14).
EDTA	Ethylenediaminetetraacetic acid. Also known as versene, sequestrene, edetic acid, and (ethylenedinitrilo)-tetraacetic acid. Usually used as its disodium salt.
Fab	Part of the immunoglobulin molecule (see Chapter 19).
FAD	Flavin adenine dinucleotide (see Chapter 16).
FAGLU	A fixative containing formaldehyde and glutaraldehyde (see Chapter 17).
Fc	Part of the immunoglobulin molecule (see Chapter 19).
FITC	Fluorescein isothiocyanate.
FMN	Flavin mononucleotide (see Chapter 16).
GBHA	Glyoxal-*bis*-(2-hydroxyanil).
H-acid	8-amino-1-naphthol-3,6-disulphonic acid.
H-chain	Part of the immunoglobulin molecule (see Chapter 19).
H. & E.	Haematoxylin and eosin (see Chapter 6).
HNAH	2-hydroxy-3-naphthoic acid hydrazide. May equally be named 3-hydroxy-2-naphthoic acid hydrazide.
HRP	Horseradish peroxidase.
IgG	Immunoglobulin-G.
L-	Indicates a compound (usually a sugar or an amino acid) of the L-series. The compound itself is not necessarily laevorotatory.
LVN	Low viscosity nitrocellulose.
M	(as in 0.1 M) Molar (= moles per litre).
m-	*meta-* (in names of benzene derivatives).
MBTH	3-methyl-2-benzothiazolone hydrazone.
mole	The molecular weight, expressed in grams.
M.W.	Molecular weight.
N	(as in 0.1 N) Normal (= gram-equivalents per litre; see Chapter 20).
N_A	Avogadro's number: 6.022×10^{23} molecules per mole.
n-	Normal, indicating an unbranched chain, as in *n*-butanol.
NAD^+	Nicotinamide adenine dinucleotide.
NADI	Naphthol-diamine (reaction).
$NADP^+$	Nicotinamide adenine dinucleotide phosphate.
NANA	*N*-acetylneuraminic acid.
Nitro-BT	Nitro blue tetrazolium.
nm	nanometre (= 10^{-9} m or 10^{-3} μm).
NQS	1,2-naphthoquinone-4-sulphonic acid.
NSM	Neurosecretory material.

o-	*ortho-* (in names of benzene derivatives).
OPT	*o*-phthaldialdehyde.
p-	*para-* (in names of benzene derivatives).
PAN	Perchloric acid–naphthoquinone (method; see Chapter 12).
PAP	Peroxidase–antiperoxidase (reagent; see Chapter 19).
PAS	Periodic acid–Schiff (method; see Chapter 11).
PCMB	*p*-chloromercuribenzoate.
pH	The logarithm (to base 10) of the reciprocal of the molar concentration of hydrogen ions.
PMA	Phosphomolybdic acid.
pg	picogram (= 10^{-12} gram).
PMS	Phenazine methosulphate.
PNS	Peripheral nervous system.
PTA	Phosphotungstic acid.
PVA	Polyvinyl alcohol.
PVP	Polyvinylpyrrolidone (also called "povidone").
R, R′	Indicate alkyl or aryl radicals, in formulae.
RNA	Ribonucleic acid.
RN-ase	Ribonuclease.
S.G.	Specific gravity (also = density, in g per cm^3).
Susa	A fixative mixture introduced by M. Heidenhain: short for *Sublimat-Säure* (see Chapter 2).
t-	Tertiary, as in *t*-butanol: $(CH_3)_3COH$.
TCA	Trichloroacetic acid.
TMB	Tetramethylbenzidine.
TPN	Triphosphopyridine nucleotide (an obsolete name for $NADP^+$, q.v.; see Chapter 16).
TPP	Thiamine pyrophosphate.
TPP-ase	Thiamine pyrophosphatase.
trans-	Indicates a geometrical isomer in which two substituents lie on opposite sides of the molecule.
TRIS	*Tris*(hydroxymethyl)aminomethane.
UDPG	Uridine-5-diphosphate glucose.
UQ	Ubiquinone.
U.S.P.	United States Pharmacopoeia.
v/v	Volume ÷ volume.
w/v	Weight ÷ volume.
w/w	Weight ÷ weight.

SPECIALIZED ABBREVIATIONS

The following are used only for special purposes and are explained in the appropriate parts of the text.

Chapter 11. Monosaccharide residues: Fuc, Gal, GalNAc, Glc, GlcNAc, GlcUA, IdUA, Man, Xyl. Lectins: A, DBA, LA, LCA_1, PHA, RCA_1, RCA_{120}, SBA, UEA_1, WGA.

Chapter 16. Tetrazolium salts: BSPT, BT, DS-NBT, INT, MTT, Nitro-BT, NT, TNBT, TTC, TV.

Chapter 17. Biogenic amines: ADR, DA, HIS, 5HT, NA.

1

Introduction to Microtechnique

1.1. Thickness and contrast 1
1.2. Staining and histochemistry 2
1.3. Some physical considerations 2
1.4. Properties of tissues 3
1.5. Books and journals 4
1.6. On carrying out instructions 4
1.6.1. De-waxing and hydration of paraffin sections 5
1.6.2. Staining 5
1.6.3. Washing and rinsing 6
1.6.4. Dehydration and clearing 6
1.6.5. Understanding the methods 7
1.7. Exercises 7

Many theoretical explanations and practical instructions are contained in this book. The present chapter concerns various aspects of the making of microscopical preparations that are fundamental to all the techniques described in the later chapters. It cannot be over-emphasized that unless the student or technician understands the rationale of all that is to be done, he will not do it properly. Some purely practical information relevant to the manipulations discussed here will be found in Chapter 4 as well as in Section 1.6 of this chapter.

1.1. THICKNESS AND CONTRAST

In order to be examined with the microscope, a specimen must be sufficiently thin to be transparent and must possess sufficient contrast to permit the resolution of structural detail. Thinness may be an intrinsic property of the object to be examined. Thus, small animals and plants, films and smears of cells, macerated or teased tissues, and spread-out sheets of epithelium or connective tissue are all thin enough to be mounted on slides directly as **whole mounts**. In the study of histology and histochemistry, one is more often concerned with the internal structure of more substantial, solid specimens. These must be cut into thin slices or **sections** in order to make them suitable for microscopical examination.

Freehand sections, cut with a razor, are rarely used in animal histology but are still sometimes employed for botanical material. Though some expertise is necessary, sectioning in this way has the advantage of requiring little in the way of time or special equipment. When sections of human or animal tissues are needed in a hurry, **frozen sections** are commonly used. The ordinary **freezing microtome** is used for fixed material, while the **cryostat**, essentially a microtome mounted in a freezing cabinet, may be used for cutting sections of fixed or unfixed tissue. More skill is called for in the operation of a cryostat than for a freezing microtome. Another advantage of cutting frozen sections, aside from speed, is the preservation of some lipid constituents, which are dissolved out during the course of dehydration and embedding in paraffin wax. A disadvantage of the freezing microtome is that the sections are commonly too thick (e.g., 20–100 μm) for the resolution of fine structural detail within cells. Much thinner sections can be obtained by using a cryostat. The **vibrating microtome** (Vibratome) can be used to cut sections of unfixed, unfrozen specimens. The blade of this instrument passes with a sawing motion through a block of tissue immersed in an isotonic saline solution. The cutting process is much slower than with other types of microtome, so it is not feasible to prepare very large numbers of sections.

When the preservation of lipids and of heat-labile substances such as enzymes is not important, the specimens are **dehydrated**, **cleared** (which means, in this context, equilibrated with a solvent which is miscible with paraffin), **infiltrated** with molten paraffin wax, and, finally, **embedded** (blocked out) in solidified wax. **Paraffin sections** are most commonly cut on a **rotary microtome**, though a **rocking microtome** or a **sledge microtome** can also be used. The sections come off the knife in ribbons, and with sufficient skill it is possible to obtain **serial sections**, which may be as little as 4 μm thick, through the whole block of tissue. **Polyester wax** (Steedman, 1960), which is miscible with 95% alcohol, is handled in much the same way as paraffin. Its lower melting point (about 40°C as compared with 55–60°C for most paraffin waxes) is an advantage for some tissues and histochemical methods. When relatively thick sections of large specimens are required, it is more convenient to embed in celloidin or low-viscosity nitrocellulose. **Celloidin sections**, 50–200 μm thick, are usually cut on a sledge microtome. Various **synthetic resins** are sometimes used as embedding media for light microscopy, though their main application is in the cutting of extremely thin sections for examination in the electron microscope. Resin-embedded tissue is usually sectioned with an **ultramicrotome**, using a glass or diamond knife. Sections 0.5–1.0 μm thick, suitably stained for optical examination, are valuable for comparison with the much thinner sections used in ultrastructural studies.

The optical contrast in a thin specimen is determined partly by its intrinsic properties but largely by the way in which it is processed. If the specimen is not stained, contrast will be greatest when the mounting medium has a refractive index substantially different from that of the specimen. Differences of this type are emphasized in the **phase-contrast microscope**. This instrument is especially valuable for the study of living cells, such as those grown in tissue-culture. In histology, the natural refractility of a tissue is usually deliberately suppressed by the use of a mounting medium with a refractive index close to that of the anhydrous material constituting the section (approximately 1.53). Almost all the contrast is produced artificially by **staining**.

Fluorescence is the property exhibited by substances which absorb light of short wavelength such as ultraviolet or blue and emit light of longer wavelength, such as green, yellow, or red. The phenomenon can be observed with a **fluorescence microscope** in which arrangements are made for the emitted (long wavelength) light to reach the eye while the exciting (short wavelength) light does not. Fluorescing materials therefore appear as bright objects on a dark background. The fluorescence microscope can be used to observe **autofluorescence** due to substances naturally present and **secondary fluorescence** produced by appropriate chemical treatment of the specimen. The autofluorescence of a tissue, which is due to various endogenous compounds, notably flavoproteins (Benson *et al.*, 1979), frequently interferes with the interpretation of secondary fluorescence.

1.2. STAINING AND HISTOCHEMISTRY

The histologist stains sections in order to facilitate the elucidation of structural details. The histochemist, on the other hand, seeks to determine the locations of known substances within the structural framework. The disciplines of histology and histochemistry overlap to a large extent, but one consequence of the two approaches is that the staining techniques used primarily for morphological purposes are sometimes poorly understood in chemical terms. Histochemists, not wanting to be thought of as mere dyers, frequently avoid calling their procedures "staining methods". Nonetheless, the science of histochemistry has contributed several techniques to the art of histology: it is obviously most desirable to be able to demonstrate a structural component by "staining" for a substance which it is known to contain. Conversely, there are many empirically derived histological techniques of immense value which do not depend upon well-understood chemical principles.

1.3. SOME PHYSICAL CONSIDERATIONS

The intelligent handling of microscopical preparations requires familiarity with the physical properties of several materials which are used in almost all techniques. All too often, the beginner will ruin a beautifully stained section by forgetting that two solvents are immiscible, or by leaving the slides overnight in a liquid which dissolves the coloured product. The following remarks relate mainly to sections mounted on slides, but are also applicable to blocks of tissue or to smears, films, wholemounts, or free-floating sections.

Water is completely miscible with the common alcohols (methanol, ethanol, isopropanol, methylated ethyl alcohol). Water is immiscible with xylene, benzene, chloroform, and most other clearing agents. These clearing agents are, however, miscible with the alcohols in the absence of water. Melted paraffin wax and the resinous mounting media (Canada balsam, Xam, Permount, DPX, etc.) are miscible with the clearing agents but not with the alcohols or with water. One mounting medium, euparal, is notable for being miscible with absolute alcohol as well as with xylene. Because of these properties of the common solvents, a specimen must be passed through a **series** of liquids during the course of embedding, staining, and mounting for examination.

For example, a piece of tissue removed from an aqueous fixative, such as a formaldehyde solution, must pass through a **dehydrating agent** (such as alcohol) and a **clearing agent** (such as chloroform) before it can be infiltrated with paraffin wax. Ribbons of paraffin sections may be floated on warm water, which will remove wrinkles, when they are being mounted on glass slides. A thin layer of a suitable **adhesive** (see Chapter 4) may be interposed between the slide and the sections, but this is not always necessary. The slides must then be **dried** thoroughly in warm air before being placed in a **clearing agent**, usually xylene, to dissolve and remove the wax. The slides will now bear sections of tissue which are equilibrated with the clearing agent. Passage through alcohol (or any other solvent miscible with both xylene and water) must precede immersion of the slides in water. Sudden changes are avoided if possible, so a series of graded mixtures of alcohol with water is used. Most staining solutions and histochemical reagents are aqueous solutions. If a permanent mount in a resinous medium is required, the slides carrying the stained sections must be **dehydrated**, without unintentionally removing the stain, in alcohol or a similar solvent, **cleared** (usually in xylene), and, finally, **mounted** in the resinous medium.

The physical properties of some commonly used solvents are given in Chapter 4, where some practical guidance for their use will also be found.

Since resinous media are themselves dissolved in clearing agents, a mounted preparation will not be completely transparent for a few hours. The resin has to permeate the section and the solvent has to evaporate at the edges of the coverslip. When these events have taken place, the specimen will be equilibrated with the mounting medium and should have almost the same refractive index as the latter. Consequently, most of the observed contrast will be due to the staining method.

Frozen sections are collected into water or an aqueous solution. They may be affixed to slides and dried in the air either before or after staining. The frozen section on the slide is, therefore, at first equilibrated with water and must be dehydrated and cleared before mounting in a resinous medium. If the products of staining would dissolve in organic solvents, as is the case with the Sudan dyes and with the end-products of some histochemical reactions, it is necessary to use a water-miscible mounting medium. Several such media are available (e.g. glycerol jelly, Apathy's, Farrant's, polyvinylpyrrolidone) but they usually do not suppress the intrinsic refractility of the specimen as completely as do the resins.

Celloidin sections require special handling owing to the properties of the embedding medium. Nitrocellulose is soluble in a mixture of equal volumes of ethanol and diethyl ether, commonly called ether–alcohol. Celloidin is hardened by 70% alcohol, chloroform, or phenol. Absolute alcohol makes celloidin swell but does not dissolve it. Aqueous solutions penetrate freely through the matrix of nitrocellulose, so it is not necessary to remove the latter in order to stain the contained section of tissue. Molten paraffin wax can also permeate a nitrocellulose matrix without dissolving it. Specimens which are expected to be difficult to section are often infiltrated with celloidin, cleared, and then infiltrated with and embedded in wax. This procedure is known as **double embedding**.

More infrequently used embedding media are the water-soluble polyethylene glycol waxes and glycol methacrylate resins. The specialized techniques for the use of these substances are also based on simple physical principles, but will not be discussed here.

Many of the dyes used in histology can be removed from stained sections by alcohol. This property is useful for the extraction of excesses of dye, a process known as **differentiation**, but it can also be a nuisance. Since a stained preparation must be completely dehydrated as well as adequately differentiated, the timing and rate of passage through graded alcohols is often critical. It is one of the arts of histological technique to obtain the correct degree of differentiation.

1.4. PROPERTIES OF TISSUES

Freshly removed cells and tissues, especially those of animals, are chemically and physically unstable. The treatments to which they are exposed in the making of microscopical preparations would damage them severely if they were not stabilized in some way. This stabilization is usually accomplished by **fixation**, which is discussed in Chapter 2. For some purposes, especially in enzyme histochemistry, it is necessary to use sections of unfixed tissues. As has already been stated, such sections may be cut with a cryostat or a vibrating microtome. Unfixed sections are stable when dried onto glass slides or coverslips but become labile again when wetted with aqueous liquids that do not produce fixation. Many histological staining methods do not work properly on unfixed tissues.

Most methods of fixation make the tissues harder than they were in the living state. Provided that it is not excessive, hardening is advantageous because it renders the tissues easier to cut into sections. However, some tissues such as bone are too hard to cut even before they have been fixed. These have to be softened after fixation but before dehydration, clearing, and embedding. Calcified tissues are softened by dissolving out the inorganic salts that make them hard, a procedure known as **decalcification** (see Chapter 3). Other hard substances such as cartilage, chitin, and wood require different treatments. Exceptionally robust microtomes equipped with massive chisel-like knives of tungsten steel have become available in recent years. These will cut sections of undecalcified bone and other hard tissues.

Even initially soft specimens sometimes become unduly hard by the time they are embedded in wax. These can be softened by cutting sections to expose the interior of the tissue at the face of the block and then immersing for a few hours in water. Although the solid wax is present in all the interstices of the tissue, materials such as collagen can still imbibe some water and be made much softer. Various proprietary "softening agents" are marketed for the same purpose, but they are, in my experience, no better than plain water. Another important factor in microtomy is the hardness of the embedding mass relative to that of the tissue. This is determined by the composition of the former and by the ambient temperature. Obviously, the proper use of the microtome is also necessary if satisfactory sections are to be cut.

1.5. BOOKS AND JOURNALS

There is a profusion of books, large and small, that give directions in practical microtechnique. Some of the more modern ones also briefly explain the rationales of the different methods described. The texts of Baker (1966), Humason (1972), Bradbury (1973), Culling (1974), Cook (1974), and Bancroft & Stevens (1977) can all be recommended, but there are many others equally valuable. Some of them will be cited in the following chapters. Books such as these should always be at hand for the practising microscopist. For histochemistry, the major treatise in English is that of Pearse (1968, 1972); other important ones are Barka & Anderson (1963), Thompson (1965), and Lillie & Fullmer (1976). In these works it is usually assumed that the reader is familiar with the chemical principles underlying the explanations of how the methods work.

Some journals are devoted largely to the publication of papers on methodology. The major ones are *Stain Technology*, the *Journal of Histochemistry and Cytochemistry*, *Histochemistry* (formerly *Histochemie*), the *Histochemical Journal* and the *Journal of Microscopy* (formerly *Journal of the Royal Microscopical Society*). Relevant papers appear in other journals too, but by scanning the ones listed above it is not difficult to keep up with the major advances in the field.

1.6. ON CARRYING OUT INSTRUCTIONS

THIS IS IMPORTANT. READ THIS SECTION BEFORE ATTEMPTING TO PERFORM ANY OF THE TECHNIQUES DESCRIBED IN LATER CHAPTERS. See also "Conventions and Abbreviations" for methods used to express concentrations of solutions and for the correct interpretation of such terms as "alcohol" and "water" and for guidance on accuracy of measurement of weight, volume, and temperature.

In this and other texts, practical schedules are given for many techniques. Since the number of methods described in this book is relatively small, it has been possible to be quite explicit, so that the methods should all work properly if the instructions are followed exactly. There are, however, some general rules applicable to nearly all staining methods. These will therefore be given now in order to avoid tedious repetition in the following chapters.

1.6.1. De-waxing and hydration of paraffin sections

Place the slides (usually 4–12 of them) in a glass or stainless steel rack and immerse in a rectangular tank containing about 400 ml of xylene. This is the most useful size of tank for most purposes. Smaller ones are available, but when they are used their contents must be renewed more often. A "commercial" or "technical" grade of xylene (mixed isomers) is satisfactory. Agitate the rack, up and down and laterally, three or four times over the course of 2–3 min. If for some reason it is inconvenient to agitate the slides, they should be left to stand in the xylene for at least 5 min. A single slide is de-waxed by moving it slowly back and forth in the tank of xylene for 1 min. Individual slides should be held with stainless steel forceps.

Lift the rack (or individual slide) out of the xylene, shake it four or five times and touch it onto bibulous paper (three or four thicknesses of paper towel, or filter paper) and place in a second tank of xylene. Agitate as described above, but this time 1 min is long enough. The purpose of this second bath of xylene is to remove the wax-laden xylene from the initial bath, thereby reducing the risk of precipitation of wax upon the sections when they are passed into alcohol, in which wax is insoluble. The removal of excess fluid by shaking and blotting is very important and is done every time a rack or slide is passed from one tank to another. If it is not done, the useful life of each tankful of xylene or alcohol will be greatly shortened. The instruction "drain slides" refers to this shaking-off of easily removed excess liquid.

After the second bath of xylene, drain the rack of slides and place in a tank containing about 400 ml of absolute ethanol. Agitate at intervals of 10–20 s for $1\frac{1}{2}$–$2\frac{1}{2}$ min. Drain slides and transfer to 95% ethanol and agitate in this for about 1 min. Drain slides and transfer to 70% alcohol. Agitate for at least 1 min. For an individual slide, it is sufficient to move it about with forceps for about 20 s in each change of alcohol. If the slides have to be left for several hours, or even for a few days, they should be immersed in 70% alcohol. This will prevent the growth of fungi and bacteria on the sections but will not make them come off the slides or become unduly brittle. If some sections do detach from the slides during de-waxing or hydration, more will certainly be lost in later processing. If attachment of the sections appears to be precarious, a film of nitrocellulose should be applied, as described in Chapter 4.

Hydration of the sections is completed by lifting the rack (or individual slide) out of the 70% alcohol, draining it, and immersing in water. Agitation for at least 30 s is necessary for removal of the alcohol. Without agitation, this takes 2–3 min. A second rinse in water is desirable if all traces of alcohol are to be removed.

It is possible to use small volumes of xylene and alcohol by carrying out the above operations in coplin jars (which usually hold up to five slides) or rectangular staining dishes (usually for 8–12 slides). The xylenes and alcohols are poured into these vessels and the slides agitated continuously with forceps. To change the liquid, pour it out (without losing the slides) and replace it with the next one in the series. Working in this way, each lot of xylene or alcohol should be used only once. When tanks holding 400 ml are used, the liquids can be used repeatedly. They should all be renewed when traces of white sludge (precipitated wax) appear in the absolute ethanol. This commonly occurs after 10–12 racks of slides have been de-waxed and hydrated. In order to minimize evaporation, contamination by water vapour and the risk of fire, all tanks containing alcohol or xylene should have their lids on when not in use.

1.6.2. Staining

An instruction such as "stain for 5 min" means that the sections must be in intimate contact with the dye solution for the length of time stated. Slides (alone, or in racks, coplin jars or staining dishes, as convenient) are immersed in the solution, agitated for about 10 s, and then left undisturbed. Free-floating frozen sections are transferred to cavity-blocks, watch-glasses, or the wells of a haemagglutination tray containing the staining solution. Folds and creases in such sections must be straightened out if uniform penetration of the dye is to occur. The best instrument for handling frozen sections is a glass hook or "hockey stick", fashioned by drawing out a piece of glass rod in the flame of a Bunsen burner. Sections of nitrocellulose-embedded material, if not affixed to slides, are manipulated in the same way as frozen sections.

In some techniques of enzyme histochemistry and immunocytochemistry, only one drop of a scarce or costly reagent can be applied to each section. This is done with the slide lying horizontally. The slide bearing the section, covered by the drop, is placed on wet filter paper in a closed petri dish: the drop will not evaporate if the air above is saturated with water vapour. Horizontal slides are also used in staining methods for blood films. In this case the reagents are not expensive, so the slides are placed, film upwards, on a pair of glass rods over a sink. The staining solution is poured on to flood the slides and later washed off by a stream of water or of a suitable buffer.

Since sections of tissue take up only minute quantities of dyes and other substances, there is no need for the volume of staining solution or histochemical reagent to be any greater than that required to cover the sections. Exceptions to this general rule are rare and are mentioned in the instructions for the methods concerned.

1.6.3. Washing and rinsing

The excess of unbound dye or other reagent is removed from the stained sections by washing or rinsing, usually with water. A "wash" is a more prolonged and vigorous treatment than a "rinse". Agitation of slides for a minute or more in each of three changes of water constitutes an adequate wash. When tap water is suitable, the slides are placed for about 3 min in a tank through which the water is running quickly enough to produce obvious turbulence. A rack of slides should be lifted out of the running tap water and then replaced every 20–30 s in order to ensure that all the slides are thoroughly washed. A rinse, rather than a wash, is prescribed when excessive exposure to water would remove some of the dye specifically bound to the sections. Rinsing is done in the same way as washing, but the slides are agitated continuously and the total time of exposure to water is only about 15 s.

With unmounted frozen or nitrocellulose sections, it is more difficult to control the process of washing. The sections are carried through three successive baths (50 ml beakers are convenient) of water and are kept in constant motion for 20–40 s in each. Free-floating sections should not be allowed to fold or to crumple into little balls. Stains that are easily extracted by water should not be applied to sections of this kind.

1.6.4. Dehydration and clearing

For stains or histochemical end-products insoluble in water and alcohol, dehydration is a simple matter. The slides are agitated continuously for about 1 min in each of the following: 70% alcohol, 95% alcohol; two changes of absolute ethanol, methanol, or isopropanol. There is no objection to taking them straight into 100% alcohol (in which case, three or four changes will be needed), but the use of lower alcohols will protect the more expensive anhydrous liquids from excessive contamination with water. **Slides must be drained** (see Section 1.6.1) as they are transferred from one tank to the next. Clearing is accomplished by passing the slides from absolute alcohol into xylene (two changes, 1 min with agitation in each). They may remain in the last change of xylene for several days if mounting in a resinous medium cannot be carried out immediately. Used as described the alcohols and xylenes, in 400 ml tanks, can be used repeatedly for 10–12 racks of slides. The contents of all the tanks must be renewed when the xylene becomes faintly turbid.

Many dyes are extracted by alcohol, especially if water is also present. When this is the case, the instruction will be to "dehydrate rapidly". For rapid dehydration, drain off as much water as possible and then transfer the slides directly to absolute alcohol. Agitate very vigorously for 5–10 s in each of three tanks of this liquid, draining between changes, and then clear in xylene as described above. The alcohol used for rapid dehydration should be renewed after processing four racks of slides. It is not possible to dehydrate free-floating sections rapidly. These should be mounted onto slides after washing, allowed to dry in the air, and then passed quickly through three changes of absolute alcohol into xylene.

Some cationic dyes are much less soluble in *n*-butanol than in lower alcohols. An alternative method of dehydration consists of blotting of the slides between three thicknesses of filter paper and then immersing them (using a clean, dry staining rack) in *n*-butanol, two changes, each 5 min with agitation once or twice in each. Resinous mounting media are miscible with *n*-butanol, but clearing in xylene allows optical clarity to develop more rapidly.

1.6.5. Understanding the methods

There is a reason for everything that is done in the making of a microscopical preparation. Before trying out a technique for the first time, the student should read about and understand the underlying physics and chemistry. He should then read through all the practical instructions and make sure that he understands the purpose of every stage of the procedure.

Some technical methods can be learned only by practice. These include the cutting of sections, the mounting of sections onto slides, and the application of coverslips to sections. In this book, no attempt is made to teach these skills, which must be acquired under the guidance of more experienced colleagues. Excellent descriptions of these procedures are to be found in the works of Krajian & Gradwohl (1952), Culling (1974), Gabe (1976), and Bancroft & Stevens (1977), but there is no substitute for practice in the laboratory. Success in microtechnique requires the integration of craftsmanship with intelligent appreciation of scientific principles.

1.7. EXERCISES

Theoretical

1. What type of preparation would you make in order to investigate the populations of cells in (a) the mesentery of the rat, (b) the articular cartilage of the head of the human femur?

2. A frozen section of skin is stained with an oil-soluble dye in order to demonstrate lipids. What would be an appropriate procedure for making a permanent microscopic preparation of the stained section?

3. If alcohol that has been used to dehydrate a fairly large piece of animal tissue is mixed with water, the resulting liquid is turbid. Why?

4. Some staining methods are applied to whole blocks of tissue which are subsequently embedded, sectioned, cleared, and mounted. To what artifacts would this type of technique be especially prone? How could the artifacts be minimized?

5. It is sometimes necessary to remove the coverslip from a section mounted in a resinous medium. Devise a reasonable procedure by which such a section could be made ready to stain in an aqueous solution of a dye.

Practical

6. Kill a rat (by overdosage with ether vapour) and make spreads of its mesentery. Stain the spreads with polychrome methylene blue (Chapter 6) and make permanent mounts in a resinous medium. Mount a stained spread from water into an aqueous mounting medium. What happens? Deliberately pass one of the slides too quickly through the dehydrating and clearing agents and observe the consequences of doing this.

7. Stain a spread of mesentery with Sudan IV or Sudan black B (Chapter 12) and make a permanent preparation. The Sudan dyes colour lipids by dissolving in them.

2

Fixation

2.1. Physical methods of fixation 8
2.2. Chemical methods of fixation 9
2.3. General properties of fixatives 9
2.3.1. Physical considerations 9
2.3.2. Coagulants and non-coagulants 10
2.3.3. Effects on staining 12
2.4. Individual fixative agents 12
2.4.1. Ethanol, methanol, and acetone 12
2.4.2. Acetic acid 12
2.4.3. Trichloroacetic acid 12
2.4.4. Picric acid 12
2.4.5. Mercuric chloride 13
2.4.6. Chromium compounds 14
2.4.7. Formaldehyde 15
2.4.8. Glutaraldehyde 17
2.4.9. Osmium tetroxide 18
2.4.10. Other fixative agents 19
2.4.11. Chemically inactive ingredients 20
2.4.12. Summary of some properties of individual fixatives 20
2.5. Choice of fixative 21
2.5.1. Carnoy's fluid 21
2.5.2. Bouin's fluid 21
2.5.3. Gendre's fluid (alcoholic Bouin) 21
2.5.4. Heidenhain's "Susa" 21
2.5.5. Helly's fluid 21
2.5.6. Zenker's fluid 22
2.5.7. Neutral buffered formaldehyde 22
2.5.8. Formal–calcium 22
2.5.9. Formal–ammonium bromide 22
2.5.10. Bodian's fixative 22
2.5.11. Altmann's fixative 23
2.5.12. Glutaraldehyde fixative 23
2.6. Methods of fixation 23
2.7. Exercises 24

It is not sufficient for the histologist that a specimen be transparent and that it possess adequate optical contrast. The cells and extracellular materials must be preserved in such a way that there has been as little alteration as possible to the structure and chemical composition of the living tissue. Such preservation is the object of **fixation**. Without being spatially displaced, the structural proteins and other constituents of the tissue must be rendered insoluble in all of the reagents to which they will subsequently be exposed. "Perfect" fixation is, of course, theoretically and practically impossible to attain.

Biological material may be fixed in many ways and some of these will now be discussed.

2.1. PHYSICAL METHODS OF FIXATION

The simplest physical method is the application of **heat**. This results in the coagulation of proteins and the melting of lipids. The resemblance to the living state is not very close after such treatment, but the method is often used in diagnostic microbiology. The shapes and staining properties of bacteria and some other micro-organisms are preserved sufficiently well to permit identification. Heating of large specimens is accomplished by immersion in boiling fixative solution or by cooking in a microwave oven.

Animal tissues are sometimes processed by the techniques known as **freeze-drying** and **freeze substitution**, though only the former is purely physical in nature.

For freeze-drying, the piece of tissue, which should be no more than 2 mm thick, is frozen as quickly as possible (usually by immersion in isopentane, cooled by liquid nitrogen) and then transferred to a special chamber. Here the tissue is maintained *in vacuo* at about −40°C until all the ice it contains has sublimed and has been condensed in a vapour trap maintained at a yet lower temperature. Alternatively, the water vapour may be absorbed into a tray of phosphorus pentoxide. The dried specimen may then be infiltrated with paraffin wax and sectioned in the usual way, though the sections cannot be flattened on water. Freeze-drying does not insolubilize proteins, so it is not, strictly speaking, a method of fixation. However, the morphological preservation is sometimes excellent, and water-soluble substances of low molecular weight are not lost. A freeze-dried block may be chemically fixed in a suitable gas, such as formaldehyde, thus combining the advantages of the two procedures.

In freeze substitution the frozen specimen is dehydrated by leaving it in a liquid dehydrating agent, usually ethanol or acetone, at a temperature below −40°C. The liquid dissolves the ice but does not, at the low temperature, coagulate proteins. When dehydration is complete, the temperature is raised to 4°C for a few hours in order to allow chemical fixation by the alcohol or acetone to take place. The block is then cleared and embedded in paraffin. Sections of unfixed tissue, cut on the cryostat, can be freeze substituted much more quickly than blocks. The relative merits of freeze-drying and freeze substitution are discussed in great detail, together with many technical aspects not mentioned above, by Pearse (1968).

2.2. CHEMICAL METHODS OF FIXATION

For most histological and histochemical purposes, liquid fixatives are used. These substances affect the tissues both physically and chemically. The principal physical changes produced are shrinkage or swelling, and many of the fixatives in common use are mixtures of different agents, formulated so as to balance these two undesirable effects. Most fixatives harden tissues. Moderate hardening is desirable for sectioning with a freezing microtome, or if embedding is to be in nitrocellulose, but can lead to difficulty in cutting wax-embedded material. Dehydration and infiltration with paraffin always produce some further shrinkage and hardening, whatever may be the state of the tissue when it comes out of the fixative. The volume of a fixed, paraffin-embedded specimen is commonly 60–70% of what it was in life. Another important property of a chemical fixative is its rate of penetration. This rate will determine the duration of fixation and the maximum permissible size of the specimen. These physical aspects are reviewed by Baker (1958). Many chemical reactions are involved in fixation, and some of these will now be described. The chemistry of fixation has been reviewed by Baker (1958), Pearse (1968), and Hopwood (1969).

2.3. GENERAL PROPERTIES OF FIXATIVES

The structure of a tissue is determined largely by the configuration of its contained proteins, especially the lipoproteins, which are major components of the plasmalemmae and membranous organelles of cells, the fibrous glycoproteins of such extracellular elements as collagen and basement membranes, and the globular proteins which are dissolved in the cytoplasm and extracellular fluid. In some tissues, extracellular mucosubstances (e.g. chondroitin sulphates) also contribute substantially to the local architecture. All of these substances must be stabilized by fixation. The nucleic acids and their associated nucleoproteins should also be preserved, as should the macromolecular carbohydrates (mucosubstances: comprising glycoproteins, proteoglycans, and at least one polysaccharide—glycogen) and, if their histochemical demonstration is required, the lipids. Fortunately, most of the commonly employed fixatives render insoluble the proteins, nucleic acids, and mucosubstances, though some may be more completely preserved than others by particular agents. Many fixatives do not directly affect lipids and the preservation of these substances depends largely upon the avoidance of agents which dissolve them.

2.3.1. Physical considerations

The rate of **penetration** will determine the size of a block to be fixed by immersion. Rapidly penetrating agents will usually fix in 24 h a specimen whose least dimension is 3–5 mm. For slowly penetrating fixatives the thickness of the block should not exceed 2 mm. The **duration** of fixation should not exceed 24 h except in the case of formaldehyde which takes a week to cause full stabilization of histological structure. Distortion due to slow penetration can be offset by perfusion of the fixative through blood-vessels or by injection into thin-walled cavities. **Shrinkage and swelling** are not, *per se*, detrimental to the quality of fixation, but must be allowed for in quantitative work. The overall change in size is easily determined by measuring appropriate dimensions of the fresh specimen and of the stained, mounted sections. It must not be assumed, however, that all the components of an organ or tissue will shrink or swell equally. Empty spaces due to unequal shrinkage of cells or larger regions of tissues are common artifacts, especially in paraffin sections. The consequences of **hardening** due to fixation have already been mentioned. When a specimen contains tissues of widely varying hardness (e.g. glands and muscle), the embedding procedure should be chosen to suit the hardest component. Double-embedding (Chapter 4) is often the best method for such specimens.

2.3.2. Coagulants and non-coagulants

Fixatives that cause **coagulation** of cytoplasmic proteins destroy or distort organelles such as mitochondria and secretory granules but do not seriously disturb the supporting extracellular materials, which are already essentially solid before being fixed. The sponge-like proteinaceous reticulum produced by coagulant fixatives is easily permeated by melted paraffin so that sectioning of wax-embedded material is facilitated. **Non-coagulant** fixatives convert the cytoplasm into an insoluble gel in which the organelles are well preserved but which is not so readily penetrated by embedding media. Artifactual shrinkage spaces and cracks commonly occur in specimens fixed by these agents. Many fixative mixtures contain both coagulant and non-coagulant compounds and combine the advantages of both. For light microscopy it is necessary to settle

TABLE 2.1. *Properties of individual fixative agents* (Arrows indicate increase (↑) or decrease (↓). Plus signs

	Ethanol, methanol, acetone	Acetic acid	Trichloracetic acid	Picric acid
Usual concentration (alone or in mixtures) (%)	70–100	5–35	2–5	0.5–5
Penetration	Rapid	Rapid	Rapid	Slow
Change in volume of tissue	↓ +++	↑ +++	↑ ++	↓ +
Hardening	+++	Nil	Nil (?)	+
Fixative effect on proteins	Non-additive; coagulant	Nil	Non-additive; coagulant	Additive; coagulant
Action on nucleic acids	Nil	Precipitation	Some extraction	Partial hydrolysis
Effects on carbohydrates	Nil	Nil	Nil	Nil
Effects on lipids	Some extraction	Nil	Nil	Nil
Effects on enzyme activities	Some preserved if cold	Inhibition (?)	Inhibition	Inhibition
Effects on organelles (especially mitochondria)	Destroyed	Destroyed	Preserved	Distorted
Staining with: anionic dyes	Satisfactory	Poor	Satisfactory	Good
cationic dyes	Satisfactory	Good	Good	Satisfactory
Value as sole fixative agent	For small blocks (for glycogen) or cryostat sections	Nil	Not used alone (but see Gabe, 1976)	Poor

for a mixture that gives adequate cytoplasmic fixation (due mainly to non-coagulant insolubilization of protein) or one that provides superior structural preservation on a larger scale (for which coagulation is necessary). In electron microscopy the cytoplasmic disruption due to coagulant fixation is totally unacceptable, and non-coagulant agents must be used. Fortunately the plastics used as embedding media for electron microscopy cause less distortion of the delicate architecture of tissues than do paraffin wax and nitrocellulose. The latter are still needed, however, for the larger specimens examined by the optical microscopist.

The nuclei of cells are deliberately stained in most histological preparations. All fixative mixtures should therefore contain a substance which either coagulates the chromatin or renders it resistant to extraction by water and other solvents (see Table 2.1, p. 10). The chromosomes of dividing cells are shown to best advantage after fixation in agents that coagulate nucleoproteins and nucleic acids. Most fixatives do not react chemically with macromolecular carbohydrates and lipids, though these substances are often protected from extraction as a consequence of the insolubilization of associated proteins.

show relative magnitudes of effects. ? indicates uncertainty)

			Dichromate ion		
Formaldehyde	Glutaraldehyde	Mercuric chloride	pH < 3.5 (CrO_3)	pH > 3.5 ($K_2Cr_2O_7$)	Osmium tetroxide
2–4	0.25–4	3–6	0.2–0.8	1–5	0.5–2
Fairly rapid	Slow	Fairly rapid	Slow	Fairly rapid	Slow
Nil	Nil	↓ +	↓ +	Nil	Nil
++	++	++	++	+	+
Additive; non-coagulant	Additive; non-coagulant	Additive; coagulant	Additive; coagulant	Additive; non-coagulant	Non-coagulant (see p. 18); some extraction
Slight extraction	Slight extraction	Coagulation	Coagulation; some hydrolysis	Some extraction	Slight extraction
Nil	Nil (?)	Nil	Oxidation	Nil	Some oxidation (?)
Nil (but see p. 17)	Similar to formaldehyde	Plasmal reaction	Oxidation of C=C	(see p. 14)	Addition to and oxidation of C=C
Some preserved if cold and short time	Most are inhibited	Inhibition	Inhibition	Inhibition	Inhibition
Preserved	Well preserved	Preserved	Considerable distortion	Very slight distortion	Well preserved
Rather poor	Rather poor	Good	Satisfactory	Good	Acidophilia changed to basophilia
Good	Satisfactory	Good	Satisfactory	Satisfactory	Satisfactory
Useful	Rarely used alone	Poor	Poor	Poor	Specialized applications only

2.3.3. Effects on staining

Another important consideration is the effect of fixation on the subsequent reactivity of the tissues with dyes or histochemical reagents. For example, a glance at Table 2.1 will show that osmium tetroxide should not be used for specimens intended for the staining of connective tissue with anionic dyes.

2.4. INDIVIDUAL FIXATIVE AGENTS

2.4.1. Ethanol, methanol, and acetone

These liquids displace water from proteinaceous materials, thereby breaking hydrogen bonds and disturbing the tertiary structure to produce the change known as denaturation. Soluble proteins of cytoplasm are coagulated and organelles are destroyed. Nucleic acids are not precipitated and remain soluble in water. At low temperatures (below −5°C) ethanol precipitates many proteins without denaturing them. A few secretory products (notably the hypothalamo-neurohypophysial neurosecretory material and the hormones of some cells of the adenohypophysis) remain soluble in water even after otherwise adequate fixation in alcoholic mixtures at room temperature. Alcohols, acetone, and other such solvents extract much lipid from tissues. Carbohydrate-containing components, however, are largely unaffected. Ethanol and methanol make hepatic glycogen insoluble, but acetone does not do this.

Methanol, ethanol, and acetone are used alone for fixing films, smears, and cryostat sections but not for blocks (unless very small) of tissue, since they cause considerable shrinkage and hardening. These undesirable properties can be offset by the presence in the fixative mixture of one or more other substances with opposing properties.

2.4.2. Acetic acid

Acetic acid does not fix proteins, and its only valuable effect is the coagulation of nuclear chromatin. The mechanism by which this change is brought about is obscure. It is a property of the acid and not of the acetate ion. Acetic acid is included in fixative mixtures to preserve chromosomes, to precipitate the chromatin of interphase nuclei, and to oppose the shrinking actions of other agents such as ethanol and picric acid.

Acetic acid is a liquid which freezes to an ice-like mass at 16.6°C (hence the name "glacial"). It has a pungent smell and is injurious to the skin.

2.4.3. Trichloroacetic acid

This acid (TCA; CCl_3COOH) is widely used by biochemists to precipitate proteins from solutions. It serves an equivalent purpose as a component of fixative mixtures but is seldom used alone. The coagulation is probably due to electrostatic interaction of trichloroacetate anions with positively charged groups ($—NH_3^+$, etc.) of proteins. The highly non-polar $Cl_3C—$ group enables the TCA to penetrate into hydrophobic domains within protein molecules. The consequent combination of hydrophobic interaction and ionic attraction by the same ion is probably responsible for the breaking of the hydrogen bonds that hold the proteins in their normal configurations. TCA extracts nucleic acids but only does so significantly under conditions more rigorous than those used in histological fixation.

Crystals of TCA are deliquescent and are often too wet to be weighed. The saturated aqueous solution, which has a concentration of 120% (w/v), is more easily and safely handled than the damp solid. Trichloroacetic acid and its concentrated solutions must be treated with respect: they will cause severe burns if they come into contact with the skin.

2.4.4. Picric acid

Picric acid is trinitrophenol. It is a much stronger acid than unsubstituted phenol in aqueous solution, owing to the electron-withdrawing effect of the three nitro groups on the hydroxyl group

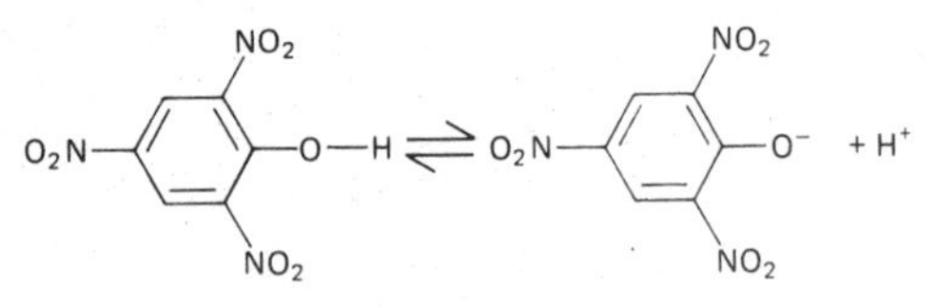

The anion causes coagulation by forming salts (picrates) with the basic groups of proteins. Because some of the precipitated protein picrates are soluble in water (though not, on account of the common ion effect, in picric acid), tissues fixed in mixtures containing picric acid are usually transferred directly to 70% alcohol in order to coagulate all the precipitated protein. Some authorities, however, state that it makes no difference whether the specimens are washed in water or alcohol. When other fixatives (e.g. formaldehyde) are mixed with picric acid it is unlikely that any proteins in the fixed tissue are still soluble in water, yet coagulable by alcohol. Picric acid is sufficiently strong (i.e. low pH due to dissociation into hydrogen ions and picrate ions) to bring about hydrolysis of nucleic acids, so fixatives containing it are avoided when DNA and RNA are to be studied histochemically. The blocks must remain in 70% alcohol (several changes) until as much as possible of the yellow colour of picric acid is removed. Prolonged contact with this reagent (months to years) causes undue maceration, even within solid paraffin wax (Luna, 1968). When the tissues are to be sectioned within a few weeks, persistence of picric acid does not matter. It is easily washed out of the sections by a dilute (e.g. 0.13% = one-tenth of saturated) aqueous solution of lithium carbonate (Li_2CO_3).

Since, used alone, picric acid makes tissue shrink, it is used in combination with other substances, notably acetic acid, which, unopposed, would cause swelling. The most commonly used fixatives containing picric acid are Bouin's and Gendre's fluids (see below). As an alcoholic fixative, Gendre's fluid preserves glycogen, though less completely than ethanol used alone.

Trinitrophenol is a bright yellow solid. It is dangerously explosive when dry and is therefore stored under water. Stock bottles should be inspected from time to time and water added as necessary to give a layer about 2 cm deep above the level of the yellow powder.

2.4.5. Mercuric chloride

Solutions of mercuric chloride contain molecules Cl—Hg—Cl and hardly any free Hg^{2+} ions. With water there is slight formation of a hydroxo complex:

$$HgCl_2 + H_2O \rightleftharpoons HO - Hg - Cl + H^+ + Cl^-$$

but this hydrolysis is reversed in acid solutions. In the presence of chloride ions, as from added sodium chloride, a complex anion is produced:

$$HgCl_2 + 2Cl^- \rightleftharpoons [HgCl_4]^{2-}$$

The chemistry of fixation by mercuric chloride is poorly understood, though some tentative deductions may be made from the known reactions of the compound with ammonium salts, ammonia, sulphydryl groups, and olefins:

$$HgCl_2 + 2NH_4^+ \rightleftharpoons Hg(NH_3)_2Cl_2\downarrow + 2H^+ \qquad (2.1)$$

$$HgCl_2 + 2NH_3 \rightleftharpoons H_2N-Hg-Cl\downarrow + NH_4^+ + Cl^- \qquad (2.2)$$

$$HgCl_2 + R-SH \rightleftharpoons R-S-Hg-Cl + H^+ + Cl^- \qquad (2.3)$$

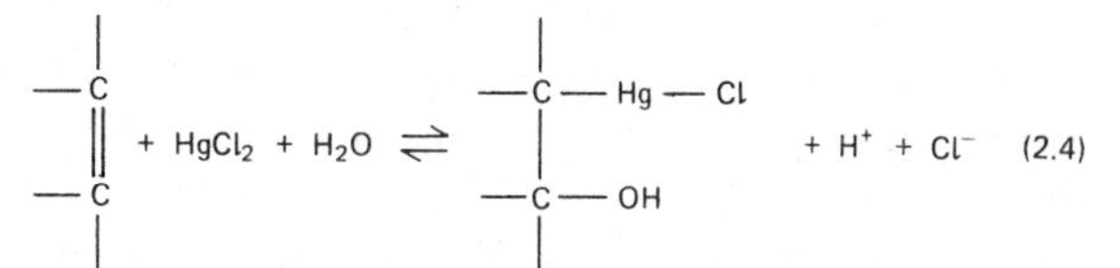

Since most fixatives containing mercuric chloride are acidified, amino groups of proteins would be more likely to participate in reaction (2.1) than in reaction (2.2):

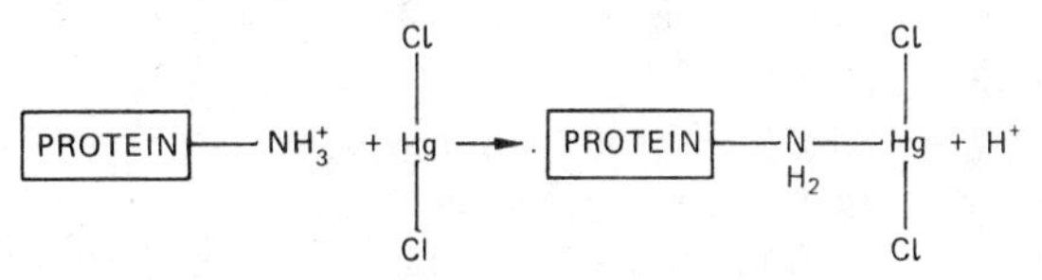

Suitably spaced amino groups could be cross-linked through atoms of mercury in structural arrangements such as

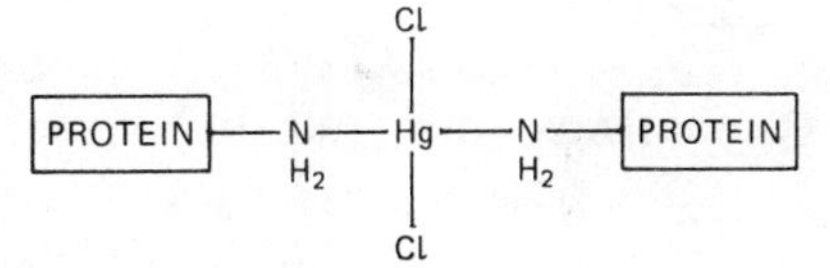

In fixative solutions containing excess chloride ions the active ingredient may be the complex anion $[HgCl_4]^{2+}$. This may be able to cross-link proteins by forming coordinate bonds with ionized amino groups in reactions such as:

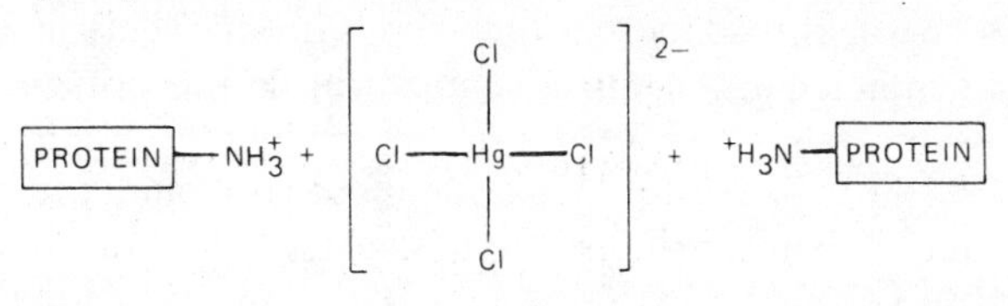

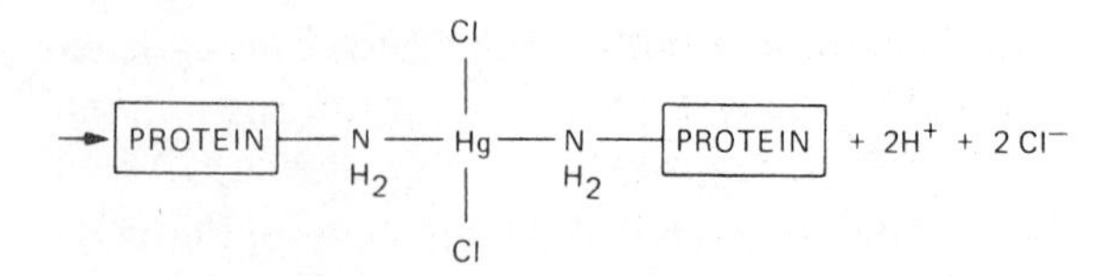

The best-known reaction of mercuric chloride with proteins is condensation with the sulphydryl groups of cysteine:

$$\boxed{\text{PROTEIN}}\text{—SH} + HgCl_2$$

$$\longrightarrow \boxed{\text{PROTEIN}}\text{—S—Hg—Cl} + H^+ + Cl^-$$

This certainly occurs during fixation, but the importance of the reaction in stabilizing fine structure is not known. Addition of mercuric chloride to unsaturated linkages in lipids is probably not important in fixation except in relation to the plasmal reaction (Chapter 12).

Fixative mixtures containing mercuric chloride are notable for enhancing the brightness of subsequent staining with dyes. The reason for this is not known. A crystalline precipitate (of uncertain chemical composition but probably mostly mercurous chloride, Hg_2Cl_2) forms within mercury-fixed tissues and must be removed before the sections are stained by treatment with a solution of iodine, followed by sodium thiosulphate (see p. 36). Another important practical point is that while $HgCl_2$ penetrates rapidly through the tissue and causes hardening, the latter effect becomes excessive if the time of fixation is unduly prolonged. In old texts of histological technique, $HgCl_2$ is usually called "corrosive sublimate" or simply "sublimate". It is a very poisonous substance.

2.4.6. Chromium compounds

The compounds of chromium used in fixation contain the metal in its highest oxidation state, +6. Chromium trioxide, CrO_3, dissolves in water to form the completely ionized dichromic acid

$$2CrO_3 + H_2O \longrightarrow H_2Cr_2O_7 \longrightarrow 2H^+ + Cr_2O_7^{2-}$$

This acid, also loosely known as "chromic acid", is contained in some fixative mixtures, but its potassium salt, $K_2Cr_2O_7$, is more frequently used. Dichromic acid (pH below about 3.5) and the dichromate ion (in less strongly acid solutions) have quite different effects on tissues. Their contrasting properties are shown in Table 2.1 (p. 11). The strongly acid solutions coagulate cytoplasmic proteins, producing a reticulated texture, and coagulate the chromatin. DNA is partially hydrolysed by chromic acid so that it gives a positive reaction in histochemical tests for aldehydes. This is an effect of any sufficiently strong acid and is discussed in Chapter 9. Less acid solutions of the dichromate ion insolubilize proteins without coagulation but do not fix nucleic acids.

The chemistry of fixation of proteins by dichromate is poorly understood, but it is possible that the atoms of chromium form co-ordinate linkages through oxygen with one another and with oxygen and nitrogen atoms in carboxyl, hydroxyl, and amino groups (see Gustavson, 1956; Pearse, 1968). This would cause both precipitation and cross-linking of protein molecules. The cross-linking of proteins by chromium has been studied principally in relation to the tanning of leather (Thorstensen, 1969) and the hardening of photographic gelatine (Pouradier & Burness, 1966). In both these industrially important processes the metal is used as a salt of the chromic ion (Cr^{3+}), not as a dichromate.

As well as fixing proteins, the dichromate ion can also react with phospholipids in such a way as to make them insoluble in non-polar solvents. The chemical mechanism of this process (see Chapter 12) is still obscure. The lipid-stabilizing action of dichromate does not occur with the usual 1 or 2 days of fixation at room temperature.

The oxidizing properties of the dichromate ion are exploited in the chromaffin reaction, whereby catecholamines (adrenaline and noradrenaline) are transformed into brown compounds (see Chapter 17).

Material fixed in dichromate must be washed for 24–48 h in running tap water before being transferred to the dehydrating alcohol in order to avoid the reaction

$$H_2Cr_2O_7 + 3C_2H_5OH \longrightarrow 2Cr(OH)_3\downarrow \cdot 3CH_3CHO + H_2O$$

which would produce an insoluble green precipitate in the tissue. Potassium dichromate is used in mixtures such as Zenker's and Helly's fluids. The first of these contains enough acetic acid to cause the fixative action of the dichromate ions to be that of a strongly acid solution. Helly's fluid is less acid and the dichromate in it acts as a non-coagulant fixative. Both these mixtures also contain mercuric chloride, which is a coagulant. Helly's fluid contains formaldehyde as well and is therefore unstable owing to the reaction

$$Cr_2O_7^{2-} + 3HCHO + 5H^+ \longrightarrow 2Cr^{3+} + 3HCOO^- + 4H_2O$$

The reaction proceeds quite slowly, however, since $[H^+]$ in Helly's fluid is not very high. Both the dichromate and the formaldehyde have time to act upon the tissue before they are themselves changed into chromic and formate ions.

2.4.7. Formaldehyde

Formaldehyde is a gas (B.P. −21°C) with the structural formula

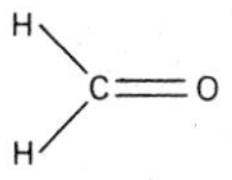

It is available to the histologist as a solution (formalin) containing 37–40% by weight of the gas in water, and as a solid polymer, **paraformaldehyde**, which is $HO(CH_2O)_nH$, n being 6 to 100. Paraformaldehyde is also seen as the white precipitate which forms in old bottles of formalin. Although formaldehyde is the simplest of the aldehydes, its chemistry is quite complicated (see Walker, 1964).

In aqueous solutions formaldehyde is present as methylene hydrate, the product of the reaction

$$H_2C{=}O + H_2O \rightleftharpoons HOCH_2OH$$

Formaldehyde Methylene hydrate

The equilibrium lies far to the right, and hardly any true formaldehyde is present in the solution. However, the chemical reactions of methylene hydrate are those of formaldehyde in the presence of water, so it is usual to speak and write of aqueous solutions as if they contained formaldehyde. Formalin also contains soluble polymers of the form $HO(CH_2O)_nH$ (where $n = 2$–8), known as lower polyoxymethylene glycols. Continued addition of methylene hydrate molecules occurs spontaneously, and this is the reason for the precipitation of paraformaldehyde in formalin that has been stored for a long time. Formalin also contains methanol (commonly about 10% v/v) which is added as a stabilizer to inhibit polymerization. Methanol does this by forming with formaldehyde a hemiacetal (methylal) which is more stable than methylene hydrate.

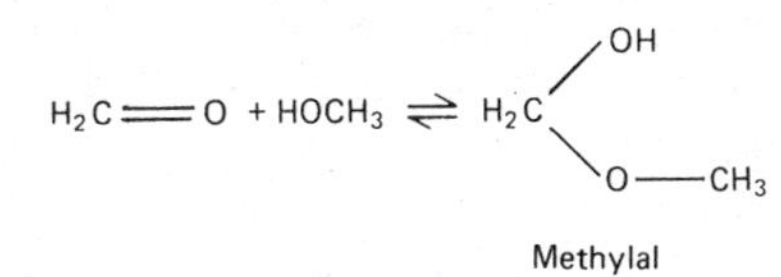

Methylal

The polymers are hydrolysed when formalin is diluted with an excess of water. Thus, for trioxymethylene glycol, $HO(CH_2O)_3H$,

$$HO{-}CH_2{-}O{-}CH_2{-}O{-}CH_2{-}OH + 2H_2O \rightleftharpoons 3HOCH_2OH$$

The equilibrium is displaced to the right because of the high concentration of water, but the reaction is very slow (taking some weeks for completion) between pH 2 and 5, as in non-neutralized formalin. The depolymerization occurs quite rapidly, however, in neutral media. If formalin is to be used as a fixative in simple aqueous solution it should be diluted several days in advance, but when buffered to approximate neutrality it may be used immediately. Neutrality may also be achieved by keeping some marble chips (calcium carbonate) in the bottom of the bottle of diluted formalin.

Formalin deteriorates during storage as the result of a Cannizzaro reaction:

$$2HOCH_2OH \longrightarrow CH_3OH + HCOOH + H_2O$$

The pH of the solution gradually falls as more formic acid is produced. With other aldehydes, Cannizzaro reactions are significant only in alkaline media. Oxidation of formaldehyde by atmospheric oxygen (which would also generate formic acid) is exceedingly slow and is not a significant cause of deterioration.

Formaldehyde solutions for use as fixatives are also made by depolymerizing paraformaldehyde. This substance dissolves very slowly in water but more quickly in near-neutral buffer solutions. Methylene hydrate made in this way does not contain methanol or formic acid and is preferred to diluted formalin as a fixative for histochemistry and electron microscopy. Paraformaldehyde is also depolymerized by heating. Small specimens, including those of freeze-dried tissues, can be fixed by exposure to the vapour, which is gaseous formaldehyde, at 50–80°C.

The content of formaldehyde in a fixative is best denoted by stating the percentage by weight of the gas rather than the amount of formalin which was used in preparing the mixture. Thus, "4% formaldehyde" is preferred to "10% formalin" (for the same solution), though the latter designation is in common use.

Formaldehyde reacts with several parts of protein molecules (see Walker, 1964; Hopwood, 1969; Pearse, 1968 for more information). The methylene glycol molecule adds to many functional groups to form hemiacetals and related adducts. For example, with primary amines (N-terminal amino acids and lysine side-chains):

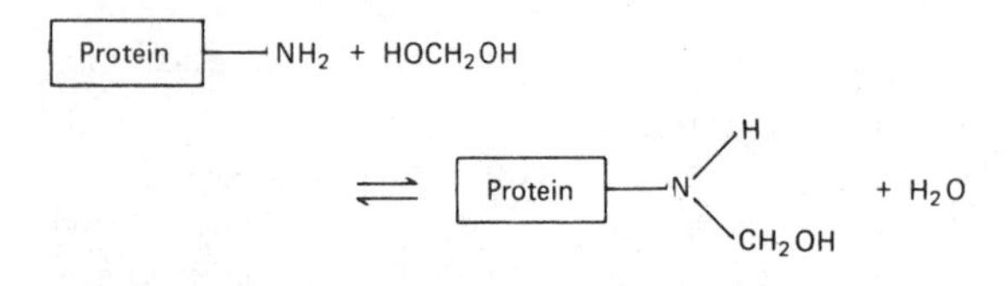

with guanidyl groups of arginine side-chains:

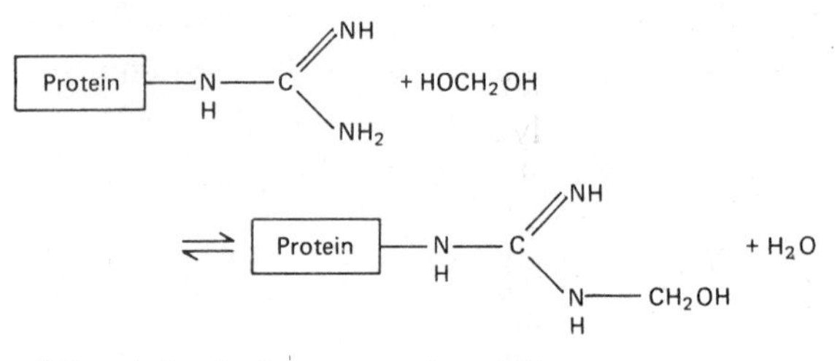

with sulphydryl groups of cysteine:

$$\boxed{\text{Protein}}—SH + HOCH_2OH$$

$$\rightleftharpoons \boxed{\text{Protein}}—S—CH_2OH + H_2O$$

with aliphatic hydroxyl groups (serine, threonine):

$$\boxed{\text{Protein}}—OH + HOCH_2OH$$

$$\rightleftharpoons \boxed{\text{Protein}}—O—CH_2OH + H_2O$$

with amine nitrogen (at accessible peptide linkages):

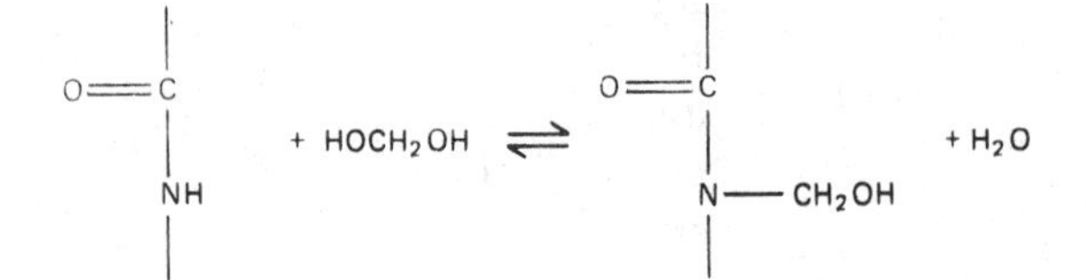

These reactions are all readily reversible by washing in water or alcohol, so the simple addition of formaldehyde to proteins does not contribute significantly to fixation for $—NH_2$, $—NHC(NH)NH_2$, and —SH groups. However, the hemiacetal-like adducts all have free hydroxymethyl groups, and these are capable of further reaction with suitably positioned functional groups of proteins:

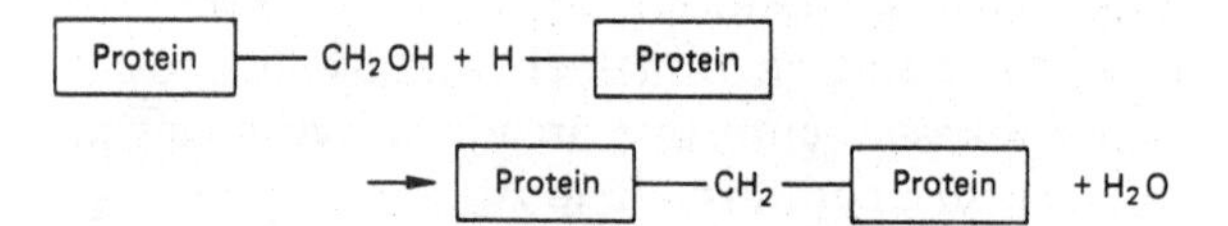

Thus, different protein molecules can be joined together by methylene bridges, which are chemically stable. Formaldehyde is used in the tanning of collagen to produce leather and investigations of this industrial process (see Gustavson, 1956) have led to the conclusion that although cross-links of many kinds are possible, the great majority are formed between ε-amino groups of lysine and the amide nitrogen atoms of peptide linkages:

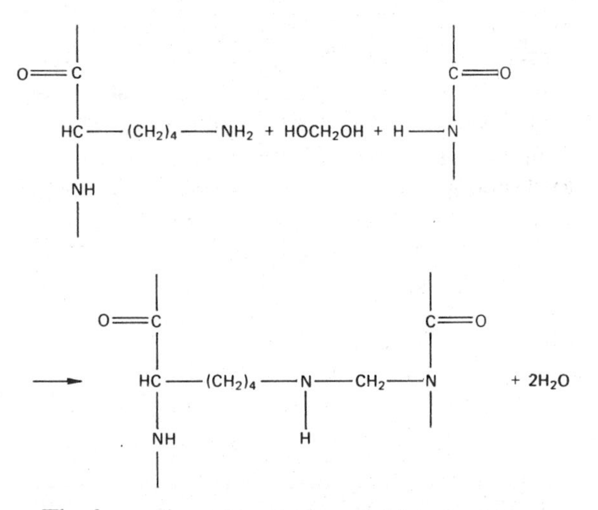

The formation of methylene bridges in this reaction is probably largely responsible for the cross-linking of protein molecules that constitutes fixation by formaldehyde. Many ε-amino groups will not be close to peptide linkages and will therefore be able to form only the unstable hemiacetal-like adducts. Thus, the histochemical reactions of primary amines (including the binding of anionic dyes) will be only slightly depressed in the fixed tissue. After prolonged storage (e.g. 6 months to 10 years) in formalin, the free $—NHCH_2OH$ groups may be oxidized by the atmosphere to the more stable —NHCOOH. This change would account for the eventual loss of stainability by anionic dyes that occurs in old specimens.

Cross-linking of protein molecules by formaldehyde is a slower process than most of the chemical reactions of other fixative agents and requires 1–2 weeks for completion at room temperature. For histochemical purposes, tissues are commonly fixed for 12–24 h at 4°C, but many non-histochemical methods, especially for constituents of the nervous system, work better after more complete fixation. Sufficient hardening for the cutting of frozen sections is usually attained after 24 h of fixation. Very long periods of storage in formaldehyde solutions result in excessive hardening, loss of stainability of nuclei, and (with acidic solutions) deposition of brown "formalin pigment". This is a haematin formed by acid degradation of haemoglobin. It can be removed by treating the sections with an alcoholic solution of picric acid or with any of a variety of oxidizing agents or alkalis (see Lillie & Fullmer, 1976). The "pigment" does not form if the formaldehyde solution is buffered to approximate neutrality.

Formaldehyde preserves most lipids, especially if the fixing solution contains calcium ions which, for ill-understood reasons, reduce the solubilities of some phospholipids in water. The only chemical reactions of formaldehyde with lipids under ordinary conditions of fixation are (i) addition to the amino groups of phosphatidyl ethanolamines, which is probably reversible by washing in water, and (ii) prevention of the histochemical reactivity of plasmalogens owing to oxidation, probably to a glycol, of the ethylenic linkage next to the ether group. The latter reaction may be brought about by atmospheric oxygen rather than by formaldehyde. With prolonged fixation in formaldehyde solutions (3 months to 2 years), other double bonds are attacked with the formation of a variety of products, all of which are more soluble in water than the original lipids (Jones, 1972).

Formaldehyde does not react significantly with carbohydrates. All the common mucosubstances can be demonstrated after fixation with formaldehyde, though appreciable quantities of glycogen are lost.

A neutral, buffered aqueous solution (pH 7.2–7.4) containing 2–5% of formaldehyde and isotonic with tissue fluids, is probably the most generally useful of all fixatives for most histological and histochemical purposes. However, when histochemical methods are not to be used, mixtures such as Bouin's fluid and Susa (see Section 2.5.4) are preferable to formaldehyde because they leave the tissue in a state more amenable to staining with dyes.

2.4.8. Glutaraldehyde

Glutaraldehyde

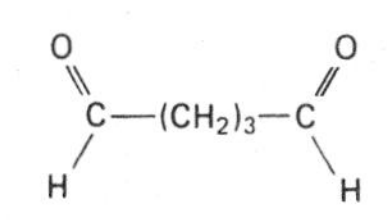

is the most widely used bifunctional aldehyde fixative. The aldehyde (—CHO) groups react in much the same way as those of formaldehyde, but a glance at the formula of glutaraldehyde shows that both ends of the molecule are potentially available to combine with reactive sites of proteins. Tissues fixed in glutaraldehyde are more strongly stabilized by cross-linking than those fixed in formaldehyde, and this is probably the reason why ultrastructural features are so well preserved by the former substance. Glutaraldehyde causes more hardening than does formaldehyde, and this can lead to difficulty in sectioning paraffin-embedded material. Some of the difficulty may also be due to failure of the wax molecules to enter the tightly cross-linked proteinaceous matrix. Glutaraldehyde penetrates tissues slowly, so it should be perfused through the vascular system if possible. The cross-linking of proteins occurs much more rapidly than with formaldehyde: fixation is complete after only a few hours.

When only one end of a glutaraldehyde molecule is involved in fixation, the other aldehyde group remains free to react with histochemical reagents. This phenomenon must be taken into account when sections of glutaraldehyde-fixed material are to be treated by the Feulgen or periodic acid–Schiff procedures, in which the detection of DNA or of carbohydrate-containing substances is made possible by the chemical production of aldehydes. Glutaraldehyde may also cause false-positive results in autoradiography. Aldehyde groups introduced by the fixative must be irreversibly chemically blocked before such methods can be used. Suitable blocking procedures are described in Chapter 10 (Sections 10.10.6 and 10.11.14.11).

The activities of some enzymes are preserved after brief fixation in glutaraldehyde, but formaldehyde usually causes less inhibition, probably because it forms fewer cross-links between protein molecules.

Acrolein ($H_2C{=}CHCHO$) is a very toxic unsaturated aldehyde occasionally used as a fixative for electron microscopy. Reaction occurs with the olefinic as well as with the aldehyde group, with resultant cross-linking of proteins.

2.4.9. Osmium tetroxide

This substance, sometimes known colloquially and very wrongly as "osmic acid", is a non-ionic solid, OsO_4, which is volatile at room temperature. The vapour is irritating and can cause corneal opacities. OsO_4 is soluble (without ionization) in water but much more soluble in non-polar organic solvents. It is somewhat unstable in solution, being reduced by traces of organic matter to the dioxide, $OsO_2.2H_2O$. This reduction is also brought about by alcohols, but not by carbonyl compounds such as formaldehyde, glutaraldehyde, and acetone, provided that these substances are pure.

Used solutions of OsO_4 should not be thrown away. The old fixatives should be collected in well-stoppered bottles and then chemically processed for recovery of the OsO_4. A simple technique for recycling is described by Kiernan (1978). If the osmium tetroxide is to be discarded, it must be safely buried at an approved site for disposal of noxious chemicals. In view of the high cost of the reagent, reclamation is well worth while. Small quantities (e.g. 100 ml) of very dilute solutions (less than 0.1%) can be safely washed down the sink with plenty of water.

Although OsO_4 can react in the test-tube with proteins (especially sulphydryl groups) and carbohydrates (Bahr, 1954), it extracts quite large amounts of these substances from tissues during the course of fixation. Protein solutions are gelated but not coagulated by OsO_4 and some cross-linking occurs. The chemical reactions involved in these changes are not yet understood (Hopwood, 1977; Nielson & Griffith, 1979). The best understood fixative action is with the unsaturated linkages of lipids (see Adams *et al.*, 1967). A cyclic ester is first formed, by addition:

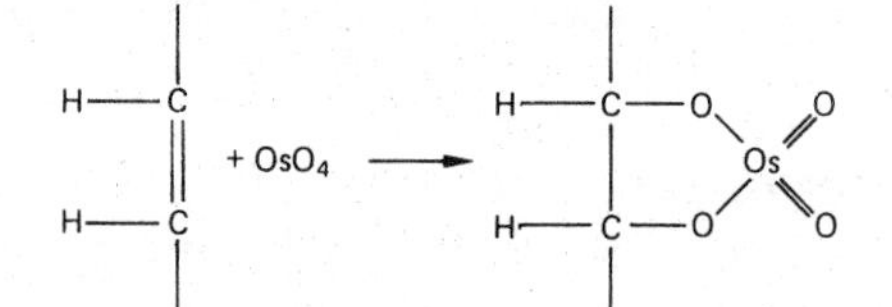

In this reaction, oxidation occurs at the carbon atoms (2 electrons lost), while the osmium is reduced, by gaining 2 electrons, so that its oxidation number changes from +8 to +6.

When two unsaturated linkages are suitably positioned they may be cross-linked (Wigglesworth, 1957). Chemical studies by Korn (1967) indicate that a diester is formed by the reaction

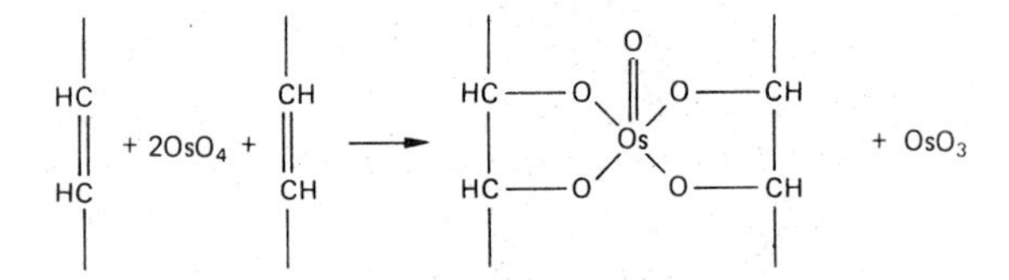

The oxide OsO_3 is unstable and disproportionates:

$$2OsO_3 \rightarrow OsO_2 \downarrow + OsO_4$$

The osmium(VI) addition compounds and diesters are colourless or brown and may be soluble in organic solvents. The precipitated osmium dioxide is black and insoluble. Blackening of tissue fixed or stained with osmium tetroxide is increased in some circumstances with passage through alcohol, which may effect a reaction of the type

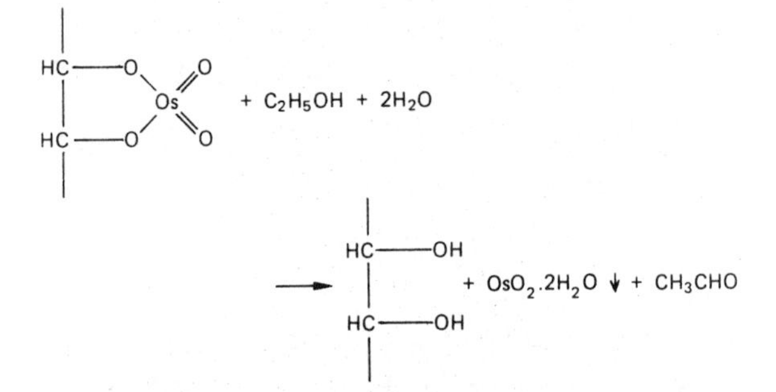

Any unreacted OsO_4 that has not been washed out of the tissue is similarly reduced:

$$OsO_4 + 2C_2H_5OH \longrightarrow OsO_2.2H_2O \downarrow + 2CH_3CHO$$

The uses of OsO_4 in the histochemical study of lipids are discussed in Chapter 12.

Ordered arrays of phospholipid molecules are present predominantly in biological membranes, and OsO_4 is most valuable as a fixative for these structures, which are rendered insoluble, black, and electron dense. Although individual membranes cannot be resolved with the light microscope, it is possible to see structures which are largely composed of membranous material, such as mitochondria and the myelin sheaths of nerve fibres. The solubility of OsO_4 in non-polar substances also causes the blackening of hydrophobic lipids such as the contents of fat cells and the products of Wallerian degeneration of myelinated axons. The staining of the latter forms the basis of the Marchi method, used in neuroanatomy.

Treatment with OsO_4 largely abolishes the affinity of tissue proteins for anionic (acid) dyes, and normally acidophilic elements become stainable by cationic (basic) dyes. This change may be due to oxidation of terminal and side-chain amino groups of proteins with concomitant formation of carboxyl groups. This effect is in need of investigation. By analogy with other oxidative deaminations the reaction would be expected to occur in three stages:

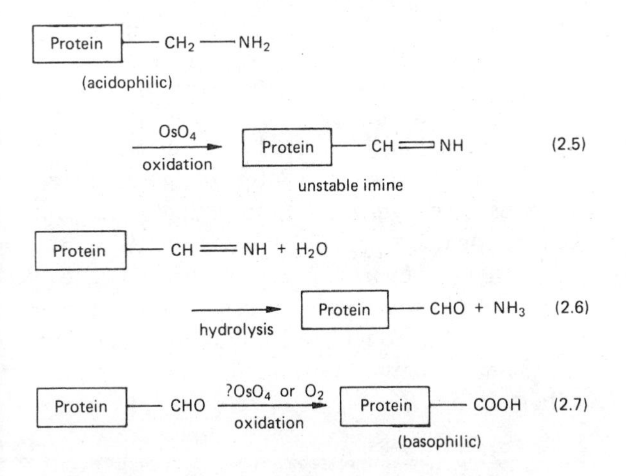

The product of reduction of OsO_4 is probably not $OsO_2.2H_2O$ since, if it were, proteins would be blackened. A soluble osmate, $[OsO_2(OH)_4]^{2-}$, or osmiamate, $[OsO_3N]^-$, may be the principal by-product of the reaction.

Osmium tetroxide penetrates blocks of tissue to a depth of only 0.5–1.0 mm, so that its use as a fixative is limited to small blocks. By stabilizing membranes and gelating dissolved proteins, OsO_4 provides lifelike fixation of the internal structures of cells. The absence of morphological artifact has been demonstrated by microscopic observation of living cells during the course of fixation by OsO_4. Unfortunately, however, pieces of tissue fixed in this agent have a crumbly consistency which is made worse by embedding in wax, and leads to the formation of cracks and shrinkage spaces. Osmium tetroxide may be used as a secondary fixative (post-fixation) after formaldehyde or glutaraldehyde, as in electron microscopy, and it may be used to stain frozen sections for unsaturated lipids. The vapour of OsO_4 is as effective as an aqueous solution, both as a fixative and as a stain.

2.4.10. Other fixative agents

The major individual fixatives have now been described, but many other substances have also been used, some of them in mixtures that are now obsolete or at least out of fashion, others in more recent years for special purposes. These minor fixative agents include:

(a) Several metal ions and complexes which cause precipitation of proteins, e.g. Cu^{2+}, Cd^{2+}, Cr^{3+}, $[PtCl_6]^{4-}$, Pb^{2+}, $[UO_2]^{2+}$, Co^{2+}. See Osterberg (1974) for a review of metal ion–protein interactions.

(b) Mineral acids, which also serve as coagulants, e.g. H_2SO_4, HNO_3.

(c) Cationic detergents and cationic dyes, which form insoluble salt-like complexes with polyanions, especially with proteoglycans (Williams & Jackson, 1956).

(d) Organic protein coagulants, e.g. tannic acid, *p*-toluenesulphonic acid (Malm, 1962).

(e) Aldehydes other than formaldehyde and glutaraldehyde, e.g. chloral hydrate, acrolein, hydroxyadipaldehyde.

(f) Bifunctional organic compounds other than aldehydes which are capable of combining with and cross-linking amino, hydroxyl, and carboxyl groups, e.g. cyanuric chloride, carbodiimides, quinones, diethyl pyrocarbonate (Pearse & Polak, 1975).

Though some of these fixatives have been shown to be potentially as useful as the more popular ones, none has yet acquired great importance, so they will not be discussed further. Recent research has been directed mainly towards the discovery of fixatives that would give adequate structural preservation at the level of the electron microscope without unduly disturbing the activities of enzymes and the antigenic properties of substances to be studied by immunohistochemical techniques.

2.4.11. Chemically inactive ingredients

In addition to one or more of the components discussed above, an aqueous fixative mixture commonly includes "indifferent" substances which influence its osmotic pressure and its pH.

Since the most rapidly penetrating component of a fixing fluid is the water in which the active ingredients are dissolved, the central parts of a specimen are likely to be bathed in a hypotonic medium before they are fixed. This would lead to swelling or rupture of cells and disorganization of their surrounding framework of connective tissue. In order to prevent such damage, unreactive salts with small, rapidly diffusing ions (e.g. sodium chloride or sulphate) are incorporated into the fixative mixture. The necessity for control of osmotic pressure is disputed, but most histologists and electron microscopists agree that the best results are obtained when the net osmotic pressure of the fixative, due to all its solutes, is slightly higher than that of the extracellular fluid. Sucrose is also used to raise the osmotic pressure, but since its molecules are fairly large it should probably be employed only in fixatives for vascular perfusion or for immersion of very small specimens.

Non-aqueous fixatives coagulate the tissue as they penetrate, so for these the control of osmotic pressure is unnecessary. The solvent penetrates more rapidly than any dissolved ingredients, however. For example, the centre of a specimen immersed in an alcoholic solution of picric acid will be fixed primarily by the alcohol alone.

The pH of a fixative must be appropriate for the chemical reactions of fixation. Formaldehyde, glutaraldehyde, and OsO_4 are all fully active at and around neutrality, so their solutions may, with advantage, be buffered to approximately the same pH as the extracellular fluid. This is pH 7.2–7.6 for mammals. The prevention of an abrupt change in pH during fixation probably minimizes fine structural disturbance within cells. Buffering is of greatest value in the case of formaldehyde, which reacts with proteins much more slowly than do other fixatives.

Most fixatives other than the three mentioned in the preceding paragraph are active only when their solutions are considerably more acid than the extracellular fluid. The correct pH is produced when the fixative mixture is made and the use of buffers is not necessary. In the design of new fixatives, care must be taken to avoid irrational combinations. For example, the advantages of mixing acetic acid with a formaldehyde or osmium tetroxide solution would be lost if the mixture were then carefully buffered to pH 7.4. The effect of pH on the type of fixation produced by the dichromate ion has already been described.

Solutions of indifferent salts, buffers, and sucrose are often used to wash fixed specimens, especially in processing for electron microscopy. Completely fixed tissues are, however, no longer responsive to changes in the osmotic pressures of the solutions in which they are immersed. Plain water is satisfactory for washing all specimens intended for ordinary histological or cytological examination with the light microscope. Isotonic buffers are recommended, however, for washing specimens that have received only partial fixation preparatory to the application of histochemical methods for enzymes or to subsequent processing for electron microscopy. Paljarvi *et al.* (1979) have shown that tissues are still susceptible to damage by osmotic stress after fixation for 24 h in buffered 4% formaldehyde, though 3% glutaraldehyde used for the same length of time confers complete stabilization.

2.4.12. Summary of some properties of individual fixatives

The more important effects of the major fixative agents upon tissues are presented briefly in Table 2.1 on pages 10 and 11.

2.5. CHOICE OF FIXATIVE

A fixative is chosen according to the structural or chemical components of the tissue which must subsequently be demonstrated. Often a mixture of different agents is employed in order to offset undesirable effects of individual substances and to obtain more than one type of chemical fixation. A few of the commoner fixatives will now be described.

2.5.1. Carnoy's fluid

Absolute ethanol:	60 ml	Mix just before using
Chloroform:	30 ml	
Glacial acetic acid:	10 ml	

A rapidly penetrating fixative which coagulates protein and nucleic acids and extracts lipids. Many carbohydrate components are also preserved. Blocks of tissue up to 5 mm thick are fixed for 6–8 h. Fixation for more than 18 h can result in hydrolysis of nucleic acids, with loss of RNA.

2.5.2. Bouin's fluid

Saturated aqueous picric acid:	750 ml	Keeps indefinitely
Formalin (40% HCHO):	250 ml	
Glacial acetic acid:	50 ml	

Not used in histochemistry but preserves morphological features, especially of connective tissue, well. This fixative is ideal for purely histological work, since physical distortion of tissues is minimal, but intracellular structures other than nuclei are poorly preserved. Specimens are usually fixed in Bouin for 24 h, but material stored in it for several months is sometimes still usable.

2.5.3. Gendre's fluid (alcoholic Bouin)

Saturated alcoholic picric acid:	800 ml
Formalin (40% HCHO):	150 ml
Glacial acetic acid:	50 ml

Keeps indefinitely

(The picric acid solution is made up in 95% ethanol or industrial methylated spirit.)

Similar to Bouin, but also immobilizes some water-soluble carbohydrates (glycogen; mast cell granules of some species) which dissolve in aqueous fixatives. Specimens are fixed overnight (16–20 h) and then washed in several changes of 95% alcohol.

2.5.4. Heidenhain's "Susa"

Mercuric chloride ($HgCl_2$):	45 g
Sodium chloride (NaCl):	5 g
Trichloroacetic acid:	20 g
Glacial acetic acid:	40 ml
Formalin (40% HCHO):	200 ml
Water:	to 1000 ml

Keeps indefinitely

Pieces of tissues are fixed for no more than 24 h and then transferred directly to 95% alcohol, supposedly to avoid swelling. Stains for nuclei, cytoplasm, and connective tissue all work brightly after fixation in Susa, as do some histochemical methods for carbohydrates, though this is primarily a "morphological" fixative. Mercurial deposits must be removed prior to staining.

2.5.5. Helly's fluid

Stock solution

Mercuric chloride ($HgCl_2$):	50 g
Potassium dichromate ($K_2Cr_2O_7$):	25 g
Sodium sulphate ($Na_2SO_4 . 10H_2O$):	10 g
Water:	to 1000 ml

Keeps indefinitely

Immediately before use, mix 5 ml of formalin (40% HCHO) with 100 ml of the stock solution.

Helly's fluid is used mainly to fix cytoplasmic elements such as mitochondria and secretory granules, as in endocrine organs and haemopoietic tissue. The non-coagulant components (formaldehyde, dichromate) offset the coagulant action of the mercuric chloride so that cytoplasm is not coarsely coagulated. The mixture goes green and murky on standing, owing to reduction of dichromate to chromic salts by formaldehyde. Fixation should not be for more than 12–24 h. See comments relating to mercury- and chromium-containing fixatives (p. 14).

2.5.6. Zenker's fluid

This is similar to Helly's except that the formalin is replaced by 5 ml of glacial acetic acid. This mixture is stable indefinitely. Because of the low pH at which the dichromate is used (see p. 14) cytoplasmic proteins are rather coarsely coagulated and most organelles are not preserved. Helly and Zenker are not generally suitable for histochemical work.

2.5.7. Neutral buffered formaldehyde

Sodium phosphate, monobasic ($NaH_2PO_4 . H_2O$):	4 g
Sodium phosphate, dibasic (Na_2HPO_4):	6.5 g

Dissolve in water to 1000 ml

Heat to boiling, add 25–40 g of paraformaldehyde powder (in fume hood), stir well, cool, and filter. Keeps for several weeks

Sometimes the dissolution of the paraformaldehyde is very slow and the mixture has to be left for about 24 h in an oven (55–60°C), in a tightly capped bottle. The concentration of formaldehyde is not critical. A 4% solution is commonly used but 2.5% is just as effective. The pH is 7.2–7.4.

Tissues may remain in this fixative for several months, but if histochemical methods for enzymes (those which resist such fixation) are to be carried out, 12–24 h at 4°C is usually the maximum permissible time before cutting frozen sections. Longer periods of fixation, such as 1 or 2 weeks, render the tissue harder. This is sometimes advantageous, especially for the nervous system. If the presence of traces of methanol and formate can be tolerated (as in most histological work), the fixative may be more conveniently made by substituting 100 ml of formalin (40% HCHO) for the paraformaldehyde, thus obviating the need for heating and filtering.

The addition of 10 ml of 25% aqueous glutaraldehyde to 90 ml of the above mixture (preferably made with paraformaldehyde rather than formalin) gives a solution which will preserve more intracytoplasmic structures, such as mitochondria. These may be further stabilized for paraffin embedding by re-fixing small pieces of the tissue in osmium tetroxide (1% aqueous; 6 h) or potassium dichromate (3% aqueous; 7 days).

2.5.8. Formal–calcium

Formalin (40% HCHO neutralized by standing for a few days over marble chips):	100 ml
Calcium chloride ($CaCl_2$):	20 g
Water:	to 1000 ml

Prepare just before using

This is a formulation devised to insolubilize phospholipids prior to cutting frozen sections. Calcium acetate, ($(CH_3COO)_2Ca.H_2O$; 20 g), may be used instead of calcium chloride. This salt will also serve to neutralize any formic acid present in previously unneutralized formalin. Specimens are fixed for 1–3 days.

2.5.9. Formal–ammonium bromide

Formalin (40% HCHO; neutralized with calcium carbonate):	15 ml	Prepare just before using
Ammonium bromide (NH_4Br):	2 g	
Water:	to 100 ml	

This mixture has long been known to enhance the subsequent staining of neuroglia by metallic impregnation methods, for reasons which are unknown. It has no other uses. Specimens are fixed for 3–10 days.

2.5.10. Bodian's fixative

Formalin (40% HCHO):	5 ml	Mix just before using
Glacial acetic acid:	5 ml	
80% aqueous ethanol:	90 ml	

This mixture ("Formula No. 2" of Bodian, 1937) is useful when axons are to be stained by silver methods in paraffin sections. It and many other similar mixtures are also valuable when rapid penetration and the preservation of mucosubstances are required. Nucleic acids are also well preserved. Fix for 12–24 h.

2.5.11. Altmann's fixative

2% aqueous osmium tetroxide:	5 ml	Mix just before using
5% aqueous potassium dichromate:	5 ml	

This is one of many chromium–osmium mixtures (others are Flemming's, Champy's, Bensley's, Dalton's, etc.). Though once used as cytoplasmic fixatives, especially for mitochondria and the Golgi apparatus (which is blackened), these mixtures are rarely used today. Fixation time is 6–18 h. Specimens must be less than 2 mm thick. The metallic compounds must be removed by prolonged washing (24–48 h) in running tap water before dehydration.

2.5.12. Glutaraldehyde fixative

25% aqueous glutaraldehyde:	10 ml	Mix just before using
0.06 M sodium phosphate buffer pH 7.2–7.4:	to 100 ml	

This simple fixative is suitable for studying the effects of glutaraldehyde fixation in light microscopy. It penetrates rather slowly, so pieces of tissue should be not more than 3–4 mm in any dimension. Fix for 6–18 h.

2.6. METHODS OF FIXATION

Specimens may be fixed by **immersion** in at least twenty times their own volumes of the appropriate solution or by perfusion of the fixative through the vascular system. The latter technique has the advantage of ensuring that the fixing agent is rapidly brought into contact with all parts of all organs. Usually a whole animal is perfused, but it is also feasible to cannulate and perfuse the vasculature of a single organ. Fixation of material removed from a perfused animal is usually continued in the same fluid by immersion.

Very small or thin pieces may be fixed by leaving them for 2–24 h in the vapour above an aqueous solution of osmium tetroxide and freeze-dried blocks may be fixed in the vapour of formaldehyde or glutaraldehyde. Despite the excellence of the latter procedures, they are rarely used except in histochemical studies of monoamines.

Films, smears, and cryostat sections of unfixed material are immersed in buffered formaldehyde, cold (0–4°C) acetone, or absolute ethanol or methanol, either before or after the application of histo-

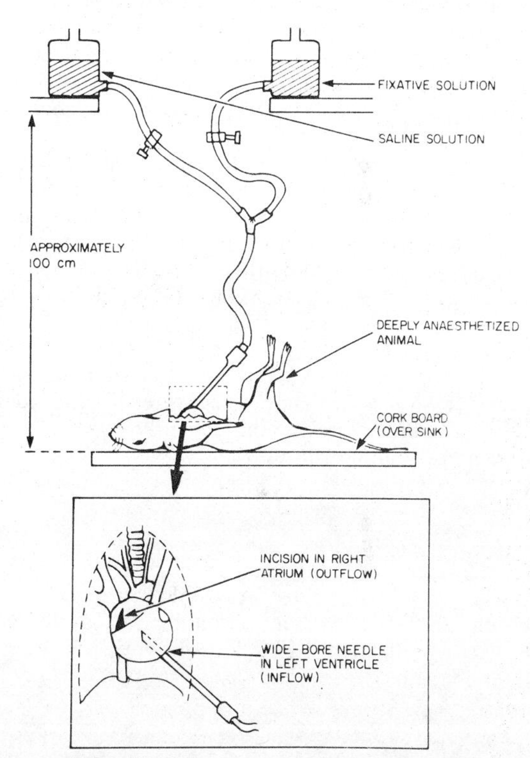

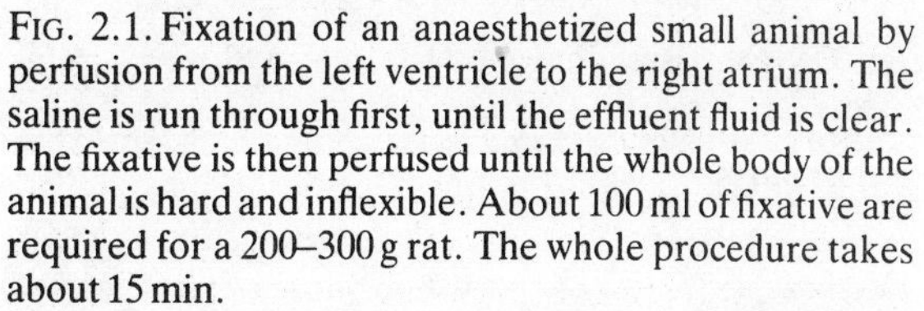

FIG. 2.1. Fixation of an anaesthetized small animal by perfusion from the left ventricle to the right atrium. The saline is run through first, until the effluent fluid is clear. The fixative is then perfused until the whole body of the animal is hard and inflexible. About 100 ml of fixative are required for a 200–300 g rat. The whole procedure takes about 15 min.

chemical methods for enzymes, according to the requirements of the particular techniques.

A simple method for perfusing small animals with fixatives is illustrated in Fig. 2.1. An **injected** anaesthetic should be used in this procedure.

2.7. EXERCISES

Theoretical

1. Deduce the differences that you would expect to see between the appearances of stained sections of animal tissue fixed in: (a) formalin, diluted ten times with water immediately before use, (b) an isotonic solution containing 4% formaldehyde and buffered to pH 7.4.

2. Why is it preferable to use Bouin's fluid rather than an aqueous solution of picric acid alone as a fixative?

3. Before the discovery of the fixative properties of glutaraldehyde, osmium tetroxide (in buffered, isotonic solutions) was the best available fixative for specimens to be examined in the electron microscope. In light microscopy, however, OsO_4 is of limited value as a fixative. Why should this be? Why is OsO_4 useful as a fixative for peripheral nerves but almost useless for central nervous tissue in light microscopy?

4. Glutaraldehyde, although it is an excellent fixative for electron microscopy, is not often used to fix tissues that are to be examined in the light microscope. Why? For which purposes would you expect glutaraldehyde to be the fixative of first choice in light microscopy?

5. Which fixative mixture would you use in order to preserve: (a) RNA, (b) neutral fat, (c) the Golgi apparatus, (d) myelin, (e) collagen and reticulin, (f) free amino groups of protein, (g) free carboxyl groups of mucosubstances, (f) an enzyme that was inactivated by aldehydes and by any protein-precipitating reagent?

6. Cetylpyridinium chloride (CPC)

$$CH_3(CH_2)_{14}CH_2N^{+}C_5H_5 \quad Cl^{-} \cdot H_2O$$

has been used as a fixative for a specific histochemical purpose. What type of substance would you expect to be insolubilized by CPC? Many morphological features are distorted after fixation in CPC. Explain why.

Practical

(See also exercises following Chapters 3 and 4)

7. Kill a rat and remove and fix pieces of the following organs:

(a) Skin, from the back after removing most of the hair (two pieces: neutral, buffered formaldehyde, and Bouin's fluid).
(b) The brain (formal–ammonium bromide).
(c) The external ear (Carnoy).
(d) Liver (Gendre's fluid).
(e) Kidney (Helly's fluid).
(f) The pituitary gland (Susa).
(g) The sternum (Helly's fluid).
(h) The adrenal glands (glutaraldehyde; formal–calcium).
(i) Submandibular salivary gland (two pieces: Carnoy and Bouin).

Keep (a), (b), and (h) for sectioning on the freezing microtome. Process (c) to (f) for paraffin embedding, with due attention to the fixatives used. Decalcify (g) and then embed in wax. Keep (b) in fixative for 2–3 weeks prior to cutting frozen sections. Embed (i) and the Bouin-fixed piece of (a) in paraffin. These specimens are for use in the practical exercises described at the ends of some of the following chapters. Other tissues (e.g. heart, lung, lymph node, intestine, bladder, testis, peripheral nerve) may also be taken if desired.

Cut paraffin sections as the blocks become ready. Obtain six slides of each with sections 7 μm thick, and six with sections 4 μm thick.

8. Anaesthetize a rat with ether and give it an intraperitoneal injection of Nembutal (solution containing 60 mg per ml of pentobarbitone sodium), 0.5 ml. This is a lethal dose for almost any rat. When the animal is deeply unconscious, perfuse it from the left ventricle to the right atrium with 0.9% NaCl (to wash out the blood) followed by 4% buffered formaldehyde (pH 7.2–7.4) for about 10 min. Remove the following organs and continue their fixation by immersion for 12–24 h.

(a) The brain (a piece 3–5 mm thick, passing transversely through the telencephalon and a similar piece passing through the pons and the cerebellum).
(b) A piece of the small intestine 10 mm long (trim to 5 mm long when in one of the dehydrating fluids).
(c) A submandibular salivary gland.
(d) A peripheral nerve, such as the sciatic. Take two pieces, for longitudinal and transverse sections.
(e) Some thick skin, together with subcutaneous tissue, from the back of the neck.
(f) The adrenal glands.
(g) The tongue (two pieces).

(a) to (d) and one of the pieces of (g) should be embedded in wax and sectioned at 7 μm and at 4 μm. About six slides at each thickness will be needed. (e), (f) and the other piece of (g) can be kept in the fixative for a few days. Frozen sections will be cut. These specimens are suitable for various histochemical methods described in later chapters.

3

Decalcification

3.1. Decalcification by acids 25
3.2. Decalcification by chelating agents 25
3.3. Decalcification in practice 26
3.3.1. Acid decalcifiers 26
3.3.2. Chelation with EDTA 27
3.3.3. End-point of decalcification 27
3.4. Exercises 28

Calcified tissues such as bone and dentine cannot be sectioned with an ordinary microtome using a steel knife. Such material is softened, after fixation, by chemically removing insoluble salts of calcium which are responsible for the hardness of the tissue. Decalcification is accomplished by treatment of the specimen with a suitable acid or with a chelating agent. Several techniques are available and are thoroughly discussed by Culling (1974), Page (1977), and Eggert & Germain (1979).

3.1. DECALCIFICATION BY ACIDS

The principal mineral component of the calcified tissues of vertebrate animals is a hydroxyapatite formally designated as $Ca_{10}(PO_4)_6(OH)_2$. Like other "insoluble" salts this exists, when wet, in equilibrium with its saturated solution, which contains very low concentrations of calcium, phosphate, and hydroxide ions:

$$Ca_{10}(PO_4)_6(OH)_2 \rightleftharpoons 10Ca^{2+} + 6PO_4^{3-} + 2OH^-$$

Continuous removal of calcium, phosphate, or hydroxide ions from the solution will prevent the system from reaching equilibrium. The reaction will therefore proceed from left to right until all the hydroxyapatite has dissolved. If the liquid surrounding the specimen has a high concentration of hydrogen ions, the reaction

$$H^+ + OH^- \rightarrow H_2O$$

will be driven from left to right, removing the hydroxide ions liberated as a result of dissolution of the hydroxyapatite. Any strong acid could serve as a source of hydrogen ions, but those that form sparingly soluble calcium salts (e.g. sulphuric acid) are, for obvious reasons, unsuitable. Hydrochloric, nitric and formic acids are the ones most often used. The overall reactions of these acids with hydroxyapatite are, respectively:

$$Ca_{10}(PO_4)_6(OH)_2 + 20H^+ + 20Cl^- \rightarrow 10Ca^{2+} + 20Cl^- + 6H_3PO_4 + 2H_2O$$

$$Ca_{10}(PO_4)_6(OH)_2 + 20H^+ + 20NO_3^- \rightarrow 10Ca^{2+} + 20NO_3^- + 6H_3PO_4 + 2H_2O$$

$$Ca_{10}(PO_4)_6(OH)_2 + 20HCOOH \rightarrow 10Ca^{2+} + 20HCOO^- + 6H_3PO_4 + 2H_2O$$

The calcium from the tissue ends up as calcium ions dissolved in the decalcifying fluid. If the latter is changed frequently the completion of decalcification can be recognized when calcium ions are no longer detectable by a simple chemical test (see below).

Decalcification by strong acids can result in hydrolysis of nucleic acids. This occurs less completely than in a deliberate Feulgen hydrolysis (see Chapter 9), but may still interfere with the interpretation of the results obtained with histochemical techniques. Most enzymes are put out of action by acid decalcifying agents, but the structure of the tissue is only slightly disrupted provided that fixation has been adequate. The crystal lattice of the hydroxyapatite of bones and teeth incorporates small numbers of carbonate ions in addition to the more abundant phosphates and hydroxides. The mineralized tissue contains, in effect, a small percentage of calcium carbonate. This is dissolved by acids:

$$CaCO_3 + 2H^+ \rightarrow Ca^{2+} + H_2O + CO_2\uparrow$$

Minute bubbles of carbon dioxide are formed within and on the surfaces of specimens being decalcified in acids, but they do not usually produce signs of damage visible under the microscope.

3.2. DECALCIFICATION BY CHELATING AGENTS

A chelating agent is an organic compound or ion which is able to combine with a metal ion to form a compound known as a metal chelate:

$$[\text{CHELATING AGENT}]^{-} + [\text{METAL}]^{+} \rightleftharpoons [\text{METAL CHELATE}]$$

In the chelate, the metal atom is covalently bound as part of a ring.

Chelates are very stable compounds and do not readily decompose to liberate metal ions. Consequently, the reaction of a metal ion with a chelating agent, although reversible, proceeds almost to completion. If the chelating agent is present in excess, virtually all the metal ions will be removed from the solution.

Ethylenediamine tetraacetic acid (EDTA) forms ordinary salts with sodium and other alkali metals, but the ethylenediamine tetraacetate ions combine with most other metal ions to form stable, soluble chelates. If a piece of calcified tissue is immersed in a liquid containing EDTA anions, calcium ions will be removed from the solution by chelation. Hydroxyapatite will therefore dissolve because it will be unable to attain equilibrium with a saturated solution.

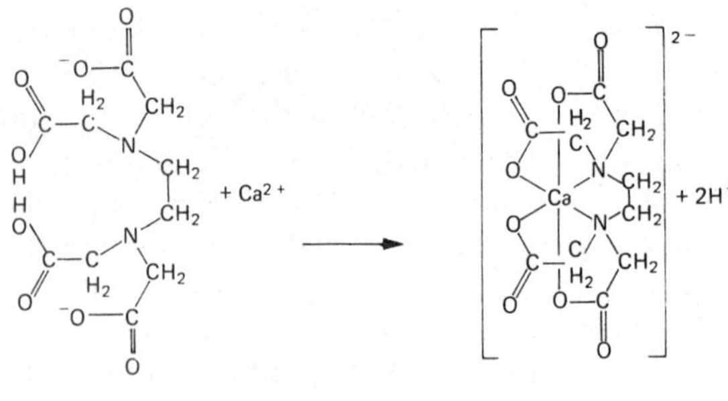

$(EDTA)^{2-}$
As in a solution of the disodium salt of EDTA

$[CaEDTA]^{2-}$
The bonds to the Ca atom are of equal length and mutually at right angles, directed as if to the vertices of a regular octahedron

For a full account of the chemistry chelation, see Chaberek & Martell (1959). The chelation of metal ions by dye molecules is described in Chapter 5 of this book. Decalcification by EDTA differs from decalcification by acids in that hydrogen ions play no part in the chemical reaction involved. The chelating agent is used at or near to neutrality, so the deleterious effects of strong acids on labile substances such as nucleic acids and enzymes are avoided. The main disadvantage of EDTA is that it acts much more slowly than the acids.

3.3. DECALCIFICATION IN PRACTICE

Specimens that are to be decalcified must be properly fixed. Swelling, with consequent disruption of structure, will occur if any undenatured protein, especially collagen, is immersed in a strongly acid fluid. The fixative must be thoroughly washed out of the tissue prior to decalcification in order to avoid undesirable chemical reactions. For example, dichromate ions would be reduced by formic acid; mercury would form a chelate with EDTA. The volume of decalcifying fluid should be at least twenty times that of the specimen. An acid mixture is changed every 24–48 h; an EDTA solution every 3–5 days. If the ammonium oxalate test (Section 3.3.3) is to be used, the anticipated last change of acid should have only five times the volume of the specimen, so that any calcium in the liquid will be present at a higher concentration than in a larger volume.

3.3.1. Acid decalcifiers

Buffered formic acid (Clark, 1954)

Formic acid (90%):	250 ml	Keeps indefinitely
Water:	750 ml	
Sodium formate ($HCOONa$):	34 g	

Alternatively, a solution with the same composition can be made by dissolving 19.8 g of sodium hydroxide in 729 ml water and adding 271 ml of 90% formic acid.

This solution, whose pH is 2.0, produces only minimal suppression of the stainability of nucleic acids. It is the decalcifier of first choice for most purposes.

De Castro's fluid

Absolute ethanol:	300 ml	**Mix in order stated.** Keeps indefinitely
Chloral hydrate:	50 g	
Water:	670 ml	
Concentrated nitric acid (70% HNO_3):	30 ml	

Caution. The nitric acid must be added last. Concentrated HNO_3 reacts explosively with absolute ethanol.

De Castro's fluid is traditionally used in conjunction with neurological staining methods. Tissues should not be left in it for longer than necessary or most staining properties will be suppressed. It is more strongly acidic (pH 1.0) than buffered formic acid and therefore decalcifies more rapidly. The alcohol and chloral hydrate in the mixture are supposed to prevent swelling of the tissue, presumably by increasing the osmotic pressure of the fluid. This is probably unimportant when adequately fixed specimens are being decalcified. However, de Castro's mixture is occasionally used as a simultaneous fixing and decalcifying agent. In this circumstance the osmotic effects, as well as the fixative properties of all three ingredients, assume greater significance.

Perenyi's fluid

Dilute nitric acid (10% aqueous HNO_3):	40 ml	Prepare just before using
Absolute ethanol:	30 ml	
0.5% aqueous solution of chromium trioxide (CrO_3):	30 ml	

This mixture is also used as a fixative, especially for insect tissues, eyes of vertebrates, and objects containing small calcareous deposits. It is a rather mild decalcifying agent, suitable only for specimens containing small amounts of calcified material. The solution turns blue soon after mixing owing to reduction of the dichromate ion (from CrO_3 and water) by ethanol:

$$Cr_2O_7^{2-} + 3C_2H_5OH + 8H^+ \rightarrow 2Cr^{3+} + 3CH_3CHO + 7H_2O$$

A simple calculation based on the stoichiometry of this equation shows that in Perenyi's fluid all the chromium is present as chromic cations and that only a tiny proportion of the ethanol is consumed in the reaction. When used as a fixative the active ingredients are therefore ethanol, nitric acid, and chromic ions.

Drury (1973) has shown that cartilage can usefully be softened by treating formaldehyde-fixed blocks with Perenyi's fluid for 24 h. This effect must occur by a mechanism different from that involved in decalcification. Perenyi's fluid also softens chitin.

3.3.2. Chelation with EDTA

Decalcification with EDTA proceeds much more slowly than with acids, several weeks often being required, but this is not injurious to the tissues. Some histochemical methods for enzymes can subsequently be performed upon frozen sections.

Either 5 or 10 g of disodium ethylene diamine tetraacetate, $[CH_2N(CH_2COOH)COONa]_2.2H_2O$, is dissolved in 100 ml of water and sufficient 4% NaOH added to bring the pH to about 6. The solution should be freshly prepared. It is changed every fourth day. Tissues treated with EDTA should be washed in water before dehydration, since the chelating agent is insoluble in alcohol.

3.3.3. End-point of decalcification

If the specimen is larger than necessary and contains calcified tissue throughout, it may be trimmed with a scalpel at intervals during decalcification. When it can be cut easily with the scalpel blade it will also be soft enough to be sectioned on a microtome. Sometimes the specimen is too small to be trimmed, or the calcified material is isolated in the middle of the block. An otherwise unwanted piece of bone of approximately the same size can then be processed alongside the specimen. When this piece of bone is fully softened the specimen for histological study should also be decalcified. Specimens should never be tested by poking needles into them; the holes will persist in the sections.

A more exact test for completeness of decalcifications makes use of the fact that calcium oxalate, though soluble in mineral acids, is insoluble in water and in aqueous solutions of alkalis.

$$Ca^{2+} + C_2O_4^{2-} + H_2O \rightarrow CaC_2O_4.H_2O$$

The test is conducted as follows:

(1) Add drops of strong ammonia solution (ammonium hydroxide, S.G. 0.9) to about 5 ml of used decalcifying fluid until the mixture becomes alkaline to litmus ($pH > 7$).

(2) Add 5 ml of a saturated aqueous solution of ammonium oxalate (approximately 3% $(NH_4)_2C_2O_4.H_2O$) and leave to stand for 30 min.

If no white precipitate has formed after this time, the fluid contains no calcium ions. This test can also be used to determine the end-point of decalcification by EDTA, even though solutions of the latter do not contain free calcium ions (Eggert & Germain, 1979). The $[CaEDTA]^{2-}$ anion presumably dissociates as the highly insoluble oxalate is formed.

If an X-ray machine is available the presence or absence of calcified deposits in a specimen can be demonstrated by radiography. Control specimens known to contain and not to contain calcified material should be X-rayed alongside the specimen being tested in order to enable the amount of radio-opacity due to soft tissues to be assessed.

3.4. EXERCISES

Theoretical

1. Following decalcification in acidic mixtures, tissues are often placed overnight in 5% sodium sulphate ($Na_2SO_4.10H_2O$; M.W. 322; atomic wt. of Na = 23). This is often said to be done in order to neutralize any acid remaining in the specimen, with the effect of preventing swelling during subsequent processing. Is this justification of the use of sodium sulphate correct? (*Hint*: a solution of NaCl (M.W. 58.5) isotonic with mammalian extracellular fluid is 0.154 M.)

2. Sections are required of a bone-containing specimen fixed in Bouin's fluid. What would you do before embedding the specimen in paraffin wax? Give reasons for all stages of the procedure. How would the procedure differ for a piece of Zenker-fixed tissue?

Practical

3. Decalcify a specimen of a rat's sternum together with associated muscle and costal cartilages, which has been fixed in Helly's fluid. Embed it in paraffin wax and cut sections at 7 μm and at 4 μm. Other instructive objects for decalcification include the posterior part of the snout and a piece of the vertebral column. These should be sectioned in the transverse plane. If the specimens are taken from the mouse rather than from the rat, they will not take so long to decalcify.

These sections of decalcified tissue may be used in the practical exercises associated with Chapters 6, 7, and 8.

4

Processing and Mounting

4.1. Processing 29
4.1.1. Gelatine embedding 29
4.1.2. Agar embedding 29
4.1.3. Dehydration, clearing, and paraffin embedding 30
4.1.3.1. Standard procedure 30
4.1.3.2. Chemical dehydration 31
4.1.4. Nitrocellulose embedding 31
4.1.5. Double embedding 34
4.1.6. Adhesives for sections 34
4.1.6.1. Mayer's albumen 34
4.1.6.2. Chrome-gelatine 35
4.1.7. Coating slides with nitrocellulose 35
4.1.8. Removal of mercury precipitates 36
4.2. Mounting media 36
4.2.1. Resinous media 36
4.2.2. Aqueous media 36
4.2.2.1. Glycerol jelly 36
4.2.2.2. Apathy's medium 37
4.2.2.3. Polyvinylpyrollidone (PVP) medium 37
4.3. Exercises 37

This chapter contains some practical instructions for the processing of specimens and the handling of sections and slides. The procedures described are basic to nearly all the staining and histochemical techniques presented elsewhere in the book. Much of the underlying theory has already been explained in Chapter 1.

4.1. PROCESSING

For comprehensive accounts of dehydration, clearing, and embedding the reader is referred to Gabe (1976) and Bradbury & Gordon (1977). The properties of several of the solvents used in these procedures are summarized in Table 4.1 (pp. 32–33).

4.1.1. Gelatine embedding

Gelatine does not penetrate the minute interstices of a specimen in the same way as paraffin or nitrocellulose. It surrounds the block of tissue and fills in the larger cavities and cracks. This amount of support permits the cutting of frozen sections of objects which would otherwise disperse into fragments when placed into water.

1. Wash formaldehyde-fixed specimens in a large excess of water at 37°C for 30–60 min.
2. Infiltrate with the following solution (which has to be melted before use) for 1–2 h at 37°C, with occasional turning of the specimen:

Gelatine powder:	30 g	Keeps for a few weeks at 4°C
Glycerol:	30 ml	
Water:	140 ml	
Thymol (as bacteriostatic):	One small crystal	

3. Orientate the specimen in the gelatine in a small petri dish and place in refrigerator (4°C) until set, about 1 h.
4. Cut out a square block, leaving 3–5 mm of gelatine around the specimen, and place the block in 4% neutral buffered formaldehyde (see Chapter 2, p. 22) at 4°C overnight. This will harden the gelatine.
5. Cut frozen sections in the usual way and mount onto slides. Albumen or chrome-gelatine (see Section 4.1.6 below) may be used as an adhesive. Dry by draining and then place slides on hotplate (45–50°C) until the gelatine begins to melt (about 15 min).

Pearse (1968, pp. 577–8) describes a rather less rigorous procedure for use in conjunction with enzyme histochemistry. The gelatine embedding mass stains quite strongly with anionic dyes, but this does not interfere with the interpretation of the appearance of the stained section.

4.1.2. Agar embedding

Agar is a macromolecular carbohydrate extracted from seaweeds. Gels made from agar melt at about 60°C and set on cooling to room temperature.

1. Thoroughly wash the fixed specimen.
2. Infiltrate for 2 h at 56–60°C in the following solution:

Water: 100 ml
Boil and then add:
Agar (bacteriological grade): 2 g
Stir until dissolved. If further heating is required, be careful not to char the undissolved agar at the bottom of the vessel. Store at 4°C.

Keeps for several months if not infected. Formalin (2 ml) may be added as an antibacterial preservative.

3. Orientate specimen in agar in a small petri dish and leave to set either at room temperature or at 4°C.
4. Cut out a block, leaving 3–5 mm of agar around the specimen.
5. Cut frozen sections and mount onto slides with an adhesive (see Section 4.1.6 below) if desired.
6. Drain the slides and dry on hotplate (45–50°C) for about 20 min.

The agar embedding mass is stained by cationic dyes and also by the periodic acid–Schiff method. Agar with formalin can be used for fixing and sectioning suspensions of protozoa and other small organisms (see Humason, 1972).

4.1.3. Dehydration, clearing, and paraffin embedding

The standard procedure is suitable for specimens no more than 5 mm thick. The times for washing, dehydration, clearing, and infiltration should be shorter for small specimens and longer (with more changes) for larger ones.

Many forms of paraffin wax are available both commercially and as extempore mixtures. The term "wax" is used below to mean any satisfactory formulation. Paraffin wax melting at 56°C and containing a little synthetic polymer is suitable for most purposes. The virtues and vices of different waxes are discussed in great detail by Steedman (1960). The wax should be kept about 2°C above its melting point.

It is often said that prolonged treatment with molten wax, especially if the temperature is too high, causes excessive hardening of the tissue. Gabe (1976) maintains that the hardening occurs only if the specimens have not been completely dehydrated. I agree with Gabe; the effects of insufficient infiltration are much more serious than those of unnecessarily long exposure to hot paraffin.

4.1.3.1. Standard Procedure

The starting point given below is water. If alcoholic fixatives have been used, the earlier stages of dehydration are omitted. Throughout the procedure, the volume of liquid should be 10–20 times that of the specimen. The "alcohol" may be ethanol, methanol, industrial methylated spirit (up to 95%), or isopropanol.

Pass the specimen through:

1. (Delicate specimens only; most solid pieces can go directly to stage 2.) Two hours in each of: 15%, 25%, and 50% alcohol

2. 70% alcohol	2 h (or overnight)	
3. 95% alcohol	2 h	
4. 100% alcohol (1)	2 h	
5. 100% alcohol (2)	1 h	
6. Terpineol (1)	2–24 h	One of these is conveniently overnight
7. Terpineol (2)	2–24 h	
8. Benzene	15–30 min (to remove terpineol from surface of specimen)	
9. Wax (1)	1 h	
10. Wax (2)	1 h	
11. Wax (3)	1 h	
12. Wax (4)	1 h	

The infiltration with wax may be carried out in the oven or in a vacuum-embedding chamber. The latter has the advantages of accelerating removal of volatile solvents and of extracting air bubbles (e.g. among hairs) which may be adherent to the specimen.

13. Block out in wax in a suitable mould.

Alternative clearing agents include chloroform and a mixture of equal volumes of chloroform and benzene. Times are the same as for terpineol, but stage 8 is not needed. Cedarwood oil is used for clearing hard tissues and is also generally considered to give superior preservation of intracellular structures. Two changes of 24 h are required, followed by a rinse in benzene (stage 8) and at least

2 h in each of the four changes of wax. The first two lots of wax must be discarded after they have been used to infiltrate specimens cleared in non-volatile solvents such as terpineol or cedarwood oil. The wax from the third and fourth changes may be used **once** again, for stages 9 and 10 with subsequent specimens. New terpineol or cedarwood oil must always be used for stage 7, but once-used solvents are permissible for stage 6. All other dehydrating and clearing agents should be discarded after using them once only. Quantities of inflammable liquids greater than 100 ml should not be poured down the sink, especially if immiscible with water. They should be collected in metal solvent drums for eventual burning.

Some liquids, such as dioxane, are miscible with water and with melted paraffin wax (Table 4.1) and may be used as combined dehydrating and clearing agents. These liquids are usually rather expensive. Dioxane is too toxic to use in large quantities.

Clearing agents of high refractive index (e.g. methyl salicylate, cedarwood oil) render pieces of tissue transparent. This change indicates that "clearing" is literally complete and is responsible for the customary but rather illogical use of the term. Several authors have advocated such expressions as "intermediate solvent", but "clearing agent" remains fixed in the practising histologist's vocabulary.

4.1.3.2. Chemical dehydration

Instead of replacing the water in a specimen with alcohol, dioxane, or the like, it is possible to use a reagent that reacts chemically with water to give liquid products which are miscible with the clearing agent. 2,2-dimethoxypropane (DMP) is such a reagent. It is a ketal and is hydrolysed by water, in the presence of an acid as catalyst, to yield methanol and acetone:

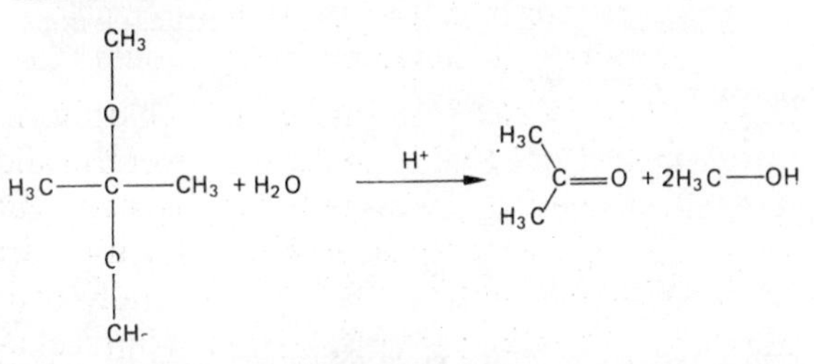

If a reasonable excess of DMP is provided, virtually all the water in a specimen will react. The hydrolysis is a strongly endothermic reaction, so it is advisable to warm the specimen and the acidified DMP to about 30°C in order to minimize the risk of formation of ice within the specimen while it is being penetrated by the reagent. When dehydration is complete, the specimen is equilibrated with a mixture of DMP, methanol, and acetone. DMP itself is immiscible with water but fully miscible with acetone, methanol, most other organic solvents, and paraffin wax. Since acetone and methanol do not mix with melted paraffin, specimens are cleared in benzene before infiltration.

Several practical procedures for chemical dehydration with DMP have been described. The following method (Prentø, 1978) is suitable for fixed pieces of tissue up to 3 mm in thickness.

1. Add 0.05 ml (2 small drops) of concentrated hydrochloric acid to 100 ml of 2,2′-dimethoxypropane (DMP) and stir thoroughly: 20 ml are needed for each specimen.
2. Warm the acidified DMP to 25–40°C in screw-capped vials. This is conveniently done by standing the vials on top of an oven or hotplate.
3. Place each specimen in 10 ml of the warm acidified DMP for 15 min. Agitate every 3–5 min.
4. Transfer to a second change acidified DMP for a further 15 min.
5. Clear in benzene or methyl salicylate (25 ml per specimen) for 15 min. Agitate every 3–5 min.
6. Infiltrate with three changes (each 45 min) of molten wax and make blocks in the usual way.

The rather slow rate of penetration of DMP precludes its use for the dehydration of large specimens, even when the time of exposure to this reagent is greatly extended.

4.1.4. Nitrocellulose embedding

Suitable types of nitrocellulose are sold under such names as celloidin, Parlodion, Necolloidin, and low-viscosity nitrocellulose (LVN). The solutions of nitrocellulose are made up well in advance and are stored in tightly screw-capped bottles. **They are dangerously inflammable.** Those required are: 8%, 4%, and 2% nitrocellulose, dissolved in a mixture of equal volumes of ethanol and diethyl ether. The times given below are for specimens approximately 15 mm thick. They may be halved for specimens 5 mm thick. The volume of liquid should be 5–10 times that of the specimen. The tissue is equilibrated with absolute ethanol and then transferred to:

Ether–alcohol mixture:	Overnight	In tightly stoppered specimen tubes
2% nitrocellulose:	1 week	
4% nitrocellulose:	1 week	
8% nitrocellulose:	1 week	

TABLE 4.1. *Properties of solvents used in histology*

Name	Refractive index (at 20°C)	Boiling point (°C)	Fire hazard[(a)]	Remarks
GROUP I. *Dehydrating agents*				
Miscible with water and with clearing agents (Group II), but not miscible with melted paraffin wax or with resinous mounting media.				
ACETONE	1.36	56	++	The "absolute" liquids usually contain about 1% by volume of water, but this can usually be ignored. Ethanol and isopropanol are preferred to the more volatile solvents for most purposes
ETHANOL (ethyl alcohol)	1.36	78	+	
ISOPROPANOL (isopropyl alcohol; propan-2-ol)	1.38	82	+	
METHANOL (methyl alcohol)	1.33	65	+	
GROUP II. *Clearing agents*				
Miscible with dehydrating agents (group I), melted paraffin wax, and resinous mounting media. Not miscible with water.				Tissues become transparent only when the refractive index of the clearing agent is higher than 1.47. Replacement of clearing agent by wax is notably slower when the boiling point of the former is above 150°C
AMYL ACETATE (isoamyl acetate)	1.40	142	+	
BENZENE	1.50	80	+	
n-BUTANOL (n-butyl alcohol; butan-1-ol)	1.40	118	+	Partially miscible with water. May be used for dehydration of blotted sections. Less extraction of dyes than with ethanol. Irritating vapour
CARBON TETRACHLORIDE	1.46	77		Toxic vapour
CHLOROFORM	1.45	61		S.G. 1.49. Cleared tissues do not sink to bottom
CEDARWOOD OIL	1.51	260 (approx.)		Only partially miscible with methanol
BENZYL BENZOATE	1.57	323		
METHYL BENZOATE (oil of niobe)	1.51	200		
METHYL SALICYLATE (oil of wintergreen)	1.54	223		
TERPINEOL (mixed isomers; synthetic oil of lilac)	1.48	200 (approx.)		Miscible with ethanol containing 15% water
TOLUENE	1.50	111	+	

(*continued*)

TABLE 2.1 (*cont.*)

Name	Refractive index (at 20°C)	Boiling point (°C)	Fire hazard[a]	Remarks
XYLENE (mixed isomers)	1.50	140 (approx.)	+	Clearing agent of choice for sections. Best avoided for blocks because it causes more hardening than any other clearing agent
GROUP III. *Solvents for combined dehydration and clearing*				
Miscible with water and with melted paraffin wax. Not necessarily miscible with resinous mounting media.				
t-BUTANOL (tertiary butyl alcohol; 2-methyl-propan-2-ol)	1.39	83	+	Freezes at 26°C
CELLOSOLVE (2-ethoxyethanol; ethylene glycol monoethyl ether)	1.41	135	+	
DIOXANE (1,4-dioxane; diethylene oxide)	1.42	101	+	Toxic vapour
TETRAHYDROFURAN (tetramethylene oxide)	1.41	66	++	
TETRAHYDROFURFURYL ALCOHOL	1.45	177		
GROUP IV. *Miscellaneous liquids*				
WATER	1.33	100		
ETHYLENE GLYCOL	1.43	198		Miscible with water and dehydrating agents (group I) but not with clearing agents (group II)
GLYCEROL (glycerine)	1.47	290		(as above)
PROPYLENE GLYCOL	1.43	188		(as above)
ETHER (diethyl ether)	1.35	35	++	(Mixed with ethanol.) Used in nitrocellulose embedding. Slight miscibility with water. The vapour softens nitrocellulose and enhances adhesion of celloidin sections to slides
LIGROIN (low boiling)		60–120	++	Sometimes used as clearing agents for specimens or slides. Only partially miscible with alcohols
PETROLEUM ETHER (mainly hexanes)	1.37 (*n*-hexane)	60 (approx.)	++	(as above)
ISOPENTANE (2-methyl butane)	1.42 (at −150°C)	28	++	Freezes at −160°C. The cooled liquid is used for rapid freezing of tissues
LIQUID NITROGEN		−196		Used in rapid freezing procedures
ACETIC ACID	1.37	118		Miscible with water and all solvents in groups I–III. Freezes at 16.6°C. Used in fixatives, buffer solutions, and staining solutions

[a] + means the liquid is inflammable and should not be used near a naked flame. ++ means the liquid is very dangerously inflammable. When these liquids are in use, all flames must be extinguished and no smoking allowed anywhere in the room. No symbol in this column means the liquid, if inflammable, is not volatile enough to constitute a serious hazard. Water, chloroform, and carbon tetrachloride are the only volatile liquids in the list whose vapours are non-flammable.

Orient the tissue in a suitable mould, which should be at least three times as deep as the specimen. Mark the level of the surface of the 8% nitrocellulose solution on the outside of the vessel. Leave covered for 12 h or until free of bubbles, then remove the lid from the mould and place in a desiccator, with the lid partly open, in a fume hood. Leave until the depth has halved as a result of **slow** evaporation of the solvent. This may take 2–10 days. Put some chloroform in the bottom of the desiccator, close the lid, and leave for a further 48 h. The nitrocellulose is hardened by the action of the chloroform vapour. Cut out a block of celloidin containing the specimen and glue it with 4% nitrocellulose to a wooden block. The block is stored in 70% ethanol, which causes further hardening of the embedding medium. The hardened nitrocellulose blocks must not be allowed to dry out.

If LVN is used, 5%, 10%, and 20% solutions are used (in ether–alcohol containing 0.5% of castor oil), the times of infiltration given for celloidin may be halved and evaporation is allowed to proceed only until a hard crust has formed. Culling (1974) gives a full account of the use of LVN.

4.1.5. Double embedding

Hardened nitrocellulose can itself be infiltrated with melted wax in the procedure known as double embedding. This should be used for specimens that contain both hard and soft tissues. *Method 1* is popular, but probably does not introduce enough nitrocellulose into the specimen to make much difference to the hardness of the softer tissues. *Method 2* (Pfühl's method, as described by Gabe, 1976) takes longer but allows a thorough infiltration with nitrocellulose, which is then hardened by chloroform and by the action of the phenol dissolved in the clearing agent. Double-embedded blocks are sectioned on a rotary microtome in the same way as ordinary wax-embedded specimens.

Method 1

The dehydrated specimens are processed as follows. Times given are for pieces 5 mm thick.

1. Ether–alcohol: 2 h.
2. 2% celloidin in ether–alcohol: four changes for a total of 48 h.
3. Methyl salicylate: overnight or until the specimen is transparent (use at least two changes).
4. Rinse in benzene: 15 min.
5. Infiltrate with wax as described previously.

Method 2

Dehydrated specimens (5–10 mm thick) are processed as follows. Times are not critical. Specimens should be moved about and turned over two or three times every day.

Clearing agent (phenolic benzene)

Phenol:	100 g Heat gently until all melted, then pour into:	Keeps for several months in a dark cupboard. Discard when dark brown. Slight brown coloration does not matter
Benzene:	1000 ml	

Procedure

1. Ether–alcohol: 4–8 h.
2. 2% nitrocellulose: 2–4 days.
3. 4% nitrocellulose: 3–6 days.
4. Wipe off excess 4% nitrocellulose and transfer specimen to chloroform: 1–2 days.
5. Clear in phenolic benzene: 12–24 h.
6. Infiltrate with wax as described previously.

4.1.6. Adhesives for sections

Properly flattened sections will usually stick to grease-free glass slides without the assistance of an adhesive. Thin sections (<7 μm) generally stick better than thicker ones. For most purposes, however, the use of an adhesive is strongly advised. It is essential when the sections are to be exposed to solutions more alkaline than about pH 8. The two adhesives given below are suitable for sections to be stained by almost any techniques. Mayer's albumen has better keeping properties than chrome-gelatine, but the latter is very much more efficacious.

4.1.6.1. Mayer's albumen

Preparation (three alternatives are available)

(a) Collect the whites of one or two eggs into a graduated 250 ml beaker. Add an equal volume of glycerol and mix thoroughly. Filter through cloth and then through cotton wool or coarse filter paper. Filtration is accelerated if carried out in an oven at 55–60°C. Add a small crystal of thymol to inhibit growth of micro-organisms.

(b) Make a 5% solution of dried egg-white (commonly called "albumen, egg" in catalogues) in 0.5% aqueous sodium chloride. It takes a day to dissolve, with occasional stirring. Filter through coarse filter paper and add an equal volume of glycerol to the filtrate. Add a crystal of thymol.

(c) Buy a ready-made Mayer's albumen solution. These solutions will all keep for several months at 4°C.

Application

Place a *small* drop on a slide and distribute it evenly over the surface with the tip of the finger. Float out the sections on water, collect onto the slide, drain, and flatten in the usual manner. Alternatively, do not put Mayer's albumen solution on the slides but add about 20 ml of it to each litre of the water used for floating out the sections.

When the slides are dry, put them in the wax oven at about 60°C for 30 min or turn up the temperature of the hotplate used for drying the slides. The heat coagulates the egg albumen. The melting of the wax probably also promotes closer contact between the sections and the glass. The albumen is lightly stained by most dyes and is most conspicuous around the sites of the edges of the wax ribbon. The layer between the sections and the slide must be extremely thin.

4.1.6.2. Chrome-gelatine

Preparation

Dissolve 1.0 g gelatine powder (a high quality bacteriological grade gelatine is advised) in 80 ml of warm water and allow to cool. Dissolve 0.1 g chrome alum (chromic potassium sulphate: $CrK(SO_4)_2.12H_2O$) in 20 ml of water. Mix the two solutions. This mixture quickly becomes infected and should never be used if it is more than 2 days old.

Application

Place a *large* drop on the slide, spread it over the surface with a finger, and leave for about 10 min to dry. Treated slides may be kept for a few days if protected from dust. Float the sections onto the slides, flatten, drain, and dry in the usual way. Further heating may be applied to melt the wax if desired.

4.1.7. Coating slides with nitrocellulose

When mounted sections have to be subjected to rough treatment, especially immersion in alkalis, it is advisable to encase the slide in a thin film of nitrocellulose. This will hold the sections in place if there is failure of the adhesion between tissue and glass. The film should cover the whole slide, both surfaces and all four edges, if it is to be effective. The procedure is as follows:

1. Take slides to absolute alcohol.
2. Immerse in ether–alcohol (equal volumes of ethanol and diethyl ether) for 30–60 s.
3. Immerse in a 0.2–0.4% solution of nitrocellulose in ether–alcohol for 30–60 s. (This solution is made by diluting one of the stock solutions kept for nitrocellulose embedding. Remember that ether is volatile and dangerously inflammable.)
4. Lift out the slides, drain, and allow them to become almost dry. A change in the reflection of light from the glass surfaces indicates the moment at which to move on to stage 5. This end-point is easily learned with a little practice.
5. Place slides in 70% ethanol for about 5 min to harden the film of nitrocellulose.
6. Carry out the staining procedure. Dehydrate as far as the 95% alcohol stage.
7. Transfer the slides from 95% alcohol into ether–alcohol, with minimum agitation. Leave in ether–alcohol for 2–3 min to dissolve the nitrocellulose film.
8. Carefully take the slides from ether–alcohol into xylene (1 min) and then into a second change of xylene (at least 1 min). Try not to agitate the slides: this could loosen the sections.
9. Apply coverslip, using a resinous mounting medium.

4.1.8. Removal of mercury precipitates

Tissues fixed in mixtures containing mercuric chloride contain randomly distributed black particles. The chemical nature of this deposit is not known with certainty, but it is generally assumed to be mercurous chloride (Hg_2Cl_2), possibly with some finely divided metallic mercury. The precipitate does not disturb the structure of the tissue and it is easily removed by the following method. Sections of all mercuric-chloride-fixed specimens must be subjected to this procedure before staining.

Solutions required

A. Alcoholic iodine

Iodine:	2.5 g	Keeps indefinitely and can be re-used many times (see *Note* below)
70% ethanol:	500 ml	

B. Sodium thiosulphate solution

Sodium thiosulphate ($Na_2S_2O_3.5H_2O$):	15 g	Keeps indefinitely. May be re-used three or four times
Dissolve in water and make up to:	250 ml	

Procedure

1. Take sections to 70% ethanol.
2. Immerse in alcoholic iodine (solution A) for approximately 3 min. With continuous agitation, 1 min is sufficient.
3. Rinse in water.
4. Immerse in 5% sodium thiosulphate (solution B) until the yellow staining due to iodine has all been removed. This usually takes about 30 s.
5. Wash in running tap water for 2 min, then rinse in distilled water.

Note: Alternatively, a 1% solution of iodine in 2% aqueous potassium iodide may be used. This stains the sections deep brown and the decolorization at stage 4 takes longer than after 0.5% alcoholic iodine.

4.2. MOUNTING MEDIA

4.2.1. Resinous media

Resinous mounting media are of three types: natural, semi-synthetic, and wholly synthetic. Canada balsam is the traditional natural mounting medium. It has a pale yellow colour which deepens with age. One of its principal components, abietic acid ($C_{19}H_{29}COOH$), actively differentiates cationic dyes and any other substances in the sections that are susceptible to attack by acids. Unsaturated compounds are also present and can behave as reducing agents towards dyes and other coloured substances. Other natural resins, such as gum dammar and gum sandarac, have similar faults. The semi-synthetic media contain natural resins mixed with synthetic polymers. In some, the carboxylic acids of the natural materials have been esterified. The entirely synthetic resinous mounting media do not have the drawbacks mentioned above. A polystyrene-based medium, DPX, is recommended for all purposes. Its refractive index (1.52) is close to that of most tissues and, since it is not fluorescent, it is suitable for fluorescence microscopy. DPX contains no acid and no reducing agents.

Instructions for making resinous mounting media are given by Lillie & Fullmer (1976), but it is generally best to buy them ready-made.

4.2.2. Aqueous media

Water-miscible media are easily made in the laboratory. Three are listed below. They should be stored in screw-capped bottles, and care is necessary to ensure that the lids do not become too firmly cemented on.

4.2.2.1. GLYCEROL JELLY

Gelatine powder:	10 g
Water:	60 ml
Dissolve by warming and add:	
Glycerol:	70 ml

Add **either** one drop of saturated aqueous solution of phenol **or** 15 mg of sodium merthiolate as an antibacterial agent

Keeps for a few weeks at 4°C. Discard when turbid or mouldy.

Glycerol jelly must be melted and freed of air bubbles before use. This is conveniently done in a vacuum-embedding chamber. Because of the low refractive index (1.42), many unstained structures remain visible in this medium.

4.2.2.2. Apathy's Medium

Gum arabic (= gum acacia):	50 g
Sucrose:	50 g
Water:	50 ml
Thymol:	One small crystal

Keeps for a few months at room temperature. Discard if it becomes infected or if the sugar crystallizes.

Dissolve the ingredients, with frequent stirring and occasional heating on a water-bath. This can take up to 3 weeks, even with continuous magnetic stirring, for some batches of gum. The final volume should be approximately 100 ml. Apathy's medium has a refractive index (about 1.5) higher than that of glycerol jelly, so it provides more transparent preparations.

4.2.2.3. Polyvinylpyrollidone (PVP) Medium

Polyvinylpyrollidone (M.W. 10,000):	25 g
Water:	25 ml

Dissolve the PVP by leaving for several hours on a magnetic stirrer. Then add:

Glycerol:	1.0 ml
Thymol:	One small crystal

Usually keeps for 2 to 3 years. Discard if it looks infected.

This mounting medium is less viscous than glycerol jelly or Apathy's and is very easy to handle. The refractive index is 1.46 (Pearse, 1968), but increases as the water evaporates at the edges of the coverslip until unstained structures are barely visible.

4.3. EXERCISES

Theoretical

1. Sections of a formaldehyde-fixed specimen are required for a histochemical study of carbohydrate-containing lipids. Which embedding procedures would and would not be satisfactory, and why?

2. Explain how a piece of tissue is dehydrated by (a) dioxane, (b) 2,2-dimethoxypropane. Devise a reasonable practical procedure for the use of dioxane in taking a specimen 5 mm^3 from Helly's fixative into paraffin wax.

3. Paraffin sections of material fixed in Heidenhain's Susa are to be stained by a histochemical method for arginyl residues which involves the use of a strongly alkaline reagent. How would you handle the slides between removal of the wax and immersion in the alkaline solution?

4. Euparal is a semi-synthetic resinous mounting medium containing gum sandarac and paraldehyde. It is miscible with absolute ethanol or xylene. Its refractive index is 1.48. For which of the following purposes would it be reasonable to use euparal?

(a) Sections stained with a cationic dye.
(b) Sections in which occasional cells contain a stable, insoluble, coloured histochemical end-product.
(c) Unstained whole mounts of small multicellular animals or plants.
(d) Sections stained with fat-soluble dyes.
(e) Stained preparations containing coloured substances that are easily bleached by reducing agents.

Practical

5. Process and embed specimens fixed according to the practical exercises at the end of Chapter 2, and at least one decalcified specimen (Chapter 3, Exercise 3). Use the techniques best suited to the individual specimens.

6. Take two slides bearing paraffin sections of a tissue fixed in a mercuric-chloride-containing mixture. Process one but not the other as described in Section 4.1.8. Stain both slides by one of the methods described in Chapter 6. Observe the effect of failing to remove mercurial deposits. This should be the first and last time that you will see this artifact.

7. Stain some paraffin sections of any tissue with (a) polychrome methylene blue (p. 85), (b) haematoxylin and eosin (p. 81); and (c) nothing. Mount in DPX, glycerol jelly, and Apathy's medium (nine slides altogether).

Examine the slides immediately, after 1 or 2 h and after 1 or 2 weeks. Account for the changes that occur. If you do anything foolish, such as mounting from xylene into an aqueous medium, take note of the result and do not do it again.

5

Dyes

5.1. General structure of dye molecules 38
5.2. Colour and chromophores 39
5.3. Auxochromic groups 40
5.4. Fluorescent compounds 40
5.5. Combination of dyes with substrates 41
5.5.1. Electrovalent attraction 42
5.5.2. Covalent combination 42
5.5.3. Hydrogen bonding 43
5.5.4. Charge-transfer bonding 43
5.5.5. van der Waals forces 44
5.5.6. Hydrophobic interaction 44
5.5.7. Mordants 44
5.6. Nomenclature and availability of dyes 46
5.7. Classification of dyes by methods of application 48
5.7.1. Basic dyes 48
5.7.2. Acid dyes 48
5.7.3. Direct dyes 48
5.7.4. Mordant dyes 48
5.7.5. Reactive dyes 49
5.7.6. Solvent dyes 49
5.7.7. Vat dyes 49
5.7.8. Sulphur dyes 49
5.7.9. Ingrain dyes 49
5.7.10. Polycondensation (condense) dyes 49
5.7.11. Azoic dyes 50
5.7.12. Oxidation bases 50
5.7.13. Disperse dyes 50
5.7.14. Pigments 50
5.7.15. Fluorescent brighteners 50
5.7.16. Food colours 50
5.8. Classification of stains by microscopists 50
5.9. Classification of dyes by chromophoric systems 51
5.9.1. Nitroso dyes 51
5.9.2. Nitro dyes 51
5.9.3. Azo dyes 52
5.9.3.1. Synthesis 52
5.9.3.2. Structure 53
5.9.3.3. Anionic azo dyes 53
5.9.3.4. Mordant azo dyes 54
5.9.3.5. Cationic azo dyes 55
5.9.3.6. Reactive azo dyes 55
5.9.3.7. Solvent azo dyes 55
5.9.3.8. Direct azo dyes 55
5.9.3.9. Azoic dyes 56
5.9.4. Indamine and indophenol dyes 58
5.9.5. Cyanine dyes 59
5.9.6. Stilbene dyes and fluorescent brighteners 59
5.9.7. Quinoline dyes 60
5.9.8. Thiazole dyes 60
5.9.9. Aminoketone and hydroxyketone dyes 60
5.9.9.1. Natural dyes 60
5.9.9.2. Synthetic dyes 62
5.9.10. Arylmethane dyes 62
5.9.10.1. Diphenylmethane dyes 62
5.9.10.2. Triarylmethane dyes 63
5.9.11. Xanthene dyes 66
5.9.12. Acridine dyes 68
5.9.13. Azine dyes 68
5.9.14. Oxazine dyes 69
5.9.15. Thiazine dyes 70
5.9.16. Indigoid and thioindigoid dyes and pigments 71
5.9.17. Naphthoquinone dyes 72
5.9.18. Anthraquinone dyes 72
5.9.19. Phthalocyanine dyes 74
5.10. Exercises 75

Many compounds are coloured but not all of them are dyes. A dye is a coloured compound that can be bound by a substrate. In histological staining dyes are used to impart colours to the various components of tissues. Sometimes the colouring process has a high degree of chemical specificity, so that the dye can be used as a histochemical reagent. In many other histochemical techniques the reagents are colourless, but coloured substances are formed by reactions involving components of the tissue. Often these end-products are insoluble compounds chemically related to dyestuffs.

An important work of reference for histochemists and other biologists who use dyes is *Conn's Biological Stains* (9th edn., ed. Lillie, 1977), which includes descriptions of many dyes and much other information. Abrahart (1968) and Allen (1971) provide accounts of the chemistry of dyes in relation to their industrial uses. References to more advanced treatises and the *Colour Index* (see p. 46) can be found in these three books.

5.1. GENERAL STRUCTURE OF DYE MOLECULES

A molecule of a dye possesses two characteristic parts. These are the **chromophore,** which is the arrangement of atoms responsible for the absorption of light in the visible part of the spectrum, and the **auxochrome†**, which is the part of the molecule that attaches to the substrate. The auxochrome is an ionizable substituent which can form an ionic bond with an oppositely charged group in the substrate, or a substituent that readily reacts to form a covalent bond with the substrate. A **chromogen** is an organic molecule (or part of a molecule) containing a chromophoric system of atoms. Many dyes contain more than one auxochrome and most contain additional radicals that can modify the colour of the complete compound. The colour is also influenced by the number and types of auxochromic groups. A compound which includes a chromophore but is unable to ionize will be only weakly coloured. The ionic charge of an auxochrome is balanced by an oppositely charged ion, most commonly H^+, Na^+, or Cl^-. The old term **gegen-ion** conveniently describes the balancing ion.

A **fluorochrome** is a dye which absorbs ultraviolet or blue (or occasionally green) light and emits light of longer wavelength. Many fluorochromes are not or scarcely coloured. They are used in fluorescence microscopy.

5.2. COLOUR AND CHROMOPHORES

Compounds are coloured because their molecules absorb quanta of electromagnetic radiation in the visible part of the spectrum. The energy absorbed by the molecule causes changes in the energy levels of electrons—those involved in covalent bonding as well as those present as lone pairs associated with some atoms. With organic compounds such as dyes the absorption of visible light is largely attributable to the electrons associated with those combinations of atoms known as chromophores. Chromophores are arrangements of the following atomic linkages:

$$C{=}C,\ {=}O,\ C{=}S,\ C{=}N,\ N{=}N,\ N{=}O,\ {=}O,\ {-}NO_2$$

† The word "auxochrome" is used here as it is by dye chemists and histologists. Organic chemists use the same word to mean an atom or group of atoms which changes the wavelength at which a chromophore absorbs maximally. Gurr (1971) used "colligator" to denote the part of a dye molecule involved in attachment to the substrate, but this term has not found wide acceptance.

The double bonds always alternate with single bonds to form what are known as **conjugated systems.** The bonding electrons are able to move from one atom to another along a conjugated system, with the effect of exchanging the positions of the atomic linkages formally designated as double and single bonds. This phenomenon is known as **resonance.** Benzene is a resonance hybrid of the extreme structures

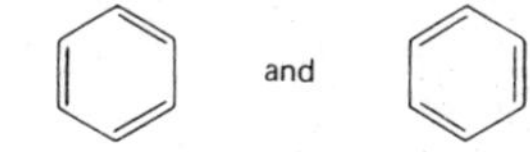

Intermediate forms cannot be represented by simple structural formulae. The equivalence of the bonds in a ring such as that of benzene confers a stability that would not otherwise be expected in a cyclic unsaturated compound. Rings stabilized by resonance are said to display **aromatic** character. The conjugated chromophoric system of a dye usually includes double bonds embodied in the formal structure of one or more aromatic rings. The resonance of the bonds in such rings commonly permits the existence of both aromatic and non-aromatic configurations in different rings within the same dye molecule.

The following chromophoric systems are those most frequently encountered in biological stains:

(a) The **nitro** group:

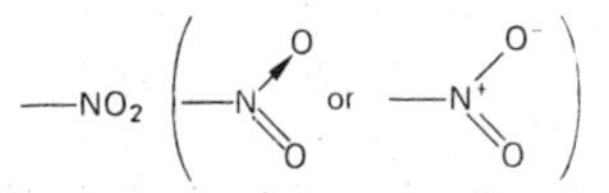

The single (dative) and double bonds between N and O are equivalent because of resonance.

(b) The **nitroso** group: —N=O.

(c) The **indamine** group: —N=. This always forms part of a larger chromophoric system.

(d) The **azo** group: —N=N—. This will be discussed at some length later (see Section 5.9.3).

(e) The **quinonoid** configuration:

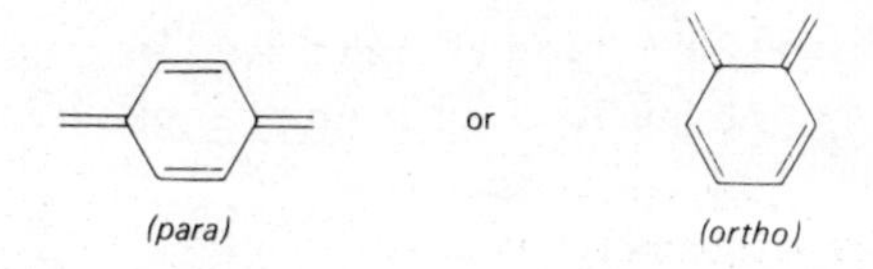

Note that the quinonoid ring itself is not aromatic. In dyes, resonance often exists among quinonoid and aromatic rings in the same molecule.

(f) Other chromophoric systems. These will be described in the accounts of the different types of dyes. Often the common chromophores (a)–(e) shown above occur in combination. Thus the **quinone–imine** configuration

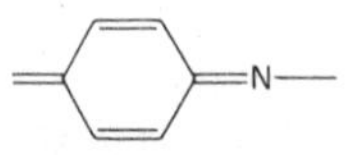

is present in many dyes such as the triphenylmethanes, the azines, the oxazines, and the thiazines.

5.3. AUXOCHROMIC GROUPS

The auxochrome or "colligator" (Gurr, 1971) is traditionally held to be responsible for attaching the chromogen to the substrate. Other factors, however, are also involved in the binding of dyes; these will be discussed later.

The **basic** auxochromes are amines, which ionize thus:

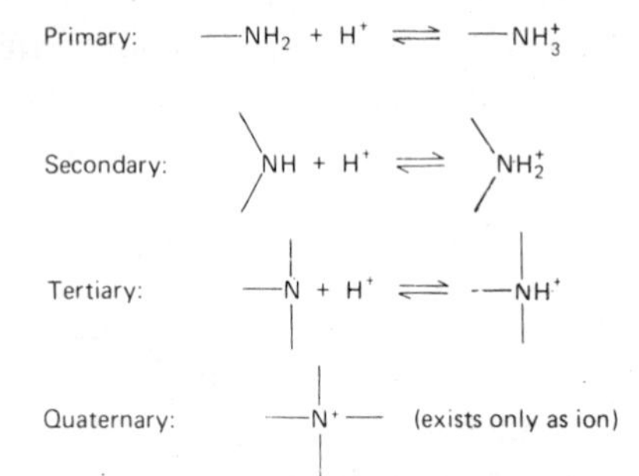

A dye with a net positive charge on the chromogen is called a **basic** or (preferably) a **cationic dye.** On treatment with strong alkali, the free bases (typically unionized amines) of such dyes will be liberated. Acid solutions will favour the formation of cations. The gegen-ions of such dyes are usually chloride or sulphate ions.

The **acid** auxochromes are derived from carboxylic or sulphonic acids or from phenolic hydroxyl groups:

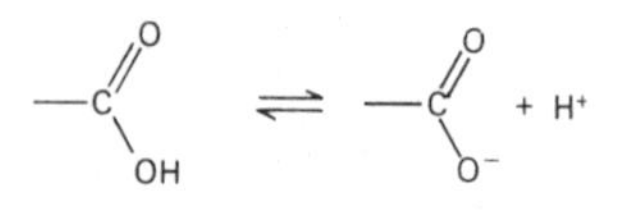

$$—SO_3H \rightleftharpoons —SO_3^- + H^+$$

$$—Ar—OH \rightleftharpoons —Ar—O^- + H^+$$

(Ar = aromatic ring)

They occur in **acid dyes** (anionic dyes), which may be encountered as the free acids or phenols or as salts, usually of sodium. The carboxylic acids are weak acids, being only partially ionized in aqueous media. Most phenols are even weaker acids. Consequently, dyes with carboxylic acid or phenolic auxochromes cannot exist as anions in the presence of high concentrations of hydrogen ions, since the equilibrium

$$Dye—H \rightleftharpoons Dye^- + H^+$$

will be pushed over to the left. On the other hand, sulphonic acids are strong, being fully ionized even at very low pH. Sulphonated dyes will therefore behave as anions even in strongly acid media.

From the foregoing paragraphs it will be seen that the pH of the solution will have profound effects on the colouring abilities of anionic and cationic dyes.

Reactive auxochromes are able to combine with hydroxyl and amino groups of the substrate to form covalent bonds. The most important reactive auxochromes are the halogenated triazinyl groups, such as dichlorotriazinyl,

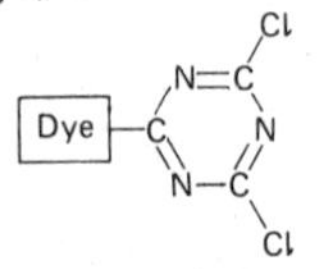

derived from cyanuric chloride. This auxochrome is usually attached to a nitrogen atom of the chromogen. In the monochlorotriazinyl group, one of the chlorine atoms in the above formula is replaced by an alkyl or amino group. The chlorine atoms confer high chemical reactivity, similar to that of the acyl halides. Condensation with hydroxyl or amino groups of the substrate, with elimination of HCl, occurs most readily in alkaline conditions. The dichlorotriazinyl group is more reactive than monochlorotriazinyl and will combine with —OH and —NH_2 at room temperature. Monochlorotriazinyl dyes have to be applied from hot solutions.

5.4. FLUORESCENT COMPOUNDS

The chemical features that enable an organic compound to fluoresçe are less easily defined than those responsible for colour. In order to be fluorescent, a molecule must contain a system of conjugated double bonds in its hydrocarbon skeleton. The double bonds may include those of aromatic rings; indeed, cyclic conjugated systems are associated with stronger fluorescence than are linear ones. Strongly fluorescent compounds, the only ones of interest in microscopy, usually have rigidly coplanar molecules. Thus, fluorene, in which planarity is stabilized by a methylene bridge, is more strongly fluorescent than biphenyl in which the two phenyl radicals are free to rotate about the single bond that joins them:

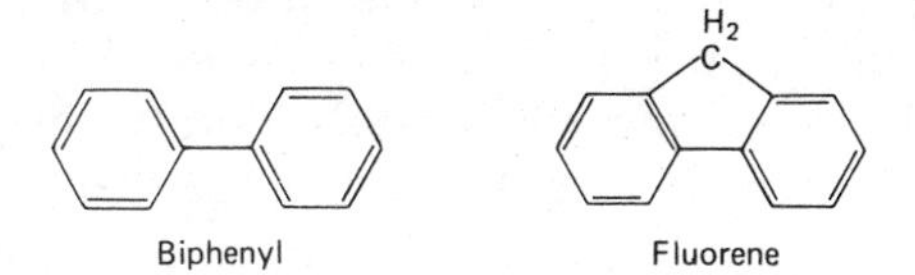

Compounds with coplanar fused ring systems are more strongly fluorescent when the rings are all joined side to side than when there is a bend in the structure. Anthracene therefore fluoresces more brightly than phenanthrene in response to the same level of exciting illumination. That is, anthracene has the higher **fluorescence efficiency** of the two compounds:

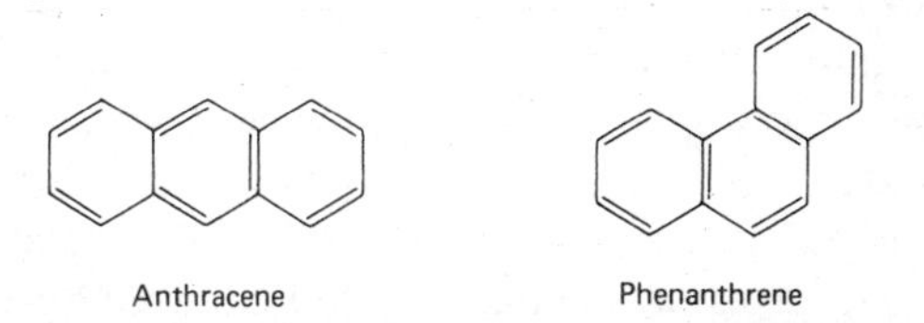

Fluorescence is also affected by substituents on the conjugated hydrocarbon skeleton of the molecule. When a single substituent is present the fluorescence efficiency is usually increased by —OH, $—OCH_3$, —F, —CN, $—NH_2$, $—NHCH_3$, and $—N(CH_3)_2$, and reduced by $>C{=}O$, —COOH, —Cl, —Br, —I, $—NO_2$, or $—SO_3H$. Alkyl side-chains generally have no effect unless they sterically hinder the assumption of a planar configuration. The above generalizations do not always apply when more than one substituent is present. Salicylic acid (*o*-hydroxybenzoic acid), for example, is fluorescent despite its carboxyl group. With heterocyclic compounds, including many dyes, the situation is complex and not fully understood. The fluorescent properties of such substances cannot be predicted by simple inspection of their structural formulae.

The fluorescence of a compound can be influenced by other substances with which it is mixed and by the pH of an aqueous solvent or histological mounting medium. For a more detailed but still introductory account of the fluorescence of organic compounds, see Bridges (1968).

5.5. COMBINATION OF DYES WITH SUBSTRATES

Dyes are not applied to tissues in the same way as to textiles. The histologist always uses a large excess of the staining solution, usually at room temperature, and the amount of dye taken up by the tissue is a negligible proportion of that present in the vessel. Staining is said to be **progressive** when a solution of a dye is allowed to act slowly until the desired effect is obtained. In **regressive** staining, the tissue is deliberately overstained, and then placed in a reagent (often water, alcohol, or acidified alcohol) that slowly removes the dye until the latter is left behind only in the components in which colour is wanted. This process of controlled removal of a dye is known as **differentiation.** The success of both the progressive and the regressive modes of staining depends on the fact that dyes generally have higher affinities for some components of tissues than for others. The textile dyer uses different methods. His dyebath is commonly used at or near boiling point, and most of the dye in it is transferred to the material being dyed. Some industrial dyeing techniques result in the deposition of quite large insoluble coloured particles; such methods would obviously be inappropriate for microscopy. Despite technological differences, however, many of the principles of textile dyeing can be applied to the use of biological stains.

The binding of dye molecules to the various textile fibres has been studied for many years, but is still not fully understood. It is realized that different mechanisms are involved for the different classes of dyes and types of fibre (see Allen, 1971, for review). The interactions between dyes and tissues have been reviewed by Baker (1958) and, more recently, by Horobin (1977). A chromogen may be attached to a dyed substrate by **electrovalent (ionic) attraction**, by **covalent combination**, by **hydrogen bonding**, by **charge-transfer bonding**, by **van der Waals forces**, by **hydrophobic interaction**, or through the mediation of a **mordant**.

5.5.1. Electrovalent attraction

The simplest mode of dyeing is that whereby there is simple electrostatic attraction of oppositely charge ionized groups in the dye and substrate. For example:

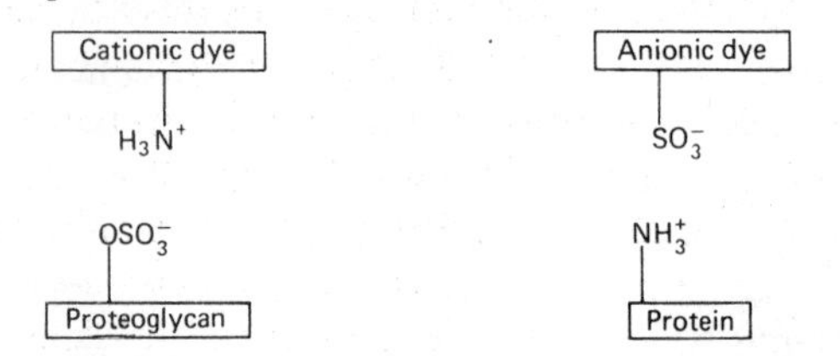

The electrovalent bonds are readily disrupted by acids or by high concentrations of electrolytes, which may therefore be included in the staining mixture to retard or limit the extent of dyeing.

Both cationic and anionic dyes are usually used in acid solutions. This enhances the ionization of amino and related groups of cationic dyes:

$$\boxed{\text{Dye}}—NH_2 + H^+ \rightleftharpoons \boxed{\text{Dye}}—NH_3^+$$

$$\boxed{\text{Dye}}—N(CH_3)_2 + H^+ \rightleftharpoons \boxed{\text{Dye}}—N(CH_3)_2H^+$$

and of basic groups of proteins:

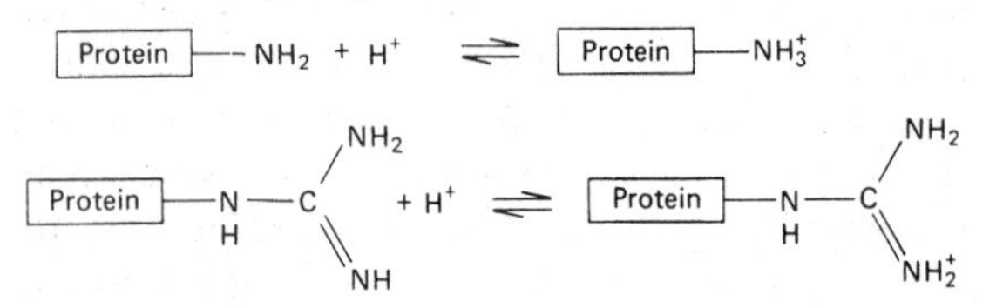

The ionization of carboxyl groups is suppressed in acid media:

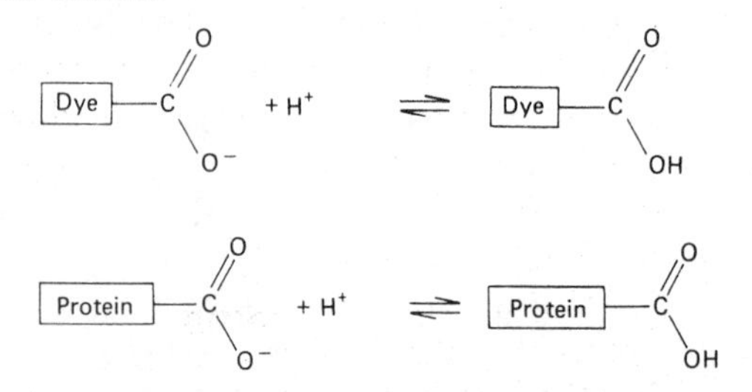

and so is that of the phosphoric acids of DNA and RNA, though in this case a higher $[H^+]$ is needed, since phosphoric acid is stronger than carboxylic acids:

$$\boxed{\text{nucleic}}—O—\overset{O^-}{\underset{O}{P}}—O—\boxed{\text{acid}} + H^+ \rightleftharpoons \boxed{\text{nucleic}}—O—\overset{OH}{\underset{O}{P}}—O—\boxed{\text{acid}}$$

These effects of acidification are simple applications of the law of mass action to the reversible chemical reactions of protonation of bases and dissociation of weak acids. Sulphonic acids are strong, so sulphonated dyes exist as anions even at very low pH. The half-sulphate ester groups of proteoglycans (see Chapter 11) are also strong acids, as are any sulphate or sulphonate radicals introduced into tissues by treatment with histochemical reagents.

The order of acidic strength of the acid (potentially anionic) groups normally present in tissues is sulphate > phosphate > carboxylate > phenol. Cationic dyes are ordinarily used at the pH sufficiently low to suppress the ionization of phenolic hydroxyl groups and carboxyl groups but high enough to allow the esters of phosphoric and sulphuric acids (i.e. nucleic acids and proteoglycans respectively) to exist as anions and therefore to be stained. The uses of cationic dyes will be discussed further in Chapters 6, 9, and 11.

Anionic dyes that are salts of sulphonic acids are used at low pH to ensure that the amino and guanidino groups of proteins in the tissue will be fully protonated and therefore stainable. As explained above, dyes of this type are fully ionized even in strongly acid media. Those anionic dyes that owe their acidity only to carboxyl or phenolic hydroxyl groups must be applied from solutions that are only slightly acidified. Anionic and cationic dyes are only rarely applied from alkaline solutions in histological practice.

Electrostatic attractions account satisfactorily for the staining properties of most cationic dyes and of the simpler anionic dyes, i.e. those with fairly small molecules. Other types of binding are also involved when anionic dyes with larger molecules are used, as will be seen below.

5.5.2. Covalent combination

Covalent bonds might conceivably be formed as the result of reactions beween anionic or cationic dyes and a substrate. For example:

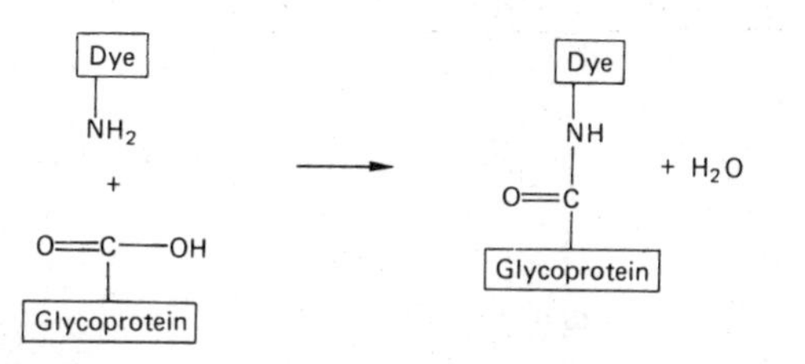

However, reactions of this type occur only very slowly. Organic chemists do not usually prepare amides by mixing amines with carboxylic acids. It is possible, however, that stains resisting extraction by acids, alkalis, and alcohol are at least partially covalently bound. Many histochemical reactions, on the other hand, certainly result in covalent bonds between chromogens and reactive groups in the tissue.

Reactive dyes (see p. 49) are, to date, not used as ordinary histological stains, probably because they are not selective enough in their reactions. Dyes with monochlorotriazinyl groups combine with free hydroxyl, amino, and other groups of the substrate. For example:

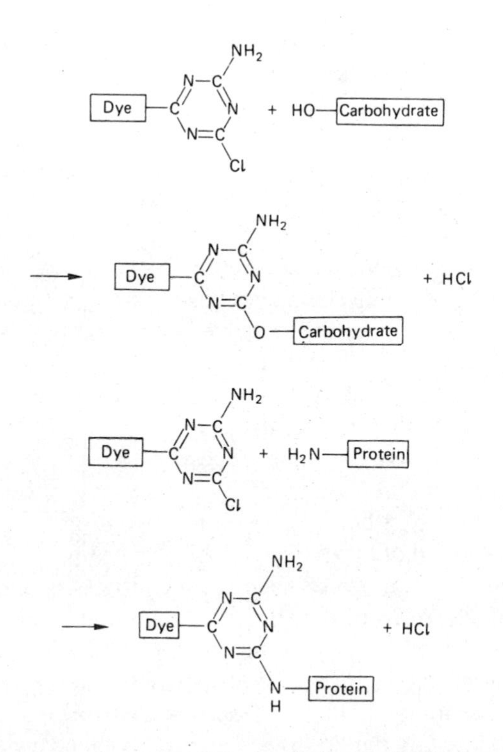

Other reactive auxochromes behave similarly.

5.5.3. Hydrogen bonding

Hydrogen bonds form between hydrogen atoms (which must be bonded covalently to oxygen or nitrogen) and bicovalent oxygen or tricovalent nitrogen. The bond is represented by a dotted line thus:

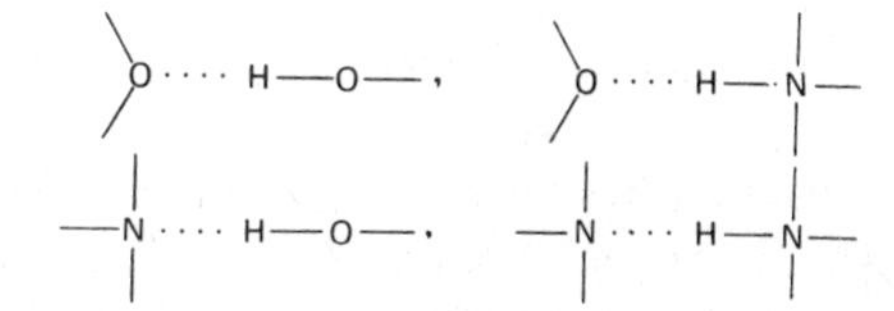

The bond is due to attraction between the lone pair of electrons (i.e. the pair of electrons not used in covalent bond) on the oxygen or nitrogen atom, and the nucleus of the hydrogen atom. This nucleus is partially exposed because the electron orbital around it is drawn towards the strongly electronegative oxygen or nitrogen atom to which the hydrogen is covalently bonded. Each of the electrical charges resulting in the hydrogen bond is of lower magnitude than the full charge of an electron or proton, so the bond is weaker than a covalent bond, though it is stronger than a van der Waals attraction.

The hydrogen atoms of water are hydrogen bonded to the oxygen atoms of other water molecules. Many dyes and components of tissues have hydrogen, oxygen, and nitrogen atoms capable of forming hydrogen bonds with water. It is therefore unlikely that bonds of this kind contribute significantly to the attachment of dyes to substrates when staining is carried out in aqueous media. The molecules of water, being much more numerous than those of the dye, would successfully compete for the available hydrogen bonding sites of the substrate. Horobin (1977) suggests that hydrogen bonding may occur when dyes are applied from non-aqueous solvents.

It has been suggested that hydrogen bonding is involved in the dyeing of cellulose by certain anionic dyes of high molecular weight (the direct cotton dyes). This explanation is unlikely, as will be seen later (pp. 55–56).

5.5.4. Charge-transfer bonding

This is another type of rather weak bonding which results in the formation of non-covalent "addition compounds". On one side of the charge-transfer bond is a molecule with electrons available for donation. These are commonly the delocalized π-electrons of systems of double bonds or of aromatic rings, though they may be lone-pair electrons of amine nitrogen atoms. On the other side of the bond is an organic molecule capable of acting as an electron-acceptor, though the exact mechanism of acceptance is not fully understood (see March, 1977, for more information). Picric acid, which is used as both a fixative and a dye, is well known for its ability to form charge-transfer complexes with aromatic hydrocarbons, olefins, and amines. Some other aromatic nitro compounds share this property.

It is conceivable that charge-transfer bonding could be partly involved in the binding of picric acid and other nitro-dyes to stained tissues, but this possibility has not been investigated.

5.5.5. van der Waals forces

The van der Waals forces are the electrostatic attractions that always exist between the electrons of one atom and the nucleus of another. This "attraction between molecules increases with increase in the number of electrons per molecule; that is, roughly also with the molecular weight" (Pauling, 1947). They are the weakest of the various forces that bind molecules together and are effective only at very short ranges. The van der Waals forces may be important in the staining of elastin (see Chapter 8) and in the attachment of some mordant-dye complexes (see below) to their substrates. The possible role of van der Waals forces in the direct dyeing of cellulose is discussed on p. 56.

5.5.6. Hydrophobic interaction

This, the affinity of non-polar molecules for one another, is responsible for the penetration of oil-soluble dyes into lipids and for their retention at hydrophobic sites in the tissue. Dyes of this type (solvent dyes; see p. 49) are applied as solutions in moderately polar organic solvents (e.g. 70% alcohol), and the stained preparations must be mounted in aqueous media. For staining to occur it is necessary for the dye to be less soluble in its solvent than in the hydrophobic substrate. The solvent should therefore be as polar as possible and saturated with the dye. Coloured, non-polar compounds are often formed as end-products in histochemical methods for enzymes (see Chapter 14). Such products may become attached by hydrophobic interaction to lipids, and thus lead to faulty localization of the enzymes. The uses of solvent dyes in lipid histochemistry are discussed further in Chapter 12.

Different non-polar molecules or parts of molecules are attracted to one another by van der Waals forces. The special term "hydrophobic interaction" merely highlights the fact that the polar (hydrophilic) parts of molecules are usually kept too far apart, by hydrogen-bonded molecules of a polar solvent such as water, for the van der Waals forces to be effective.

5.5.7. Mordants

A mordant is a substance that serves to bind a dye to a substrate. Thus cotton can be impregnated with tannic acid, with which cationic dyes form insoluble salts. The tannic acid serves as a mordant for the dye, which would not adhere to the uncharged cellulose molecules of the cotton. In histological parlance, however, the term is restricted to metal ions that are able to bind covalently to suitable dye molecules, forming **complexes** (also known as "coordination compounds" or "dye–metal complexes"). An ion or molecule that combines with a metal ion is called a **ligand**. A ligand must contain at least one atom, the **donor**, with an unshared pair of electrons in its outermost shell. These electrons become shared by the metal ion and the donor to produce the covalent bond between the two atoms.

A bond of this type resembles an ordinary covalent bond, in which each atom contributes one of the electrons of the shared pair, in that it may be very strong and resistant to harsh physical and chemical treatments, or it may be weaker, having some ionic character. Bonds with two electrons donated by the same atom are often called **coordinate**, **dative**, or **semi-polar** bonds, and several conventional representations are used in structural formulae. These include such symbols as: Cu—NH_3 (indistinguishable from any other covalent bond), $Cu^{\ominus}$—$^{\oplus}NH_3$ (indicating the acquisition of the electron by the copper atom, which consequently has a higher density of negative charge surrounding its nucleus than does the nitrogen), Cu←NH_3 (showing clearly that the nitrogen atom was the donor; relative electronegativity at the head of the arrow) and Cu---NH_3 (in which the covalent or ionic character of the bond is left deliberately vague. This should not be confused with a hydrogen bond). None of these notations are entirely satisfactory. The first one shown (representation as a simple covalency) will be used in this book.

The total number of atoms covalently bonded to a metal atom or ion is the **coordination number** of the latter. The charge remaining on an atom stripped of all its ligands is its **oxidation number**. For the rules governing the assigning of oxidation numbers to elements, the reader should consult a textbook of general and inorganic chemistry. It is important to remember that the oxygen atom of the water molecule has an unshared pair of electrons and therefore serves as a donor. Thus the copper(II) or cupric ions in an aqueous solution of $CuSO_4$ exist as $[Cu(H_2O)_4]^{2+}$, or tetra-aquocopper(II) ions. The water molecules of this complex ion can be displaced by other ligands, such as ammonia in the copper(II) tetrammine ("cuprammonium") complex $[Cu(NH_3)_4]^{2+}$. In these complexes the coordination number of the cupric ion is 4 and the oxidation number, represented by the roman numeral, is +2.

When a ligand has two or more atoms suitably positioned to form coordinate bonds with the same metal ion, the latter will be incorporated into a ring. This arrangement commonly confers stability upon the complex, which is then known as a **chelate**. Thus, cupric ions combine with sulphosalicylate ions:

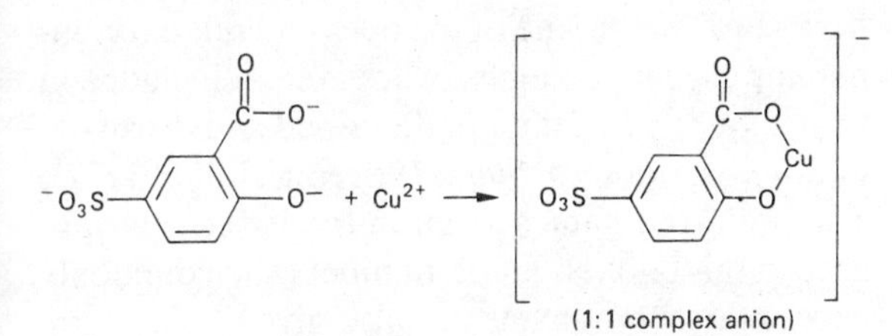

(1:1 complex anion)

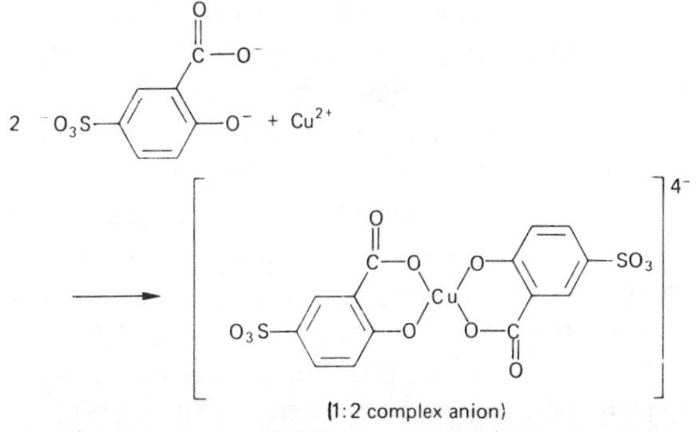

(1:2 complex anion)

Complexes of the chelate type are involved in the interactions of mordants with dyes. The cations most often encountered in conjunction with histological stains are those of aluminium, iron, and chromium. Chelation of many other metals is important in a wide variety of histochemical techniques.

The most important donor atoms of dyes are oxygen (of phenolic hydroxyl and carboxyl groups and of quinones) and nitrogen (of azo, amine, and nitroso groups). These donor atoms must, of course, be suitably positioned on the dye molecule for the latter to be able to serve as a chelating ligand. Substituents *ortho* to one another on a benzenoid ring or at the 1 and 8 positions of naphthalene, or at sterically equivalent positions on other fused aromatic rings, are usually present in dyes that form complexes with metals.

For a more thorough account of the chemistry of dye–metal complexes, with special reference to those of azo dyes, the reader is referred to Zollinger (1961).

When a metal ion acts as a mordant, coordinate bonds are formed with both the dye molecule and the substrate. The donor atoms in tissues stained by mordant dyes cannot be identified with certainty. It may reasonably be presumed, however, that any suitably positioned oxygen or nitrogen atoms that are not already fully coordinated with parts of other molecules in the tissue will be potentially available for binding a mordant metal ion. Dye–mordant–substrate complexes have such forms as

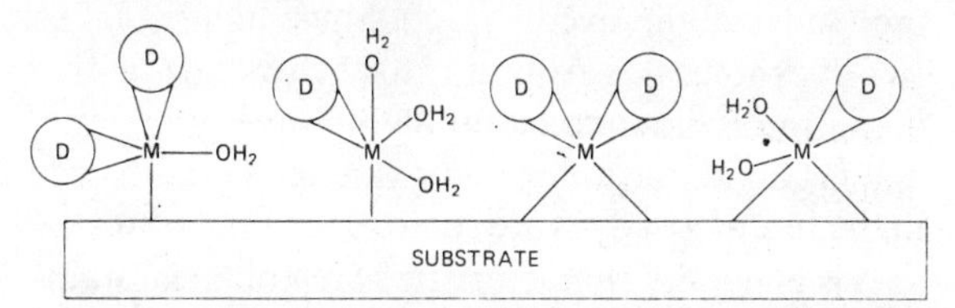

(D = dye molecule; M = metal ion capable of forming six covalent bonds; H_2O, OH_2, coordinately bound water molecules.)

The mordant is sometimes applied before the dye, but is usually mixed with it. Occasionally a metal salt is applied after the dye (see also p. 54). In a mixture of dye and mordant there will be competition among the various potential ligands (dye, water and substrate) for the metal ions. Such a mixture also contains fully coordinated dye–metal (often 2 : 1) complex molecules, and these may have their own characteristic staining properties. A complex in which the metal ions are outnumbered by dye molecules is known as a **lake**. In practical staining mixtures, an excess of mordant ions over dye molecules is nearly always present in order to ensure that the former will be in sufficient abundance to combine with the tissue as well as with the dye.

True mordant dyeing probably occurs in the staining of nuclei by iron–, aluminium–, and chromium–haematein lakes (see Chapter 6). It is known that nucleic acids are not the only substances involved in the binding of these dyes, and experimental evidence supports the contention that metal ions are interposed between the molecules of the dye and those of the tissue. For example, stained preparations can be differentiated by treating with aqueous solutions of the salts used as mordants. Nuclear staining by mordant dyes can also be differentiated by treatment with acids, which work by cleaving the bond between mordant and tissue (see Baker, 1962).

Other mordant–dye complexes (e.g. chromium–gallocyanine) behave simply as large coloured cations, there being no evidence for the interposition of a mordant ion between the tissue and the chromogen (see Lillie, 1977). It has also been suggested (Marshall & Horobin, 1973) that preformed dye–metal complexes may be held to their substrates by van der Waals forces rather than by electrovalent attraction or by covalent bonds. For such a mechanism to be effective it is necessary that the molecules of the dye be large. Since in most lakes two dye molecules are bound to each metal ion, this condition is likely to be fulfilled. Some arguments against an important role for van der Waals forces in the binding of dyes are set forth on pp. 56 and 79.

It is generally characteristic of histological staining by mordant dyes that the colours imparted to the tissues are not extracted by water, alcohols, or weakly acidic solutions used in the counterstaining and later processing of the preparations. Stronger acids decolorize the stained sections and are used for differentiation. Solutions of the mordant salt alone may also be employed for this purpose.

5.6. NOMENCLATURE AND AVAILABILITY OF DYES

The formal chemical names of most dyes are so long and cumbersome that it would be inconvenient to have to say or write them frequently. Trivial names are used instead. Some of these have been in use long enough to be considered "traditional" (e.g. haematoxylin, methylene blue, acid fuchsine). Others are trade names of manufacturers (e.g. alcian blue 8GX, procion brilliant red M2B). Letters or numbers following names are called "shade designations" and form integral parts of the names. Thus, eosin and eosin B (p. 67) are different, though related, compounds. Unfortunately, the trivial names of dyes give no clues as to their chemical natures or their uses. Azocarmine (p. 69), for example, does not contain an azo group and has no chemical or applicational similarity to the natural dyestuff carmine.

In the *Colour Index* (3rd edn., 1971), published by the Society of Dyers and Colourists, dyes and other useful coloured compounds are given numbers (the "C.I. numbers") and also names that indicate their modes of industrial application and their colours. For example, methyl blue (p. 66) is C.I. 42780, Acid blue 93. Chromoxane cyanine R (p. 66) is C.I. 43820, Mordant blue 3. Natural dyes are named as a separate class, so that haematoxylin is C.I. 75290, Natural black 1, despite the fact that it is a mordant dye. The C.I. numbers are used in other works as well as the *Colour Index* itself and they should always be specified when ordering dyes from suppliers. Some of the dyes and related compounds used in biological work are not included in the *Colour Index*, but are described and discussed in *Conn's Biological Stains* (9th edn., Lillie, 1977). The preferred names given in the latter work are used in this book, but C.I. numbers and commonly encountered synonyms are also given.

Authors have been inconsistent in their spelling of the names of dyes and in the use of initial capital letters. In this book, initial capitals are used in the preferred names only for words that are also proper names of people or places. Other words, including trade names, are printed entirely in lower case letters, but shade designations are capitalized (e.g., Victoria blue 4R; methylene violet, Bernthsen; toluidine blue O). The C.I. applicational names are treated as a botanist or zoologist would treat the generic and specific names of a species, with an initial capital for the class and lower case for the colour (e.g. Basic violet 10; Mordant black 37). The policy of Baker (1958) is followed with respect to the spelling of words in traditional preferred names that end in -in or -ine. Thus the ending -ine is used when the dye is an amine or a derivative of an amine (e.g. acid fuchsine, **not** acid fuchsin; or safranine, **not** safranin). For a dye that is not an amine, there is no terminal e (e.g. eosin, **not** eosine; haematein, **not** haemateine). This convention results in differences between some of the spellings in this book and those used by Lillie (1977). Other differences arise from the use of diphthongal ae (e.g. haematoxylin rather than hematoxylin), and that of ph rather than f in such words as "sulphur" and "sulphonic". The latter discrepancies merely reflect differences between English and American spellings.

Dyes for use as stains and histochemical reagents can be purchased from many suppliers of laboratory chemicals. A few firms specialize in biological stains, but their products are not necessarily superior to those obtained from more ordinary sources. A few dyes (e.g. aldehyde–fuchsine, p. 102) are easy to synthesize in the laboratory. Dyes are not usually manufactured by the firms that sell them. They are made in factories for the primary purpose of satisfying the requirements of the textile industry, though some stains are obsolete as industrial dyes and are now produced solely for use by microscopists.

Because of the different requirements of the dyer and the histologist, most dyes are not marketed as pure compounds. The dye is nearly always mixed with a substantial proportion of inactive filler (often sodium chloride) and several different coloured substances are often present in a sample sold as a single, pure dyestuff. Inert fillers are added in order to standardize the potency of different batches of dye so that they will behave similarly towards textiles. Some salt is often inevitably present as a result of the salting-out process used to recover the dye in solid form from the solution in which its synthesis was accomplished. The chemical reactions by which dyes are generated are usually accompanied by side reactions. These often lead to the production of other dyes, which end up as contaminants in the final products. After manufacture and packaging, some dyes undergo chemical changes in the solid state. Methyl green (see p. 64) is an important biological stain that deteriorates in this way. Changes also occur after dissolving some dyes, and this process of "maturation" may lead to deterioration of a staining solution (as with gallocyanine, see p. 69) or to improvement in its properties (as with haematoxylin as explained on p. 61 and with methylene blue as explained on pp. 70–71). Occasionally a dye with a single name is a deliberate mixture of two quite different substances (e.g. alcian green; see p. 75). In recent years there have been many investigations of the purity of dyes using paper and thin-layer chromatography. Nearly every histological stain tested has been shown to contain several coloured components. Often one of these predominates and probably corresponds to the name on the bottle, but in some cases commonly used dyes have been shown to be mixtures of two or three major components with ten or more minor contaminants. It is hardly surprising that the properties as a biological stain may vary from one batch to another of what is supposed to be the same single dyestuff.

It is necessary for the histologist to have dyes that can be relied upon to give satisfactory and repeatable results. Such constancy of performance has assumed even greater importance in recent years than in the past with the advent of automated screening devices for the examination of blood and exfoliated cells in pathology laboratories. In the early days of staining, microscopists depended upon a few suppliers, notably Dr. Georg Grübler & Co. of Berlin, who tested the products of various manufacturers and then bought up and packaged batches of dyes that proved satisfactory as stains. Grübler's dyes ceased to be available outside Germany in 1914, and most biologists became dependent upon the dye industries of other countries. The difficulty in obtaining usable stains led to the founding of the Biological Stains Commission in the United States in 1920. This body tests batches of dyes submitted by manufacturers and suppliers and certifies those suitable for their intended uses. The label on a bottle of a certified stain shows the percentage by weight of dye in the material, the C.I. number, and a batch number. Newly certified batches of dyes are regularly reported in the Commission's journal, *Stain Technology*. Since the main purpose of certification is to serve pathologists, many dyes used in other biological sciences are not tested. All samples of such dyes should be viewed with suspicion and carefully tested by the user before applying them to valuable specimens. The history of the availability and standardization of biological stains is recounted in detail by Lillie (1977), who also provides practical instructions for assaying and testing many dyes. Satisfactory assay methods are not available for some dyes. These must be tested empirically by standardized staining procedures.

When a weight or concentration of any dye is specified in the instructions for a method, in this or any other text, it refers to the material constituting an acceptable sample of the dye—not to the pure dyestuff. For the more important stains the acceptable percentages of active substances present in the commercial products are given with the descriptions of dyes in Section 5.9. The data are mostly from Lillie (1977). If the content of dye stated on the label of a bottle differs by more than a few per cent from the recommended acceptable value, appropriate correction must be made in preparing a working solution. With samples of dyes not certified by the Biological Stains Commission, the dye content is not usually stated on the label. Such materials should be carefully tested before using them on a large scale. This same stricture applies to old stock. Bottles bearing the revered name of Grübler still decorate the shelves of many laboratories. The stuff inside may still perform as well as it did in the hands of the Master, but it is just as likely to have matured to a dignified but impotent senility.

5.7. CLASSIFICATION OF DYES BY METHODS OF APPLICATION

Dyes and related colouring agents may be arranged in sixteen groups according to the different ways in which they are applied to textile fibres and other substances. These techniques are not all used in histological practice but they are briefly set out below to give the reader some idea of the variety of colouring methods employed in industry.

5.7.1. Basic dyes

These are coloured cations, which attach by electrostatic forces to anionic groups in the substrate. They are used mainly on proteinaceous fibres, cellulosic fibres that have been mordanted with tannic acid and polyacrylonitrile fibres. Cationic dyes form an important group of histological stains and histochemical reagents.

5.7.2. Acid dyes

These are coloured anions, with sulphonic acid, carboxylic acid, or phenolic auxochromes. Many dyes in this group are amphoteric, having amine as well as acid groups. An excess of anionic substituents confers a net negative charge on the molecule. The sulphonic acids are the most important industrial dyes in this group since they are strong acids, fully ionized even at low pH. They are applied from acidic dyebaths, mainly to proteinaceous fibres such as wool and silk with free amino and guanidino groups. As will be seen later, ionic interactions are not the only forces involved in the binding of acid dyes to their substrates. Anionic dyes are important as stains for cytoplasm and extracellular structures.

5.7.3. Direct dyes

These are anionic dyes with large molecules. Nearly all of them are azo dyes. They bind to cellulosic fibres directly (i.e. without the need for a mordant). Some are used as histological stains. The mode of dyeing by direct dyes will be discussed later (p. 55).

5.7.4. Mordant dyes

Mordants and their functions have already been discussed (p. 44). By convention (*Colour Index*), the mordant dyes are defined as those used in conjunction with metal salts. The mordant may be applied before the dye, together with the dye as a soluble dye–metal complex (the "metachrome" process) or after the dye (the "afterchrome" process). In some instances a mordant dye is also an anionic or, rarely, a cationic dye in its own right. Mordant dyes have many uses in histology and histochemistry.

5.7.5. Reactive dyes

These dyes combine covalently with their substrates. The most important are those with auxochromes capable of uniting with the hydroxyl groups of cellulose. Isothiocyanates, sulphonyl chlorides, diazonium salts, and a few other derivatives of ordinary dyes and fluorochromes are used for making covalently labelled proteins for use as tracers and as reagents in immunohistochemical techniques (Chapter 19). This labelling process is chemically akin to the industrial use of reactive dyes.

5.7.6. Solvent dyes

The simple solvent dyes are coloured substances which dissolve in hydrophobic materials but not in water. Solvent dyes are used in wood stains, lacquers, polishes, printing inks, inks for ball-point pens, and in the mass colouring of wax, some plastics, and soap. Other solvent dyes are salts formed by the interaction of certain anionic chromogens with hydrophobic cations. The latter are released during dyeing, leaving the water-insoluble coloured anions behind in the substrate. Biologists often call solvent dyes **lysochromes**, a term introduced by Baker (see Baker, 1958). Several lysochromes are used in the histochemical study of lipids (see Chapter 12).

5.7.7. Vat dyes

These are important in the textile industry but are not used by biologists. A vat dye is applied to cloth as its leuco compound (which is not usually colourless in practice). This is then oxidized, commonly by treatment with air and steam, to the fully coloured form of the dye, which is insoluble. Vat dyes are used mainly on cotton. The oldest vat dye is indigo (see p. 71). Chemical reactions similar to those involved in the production of indigo are involved in the indigogenic techniques of enzyme histochemistry (see Chapter 15).

5.7.8. Sulphur dyes

These are mixtures of uncertain composition made by heating a variety of organic compounds with sulphur or with alkali metal polysulphides. A sulphur dye is applied (mainly to cellulosic fabrics) mixed with aqueous sodium sulphide, which causes reduction to the leuco compound. The insoluble coloured dye is regenerated by exposure of the dyed material to air. The process is similar in principle to vat dyeing. Sulphur dyes are not used in histology.

5.7.9. Ingrain dyes

Originally this term included all dyes generated within the substrate, but it is now used specifically for temporarily solubilized phthalocyanines that change into insoluble pigments after application to their substrates. The phthalocyanines are discussed on p. 74. Alcian blue, used in carbohydrate histochemistry, is the most important biological stain in this class.

5.7.10. Polycondensation dyes (condense dyes)

This category includes dyes with a variety of chromophoric systems. They are applied as soluble monomers, the molecules of which then react with one another to form insoluble coloured polymers. Such dyes are not used as biological stains, but some histochemical methods for enzymes result in the production of coloured polymeric end-products.

5.7.11. Azoic dyes

These are formed within the substrate by reaction of diazonium salts (**azoic diazo components**) with suitable phenols or aromatic amines (**azoic coupling components**). They are important to the histochemist and are discussed in Section 5.9.3.9 below.

5.7.12. Oxidation bases

Certain colourless bifunctional amines and aminophenols are used as dyes, especially for fur and hair. Oxidation, usually by hydrogen peroxide, produces insoluble brown or black polymers containing the quinonoid chromophore. The colour that develops may be modified by adding metal salts or phenolic compounds to the oxidation base. Similar chemical reactions are encountered in several histochemical techniques.

5.7.13. Disperse dyes

These are insoluble coloured compounds used as aqueous suspensions (particle size 1–10 μm) for dyeing cellulose acetates and polyesters. The colouring process is enhanced by the presence in the dyebath of suitable oily agents (**carriers**), which are believed to promote the entry of the particles of dye into the interstices of the hydrophobic fibres. The physical nature of disperse dyes precludes their use as stains for microscopy.

5.7.14. Pigments

These are finely divided white, black, or coloured materials, insoluble in water and organic solvents. They are used as suspensions or emulsions in liquid media, as paints, and for the mass coloration of plastics, rubber, and synthetic fibres. There are many inorganic and organic pigments. They are often formed as the visible end-products of histochemical reactions.

5.7.15. Fluorescent brighteners

These compounds serve to increase the amount of visible light reflected from white or coloured objects. Invisible ultraviolet radiation absorbed by the brightening agent is emitted at a higher, visible wavelength. Most fluorescent brighteners contain the stilbene configuration (see p. 59). Some of them are used in microscopy as fluorochromes.

5.7.16. Food colours

This group includes dyes of various kinds whose principal industrial applications are in the manufacture of foodstuffs. Fast green FCF, an anionic triphenylmethane dye (p. 65), is the only one commonly used as a histological stain.

5.8. CLASSIFICATION OF STAINS BY MICROSCOPISTS

Biological stains have traditionally been designated by their users as basic, acid, or mordant dyes or as "fat stains", the last-named being the lysochromes or solvent dyes. This division of dyestuffs into four groups reflects in a rough and ready way the purposes for which the dyes are used. Thus basic (cationic) dyes are used to stain nuclei and the major anionic compounds of cytoplasm and extracellular structures. Acid (anionic) dyes are used mainly to demonstrate cytoplasmic and extracellular proteinaceous material. Mordant dyes are used mainly as nuclear stains and lysochromes are used for fats and other lipids.

This classification has the advantages of simplicity and widespread usage, but it is not adequate for the student trying to understand the reasons why different dyes are used, alone or in combination, for the demonstration of the various structural and chemical features of tissues. The uses of the dyes are determined by their physical and chemical properties, and these cannot be made to fit into any simple scheme of classification. Any dye used as a stain should be thought of in terms of its colour, its molecular size and shape, its solubility, and its potential chemical reactivity. All these properties affect the method of use and the results obtained.

5.9. CLASSIFICATION OF DYES BY CHROMOPHORIC SYSTEMS

We shall now consider the major groups of dyes following a chemical scheme of classification based on the structures of the major chromophoric systems. The more important members of each group, including all the dyes mentioned in later parts of this book, will be described. Some of the groups of dyes have little or no importance as biological stains, so for these only single representative examples are given. The number of dyes used in histology and histochemistry is likely to increase in the future, so the potential user should be aware of the variety available.

5.9.1. Nitroso dyes

Nitrous acid reacts with phenols to give nitrosophenols in which the radical —N=O replaces hydrogen *ortho* or *para* to the phenolic hydroxyl group. There is tautomerism between the nitroso compounds and the corresponding quinone oximes. Thus for *p*-nitrosophenol

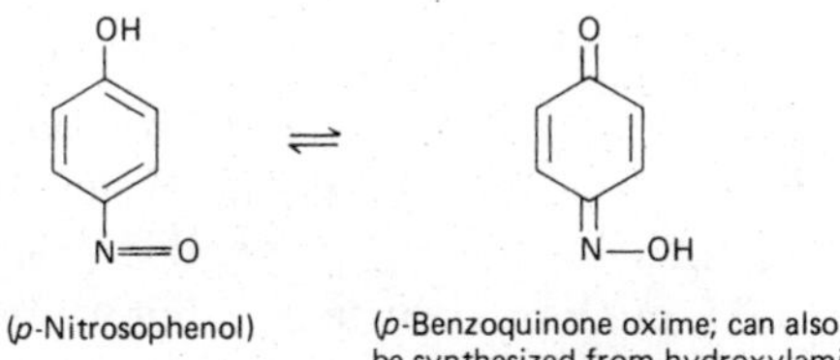

(*p*-Nitrosophenol) (*p*-Benzoquinone oxime; can also be synthesized from hydroxylamine and *p*-benzoquinone)

The *o*-nitroso compounds can form coloured chelates with metal ions and so function as mordant dyes. Nitroso dyes have found hardly any uses in histology, though they have properties that might well be exploited.

Example

NAPHTHOL GREEN Y (C.I. 10005; Mordant green 4; M.W. 173)

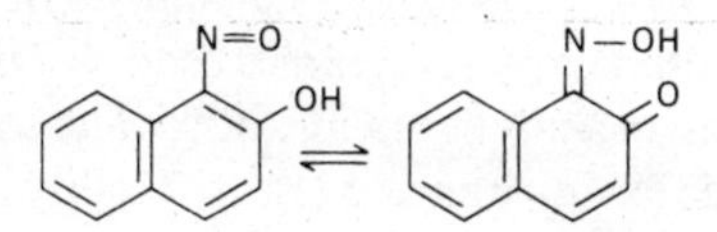

This dye gives yellowish green with iron and brownish green with a chromium mordant. Lillie (1977) states that a scarlet colour is formed with cobalt salts and suggests that this dye might be used for making visible insoluble salts of cobalt formed as the products of histochemical reactions.

5.9.2. Nitro dyes

The nitro group may be present as a substituent in dyes with other chromophoric systems, but there are some in which it may be considered the only chromophore. Hydroxyl or amino groups are also present in such dyes, and aromatic and quinonoid tautomers exist as is shown below for picric acid. The nitro group, formally represented as $—NO_2$, is a resonance hybrid of the two structures

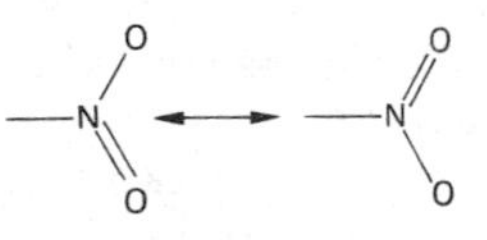

the two bonds between nitrogen and oxygen being equivalent to one another. The single N—O bond can alternatively be represented as N→O or $\overset{+}{N}—\overset{-}{O}$, the symbols having the same meanings as when used in formulae for metal complexes (see p. 45).

Examples

PICRIC ACID (2,4,6-trinitrophenol) (C.I. 10305; M.W. 299)

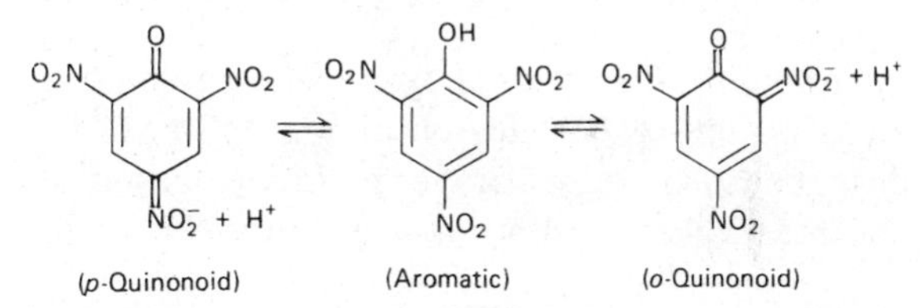

(*p*-Quinonoid) (Aromatic) (*o*-Quinonoid)

The phenolic hydroxyl group is readily ionized because of the electron-withdrawing effect of the three nitro groups. Consequently trinitrophenol is a much stronger acid than phenol. In histological practice picric acid is used as a yellow anionic dye of low molecular weight. It can also form "addition compounds" with a variety of substances. In these, the bonding is neither ionic nor covalent, its nature being incompletely understood (see March, 1977, for discussion). The addition compounds are charge-transfer complexes (see pp. 43–44).

Picric acid is also used as a fixative (Chapter 2, where precautions attendant on its use are also mentioned). It is soluble in water (1.3%) and more soluble in alcohol and aromatic hydrocarbons, but xylene does not remove the dye from stained sections.

MARTIUS YELLOW (C.I. 10315; Acid yellow 24; Naphthol yellow; 2,4-dinitro-1-naphthol; M.W. 256 (sodium salt), 251 (ammonium salt), 273 (calcium salt), 234 (the unionized phenol)).

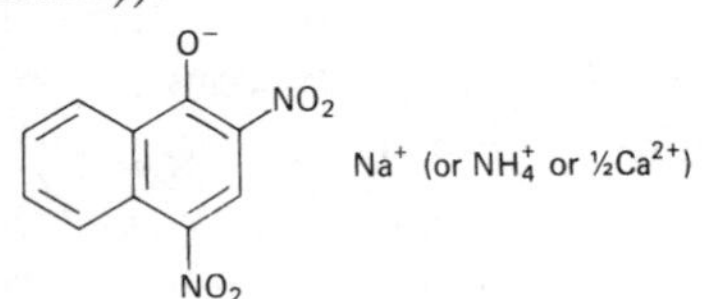

This dye is more deeply coloured (approaching orange) than picric acid and is somewhat more soluble in water and less soluble in alcohol. The ammonium and calcium salts (the latter less soluble in water) are also available. Like picric acid, Martius yellow is used as a cytoplasmic stain in conjunction with other anionic dyes.

A related compound, 2,4-dinitro-naphthol-7-sulphonic acid, is **flavianic acid**, well known to biochemists as a precipitant for the amino-acid arginine. The sodium salt of flavianic acid is **naphthol yellow S** (C.I. 10316), which is occasionally used as an anionic dye.

5.9.3. Azo dyes

This very large group of dyes is of great importance in industry and to the biologist. It will be described in some detail; many points of theoretical and practical interest are conveniently illustrated by members of the azo series.

The chromophore is the azo group (—N=N—), which connects two aromatic ring systems. Dyes may contain one (monoazo), two (bisazo), three (trisazo), or, rarely, four (tetrakisazo) or more (polyazo) azo groups. The aromatic rings are usually benzene or naphthalene, and they may bear a variety of substituents.

5.9.3.1. SYNTHESIS

Azo dyes are made by coupling diazonium salts with phenols or aromatic amines. The diazonium salts are made by reacting primary aromatic amines with nitrous acid, in acid solution, at temperatures a little above freezing:

$$Ar{-}NH_2 + HNO_2 + H^+ \rightarrow Ar{-}N^+{\equiv}N + 2H_2O$$

Most diazonium salts are unstable and cannot be isolated in pure form as solids, though some (see p. 57) are available as stable complexes. Usually the solution containing the diazonium salt is used immediately. Excess nitrous acid can be decomposed by adding a little urea:

$$2HNO_2 + (NH_2)_2CO \rightarrow 3H_2O + 2N_2\uparrow + CO_2\uparrow$$

Coupling with penols and amines occurs preferentially in the *para* position, but in the *ortho* position if another substituent or the site of fusion of rings is present *para* to the hydroxyl or amino group. With naphthols and naphthylamines, coupling is usually *ortho* to the hydroxyl or amine group. For example:

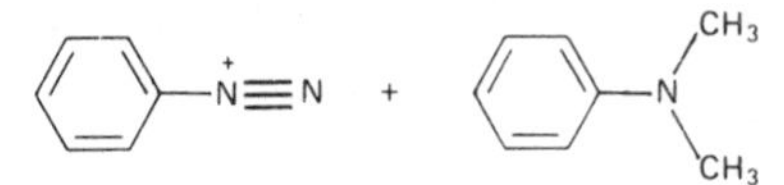

(Benzene diazonium salt) (Dimethylaminobenzene)

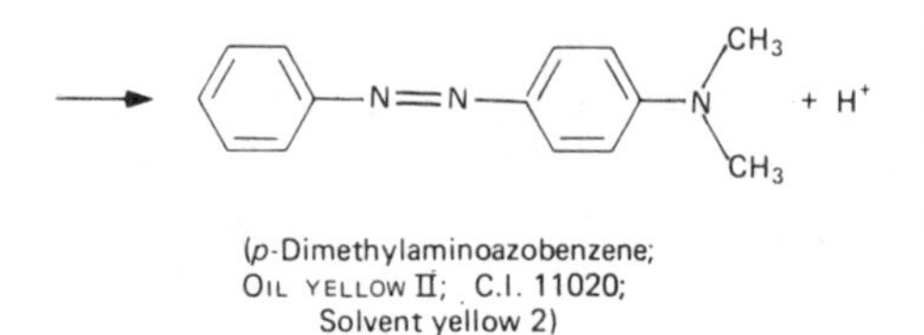

(*p*-Dimethylaminoazobenzene; OIL YELLOW II; C.I. 11020; Solvent yellow 2)

The product of this reaction is the well-known carcinogen "butter yellow" (which is neither a constituent of nor an additive to butter). It is insoluble in water and is not used in histology, but it is a dye on account of the auxochromic tertiary amine group. Solubility in water can be conferred by introducing a hydrophilic substituent. The sulphonic acid group is the one most often chosen and the sodium salt of the *para*-sulphonic acid derivative of butter yellow (formed from diazotized sulphanilic acid and *p*-dimethylaminobenzene)

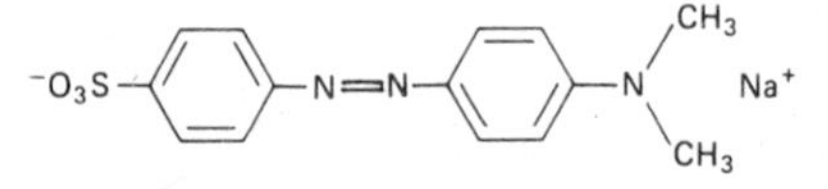

is a water-soluble dye. It is rarely used as a stain but is familiar as an acid–base indicator, methyl orange (C.I. 13025; Acid orange 52).

In coupling with diazonium salts, phenols react as their anions while amines react as the unionized bases. Both these species exist in alkaline solutions. The diazonium ion, however, is changed in the presence of excess alkali into a diazotate anion, Ar—N=N—O^-, which does not participate in coupling reactions. Consequently, there is an optimum pH for every reaction between a diazonium salt and a phenol or amine. Usually a more strongly alkaline medium is required for coupling with phenols than with amines.

Azo dyes can also be produced by reactions of quinones with aryl hydrazines, but they are always manufactured by the coupling reactions of diazonium salts. For more complete accounts of the chemistry of azo compounds, the reader should consult a textbook of organic chemistry or the monograph of Zollinger (1961).

5.9.3.2. Structure

The azo linkage can exist in two isomeric forms:

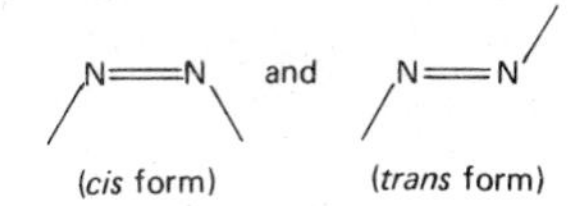

In dyes, the *trans* configuration is usually present because of intramolecular hydrogen bonding between one of the azo nitrogen atoms and a substituent in the *ortho* position on at least one of the aromatic rings:

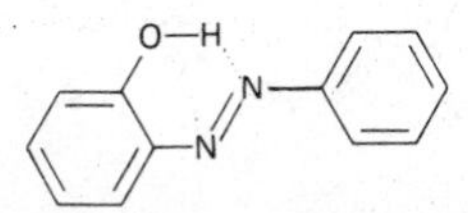

In some dyes, as will be seen, this arrangement provides a molecular shape conducive to the chelation of metal ions.

Phenolic azo dyes exist in tautomeric equilibrium with quinonoid forms:

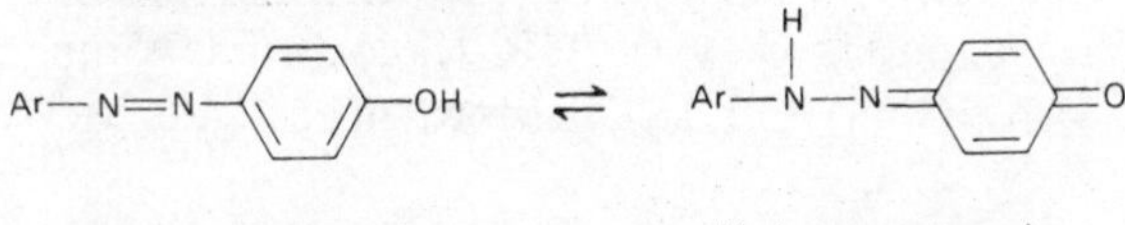

The dyes that are amines, however, usually occur only in the aminoazo form because of intramolecular hydrogen bonding.

5.9.3.3. Anionic azo dyes

The simpler anionic azo dyes are the "levelling" dyes. They are mostly monoazo compounds with sulphonic acid and phenolic substituents. An example much used in histology is orange G.

Orange G (C.I. 16230; Acid orange 10; M.W. 452)

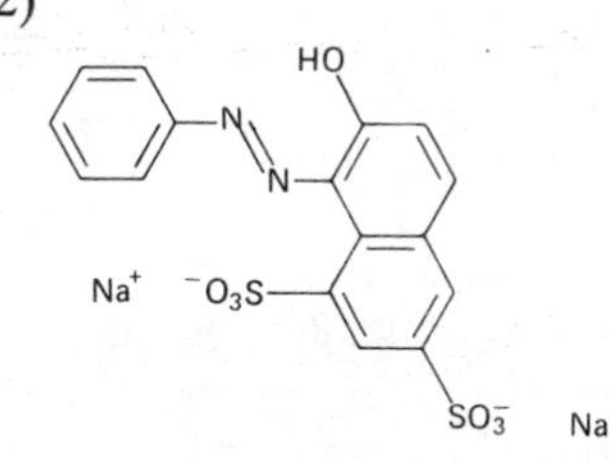

This dye is used as a cytoplasmic stain, usually in conjunction with other anionic dyes. It is very soluble in water but much less soluble in alcohol. Samples should contain at least 80% by weight of the anhydrous dye.

Chromotrope 2R (C.I. 16570; Acid red 29; M.W. 468)

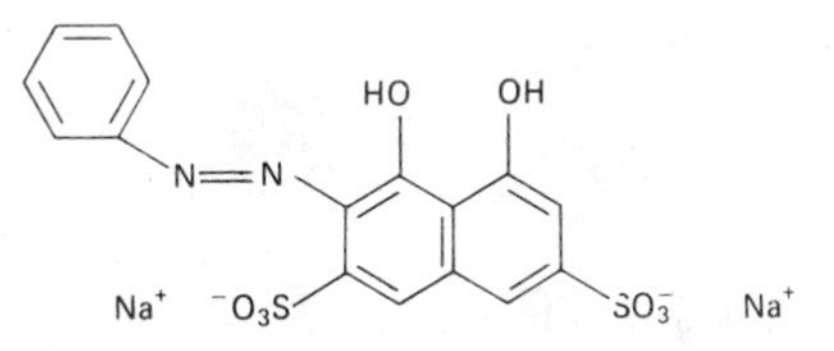

This red dye is used, in conjunction with other anionic dyes, in various methods for cytoplasm and connective tissue. It is freely soluble in water and slightly soluble in alcohol. The name "chromotrope" indicates that the phenolic coupling component chromotropic acid (1,8-dihydroxynaphthalene-3,6-disulphonic acid) was used in its preparation. Textiles dyed with "chromotrope" dyes change colour from red to blue when afterchromed with Cr^{3+} salts.

The "milling" dyes have larger molecules. They are bis- and tris-azo compounds. The "milling" dyes are used as anionic stains of large molecular size in methods for connective tissue (see Chapter 8). Some of the triphenylmethane dyes (see pp. 63–66) are used for the same purposes. Another group of anionic dyes with large molecules is that of the direct dyes, to be described below.

Example

Biebrich scarlet (C.I. 26905; Acid red 66; M.W. 557)

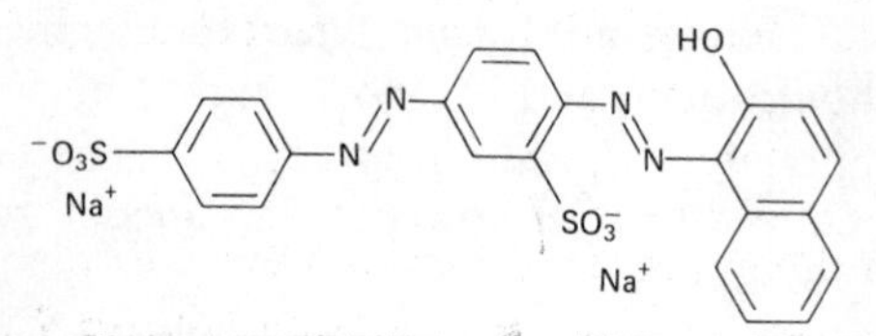

This dye is soluble in water but only slightly soluble in alcohol.

5.9.3.4. Mordant azo dyes

These have functional groups capable of chelating chromium or similar metals. The commonest arrangements are:

(a) Hydroxyl groups *ortho* to both azo nitrogens:

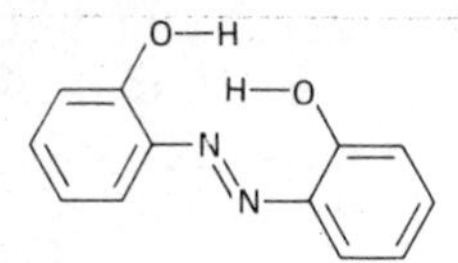

The chelate formed with an ion of a metal M has the structure

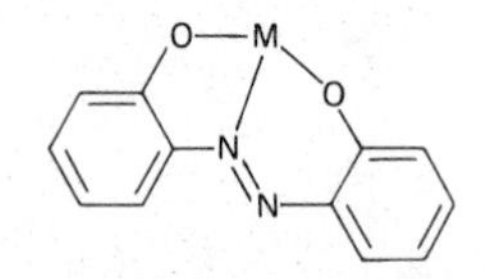

(Only three coordinate bonds to M are shown, but M will usually have coordination number 4 or 6)

It is possible for only one of the azo nitrogen atoms to be linked with one metal ion, the other being too far away as a consequence of the *trans* configuration of the azo group. However, it can be seen from the structural formula above that a metal with a coordination number higher than 3 could combine with another molecule of dye or, alternatively, with water molecules or hydroxide ions. In this way large aggregate molecules of the dye–mordant complex may be formed.

(b) Hydroxyl and amino groups *ortho* to both azo nitrogens:

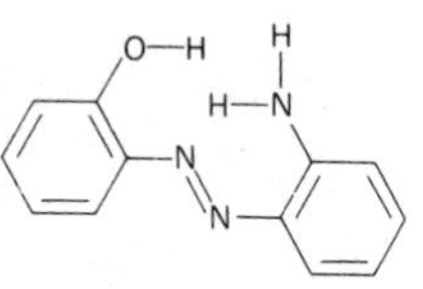

Chelates have structures analogous to that shown for (a) above.

(c) The salicylic acid arrangement on one of the aromatic rings:

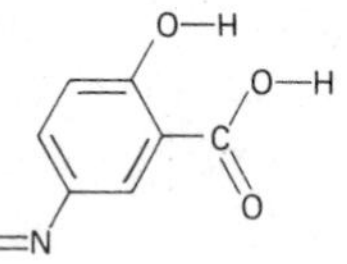

Chelation with the salicylic acid moiety of such a dye commonly results in 1:2 or 1:3 metal:dye ratios. Thus for chromium (Cr^{3+}, coordination number 6), the complexes include

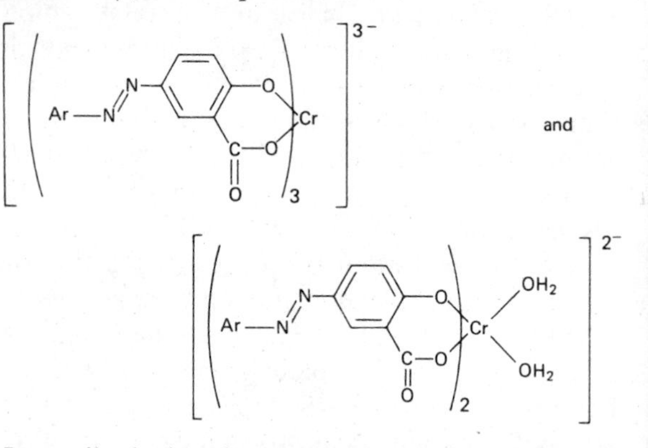

In textile dyeing the metal salt may be mixed with the dye (the **metachrome** process) or applied after the dye (the **afterchrome** process). Mordant azo dyes are but little used in histology. Afterchroming is sometimes practised in order to stabilize azo compounds formed as the end-products of histochemical methods for enzymes.

Examples

Salicin black EAG (Solochrome black A; C.I. 15710; Mordant black 1; M.W. 461)

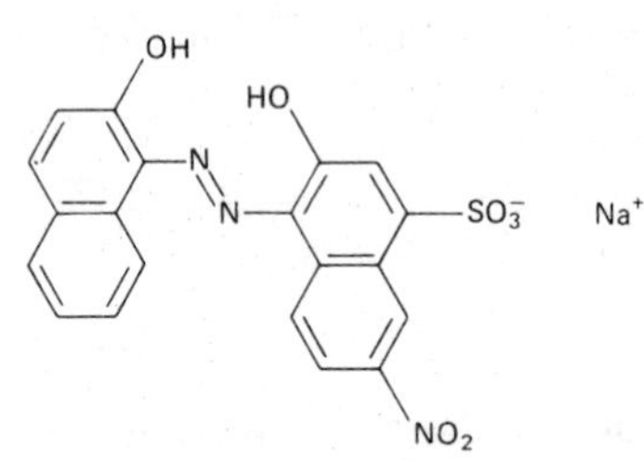

Soluble in water; slightly soluble in alcohol and acetone. The aluminium lake is fluorescent (orange emission) and is sometimes used as a fluorochrome.

Chrome orange GR (Solochrome orange GRS) (C.I. 26520; Mordant orange 6; M.W. 470)

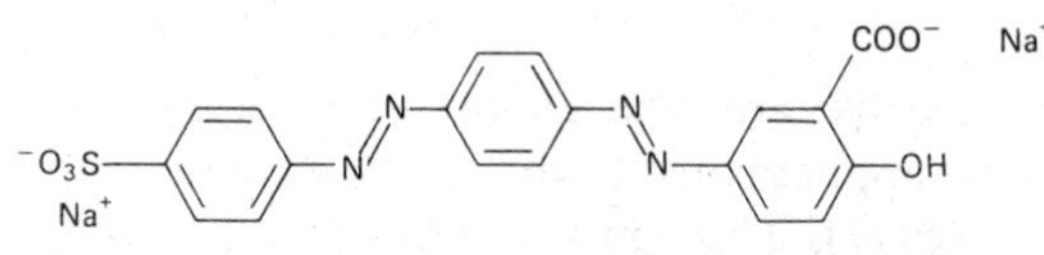

Soluble in water; slightly soluble in alcohol and acetone. Metal ions are complexed by the salicylate arrangement shown at the right-hand side of the

molecule. Chrome orange GR is not used as a stain but it is an indicator, changing from yellow to red over the pH range 10.5–12.0.

5.9.3.5. CATIONIC AZO DYES

Not many of these are used in histology. One that is encountered in a variety of techniques is Bismarck brown Y, which was the first commercially produced azo dye and is still used for colouring leather.

BISMARCK BROWN Y (Bismarck brown G; vesuvin; leather brown; C.I. 21000; Basic brown 1; M.W. 419)

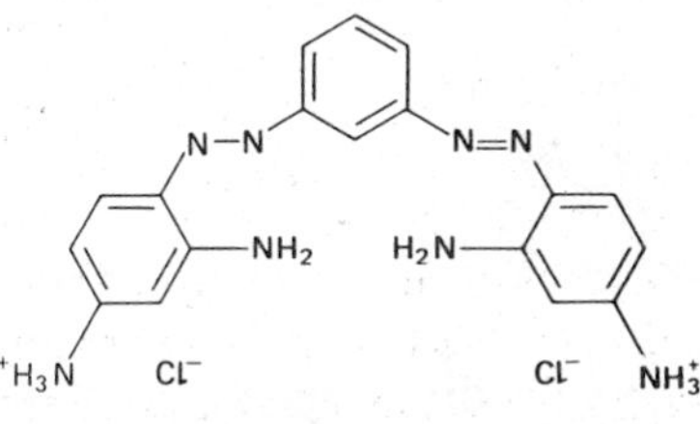

This dye is soluble in water and alcohol but insoluble in acetone and most other organic solvents. Samples should contain at least 45% of the anhydrous dyestuff.

5.9.3.6. REACTIVE AZO DYES

Reactive dyes form covalent bonds with the dyed substrate and are used mainly on cotton (see p. 49), where the attachment is to the hydroxyl groups of cellulose. The commonest reactive auxochromes are the dichlorotriazinyl and monochlorotriazinyl groups. The following is a dichlorotriazinyl azo dye.

Example

PROCION BRILLIANT RED M2B (C.I. 18158; Reactive red 1; M.W. 717)

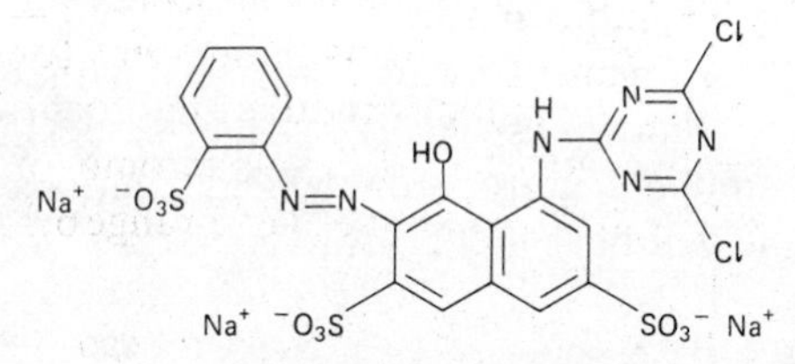

Reactive dyes are not used as stains for sections, but several applications in vital staining are cited by Lillie (1977). The constitutions of many reactive dyes, including PROCION YELLOW 4MR, which is extensively used for injection into neurons through microelectrodes (see Kater & Nicholson, 1973), have not been revealed by the manufacturers.

5.9.3.7. SOLVENT AZO DYES

These coloured substances are more soluble in lipids than in the solvents from which they are applied. Chemical dyeing does not occur. Solubilizing groups such as $—SO_3^-$ and $—NH_3^+$ are absent. Phenolic hydroxyl groups are usually present, but may be esterified without affecting the staining properties. Sudan black B is the most important histochemical reagent in this group. Its description is deferred to Chapter 12.

Examples

SUDAN IV (C.I. 26105; Solvent red 24; M.W. 380)

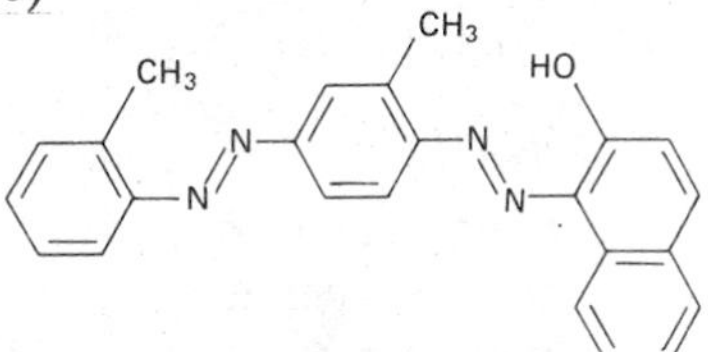

Sudan IV is suitable only for staining conspicuous accumulations of lipids, as in adipose tissue. The dye content should be at least 80% by weight.

OIL RED O (C.I. 26125; Solvent red 27; M.W. 409)

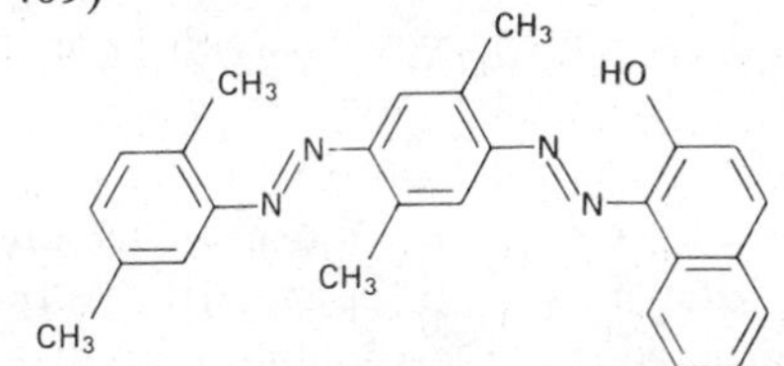

This dye is more intensely coloured than Sudan IV and therefore allows the resolution of smaller lipid-containing structures, such as intracellular droplets.

The properties and uses of these and other solvent dyes will be discussed in greater detail in Chapter 12.

5.9.3.8. DIRECT AZO DYES

These, which are also called **cellulose-substantive dyes**, are capable of colouring cellulosic fibres (cotton and linen) without the help of mordants or special reactive auxochromic groups. The ordinary auxochromes of azo dyes (the ions $—SO_3^-$, $—NH_3^+$, $Ar—O^-$) cannot be attracted by unionized hydroxyls, which are the only functional groups present in cellulose. The mode of action of the direct dyes has puzzled chemists for many years. The dyes are all anionic. Most are dis- or tris-azo dyes and all have long molecules with usually at least five aromatic rings (naphthalene counting as two). The rings must not be sterically hindered from assuming a coplanar configuration. In aqueous solution, direct dyes form molecular aggregates at room temperature but not at 90–100°C, the temperature at which they are used for dyeing textiles. This property of reversible aggregation, together with the fact that the dye can all be removed from the fabric by sufficiently vigorous and prolonged washing, suggests that the same mechanism may be involved in the attachment of dye molecules both to cellulose and to one another.

Hydrogen bonding between the hydroxyl groups of cellulose and nitrogen (azo, amino, or amide) or oxygen (hydroxyl) of the dye has been proposed as the principal force holding the dye to its substrate. However, in an aqueous medium hydrogen bonds would also be present between molecules of water and those of both the dye and the cellulose. Protection of the latter by a layer of hydrogen-bonded water might also prevent molecules of a direct dye from approaching close enough for van der Waals forces to be sufficiently strong to assist in the binding. van der Waals forces between hydrophobic regions (where hydrogen-bonded water would not be in the way) of different dye molecules might account for aggregation in cold aqueous solutions. A special kind of attraction between the conjugated array of aromatic rings and double bonds of the dye, with its extended, delocalized π-electron system, and the hydrogen atoms of the substrate's hydroxyl groups has also been suggested as a cause of attachment of direct dyes to cellulose (see Allen, 1971). Such a mode of attachment would resemble charge-transfer bonding (see p. 43).

Some of the direct azo dyes are used in histological work, especially in techniques for the demonstration of amyloid, fibrin, and elastin (see Chapter 8). The various possible modes of dyeing of cotton may or may not be involved in these procedures, which are carried out at room temperature. The dyes can, of course, behave as ordinary anionic dyes towards cationic components of a tissue. In this respect the direct dyes resemble anionic azo dyes of the "milling" class described earlier.

Examples

Congo red (C.I. 22120; Direct red 28, M.W. 697)

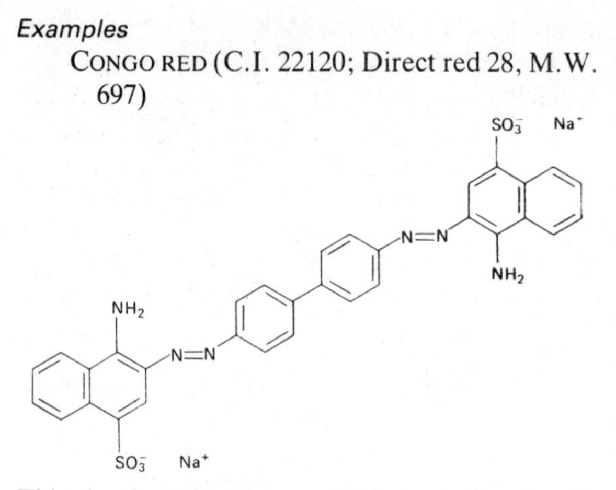

This dye is much more soluble in water than in alcohol. It is an indicator, changing from red to blue when the pH is less than about 3.0. It has several uses in histology, the best known being as a stain for amyloid. Samples should contain at least 75% by weight of the anhydrous dye.

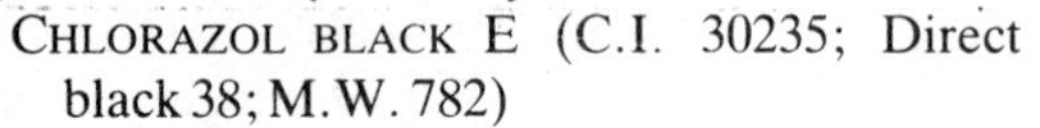

Chlorazol black E (C.I. 30235; Direct black 38; M.W. 782)

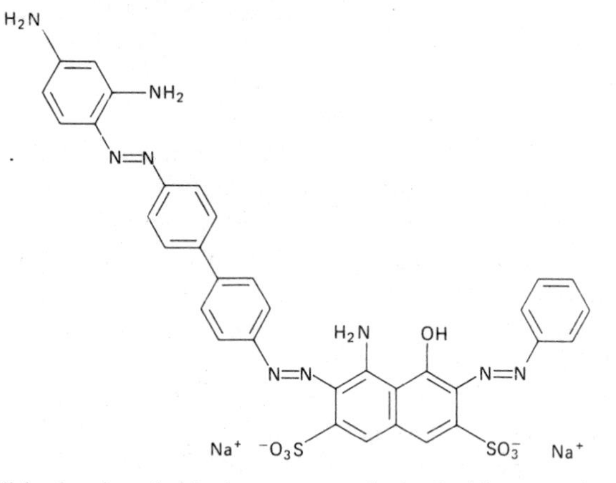

This dye is soluble in water and alcohol but not in most other organic solvents. It is used alone as a stain for sections, especially of plant tissues. The dye is amphoteric. In strongly acid solutions the molecule will bear a net positive charge.

5.9.3.9. Azoic dyes

Azoic dyes are insoluble azo compounds synthesized within or upon substrates by the combination of diazonium salts with suitable aromatic amines or phenols (known as "azoic coupling components"). In histological sections, attachment of the dye to protein, or solution in lipids, may occur as the dye is being formed. The importance of azoic dyes in microtechnique is due to their formation as the end-products of many histochemical methods. Some of these will be discussed in later chapters.

Diazonium salts and azoic coupling components are commonly used reagents in histochemistry. The diazonium salts are occasionally prepared immediately before use but more often they are bought ready-made as stabilized diazonium and tetrazonium salts. These are commonly zinc chloride double salts of the form

$$(R—N^+{\equiv}N)_2\ [ZnCl_4]^{2-}$$

Others are salts of sulphonic acids, such as the naphthalene-1,1-disulphonates,

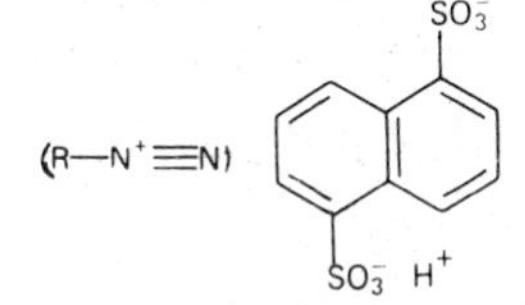

while yet others are borofluorides:

$$(R—N^+{\equiv}N)\ BF_4^-$$

A few are simple chlorides, $R—N^+{\equiv}N\ Cl^-$ or bisulphates, $R—N^+{\equiv}N\ HSO_4^-$. The diazonium salt is mixed with an inert diluent such as sodium sulphate, and other substances may also be present as stabilizers. The original amine usually accounts for less than half the weight of the commercially obtained powder. Stabilized diazonium salts are also called "azoic diazo components". They should **not** be referred to as "stable diazotates"; diazotate ions have the structure $R—N{=}N—O^-$ and are formed when strong alkalis react with diazonium ions. The terms "tetrazonium" and "hexazonium" are used for salts with two and three diazotized amino groups per molecule.

(a) *Examples of commonly used stabilized diazonium and tetrazonium salts*

FAST RED B SALT (Diazotized 5-nitroanisidine acid-1,5-naphthalene disulphonate; C.I. 37125; Azoic diazo 5; M.W. 467)

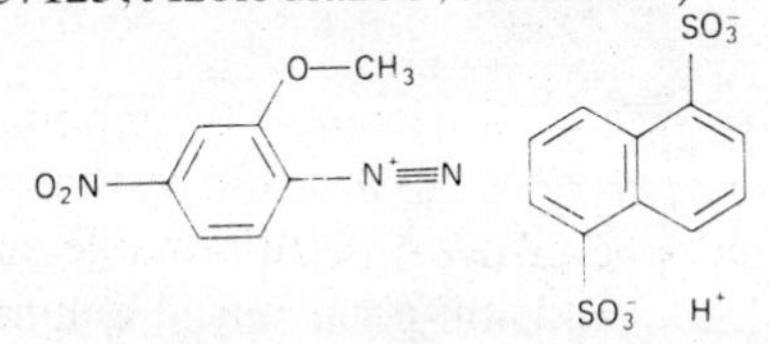

The primary amine represents 20% by weight of the commercial product (theoretical: 36% of formula weight). This diazonium salt is the one most often used for staining argentaffin cells (see Chapter 17). It does not give satisfactory results in histochemical methods for hydrolytic enzymes.

FAST BLUE RR SALT (C.I. 37155; Azoic diazo 24; M.W. 776 including two molecules of diazo)

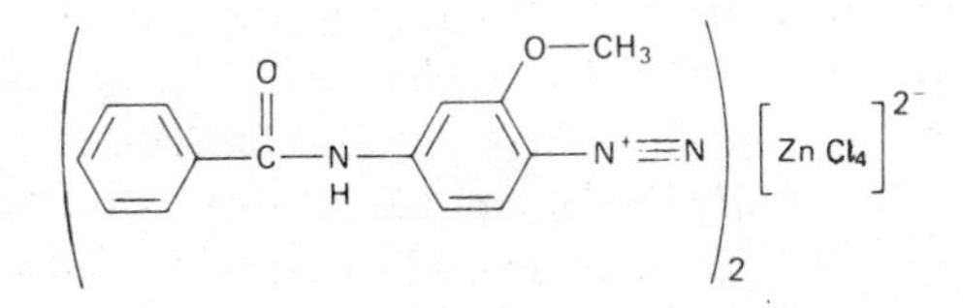

The primary amine accounts for 36% of the weight of the commercial product (theoretical: 71%). This diazonium salt is used in many methods for hydrolytic enzymes (see Chapter 15).

FAST BLUE B SALT (Tetraazotized *o*-dianisidine zinc chloride double salt; C.I. 37235; Azoic diazo 48; M.W. 476)

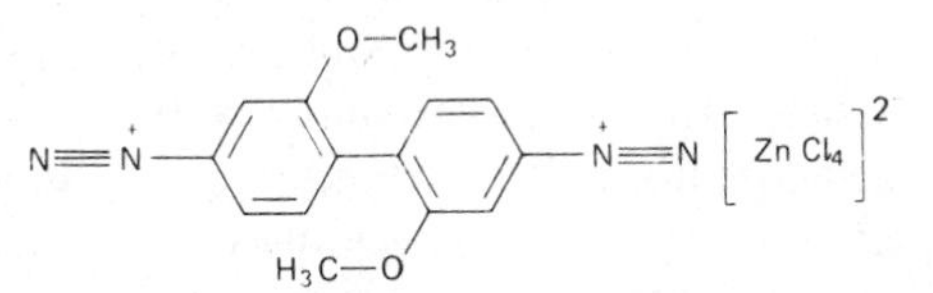

The theoretical content of the bifunctional primary amine (*o*-dianisidine) is 51% by weight but the commercial product contains about 20%. This is a "tetrazonium" salt with a diazotized amine group at each end of the molecule. It is used in the coupled tetrazonium reaction for proteins (Chapter 10).

FAST GARNET GBC SALT (C.I. 37210; Azoic diazo 4; M.W. 334)

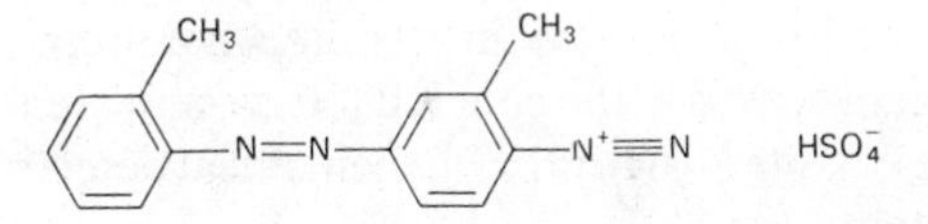

Contains 15–20% by weight of primary amine (theoretical: 67%). Used in some histochemical methods for enzymes. See also note under fast black K, below.

FAST BLACK K SALT (C.I. 37190; Azoic diazo 38; M.W. 836, including two molecules of diazo)

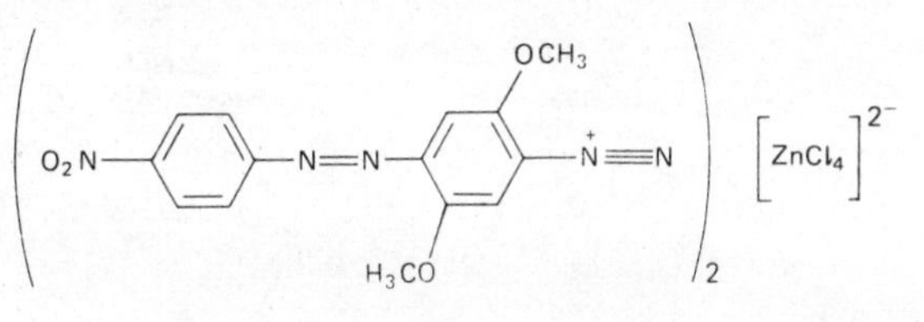

Lillie (1962) assumed that the primary amine represented 20% by weight of the commercial product (theoretical, from formula above, is 72%).

By inspection of the formulae of fast garnet GBC salt and fast black K salt, it is seen that in both cases an azo linkage is already present in the diazonium ion. These salts are therefore coloured. Coupling with a phenol or amine produces a dis-azo dye, which usually has a very dark colour. Many dyes that are primary amines can be diazotized with nitrous acid to give coloured diazonium salts. Subsequent coupling with suitable aromatic molecules in sections of tissues results in the formation of new azo compounds with colours different from and often darker than those of the parent dyes. Pararosaniline and safranine O (Lillie *et al.*, 1968) are two dyes from which histochemically useful diazonium salts are often prepared.

(b) *Azoic coupling components*

Most of the coupling components used in industry are derivatives of naphthalene, with hydroxyl, amine, and sulphonic acid groups substituted in various positions. An example is:

H-ACID (8-amino-l-naphthol-3,6-disulphonic acid; M.W. 319)

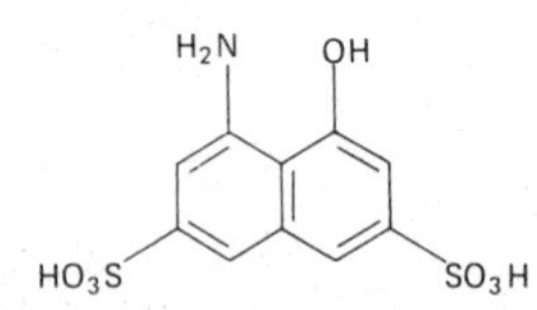

This is also available as its sodium salt (M.W. 341). Coupling occurs principally at position 2 (*ortho* to the hydroxyl group) in alkaline solution. This reagent is used in the coupled tetrazonium reaction (see Chapter 10) for the histochemical detection of proteins.

In enzyme histochemistry, azoic coupling components are released by enzymatic hydrolysis of certain of their derivatives, notably esters and amides. Such methods are discussed in Chapter 15. The coupling components most often liberated in histochemical methods for enzymes are naphthols:

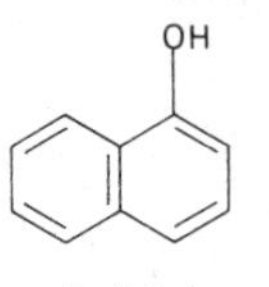

α-Naphthol
(1-naphthol)

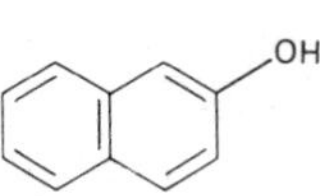

β-Naphthol
(2-naphthol)

More intensely coloured azo dyes are formed from β-naphthol, but these are more soluble and form larger particles than the coupling products of α-naphthol. The advantages of both the simple naphthols are found in naphthol-AS (also known as benzosalicylanilide):

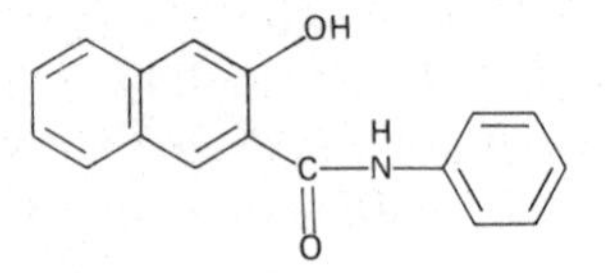

This compound couples with diazonium salts to give finely granular, brightly coloured insoluble products, but the rate of coupling is slower than with α- and β-naphthols.

5.9.4. Indamine and indophenol dyes

The chromophore consists of an indamine group (—N=) joining two rings—one aromatic and the other quinonoid. In the indamine dyes these rings bear, respectively, amine (—NRR′) and iminium (=N⁺RR′) groups *para* to the bridge. In the indophenol dyes at least one of the rings carries an oxygen atom (phenol or quinone) *para* to the indamine linkage. In the general formulations, R and R′ may be hydrogen, methyl, ethyl, or similar radicals.

Dyes and pigments of this type are formed when the products of oxidation of amines react with other amines or with phenols. They are formed in the development of colour photographs and as end-products of some histochemical reactions (see Chapter 16). The chemistry of the syntheses of indamine and related dyes is described by Weissberger (1966).

None of the synthetic indamine or indophenol dyes are important as histological stains, so only one example will be given:

FAT BLUE Z (C.I. 49705; Solvent blue 22; M.W. 304)

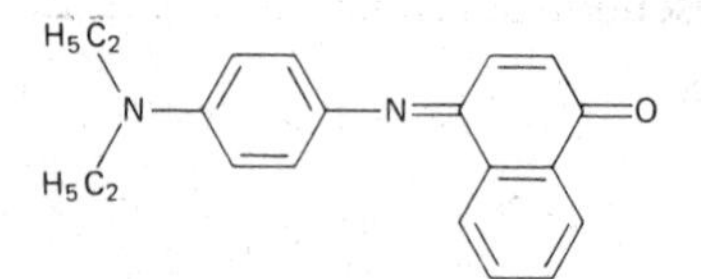

A blue indophenol dye, insoluble in water but soluble in alcohol and non-polar solvents. The tertiary amine function is potentially ionizable to give a cationic dye.

5.9.5. Cyanine dyes

These dyes are used principally as sensitizing ingredients in photographic emulsions, but some are employed as microscopic stains and as textile dyes. The chromophoric system is

$$-\overset{|}{C}(=\overset{|}{C}-\overset{|}{C})_n= \;\longleftrightarrow\; =\overset{|}{C}(-\overset{|}{C}=\overset{|}{C})_n-$$

(n = zero or an integer). The methine group (=CH—; $n = 0$) or the chain of vinylene (—($\overset{|}{C}=\overset{|}{C}$—)) groups connects two nitrogen-containing heterocyclic rings between which a positive charge resonates.

Related to the cyanines are the **hemicyanines** (with only one heterocyclic ring), the **trinuclear cyanines** (with branched vinylene chain), and the **merocyanines** (with both nitrogen- and oxygen-containing heterocycles). In the **azamethine** dyes, one or more methine linkages are replaced by indamine (=N—) groups. These dyes will not be discussed further. The reader is referred to Brooker (1966) and Allen (1971) for more information. It is possible that these groups of dyes may include valuable histological reagents that have not yet been recognized as such.

Examples of cyanines

PINACYANOL (C.I., first edition, No. 808; M.W. 480)

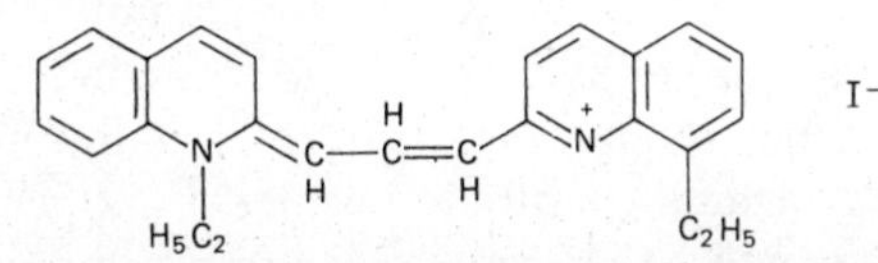

This red cationic dye has been used occasionally for vital staining and for sections. It dissolves in water and alcohols.

PSEUDOISOCYANINES (No C.I. numbers)

These red cationic dyes can also be used as fluorochromes. An example is 6,6′-dichloro-*N*,*N*-diethylpseudoisocyanine:

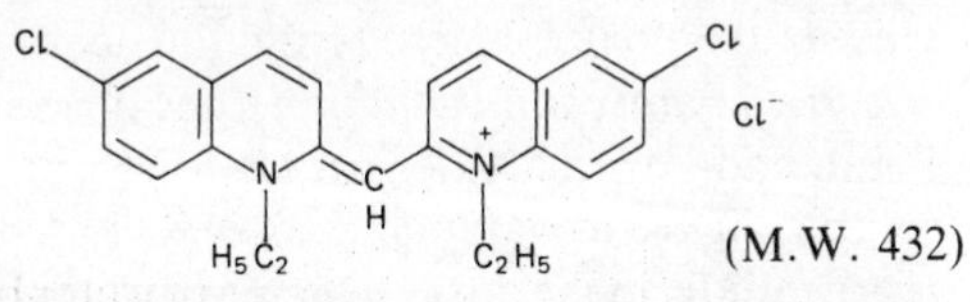

(M.W. 432)

Pseudoisocyanines have been used as fluorochromes for demonstrating cysteic acid formed by oxidation of neurosecretory material (Jakovleva *et al.*, 1968; see also Chapter 18) and for staining the secretory cells of various endocrine organs (Solcia *et al.*, 1968).

CARBOCYANINE DBTC (No C.I. number; M.W. 560)

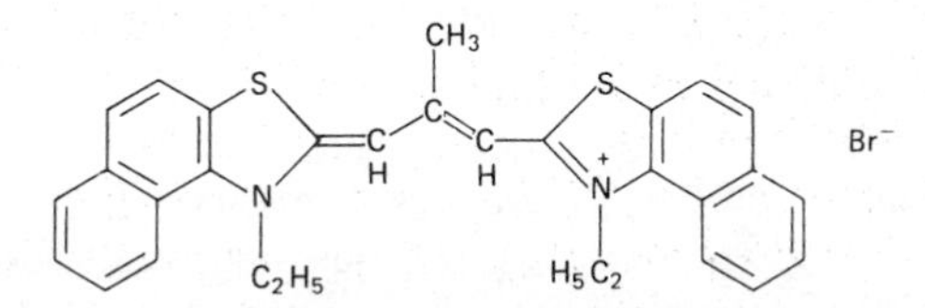

This dye, also known as "Stains-all", is used alone to stain sections in various shades of blue, purple, and red (Green & Pastewka, 1974a,b, 1979), but its usefulness is unfortunately rather limited because the stained sections fade rapidly on exposure to light.

5.9.6. Stilbene dyes and fluorescent brighteners

Stilbene is

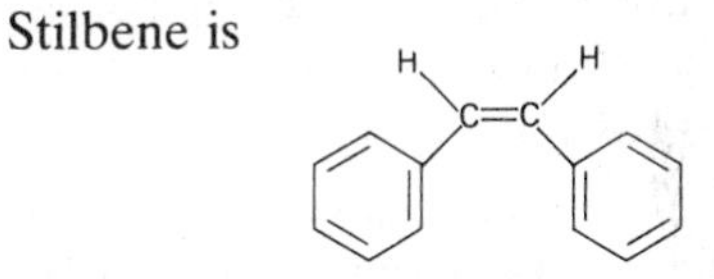

(the *cis*-isomer is shown). Several dyes include the stilbene configuration in conjunction with azo groups. Most of these dyes are mixtures of large molecules of uncertain structure and none of them are important as biological stains.

Fluorescent brighteners absorb ultraviolet radiation and emit blue or green light. A fabric containing one of these compounds appears more vividly white than it otherwise would. Thus fluorescent brighteners can offset the tendency of white materials to turn yellow with age. Most fluorescent brighteners are derivatives of 4,4′-diaminostilbene-2,2′-disulphonic acid:

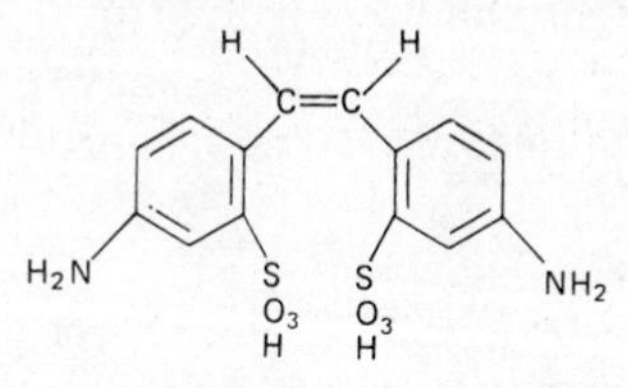

An example is C.I. 40620; Fluorescent brightening agent 32:

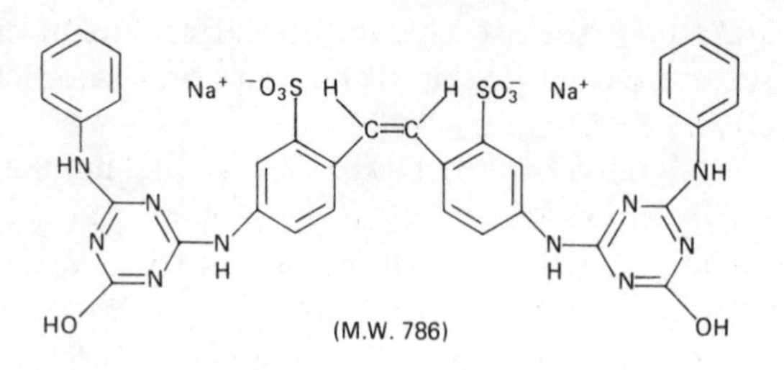

Fluorescent brighteners have occasionally been employed in microscopy as fluorochromes, especially for plant tissues. They might become more important in the future.

5.9.7. Quinoline dyes

In this small group of dyes the chromophoric system is quinophthalone:

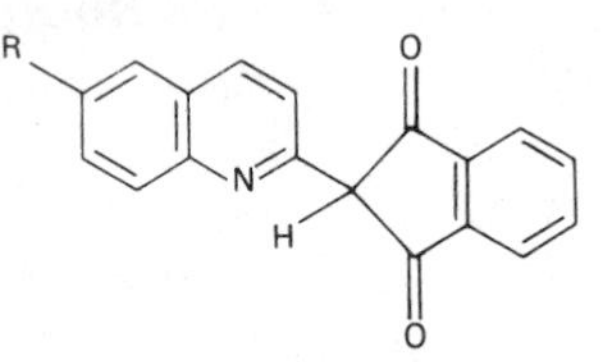

which is derived from the condensation of quinaldine with phthalic anhydride. Sulphonic acid substituents confer water-solubility. The quinoline dyes are of no importance in microtechnique. The quinoline ring is present in other types of dyes, notably in some of the cyanines and their congeners.

5.9.8. Thiazole dyes

The benzothiazole ring 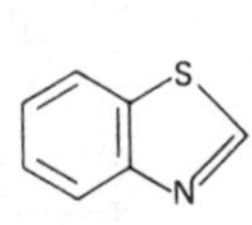

is the sole chromophore in dyes of this group. Dyes with other chromophoric systems (especially azo dyes) sometimes also contain benzothiazole rings.

Examples

PRIMULINE (C.I. 49000; Direct yellow 59)
This greenish-yellow dye (amphoteric: here shown as anionic), which is also a fluorochrome, is a mixture, the principal component of which is

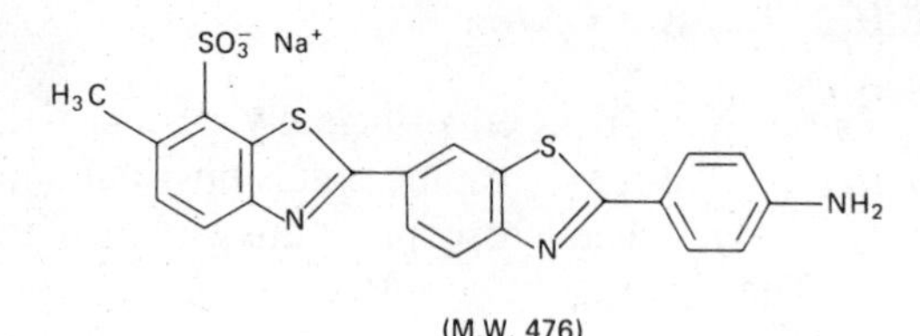

It has been used as a vital stain and fluorochrome (e.g. Kuypers *et al.*, 1977). With textiles, primuline can function as a "direct" dye for cellulosic fibres. It can be diazotized on the fibre and then coupled with suitable phenols or amines to give more intense, faster colours.

THIOFLAVINE TCN (Thioflavine T) (C.I. 49005; Basic yellow 1; M.W. 319)

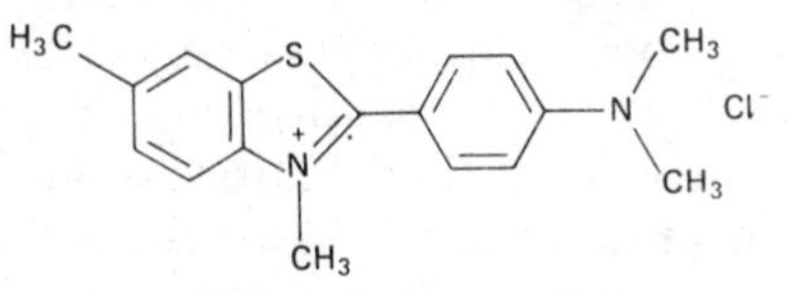

This yellow cationic dye can also serve as a fluorochrome. An insoluble yellow pigment is formed with phosphotungstomolybdic acid.

5.9.9. Aminoketone and hydroxyketone dyes

In these dyes, the chromophoric system is an array of conjugated double bonds including at least one ketone carbonyl group. An amino (or substituted amino) or hydroxyl group is also present and is attached to an aromatic or quinonoid ring. The group includes naturally occurring and synthetic dyes, of which the former are the more important in histology.

5.9.9.1. NATURAL DYES

Brazilein and haematein are hydroxyketones formed by oxidation of the naturally occurring compounds brazilin and haematoxylin.

HAEMATOXYLIN (C.I. 75290; Natural black 1)
Haematoxylin is extracted from the heart-wood of *Haematoxylon campechianum*, the logwood tree of South and Central America. Haematoxylin is very soluble in both water and alcohol but dissolves more rapidly in the latter. Haematein (see below) is only sparingly soluble in these solvents but is much more soluble in ethylene glycol and glycerol, which are also solvents for haematoxylin.

Upon partial oxidation, haematoxylin is changed into **haematein**, which is the active ingredient of the "haematoxylin" stains used in histology. The oxidation may be brought about slowly by exposure to air (ripening) or rapidly by adding an oxidizing agent such as sodium iodate or ferric ions. One gram of haematoxylin crystals (with $3H_2O$ per dye molecule) is fully oxidized to haematein by 185 mg of $NaIO_3$. In practice, when the mordant is to be aluminium, smaller amounts of oxidizing agents are used since haematein itself is slowly oxidized by atmospheric oxygen to give useless products. A reservoir of initially unoxidized haematoxylin in a staining solution is believed to allow for the continued production of haematein by slow atmospheric oxidation and so prolong the useful life of the reagent.

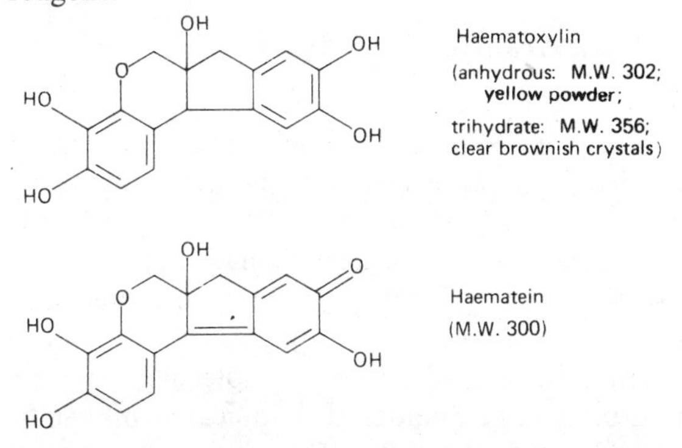

Haematoxylin
(anhydrous: M.W. 302; yellow powder;
trihydrate: M.W. 356; clear brownish crystals)

Haematein
(M.W. 300)

Ferric salts also serve to oxidize haematoxylin to haematein, so that iron–haematoxylin mixtures do not need to be "ripened". However, since the concentration of ferric ions in such stains is usually greatly in excess of that needed to oxidize the dye and form the dye–metal complex, the solutions are not indefinitely stable.

The main product of over-oxidation of haematein is an orange–yellow substance known as oxyhaematein. It is formed much more readily by oxidation with hydrogen peroxide, potassium permanganate, or sodium iodate than by exposure to oxygen. Marshall and Horobin (1972) have provided strong (though not conclusive) evidence that oxyhaematein has the structure

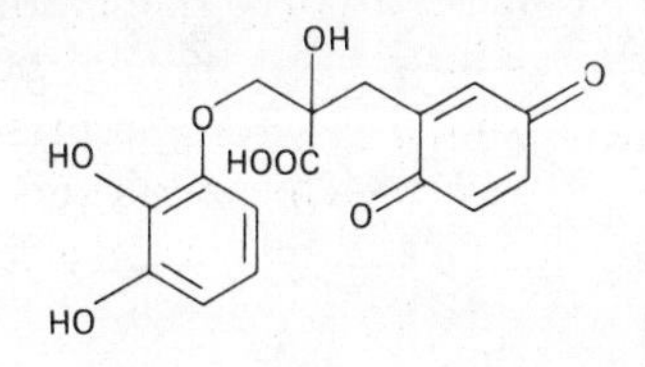

and have shown that it can behave as an ordinary anionic dye whose properties are unaffected by mixing with aluminium salts.

Staining with haematein is carried out in conjunction with a mordant, most often aluminium (for staining nuclei blue–purple) or iron (for staining nuclei, myelin or other elements blue–black, according to technique). Preformed dye–metal complexes are used for most purposes, though in the classical method of Heidenhain the mordant and the dye are applied sequentially to the sections. Many other mordants are used for special purposes. A list of mordants and structures demonstrated is given by Culling (1974, p. 160), and several recipes for mixtures containing haematoxylin are given by Stevens (1977). Nuclear staining with haematein lakes is described in Chapter 6, a method in which the dye is used for the localization of certain lipids is given in Chapter 12, and the staining of myelin is discussed in Chapter 18. For the many other uses of haematoxylin, see Gabe (1976), Lillie & Fullmer (1976), and Lillie (1977).

Spectrophotometric studies have shown that commercial samples of haematoxylin are usually quite pure (90–95% of anhydrous dyestuff). Materials sold as "haematein", however, are very variable, with dye contents ranging from 1.2 to 75.1% (Marshall & Horobin, 1974). When haematein is required, it is probably better to make it by oxidizing haematoxylin in the laboratory than to buy it.

BRAZILIN (C.I. 75280; Natural red 24)
Brazilin is extracted from brazil wood (*Caesalpina sappan* from Indonesia and *C. brasiliensis* from Central and South America). The country of Brazil was named after the discovery there of trees yielding the dyestuff, which had been imported into

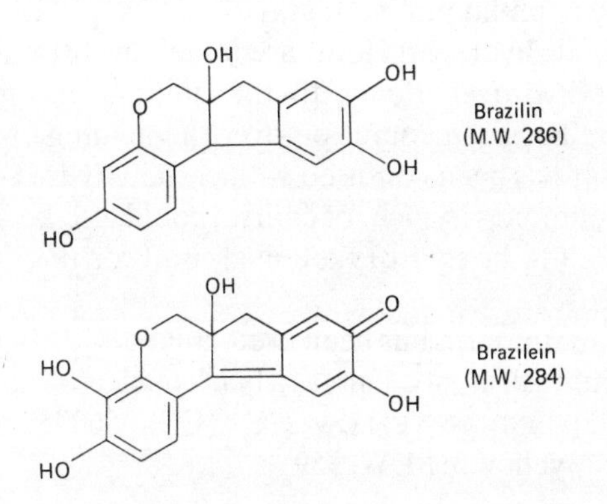

Brazilin
(M.W. 286)

Brazilein
(M.W. 284)

Europe for several centuries previously from oriental sources. Brazilin differs from haematoxylin only in lacking one hydroxyl group. Upon oxidation, a red dye, **brazilein,** is produced.

Like haematein, brazilein is used mixed with a metallic mordant. The use of the aluminium lake as a red nuclear stain is described in Chapter 6.

5.9.9.2. SYNTHETIC DYES

Very few synthetic dyes of this group have been used as biological stains (Lillie, 1977). A few examples are shown below to illustrate the variety of chemical structures available.

BRILLIANT SULPHOFLAVINE FF (C.I. 56205; Acid yellow 7; M.W. 404)

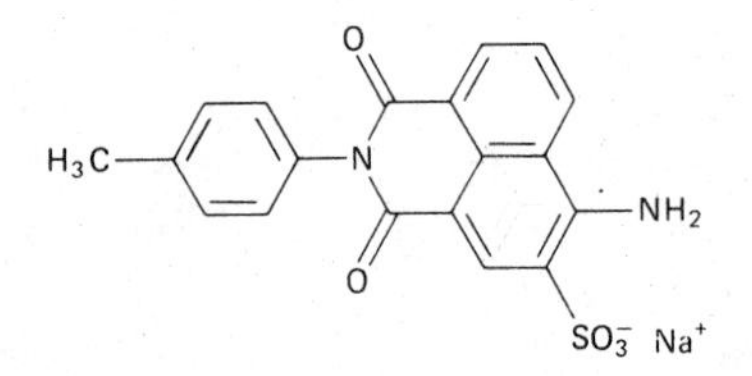

A yellow, anionic, fluorescent aminoketone dye. Two reactive dyes with similar structures are described by Lillie (1977).

NAPHTHAZARIN (C.I. 57010; Mordant black 37; M.W. 190)

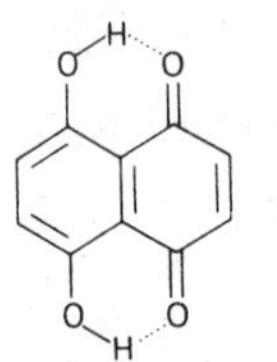

This anionic hydroxyketone dye can also be placed in the naphthoquinone class (see below, p. 72). The phenolic hydrogen atoms are joined by intramolecular hydrogen bonds to the quinone oxygens. There has been confusion over the naming of this dye. It has been confused with certain hydroxyanthraquinones (which, of course, could also be considered to be hydroxyketones) used for the same purposes.

Naphthazarin has been used, mixed with an aluminium salt as mordant, as a black nuclear stain.

HELINDON YELLOW CG (C.I. 56005; Vat yellow 5; M.W. 359)

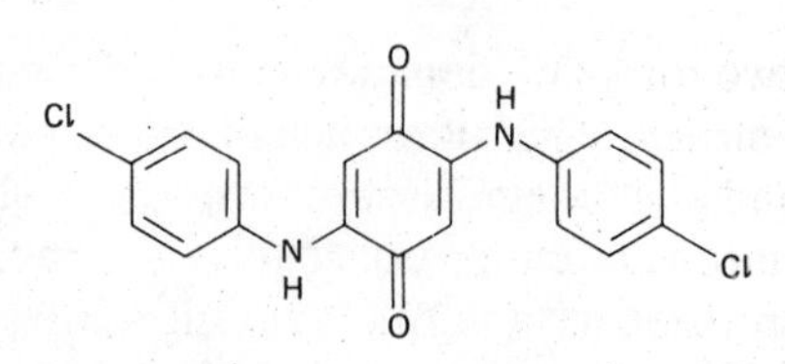

This aminoketone is a vat dye. It is reduced to a soluble leuco compound (presumably a *p*-diphenol), applied to a textile and then reoxidized to the insoluble coloured form.

RHODIZONIC ACID

This histochemical reagent, used for the detection of lead and other metals, is an orange hydroxyketone (1,2-dihydroxy-3,4,5,6-tetraoxocyclohex-l-ene), but is not usually thought of as a dye. It is discussed in Chapter 13.

5.9.10. Arylmethane dyes

These are compounds in which three or four of the hydrogen atoms of methane are replaced by aromatic ring systems. The latter bear ionizable substituents and the rings resonate between quinonoid and truly aromatic configurations. The chromophoric system of conjugated double bonds extends through all parts of the molecule, so that an unusually large number of canonical forms can be formulated for each dye. Diarylmethanes are not very important as biological stains, but the triarylmethane class includes many dyes of great value to the histologist and histochemist.

5.9.10.1. DIPHENYLMETHANE DYES

The only member of this group used in histology is auramine O.

AURAMINE O (C.I. 41000; Basic yellow 2; Anhydrous M.W. 304; commercial product has one molecule water of crystallization; M.W. 322)

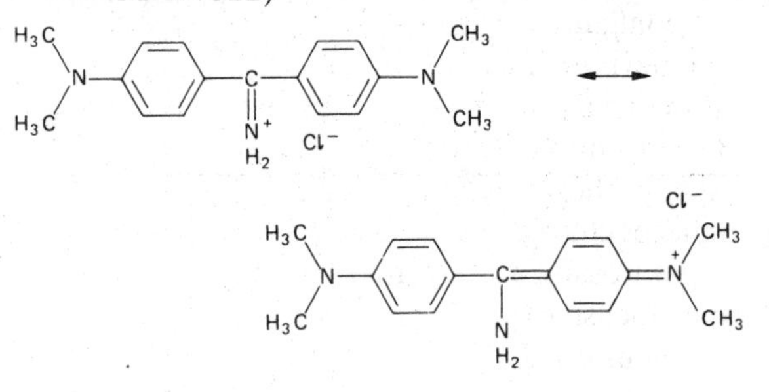

These two formulae depict two of the possible resonance structures of the chromophoric system: an imine and a quinonoid form. Auramine O is a yellow, fluorescent cationic dye used mainly in bacteriology (see Clark, 1973). Samples should contain at least 80% by weight of the **anhydrous** dye.

5.9.10.2. TRIARYLMETHANE DYES

The simplest members of this class are the triphenylmethanes. Of these, pararosaniline, a red cationic dye, has the most easily understood structure. This dye, which has many uses in histology and histochemistry, will therefore be described at some length. Other important triarylmethane dyes will be treated more briefly. Examples of cationic dyes will be followed by examples of anionic and mordant dyes.

PARAROSANILINE (C.I. 42500; Basic red 9; M.W. 324)

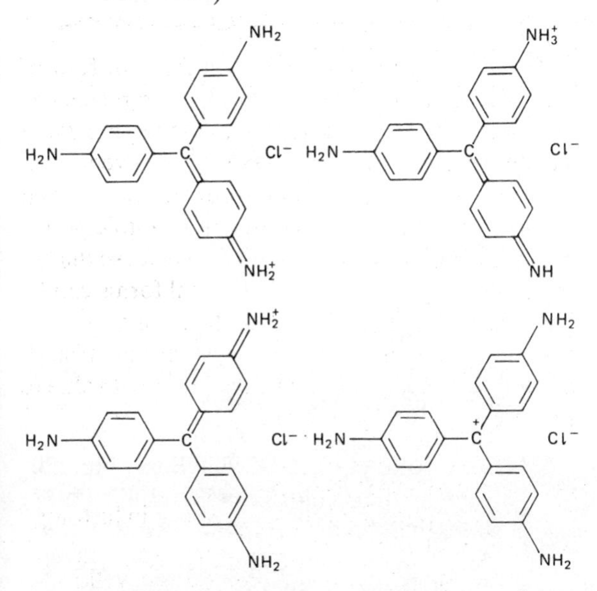

Four of the possible resonance structures are shown above. It can be seen that any one of the three rings may transiently assume the aromatic or the quinonoid configuration and any may bear a positively charged nitrogen atom. Similar resonance occurs in all the triarylmethane dyes, anionic and cationic. By convention the iminium ion forms (top left) are usually shown in structural formulae. Acetate replaces chloride as the gegen-ion in some samples of pararosaniline. For notes on purity, see under basic fuchsine (p. 64).

Colourless derivatives of three kinds are easily prepared from pararosaniline and related dyes. Reduction destroys the chromophoric structure and gives a **leuco compound**:

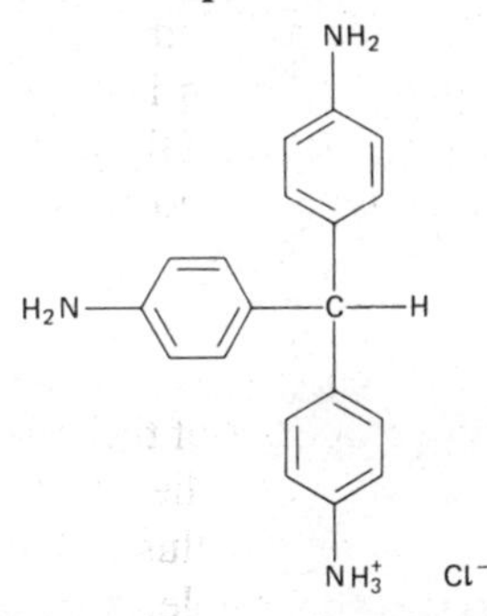

This is a salt of an amine, the **leuco base**:

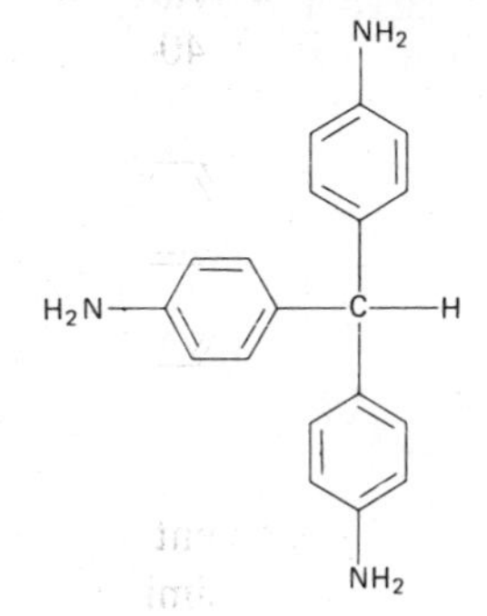

When pararosaniline is treated with strong alkali, an unstable **colour base** is first formed:

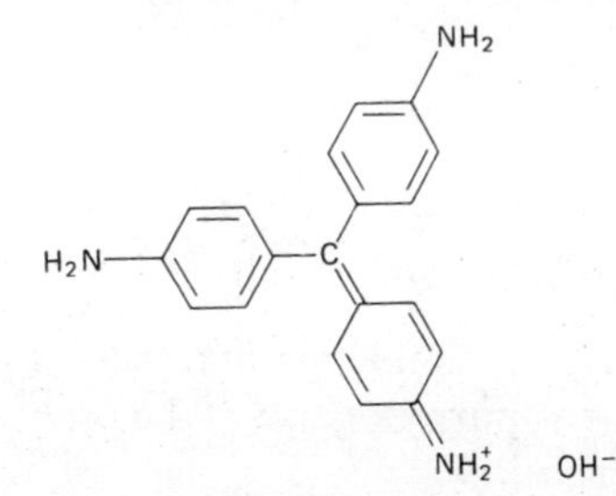

The unstable colour base rapidly isomerizes in solution to give the stable, colourless **carbinol base**:

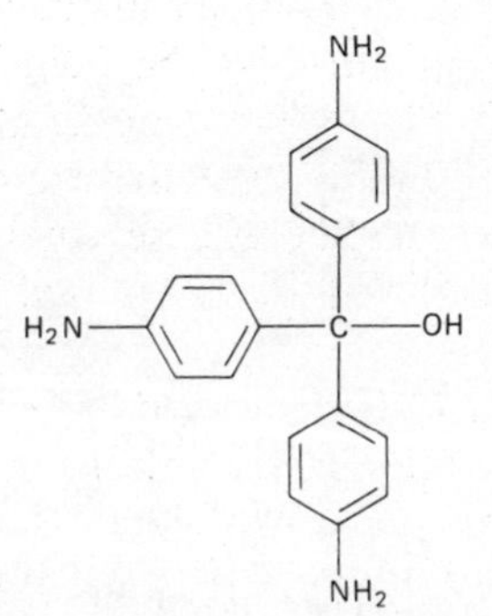

It can be seen that the carbinol base is a tertiary alcohol. It is named as a derivative of methanol (= carbinol; H_3COH).

The third way of decolorizing pararosaniline involves treatment with sulphurous acid, to give Schiff's reagent, the chemistry and uses of which are described in other chapters.

Other derivatives of pararosaniline are prepared in the laboratory for use in some special staining methods. These include a hexazonium salt (used in azo coupling procedures) and aldehyde–fuchsine (see Chapter 8).

BASIC FUCHSINE

This is a mixture of triphenylmethane dyes. Modern samples of basic fuchsine consist almost entirely of either **pararosaniline** or **rosaniline.** Older samples usually contained both dyes, together with smaller amounts of **new fuchsine** and **magenta II**. The gegen-ion is sometimes acetate rather than the chloride shown in the formulae below.

ROSANILINE (C.I. 42510; Basic violet 14; Anhydrous M.W. 338; crystals with $4H_2O$, M.W. 410)

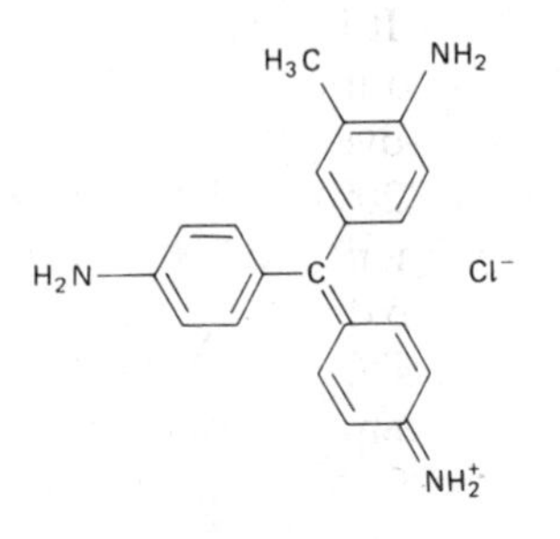

This dye is closely similar to pararosaniline and can replace it for all purposes except the preparation of aldehyde–fuchsine.

NEW FUCHSINE (Magenta III) (C.I. 42520; Basic violet 2; M.W. 366)

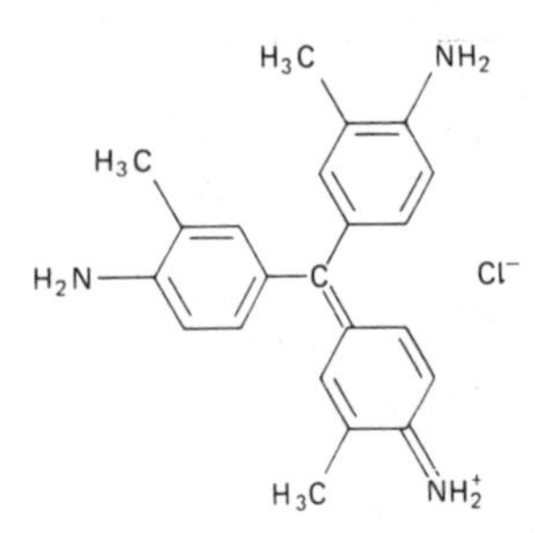

MAGENTA II (No C.I. number; M.W. 352)

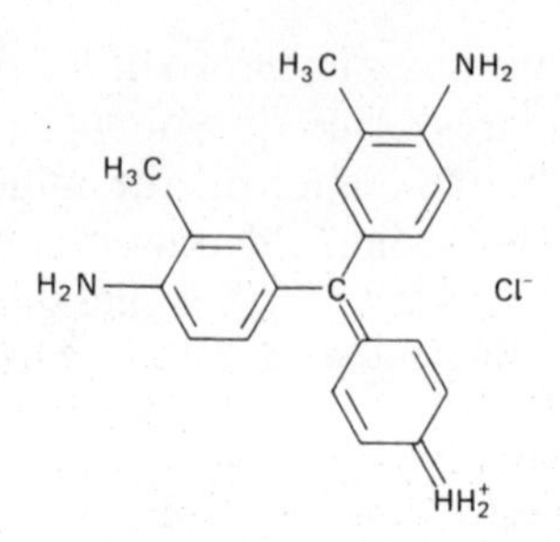

All the components of basic fuchsine are soluble in water and more soluble in alcohol. For critical work it is desirable to use pure pararosaniline. To be certified by the Biological Stains Commission, samples of basic fuchsine must contain at least 88% anhydrous dye (estimated as pararosaniline or rosaniline) and must perform satisfactorily in a variety of staining procedures.

CRYSTAL VIOLET (C.I. 42555; Basic violet 3; M.W. 408)

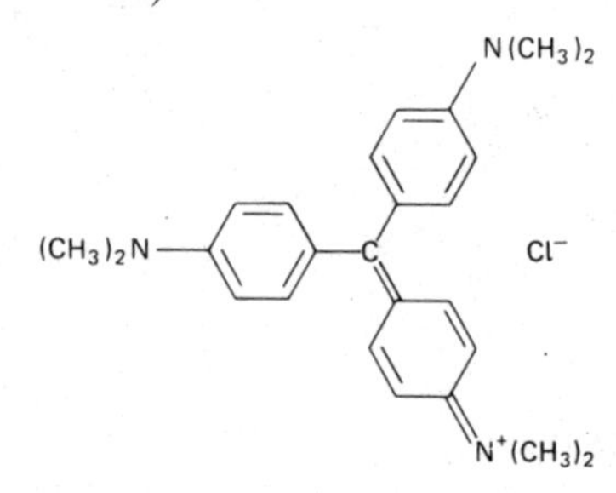

This *N*-hexamethylated derivative of pararosaniline is used in many bacteriological techniques as well as in histology. Samples should contain at least 88% by weight of the anhydrous dye.

METHYL GREEN (C.I. 42585; Basic blue 20; M.W. of the dichloride: 458.5. Lillie (1977) states that this dye is supplied as a zinc chloride double salt. This is $[C_{26}H_{33}H_3]^{2+}$ $[ZnCl_4]^{2-}$, M.W. 594)

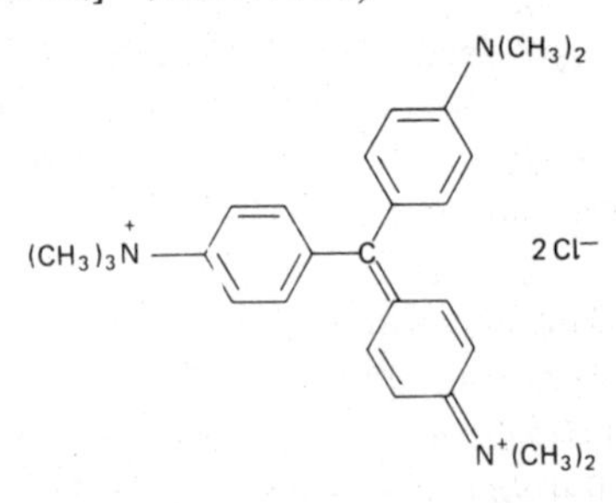

This bluish-green dye is used in several techniques, including the methyl green–pyronine method for nucleic acids. One of the methyl groups on the quaternary nitrogen (left-hand side of formula above) detaches easily, so methyl green is always contaminated with crystal violet. The latter dye can be extracted from aqueous solutions of methyl green by shaking with chloroform in which the colour base of crystal violet is very soluble. Methyl green is rather easily extracted from stained sections by water and water–alcohol mixtures, so special care is necessary when dehydrating (see p. 109, also Gabe, 1976, p. 544). Samples should contain at least 65% of the anhydrous, zinc-free dye (M.W. 458.5).

VICTORIA BLUE 4R (C.I. 42563; Basic blue 8; M.W. 520)

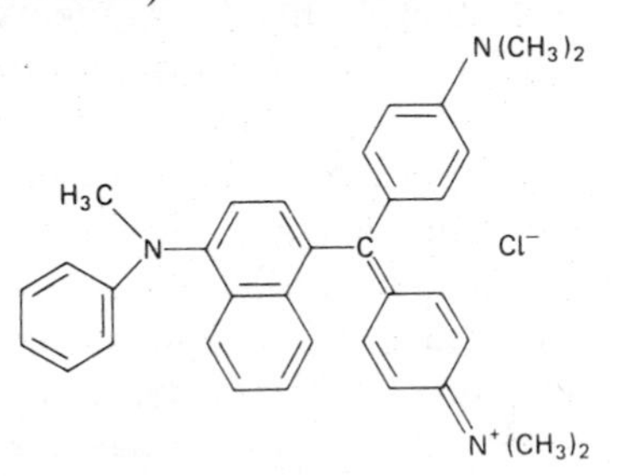

This diphenylnaphthylmethane is a cationic dye with a fairly bulky molecule. It is soluble in water and more so in ethanol. Related dyes are Victoria blue R (C.I. 44040; Basic blue 11) and B (C.I. 44045; Basic blue 26).

ACID FUCHSINE (C.I. 42685; Acid violet 19)
This is a di- or trisodium salt of the trisulphonic acid derivative of basic fuchsine. The acid fuchsine derived from rosaniline is

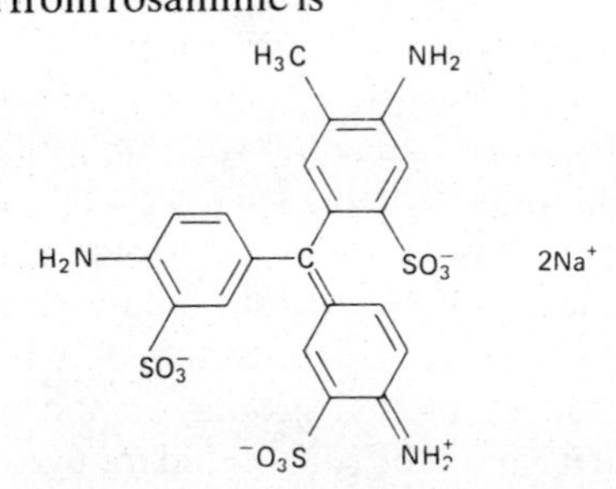

The one related to pararosaniline lacks the methyl group.

This red anionic dye has many uses, especially in conjunction with other anionic dyes, as will be explained in Chapter 8. It is very soluble in water but only sparingly so in alcohol. Stained tissues are differentiated by washing in water but are stable in 95% and absolute ethanol.

For analytical purposes acid fuchsine is assumed to consist of equal parts of the anhydrous disodium salts of trisulphonated pararosaniline and rosaniline (average M.W. 578.5). Samples should contain at least 55% by weight of this hypothetical dyestuff.

FAST GREEN FCF (C.I. 42053; Food green 3; M.W. 809)

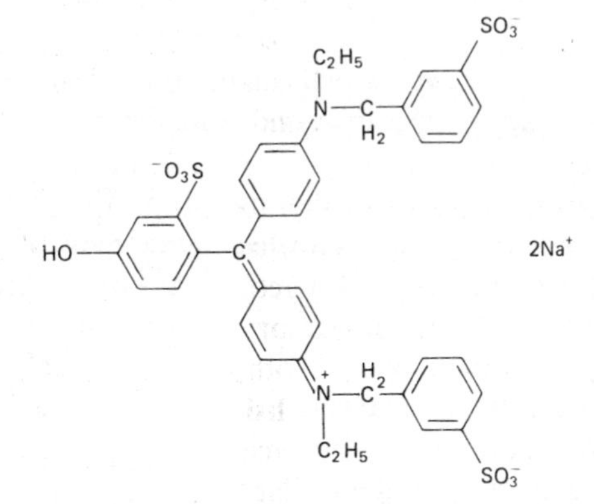

This bluish-green anionic triphenylmethane dye is used in methods for connective tissue and as a counterstain to methods that produce strong blues, purples, and reds. It is very soluble in water and considerably less so in alcohol.

A closely related dye is **light green SF** (light green SF yellowish; C.I. 42095; Acid green 5; M.W. 793), which lacks the phenolic hydroxyl group of fast green FCF. The two dyes can be used for the same purposes, but C.I. 42053 is preferred since it is less prone to fading. Samples of fast green FCF should contain at least 85% by weight of the anhydrous dye. Light green SF should be at last 65% pure.

ANILINE BLUE WS (water blue I; C.I. 42755; Acid blue 22; M.W. 738)

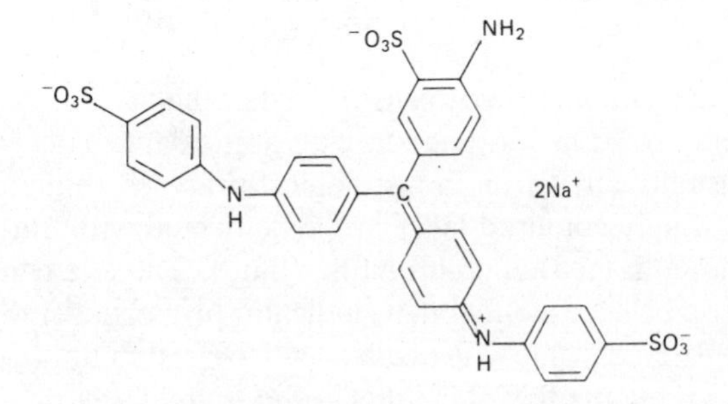

METHYL BLUE (soluble blue; ink blue) (C.I. 42780; Acid blue 93; M.W. 800)

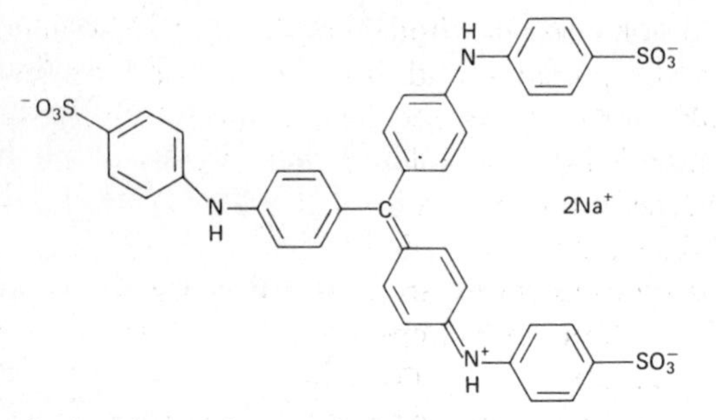

Both the above dyes and also a mixture of the two have been known in the past as "water blue", "soluble blue", and "aniline blue, water-soluble". The two dyes have identical staining properties. Both are very soluble in water and insoluble in alcohol. The colour changes to red with strong alkalis. They are used as anionic dyes of large molecular size, especially in methods for connective tissue. These dyes cannot be assayed very precisely. The criteria for certification by the Biological Stains Commission are given by Lillie (1977).

SPIRIT BLUE (C.I. 42775; Solvent blue 3), also known as "aniline blue, alcohol-soluble" is a mixture of diphenyl and triphenyl pararosanilines (i.e. aniline blue WS and methyl blue without the sulphonic acid groups). Spirit blue is a cationic dye but is insoluble in water. Its alcoholic solutions are occasionally used in histology.

CHROMOXANE CYANINE R (solochrome cyanine R; eriochrome cyanine R; C.I. 43820; Mordant blue 3; M.W. 536)

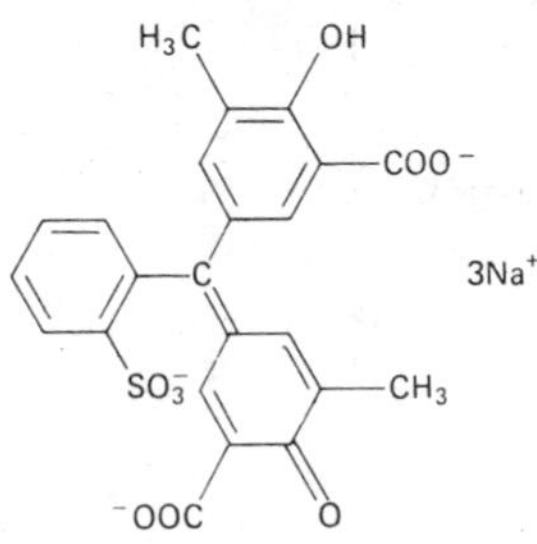

This anionic hydroxytriphenylmethane dye is employed in conjunction with a mordant. This is usually ferric iron in histological practice, though strongly coloured lakes are also formed with aluminium (red) and chromium (blue). The dye can also be used alone, but its indicator properties (red with acids, blue with alkalis) lead to a rather unpredictable medley of colours in the stained sections. A ferric lake is used, in different techniques, for staining nuclei (p. 80) and myelin (p. 266).

5.9.11. Xanthene dyes

The chromophoric system is

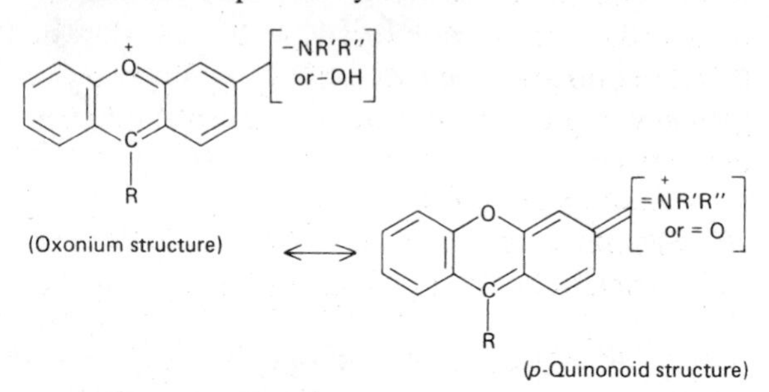

R, R′, and R″ may be hydrogen or any of a variety of alkyl or aryl radicals. Other substituents may also be present. If R is a phenyl group, the dye will also be a triphenylmethane derivative in which two of the benzene rings are joined by an ether linkage. Many of the xanthene dyes are fluorescent and some are used primarily as fluorochromes. The *p*-quinonoid structure is (arbitrarily) used in the structural formulae for the examples of xanthene dyes given below. The dyes are classified by ionic charge rather than as hydroxy- and amino-xanthenes.

Examples

Cationic xanthene dyes

PYRONINE Y (pyronine G) (C.I. 45005; M.W. 303)

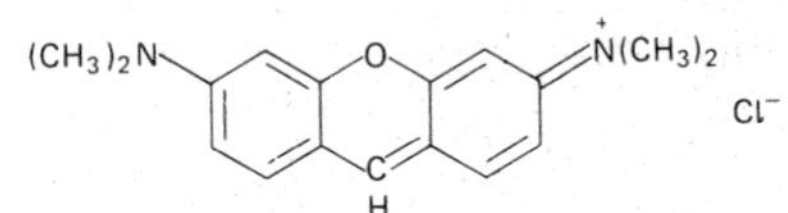

This red dye is used, together with methyl green, in a method for nucleic acids (see Chapter 9). The dye is very soluble in water and alcohol. Samples should contain at least 45% by weight of anhydrous dye.

PYRONINE B (C.I. 45010) differs only in having ethyl instead of methyl groups on the nitrogen atoms, but it cannot be substituted for pyronine Y.

RHODAMINE B (C.I. 45170; Basic violet 10; M.W. 479)

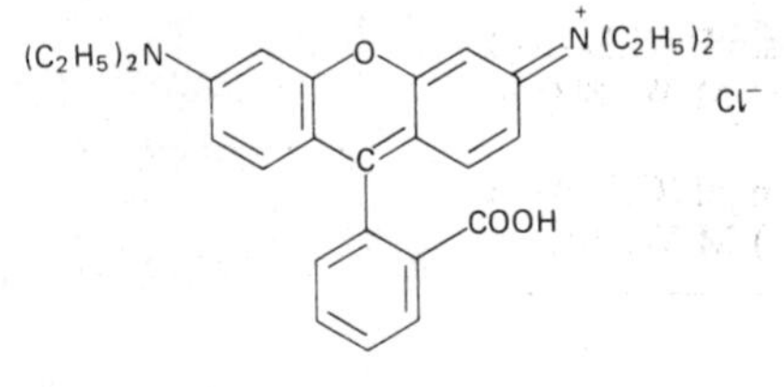

This red–violet emits a strong orange fluorescence when excited by either ultraviolet or green light. An isothiocyanate radical (S═C═N—) can be substituted in the carboxyphenyl group to give **rhodamine B isothiocyanate,** a reactive dye capable of conjugating with proteins to yield fluorescently labelled derivatives.

Anionic xanthene dyes

FLUORESCEIN SODIUM (uranin; C.I. 45350; Acid yellow 73; M.W. 376)

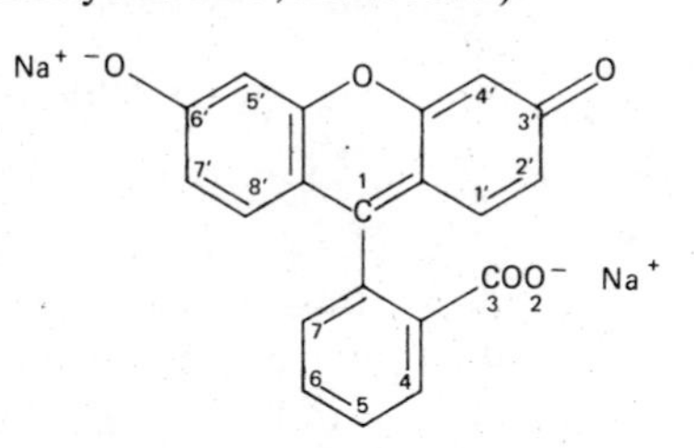

The numbering is as given by Lillie (1977).

This yellow anionic dye is used only as a fluorochrome. As with rhodamine B, an isothiocyanate can be prepared and this is used as a fluorescent label for proteins (see Chapter 19). Optimal excitation is by blue–violet or ultraviolet light and the fluorescent emission is green.

The halogenated derivatives of fluorescein are also fluorescent but are used principally as red dyes. Eosin is the most important.

EOSIN AND ITS CONGENERS

These red dyes are substituted fluoresceins with the structures indicated in Table 5.1. All the dyes in the table are red, in various shades. All are freely soluble in water and rather less soluble in alcohol and all can be used for similar purposes, though certain dyes are traditionally associated with particular techniques. For certification by the Biological Stains Commission, the percentage by weight of anhydrous dyes in a sample must not be less than 80% (eosin), 85% (eosin B), 80% (phloxine B), or 80% (rose bengal).

TABLE 5.1. *Eosin and related dyes*

Name and molecular weight	C.I. number	Substituent on fluorescein skeleton at carbon number									
		1′	2′	4′	5′	7′	8′	4	5	6	7
Eosin (eosin Y, yellowish; C.I. Acid red 87) M.W. 692	45380	H	Br	Br	Br	Br	H	H	H	H	H
Eosin B (C.I. Acid red 91) M.W. 624	45400	H	NO_2	Br	Br	NO_2	H	H	H	H	H
Phloxine (C.I. Acid red 98) M.W. 761	45405	H	Br	Br	Br	Br	H	Cl	H	H	Cl
Phloxine B (C.I. Acid red 92) M.W. 830	45410	H	Br	Br	Br	Br	H	Cl	Cl	Cl	Cl
Erythrosin (erythrosin Y; C.I. Acid red 95) M.W. 628	45425	H	H	I	I	H	H	H	H	H	H
Erythrosin B (C.I. Acid red 51) M.W. 880	45430	H	I	I	I	I	H	H	H	H	H
Rose Bengal (C.I. Acid red 94) M.W. 1018	45440	H	I	I	I	I	H	Cl	Cl	Cl	Cl

ETHYL EOSIN (eosin, alcohol soluble; C.I. 45386; Solvent red 45; M.W. 698)

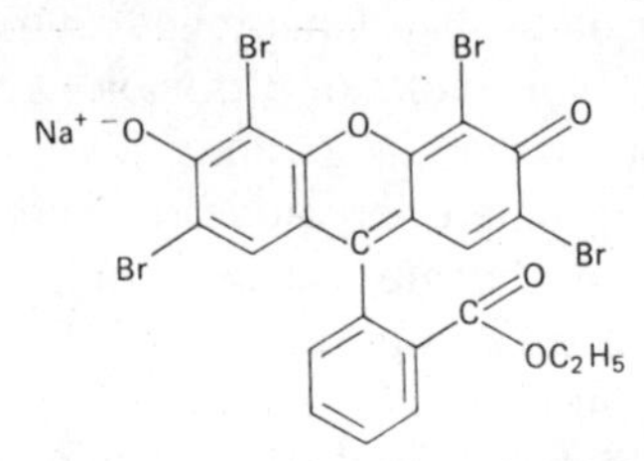

This, the ethyl ester of eosin, is only slightly soluble in cold water but dissolves freely in alcohol. It is used when staining with eosin is to be done from an alcoholic solution. Unlike the water-soluble red derivatives of fluorescein, ethyl eosin must be differentiated by alcohol rather than by water. Samples of ethyl eosin should contain at least 78% by weight of the anhydrous dye.

5.9.12. Acridine dyes

These dyes are similar to the xanthenes, but with a nitrogen rather than an oxygen atom linking the benzene rings. As with the xanthenes, the chromophoric system resonates between *o*- and *p*-quinonoid forms.

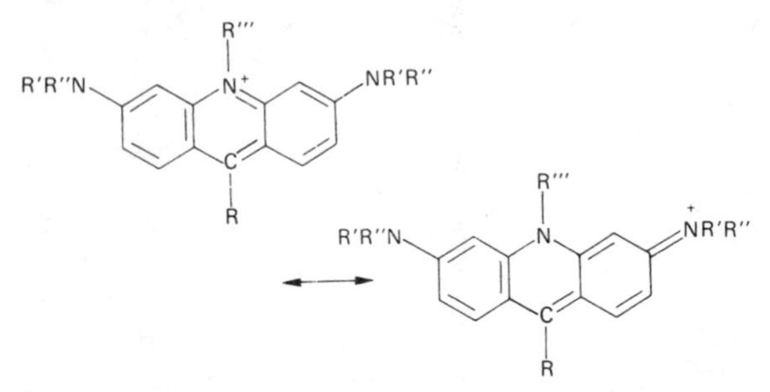

R, R′, R″ and R‴ may be hydrogen atoms or alkyl or aryl radicals. If R is a phenyl group the dye will also be a triphenylmethane derivative with two of its benzene rings linked by a nitrogen atom. The acridine dyes used in histology are all strongly fluorescent.

Examples

ACRIFLAVINE (trypaflavine) (C.I. 46000; M.W. 260)

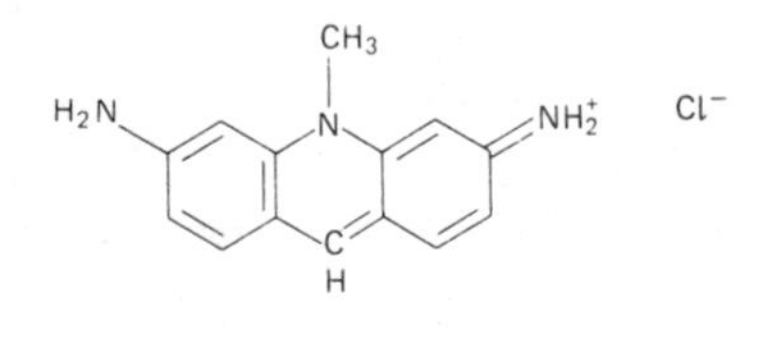

This may be used on the rare occasions when a yellow cationic dye is needed in ordinary light microscopy, and also as a fluorochrome. It is soluble in water and in alcohol. Acriflavine is better known as an antiseptic than as a biological stain.

ACRIDINE ORANGE (C.I. 46005; Basic orange 14; M.W. 302; Lillie (1977) states that this dye is usually supplied as a zinc chloride double salt. This would be $[C_{17}H_{20}N_3]_2^+$ $[ZnCl_4]^{2-}$, M.W. 438)

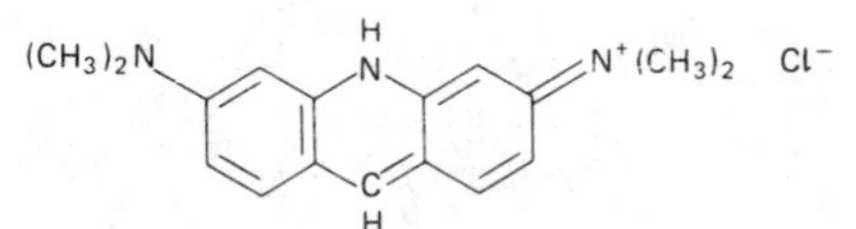

Used only as a fluorochrome, this dye is well known for its ability to impart fluorescent emissions of different colours to DNA and RNA (Bertalanffy & Bickis, 1956). It has also been used in carbohydrate histochemistry and as a vital stain.

5.9.13. Azine dyes

In these dyes the chromophore is a pyrazine ring

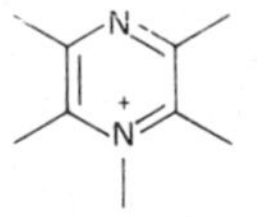

sandwiched between two aromatic systems, or *o*- and *p*-quinonoid configurations contributing to the resonance. These two structures are shown below for neutral red. For the other examples only the *p*-quinonoid forms are shown.

Examples

NEUTRAL RED (C.I. 50040; Basic red 5; M.W. 289)

Two of the canonical forms are shown. In another the positive charge is on the primary nitrogen atom.

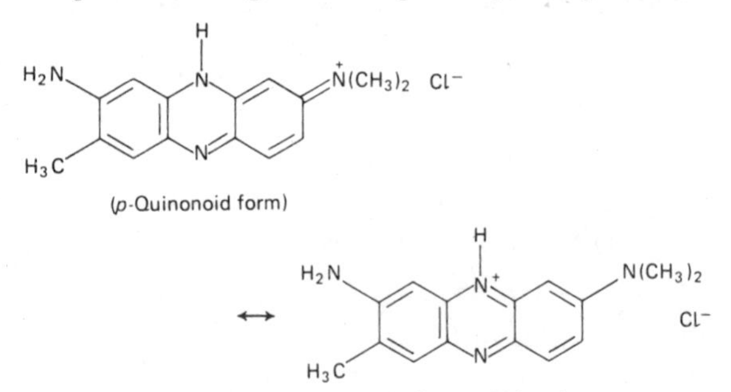

Neutral red is a cationic dye, freely soluble in water and alcohol. Aqueous solutions turn yellow on addition of alkali, the change occurring over the range pH 6.8–7.0. Very strongly acidic solutions are blue–green, and a brown precipitate is formed with an excess of alkali. The dye is valuable as a red cationic dye and is also used in some vital staining techniques. An acceptable sample of neutral red contains at least 50% by weight of the anhydrous dye.

AZOCARMINE G (C.I. 50085; Acid red 101; M.W. 580)

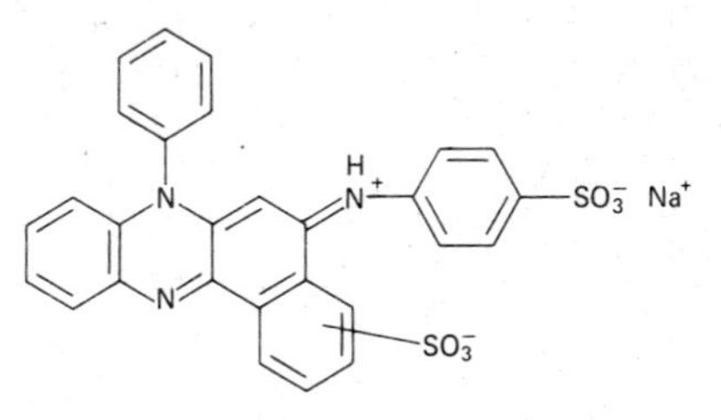

(The exact position of one of the sulphonic acid groups is unknown)

Despite its name, this red anionic dye is not an azo compound and is not related to carmine. Its principal use is in Heidenhain's "AZAN" method, a general-purpose and connective tissue staining procedure. Detailed instructions for this rather difficult technique (which can, however, yield beautiful preparations) are given by Gabe (1976). AZOCARMINE B (C.I. 50090; Acid red 103) has a third sulphonic acid group (*meta* to the one shown on the extreme right of the above formula) and is more soluble in water than azocarmine G. Although the two dyes are interchangeable, Gabe (1976) prefers azocarmine G for the AZAN method, stating that it gives a brighter shade of red.

SAFRANINE O (safranine, safranine T; C.I. 50240; Basic red 2; M.W. 351)

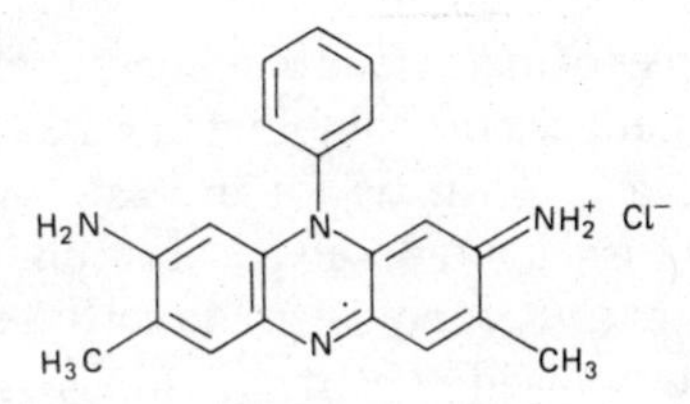

This is always a mixture of the compound formulated above with a variable amount of a related dye in which a methyl group is substituted on the phenyl radical, *ortho* to the site of attachment to nitrogen. It may have a yellowish or bluish cast, depending on the proportions of the two constituents. Safranine O is valuable as a red cationic dye. It dissolves in water and alcohol. Samples should contain at least 80% by weight of the anhydrous dye.

5.9.14. Oxazine dyes

The oxazine chromophore

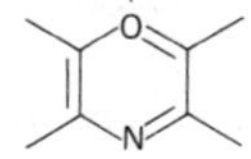

has obvious similarity to those of the azine and thiazine dyes and exists in *o*-quinonoid form (with positive charge attributed to the oxygen) and *p*-quinonoid form (as shown for the dyes described below). The natural dyes **orcein, litmus,** and **azolitmin** belong to the oxazine series. Orcein is used as a biological stain, litmus and azolitmin as pH indicators (see Lillie, 1977, for chemistry and properties). Only synthetic oxazines are considered below.

Examples

CRESYL VIOLET ACETATE (No C.I. number; M.W. 321)

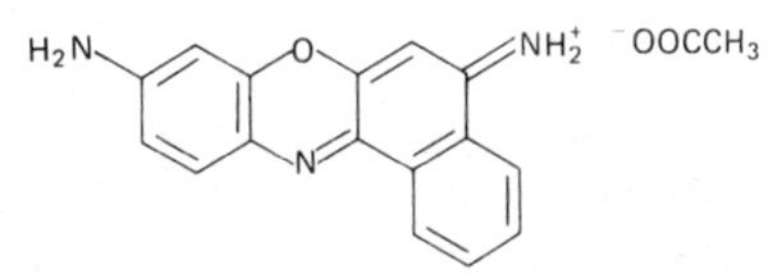

This dye is the modern equivalent of **cresyl fast violet (cresylecht violet).** Both are often loosely termed "cresyl violet". The cresyl violets are useful as violet cationic dyes, soluble in water and alcohol. They are popular as "Nissl stains" for nervous tissue.

GALLOCYANINE (C.I. 51030; Mordant blue 10; M.W. 337)

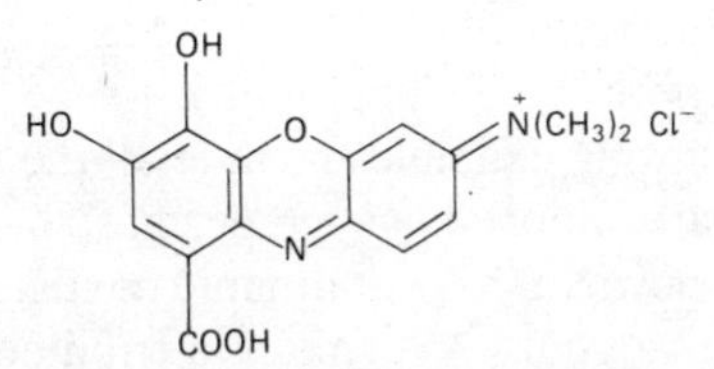

Gallocyanine is not used alone but as a preformed dye–metal complex made by boiling with chrome alum. The complex behaves as a cationic dye and is used principally to stain nucleic acids. The blue–grey colour is more resistant to extraction by water and alcohols than are simple cationic dyes. The chromium–gallocyanine staining solution is stable for only about 1 week.

CELSTINE BLUE B (C.I. 51050; Mordant blue 14; M.W. 655)

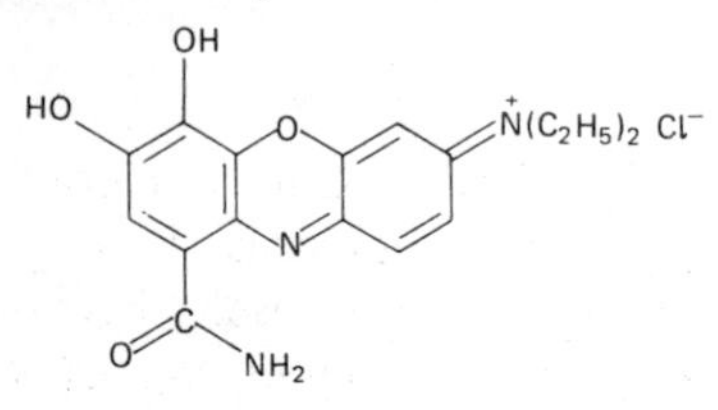

This dye differs from gallocyanine in having *N*-ethyl instead of *N*-methyl groups and in being an amide rather than a carboxylic acid. It is used as a preformed dye–metal complex with a ferric salt. The principal use is as a blue stain for nuclei, similar to alum–haematoxylin but more resistant to counterstaining in acidic reagents.

NILE BLUE (C.I. 51180; C.I. Basic blue 12; M.W. (of sulphate) 733)

This dye is discussed in Exercise 5 at the end of Chapter 12 (see p. 189). Samples should contain at least 70% by weight of anhydrous Nile blue sulphate.

5.9.15. Thiazine dyes

The chromophore of the thiazines is like that of the oxazines, with sulphur replacing oxygen. The thiazine dyes used in histology are all cationic and all except methylene green are blue or violet. All except methylene violet (Bernthsen) are soluble in water. In addition to their general usefulness as blue cationic dyes, some of the thiazines, especially the oxidation products of methylene blue, are used as eosinates in haematological staining. This will be discussed in Chapter 7. Most of the thiazine dyes can be used for metachromatic staining (see Chapter 11). Methylene blue is also used as a vital stain for nerve fibres (see Chapter 18).

The following examples include all the thiazine dyes that are important as biological stains. Structures are shown in the *p*-quinonoid configurations, with positive charges formally attributed to the most strongly basic nitrogen atoms.

THIONINE (C.I. 52000; M.W. 264)

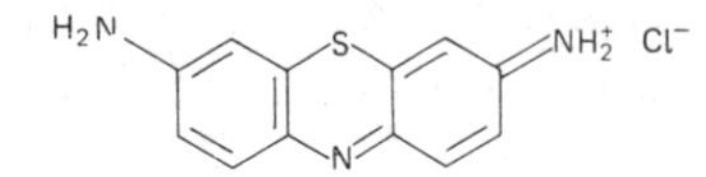

Thionine is a blue dye, soluble in water and in alcohol, but less so than most other thiazine dyes. An acceptable sample should contain at least 85% anhydrous dye. This dye must not be confused with thionine blue (see p. 71).

AZURE C (C.I. 52002; M.W. 278)

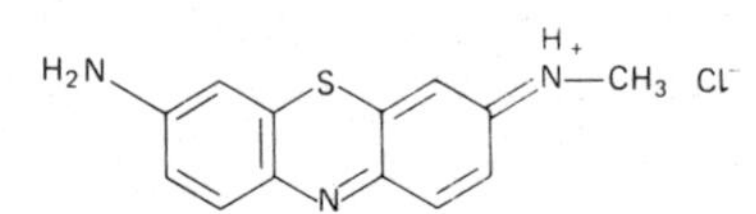

AZURE A (C.I. 52005; M.W. 292)

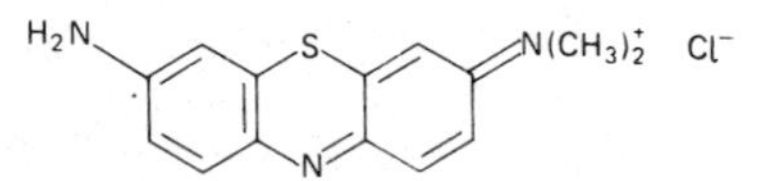

This is another blue cationic dye, soluble in water and alcohol. Samples should contain at least 55% weight of the anhydrous dye.

A symmetrical dimethylthionine (with one methyl group on each amine nitrogen atom) also exists and is said to have staining properties almost identical to those of azure A (see Lillie, 1977). Toluidine blue O (see p. 71) is also closely similar.

AZURE B (methylene azure; azure I; C.I. 52010; M.W. 306)

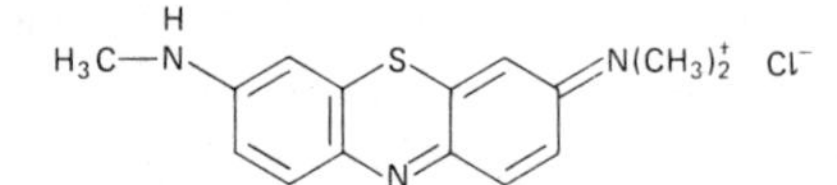

This blue dye, a product of the oxidation of methylene blue, is the most important coloured cationic component of the dye mixtures commonly used as haematological stains (see Chapter 7). It is soluble in water and in alcohol.

METHYLENE BLUE (C.I. 52015; Basic blue 9; M.W. 320)

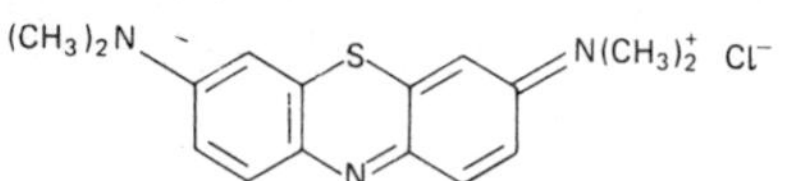

Many samples of methylene blue contain zinc chloride, which reduces the solubility in water and may render the dye too toxic for use as a vital stain (though this is disputed; see Richardson, 1969). If possible, a pharmacopoeial grade certified as zinc-free should be used. The dye is freely soluble in water and in alcohol. With time and exposure to air, methylene blue is oxidized to azure B, azure A, and other thiazine dyes. For use in bacteriology and haematology the dye is deliberately oxidized to give the mixture of dyes known as "polychrome methylene blue".

For certification by the Biological Stains Commission, a sample must contain at least 82% by weight of anhydrous methylene blue chloride and must be satisfactory as a nuclear and as a bacterial stain.

METHYLENE GREEN (C.I. 52020; Basic green 5; M.W. 365)

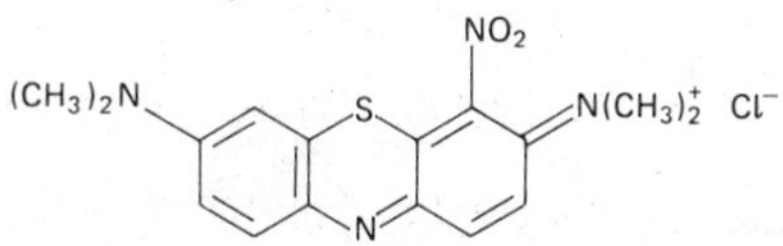

The exact position of the nitro group is uncertain. The dye is soluble in water and slightly soluble in alcohol.

THIONINE BLUE (C.I. 52025; Basic blue 25; M.W. 348)

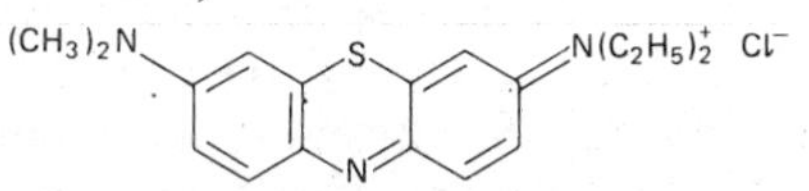

This is mentioned only to avoid confusion with thionine. Thionine blue, unlike thionine, does not give metachromatic effects.

NEW METHYLENE BLUE (C.I. 52030; Basic blue 24; M.W. 348)

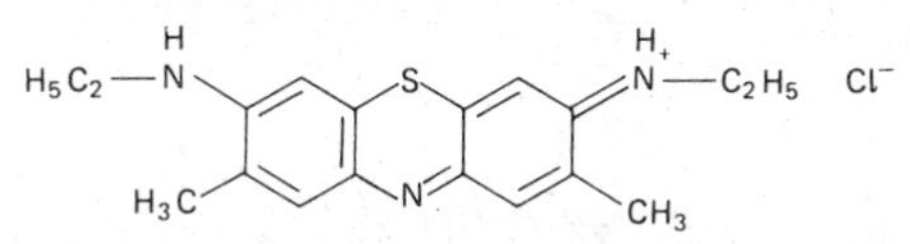

Metachromatic effects can be obtained with this dye, but it cannot replace methylene blue in vital staining methods. It is much more soluble in water than is methylene blue, but the two dyes are about equally soluble in alcohol.

TOLUIDINE BLUE O (toluidine blue; C.I. 52040; Basic blue 17; M.W. 306)

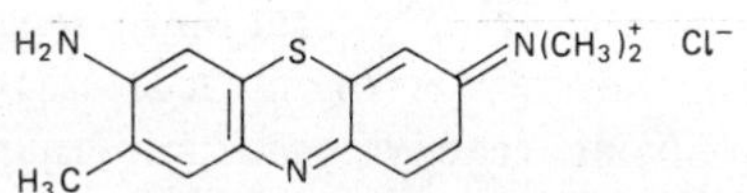

This is perhaps the most useful of all the blue cationic dyes, being similar to but cheaper than azure A. Metachromatic effects are easily obtained. Lillie (1977) points out that older samples of this dye are of the zinc chloride double salt (dye content about 60%) but that toluidine blue O is now sold in the zinc-free state (over 90% dye). An acceptable sample must contain at least 50% by weight of the anhydrous dye (M.W. 306). It is freely soluble in water and slightly soluble in alcohol.

METHYLENE VIOLET (BERNTHSEN) (C.I. 52041; M.W. 256)

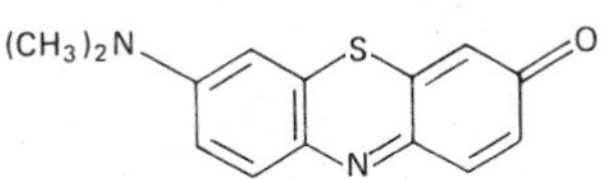

This dye is formed by oxidation of methylene blue and is a component of traditional staining mixtures for blood. The pure dye is insoluble in water but dissolves in the presence of other thiazine dyes. It is more soluble in chloroform. **Methyl thionoline** (with —$NHCH_3$ replacing —$N(CH_3)_2$ in formula above) and **thionoline** (with —NH_2 replacing —$N(CH_3)_2$ of methylene violet, Bernthsen) are related dyes also formed in the polychroming of methylene blue (Marshall, 1978).

Methylene violet (Bernthsen) should not be confused with methylene violet RR (C.I. 50205; Basic violet 5), which is an azine dye related to safranine O.

5.9.16. Indigoid and thioindigoid dyes and pigments

Natural **indigo** is the classical example of a vat dye (see p. 49): a soluble precursor, typically a colourless leuco compound, is applied to a fabric and then oxidized to an insoluble coloured substance. Indigo is manufactured by oxidizing indoxyl, which is now a synthetic product.

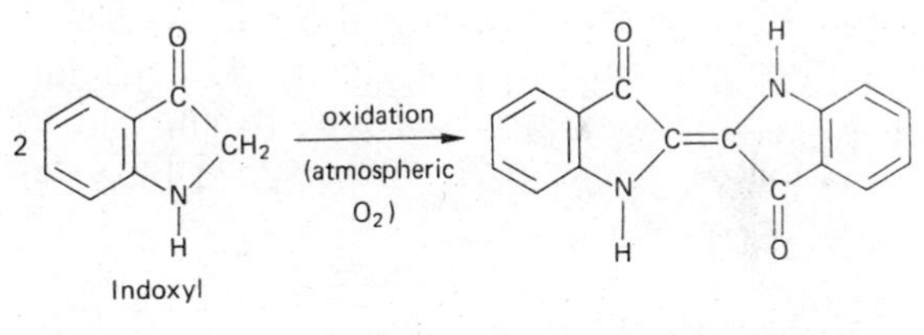

(traditionally prepared by hydrolysis of *indican*, a glycoside occurring in *Indigofera* species)

Indigo blue (indigotin) (C.I. 73000; Vat blue 1)

For use in dyeing, the indigo blue is reduced, usually by sodium dithionite ($Na_2S_2O_4$), to its soluble leuco compound, **indigo white** (C.I. 73001):

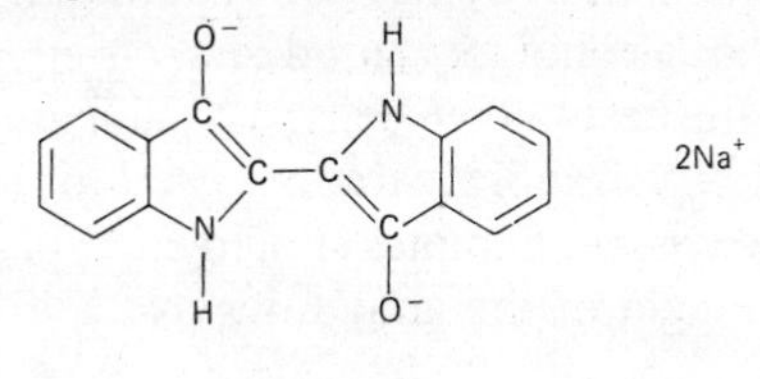

This, after application to the textile, is reoxidized by air to the insoluble indigo blue.

The chromophoric system of the indigoid dyes is the conjugated chain

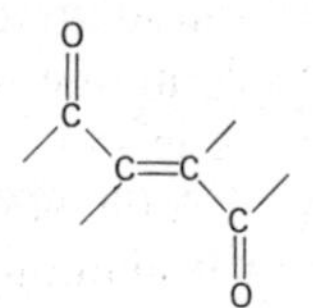

The colour is due to resonance among structures such as

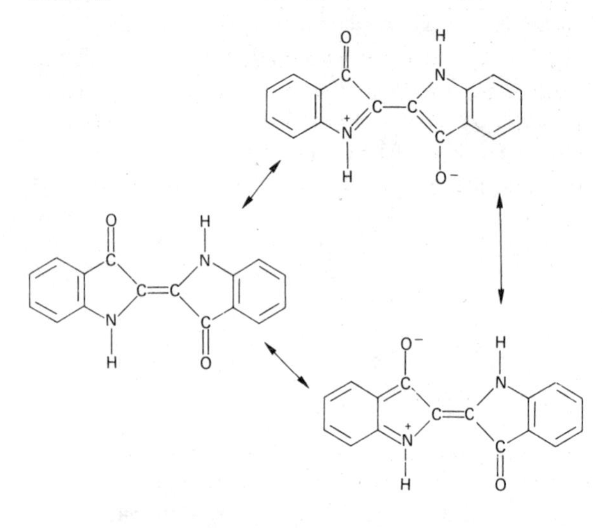

In the **thioindigoid** compounds the nitrogen of the indigo structure is replaced by sulphur.

Insoluble indigoid and thioindigoid compounds are produced as the coloured end-products of some histochemical reactions (see Chapter 15). In addition, one soluble indigoid dye is used in histology.

INDIGOCARMINE (C.I. 73015; Acid blue 74; M.W. 466)

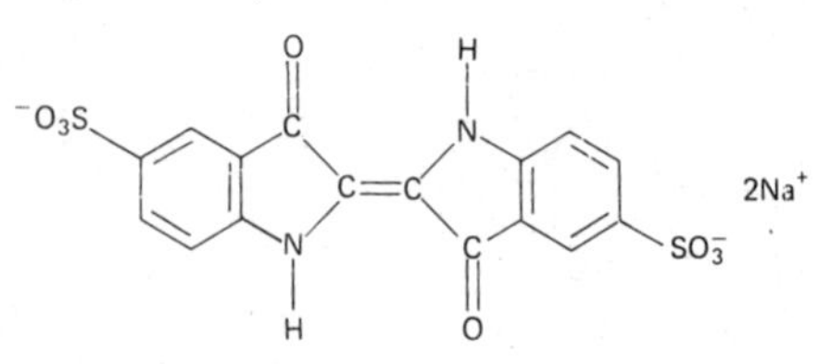

This blue anionic dye is soluble in water but almost insoluble in alcohol. It is used in a variety of procedures, the best known being a combination with picric acid (picroindigocarmine) used for staining connective tissue. Samples should contain at least 80% by weight of the anhydrous dye.

5.9.17. Naphthoquinone dyes

The chromophore in these dyes is 1,4-naphthoquinine

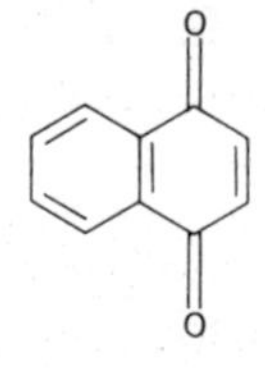

which is usually fused with or otherwise joined to other aromatic rings. The main starting material for the synthesis of naphthoquinone dyes is 2,3-dichloro-1,4-naphthoquinone

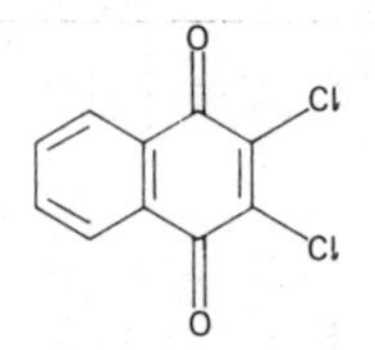

The chlorine atoms are replaced by substitution reactions with phenols, amines, mercaptans, and many other compounds. Apart from naphthazarin (p. 62; here classified as a hydroxyketone), the naphthoquinone dyes have not been used as biological stains. Neither do they form an important group of textile dyes, though some are useful as pigments (Tilak, 1971).

5.9.18. Anthraquinone dyes

Anthraquinone

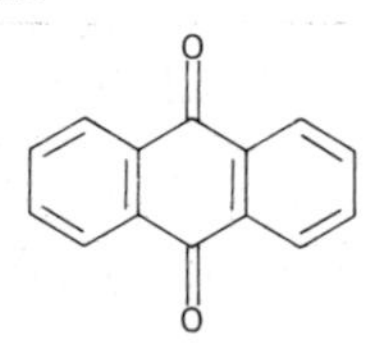

with its *p*-quinonoid configuration, is the chromophore of all the members of this large class of dyes. Anionic, solvent, reactive, and mordant anthraquinones may all have potential value as biological stains, but only those of the last-named type have achieved any importance in microtechnique. Other anthraquinones, of great industrial importance, include vat dyes, disperse dyes, and pigments (Allen, 1971; Greenhalgh, 1976).

Examples

ALIZARIN RED S. (alizarin red, water soluble;

sodium alizarin sulphonate; C.I. 58005; Mordant red 3; M.W. 342)

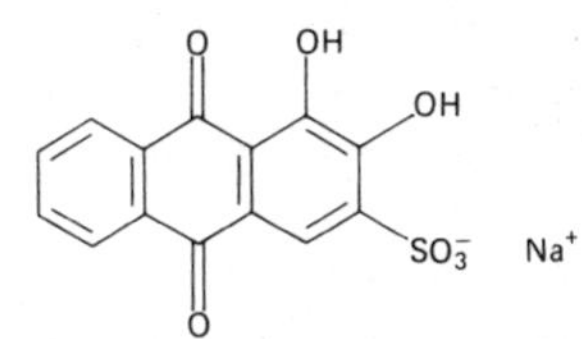

As an anionic dye this will stain tissues red. Its principal use, however, is as a histochemical reagent for calcium, with which an orange chelate is formed under appropriate conditions of use (see Chapter 13). The dye cannot be assayed chemically so the Biological Stains Commission tests it as a stain for bone in whole mounts of small animals. Samples unsatisfactory for histochemical purposes are quite frequently encountered.

ALIZARIN (C.I. 58000; Mordant red 11) lacks the sulphonic acid radical. It was extracted from the roots of madder (*Rubia tinctoria*), but is now a synthetic product. The aluminium lake of alizarin is bright red and was once used for dyeing soldiers' uniforms. Alizarin has been used only rarely as a biological stain (Lillie, 1977).

KERNECHTROT (nuclear fast red; calcium red; C.I. 60760; M.W. 357)

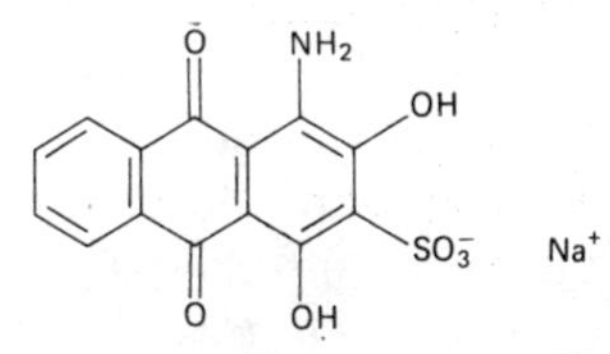

Like alizarin red S, this dye is sometimes used as a histochemical reagent for calcium. The aluminium lake is used as a red stain for nuclei. For the latter purpose kernechtrot is comparable to carminic acid and brazilein.

CARMINIC ACID (C.I. 75470; Natural red 4; M.W. 492)

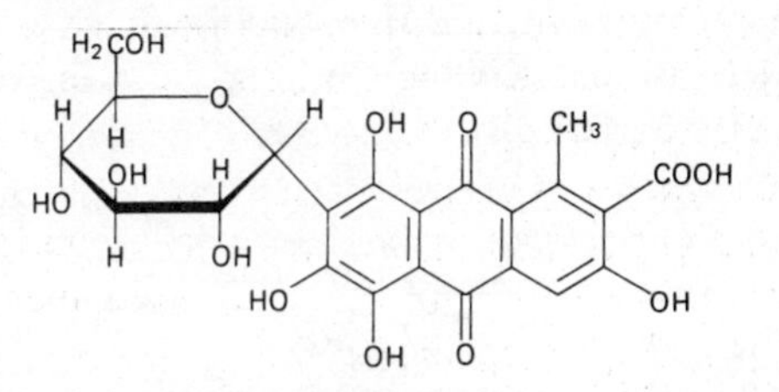

This red glycoside-like substance is the principal colouring matter of **cochineal.** Cochineal is made by grinding the dried bodies of female *Coccus* insects (from Central America). The cyclic side-chain shown on the left side of the structural formula is derived from D-glucose, which is joined by a C-glycosyl linkage (without intervening oxygen) to the polyhydroxyanthraquinone carboxylic acid. The dye is used with aluminium as mordant as a red nuclear stain for sections and whole mounts. Carminic acid is slightly soluble in water and alcohol, but in the presence of an aluminium salt it dissolves in larger amounts, with the formation of the dye–metal complex. There is no satisfactory chemical assay for the dye. Solutions of the aluminium complex deteriorate after a few weeks of storage.

Carmine, which is also made from cochineal, is used in general histology and in a staining method for glycogen. It contains 20–25% of protein (some from the insects and some added during manufacture), together with calcium ions and an anionic

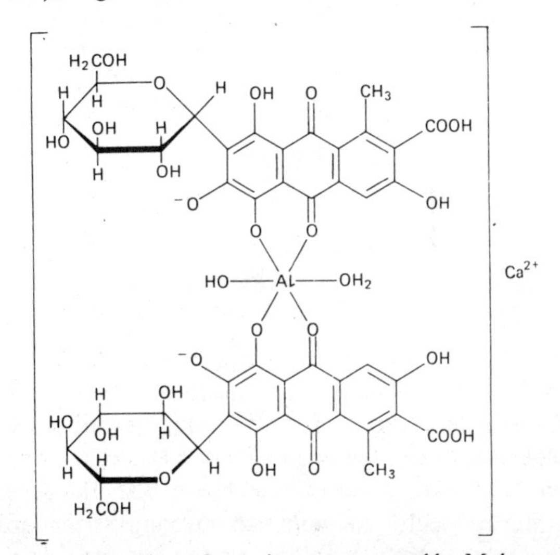

FIG. 5.1. Structure of carmine as suggested by Meloan *et al.* (1971). Their formula has been slightly modified to conform with the conventions used in this book. Ionization of the phenolic hydroxyl groups *ortho* to the sugar substituents is said to be necessary for chelation of Al^{3+} with the nearby oxygen atoms. Presumably the carboxyl groups are also at least partially ionized in solution, but it is not known whether carmine contains enough Ca^{2+} to balance these. Al^{3+} could also form complexes with the free quinonoid and α-phenolic oxygens shown on the sides of the carminic acid molecules facing away from the Al atom. This would give the dye–metal complex a net positive charge and might occur when Al^{3+} is present in excess, as in mixtures used for nuclear staining. The mode of attachment of the dye–metal complex to tissues has not been determined.

dye–metal complex of carminic acid and aluminium. The structure of carmine is uncertain and probably varies according to the method of manufacture. Calcium may not always be present. A suggested formula is shown in Fig. 5.1. Since carmine is a material of uncertain composition it should not be used as a histological stain without preliminary testing, and supposedly histochemical methods involving its use must be viewed with suspicion.

5.9.19. Phthalocyanine dyes

Most of these are derivatives of **copper phthalocyanine**, a blue pigment of high physical and chemical stability.

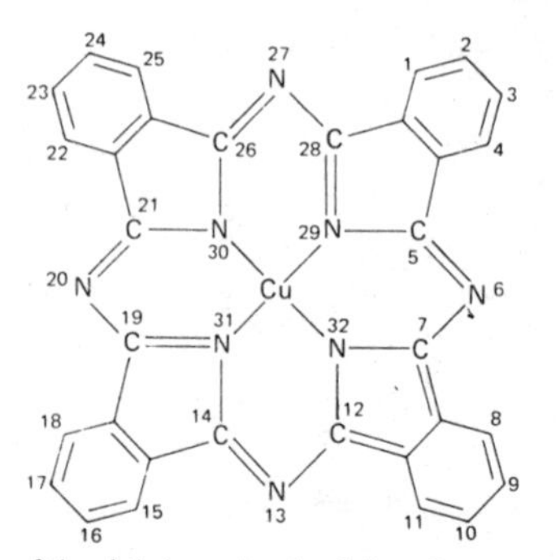

(Numbering of the phthalocyanine ring follows Patterson *et al.*, 1960)

The carbon and nitrogen atoms form a fully conjugated aromatic system, stabilized by resonance so that the *o*-quinonoid benzene ring (here shown at the lower right-hand corner) has no fixed location in the molecule. The bonds to the copper atom are shown as simple covalencies, though other representations are permissible (see p. 45).

Copper phthalocyanine is formed by the condensation of four molecules of 1-amino-3-iminoisoindolenine

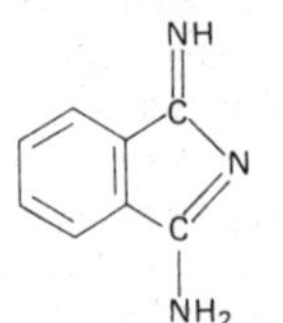

in the presence of cupric chloride. This precursor is itself a transient intermediate in the usual process of manufacture, which involves the heating together of phthalic anhydride, urea, and cupric chloride.

In dyes of the phthalocyanine series, solubilizing groups are introduced as substituents on the macrocyclic ring. The simplest dye of this type is SIRIUS LIGHT TURQUOISE BLUE GL (C.I. 74180; Direct blue 86; M.W. 780), in which sulphonic acid radicals (as their sodium salts) are present on two benzene rings at opposite corners of the molecule (see Lillie, 1977). Cationic and reactive phthalocyanine dyes are also manufactured, and phthalocyanine pigments can be synthesized on textile fibres by application of the "phthalogen" dyestuff intermediates (Vollmann, 1971). In some of the dyes, copper is replaced by cobalt or other metals.

Examples

ALCIAN BLUE (alcian blue 8GS, 8GX or 8GN; C.I. 74240; Ingrain blue 1)

The exact chemical compositions of these dyes have not been revealed by Imperial Chemical Industries Ltd., the manufacturers, but Scott (1972a) has analysed alcian blue 8GX and found that it contains four tetramethylisothiouronium groups,

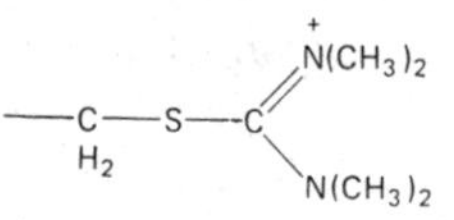

one on each of the benzenoid rings of the copper phthalocyanine structure. The tetramethylisothiouronium groups are attached at positions 2 or 3, 9 or 10, 16 or 17, and 23 or 24. The dyestuff is a mixture of the four possible geometrical isomers. The other alcian blues are believed to bear either two or four thiouronium substituents on either two or four of the benzenoid rings of copper phthalocyanine, but their full chemical formulae are not known. All these dyes are supplied mixed with boric acid, sodium sulphate, and dextrin.

In a closely related dye, NATIONAL FAST BLUE 8XM (National Aniline Co., Inc., not in *Colour Index*), two radicals

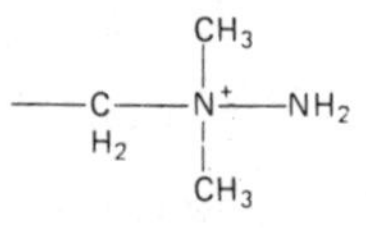

are attached at unspecified sites (Mowry & Emmel, 1966).

The dye **alcian yellow** is not a phthalocyanine but a monoazo dye which also contains the benzothiazole chromophore and a thiouronium auxochrome. **Alcian green** is a mixture of alcian blue with alcian yellow.

The most important application of alcian blue is in carbohydrate histochemistry (see Chapter 11). The other dyes mentioned above have been used for similar purposes. Unsatisfactory samples of alcian blue are often encountered, so it is necessary to test every new batch before applying the dye to important material.

LUXOL FAST BLUE MBS (C.I. Solvent blue 38; no C.I. number)

The "luxol" dyes are anionic chromogens which are balanced not by H^+ or Na^+ but by arylguanidinium cations. With gegen-ions of this type, of which diphenylguanidine is an example, the dyes are insoluble in water but soluble in alcohols. The structures of the dyes have not been disclosed by the manufacturers, but it is known that luxol fast blue MBS has a copper phthalocyanine chromogen, while LUXOL FAST BLUE ARN (C.I. Solvent blue 37) is an azo dye (Salthouse, 1962, 1963). Dyes with similar properties but of known chemical constitution can be produced in the laboratory from suitable anionic dyes and diphenylguanidine (Clasen *et al.*, 1973).

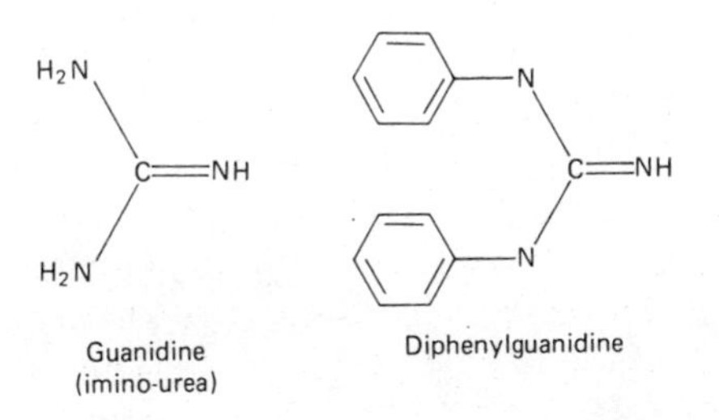

Guanidine (imino-urea) Diphenylguanidine

The "luxol" dyes are used for staining myelin and will be further discussed in Chapter 18.

5.10. EXERCISES

Theoretical

1. Which of the following compounds are coloured and which could function as dyes?

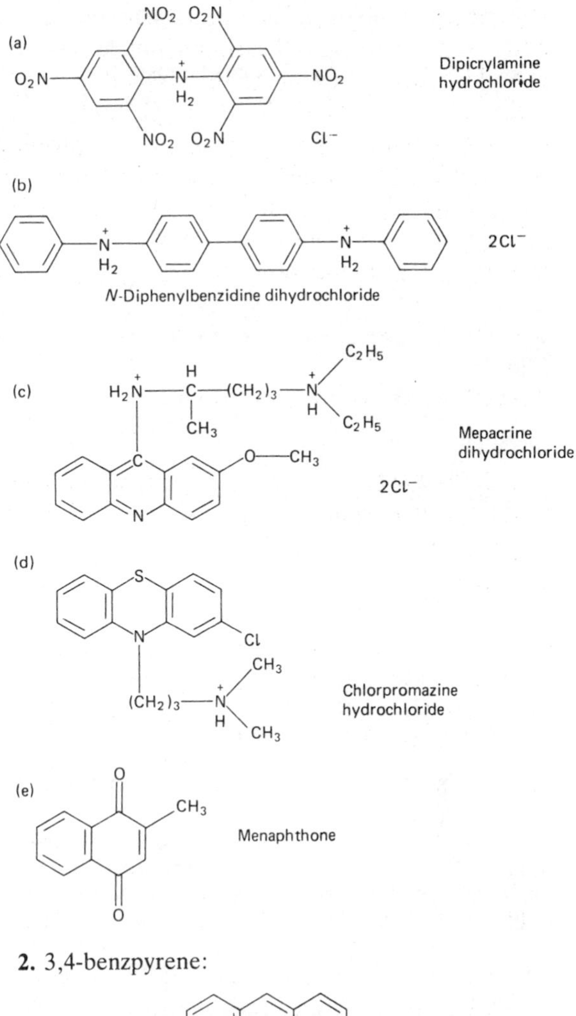

2. 3,4-benzpyrene:

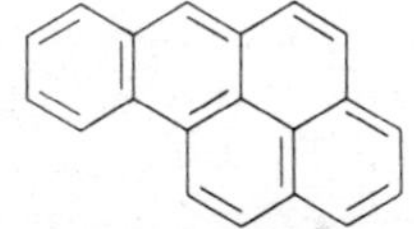

can be used as a fluorochrome. Is it a dye? What would you expect it to stain when applied to sections as a saturated solution in a saturated aqueous solution of caffeine? The caffeine serves only to enable the 3,4-benzpyrene to be dissolved in an aqueous medium.

3. Weigert's iron–haematoxylin contains 500 mg per ml of haematoxylin and 0.58 mg per ml of $FeCl_3.6H_2O$ (M.W. 270). This solution stains nuclei. Overstaining does not occur and the sections usually do not require differentiation.

If sections are mordanted in 5% iron alum, $NH_4Fe(SO_4)_2.12H_2O$ (M.W. 482) and then stained in 0.5% ripened haematoxylin, they go completely black (Heidenhain's method). Differentiation in 5% iron alum, if carefully controlled, permits the demonstration of many structures other than nuclei.

Explain the differences between the Weigert and Heidenhain methods of staining.

4. Urea readily forms hydrogen bonds with other substances. If aqueous solutions of some dyes are saturated with urea they lose their ability to stain sections. Which of the following dyes would you expect to be ineffective in a urea-containing solution: (a) Toluidine blue O, (b) Eosin, (c) Methyl blue, (d) Oil red O? Give reasons for your answers.

5. The binding of dyes by electrostatic forces can be suppressed by including a high concentration of a neutral salt such as NaCl or $MgCl_2$ in the staining solution. Which of the following dyes would you expect to be inhibited by salts: (a) Neutral red, (b) Procion brilliant red M2B, (c) Chlorazol black E, (d) Alcian blue?

6. What structural features of a dye molecule confer the ability to bind metal ions? Draw possible structures for dye–metal complexes of chromium (III) with (a) salicin black E, (b) gallocyanine, (c) brazilein.

Practical

7. Dissolve about 10 mg of basic fuchsine in 100 ml of water. Add 4% aqueous sodium hydroxide in aliquots of 0.5 ml. What happens and why?

8. Add a few drops of 0.5% neutral red to (a) 100 ml distilled water, (b) 100 ml tap water, (c) 100 ml 1% acetic acid, (d) 100 ml 0.5% sodium hydroxide. Explain the observed changes in colour.

9. Add single drops of ripened 5% alcoholic haematoxylin to: (a) slightly alkaline water, (b) slightly acid water, (c) water containing approximately 5 mg per ml potassium alum ($KAl(SO_4)_2 . 12H_2O$), (d) water containing approximately 5 mg per ml iron alum ($NH_4Fe(SO_4)_2 . 12H_2O$) or ferric chloride ($FeCl_3 . 6H_2O$). For (c) and (d) follow the addition of the haematoxylin with addition of two or three drops of 4% sodium hydroxide.

Observe the colours of the different mixtures. What do you deduce from these observations about the properties of haematein and of its complexes with Al^{3+} and Fe^{3+}?

10. Mix equal volumes of aqueous solutions of: (a) 1% neutral red with 1% fast green FCF: (b) 1% neutral red with 1% toluidine blue O; (c) 1% toluidine blue with 1% eosin; (d) 1% eosin with 1% fast green FCF.

Account for the observation that precipitates form with some but not all of the above mixtures.

11. Add 5 ml of 1% aqueous toluidine blue or methylene blue to 100 ml water and divide the solution into two parts, A and B, in 200 ml beakers.

To A add 10 ml of saturated aqueous picric acid. Observe what happens and then add 100 ml of ethanol.

To B add 10 ml of freshly dissolved 5% aqueous ammonium molybdate. Observe what happens and then add 100 ml of ethanol.

What do these observations tell you about the picrates and molybdates of thiazine dyes? Suggest a practical application of the findings.

12. Dissolve 50 mg of fast blue B salt (tetraazotized *o*-dianisidine) in 30 ml of water at 4°C. Dissolve 30–40 mg of thymol in 2–3 ml of 95% ethanol. Pour the thymol solution into the solution of the tetrazonium salt and add three or four drops of 4% aqueous sodium hydroxide. Mix well and leave at 4°C for 20 min. Allow the reddish-brown precipitate to settle and decant off as much as possible of the supernatant liquid. Add 20 ml of ethanol to the precipitate, which will dissolve.

What is the probable structural formula of the dye you have made? Thymol is 2-isopropyl-5-methyl phenol (M.W. 150). In view of its structure and the properties you have observed, how might this dye be used as a biological stain?

6

Histological Staining in One or Two Colours

6.1. Nuclear stains 77
- 6.1.1. Cationic dyes 77
- 6.1.2. Anionic dyes 78
- 6.1.3. Mordant dyes 79

6.2. Anionic counterstains 80
6.3. Single solution methods 80
6.4. Individual methods 81
- 6.4.1. Haematoxylin and eosin 81
- 6.4.2. Other nuclear stains 82
  - 6.4.2.1. Chromoxane cyanine R 82
  - 6.4.2.2. Iron–haematoxylin 83
  - 6.4.2.3. Brazilin 83
  - 6.4.2.4. Mayer's carmalum 83
  - 6.4.2.5. Alum–tetracycline 83
  - 6.4.2.6. Cationic dyes as nuclear stains 84
- 6.4.3. Mann's eosin–methyl blue 85
- 6.4.4. Chlorazol black E 85
- 6.4.5. Azure–eosin method 86

6.5. Exercises 87

For the study of microscopic anatomy and of pathological material it is usual to stain sections of tissue in such a way as to impart a dark colour to the nuclei of cells and a lighter, contrasting colour to the cytoplasm and extracellular structures. With simple methods of this type, of which the most widely practised is alum–haematoxylin and eosin ("H. & E."), it is not possible to make deductions concerning the chemical natures of any of the components of the tissue. Identification of the different types of cells and of the fibres and matrices of the intervening connective tissue must be based solely on morphological criteria. The H. & E. method is perhaps the least revealing of the two-colour techniques since the alum–haematoxylin component colours only the nuclei of most cell types. Nevertheless, this procedure is used more than any other by pathologists and in the teaching of normal histology. Other simple staining methods provide more informative preparations than does H. & E., but familiarity with the latter has not yet led to contempt.

Nuclear stains and simple counterstaining procedures are considered in this chapter. A few methods for staining in various shades of one colour are also described. The histochemical rationales of the techniques are discussed, but methods of higher chemical specificity (e.g. for nucleic acids, carbohydrates, and functional groups of proteins) are not covered. The special methods for blood and for connective tissue are likewise described elsewhere (Chapters 7 and 8).

6.1. NUCLEAR STAINS

The nucleus of a eukaryotic cell contains the two **nucleic acids**—DNA (in the chromosomes) and RNA (in the nucleolus). Both are associated with **nucleoproteins**, which are rich in the basic amino acids arginine and lysine. The cations of these amino acids are bound electrovalently to phosphoric acid residues of the nucleic acids (see Chapter 9). The DNA and nucleoprotein of the chromosomes together constitute the material known as **chromatin**. In interphase cells the chromosomes are extended and cannot be seen individually. The chromatin seen in stained preparations of interphase nuclei may be evenly distributed through the nucleoplasm or aggregated in a pattern characteristic of the cell type. The dyes used as nuclear stains impart colour to the chromatin by binding either to the nucleic acids or to the nucleoprotein.

6.1.1. Cationic dyes

The modes of binding of cationic dyes to substrates were discussed in Chapter 5, where descriptions of several of the dyes of this type can be found. The dyes are applied from acidified solutions in order to ensure the ionization of their primary, secondary, or tertiary amine groups. Three types of anions capable of binding cationic dyes occur in tissues. These are the phosphates of nucleic acids, sulphate esters of certain macromolecular carbohydrates (see Chapter 11), and carboxylate ions of carbohydrates and proteins. As explained in Chapter 5, the ionization of carboxyl groups is suppressed if the concentration of hydrogen ions in the staining solution is too high (in practice, pH 3–4), while sulphate groups remain ionized even in strongly acidic media. Phosphoric acid is of intermediate strength. A strongly basic protein, such as haemoglobin, contains a large excess of free amino over carboxyl groups, so that the latter can be made available for binding cationic dyes only from alkaline solutions (e.g. at pH 8). The staining properties of cationic dyes are profoundly influenced by the pH of their solutions. If they are too acid, only the structures rich in sulphated carbohydrates will be coloured; if they are neutral or alkaline, everything will be stained.

The interpretation of the results obtained with cationic dyes at various pH levels has been studied in some depth (see Gabe, 1976; Lillie & Fullmer, 1976, for reviews and discussion). The following list will serve as a rough guide to the use of dilute (e.g. 0.05–0.2%) solutions of cationic dyes. It applies especially to those of the thiazine series.

- pH 1.0 — Only sulphated carbohydrate components are stained. Sulphonic acid groups or sulphuric acid esters can be artificially produced in the tissue (see Chapters 10 and 11) and these, too, will bind cationic dyes from strongly acid solutions.
- pH 2.5–3.0 — Staining of the above plus phosphate of nucleic acids (DNA and RNA).
- pH 4.0–5.0 — Staining of the above plus carboxyl groups of carbohydrates and the more acid proteins. Carboxyl groups of free fatty acids may be stained in frozen sections.
- pH $>$ 5.0 — Staining of the above with increasing staining of neutral and basic proteins with increasing pH.

With higher concentrations of dye, the rules set out above are not followed, and staining is generally less specific. The duration of exposure to the dye has only minor effects on the results, though longer times are needed to achieve the full depth of staining if very dilute solutions are used. The end result will also be affected by treatments after staining. Most cationic dyes are rather readily extracted from sections by 70% alcohol. Absolute ethanol, *n*-butanol, and acetone are usually less active in this respect. In any critical work with cationic dyes the conditions of rinsing and dehydration must be carefully standardized. The water used for rinsing the sections after staining should be buffered to the same pH as the dye solution. Fixatives generally do not directly affect sulphate, phosphate, and carboxylate groups, but combination with or removal of amine groups has the effect of lowering the pH at which proteins are stained. Thus fixation by osmium tetroxide renders virtually all components of tissues stainable by cationic dyes at pH 4 or higher. The reason for this will be obvious from a consideration of one of the chemical actions of osmium tetroxide (see Chapter 2).

For general-purpose staining of nuclei, cationic dyes are used in dilute solution (e.g. 0.5%) at pH 3–5.5. The aqueous solution of the dye is usually acidified by addition of acetic acid or a suitable buffer. Objects that are coloured by cationic (basic) dyes are said to be **basophilic** (Baker, 1958, argues in favour of the term "basiphil"). When nuclei are stained by cationic dyes, basophilia will also be evident at sites of accumulation of cytoplasmic RNA such as the Nissl substance of neurons, and in structures containing sulphated carbohydrates such as mast cell granules, cartilage matrix, and many secretory products. In this propensity cationic dyes differ importantly from some of the mordant dyes used as nuclear stains.

6.1.2. Anionic dyes

For reasons given in Chapter 5 (p. 42), anionic dyes are applied from acidic solutions. Used alone, an anionic dye will colour almost all the components of a tissue, but differential staining effects can sometimes be obtained when two or more dyes are applied simultaneously or in sequence. A method of this type is Mann's eosin–methyl blue (p. 85). Nuclei, cartilage, and collagen acquire the blue colour, while cytoplasms are stained pink or red by eosin. The reasons for these results are not fully known, but it is widely held that the larger molecules of methyl blue are excluded by the smaller ones of eosin from the supposedly "dense" network of cytoplasmic protein molecules. Collagen and chromatin, with their supposedly looser textures, would admit molecules of both dyes, but the eosin would diffuse out more rapidly than the methyl blue when the sections were washed following immersion in the mixture of dyes. The influence of molecular size on the staining properties of dyes will be discussed in more detail in Chapter 8. Nuclear staining by methyl blue was studied by McKay (1962) who found that the presence of nucleic acids was necessary. He was unable to obtain evidence for binding of the dye by ionic attraction and suggested that the staining of nuclei was analogous to the direct dyeing of cotton (see p. 56). McKay compared methyl blue with chlorazol black E, a direct azo dye. Chlorazol black E was found to behave like most other anionic dyes, including eosin, and was, therefore, thought to be bound to tissue principally by electrovalent forces. Useful staining of nuclei and cytoplasms can be obtained with chlorazol black E (see p. 85). This dye also stains elastin, but probably by a non-ionic mechanism (Goldstein, 1962).

6.1.3. Mordant dyes

For nuclear staining, mixtures of mordant dyes with appropriate metal salts are applied to sections of tissue. The modes of action of mordants in binding dye molecules to their substrates have already been discussed in some detail in Chapter 5. There it was pointed out that some dye–metal complexes, such as that of chromium with gallocyanine, behave as if they were simple cationic dyes. Others, most notably haematein, are evidently true mordant dyes. These latter dyes are used in histology to obtain selective staining of chromatin. Nuclear staining by the commonly used aluminium–haematein mixtures is only partially inhibited by prior extraction of DNA from the tissue, and staining by iron–haematein is unaffected by such extraction. These and related dye–mordant complexes probably attach mainly to nucleoproteins, though staining is only partially prevented by chemical blocking of lysine and arginine residues (Lillie *et al.*, 1976a). Horobin (1977) considers it likely that the dye–metal complexes are bound to the nucleoproteins by van der Waals forces, but this explanation may not account adequately for classical evidence for the interposition of metal ions between dye and substrate molecules (see Chapter 5). In acid solutions (which are used for nuclear staining), the dye–metal complexes of haematein with iron and aluminium are predominantly anionic, so they would be electrostatically attracted to the positively charged ions formed by the lysine and arginine side-chains of nucleoproteins. Favourably conforming shapes of the nucleoprotein molecules and the dye–metal complex anions may well result in reinforcement of the ionic attractions by the weaker, short-range van der Waals forces, and perhaps also by hydrogen bonds. The binding of the dye certainly cannot be due only to electrostatic forces, since cationic components of the tissues other than nucleoproteins are not stained.

Haematein (often, but erroneously, referred to as "haematoxylin") is used in solutions containing ferric, aluminium, or, more rarely, chromium ions. In other mixtures, used only for specialized purposes, the mordant may be a lead or copper salt or phosphotungstic or phosphomolybdic acid. For nuclear staining, the solution must contain a considerably greater proportion of ferric or aluminium ions than of haematein and it must be acidified.

Staining with alum–haematoxylins may be progressive or regressive. The former is usually preferred, but if overstaining has to be corrected the sections are differentiated in dilute acetic acid or in alcohol containing a little hydrochloric acid. Further acidification of the staining bath will also suppress the tendency to overstain, thereby making the mixture more selective as a progressive nuclear stain. Acids used for differentiation probably disrupt the bonding between metal and tissue rather than that between metal and dye (Baker, 1960, 1962). The acid-lability of nuclei stained by alum–haematoxylin precludes counterstaining with solutions that are more than slightly acidic. Eosin is suitable but the van Gieson mixture (see Chapter 8) is not. The aluminium–haematein complex changes its colour from reddish brown to bluish purple at about pH 7. The latter colour is the one desired, so stained sections are washed in tap water, to which a trace of a mild alkali may have to be added, after staining. This process is called "blueing".

The iron–haematein complex has a deep blue–black colour and can be removed from stained sections only by strongly acid differentiating solutions.

The progressive mode is the method of choice for nuclear staining, so the dye solutions are made with large excesses of ferric salt and acid (usually hydrochloric). Weigert's haematoxylin is a typical mixture of this type (see p. 99). The ferric ions eventually over-oxidize the haematein, so the stain deteriorates gradually, over a period of several days, until it is no longer usable. Some iron–haematoxylins have better keeping properties than Weigert's (see Lillie & Fullmer, 1976). Almost any counterstain can be used after iron–haematoxylins.

The acid-labile colour of nuclei stained by alum–haematoxylin can be converted to an acid-stable black by treatment with a solution of a ferric salt. This simple manoeuvre might be valuable to offset the expense involved in the wastage of unstable iron–haematoxylins prepared only for occasional use. However, it is also possible to use synthetic dyes that are cheaper than haematoxylin. A suitable substitute is chromoxane cyanine R, used with a ferric salt as mordant (see below, p. 82). Several other alternatives to haematoxylin have been described by Lillie *et al.* (1976b).

Often a red nuclear stain is preferred to the colours imparted by the metal–haematein complexes. This may be achieved by using an aluminium mordant with carminic acid ("carmalum", see p. 83) or brazilein ("brazalum", see p. 83). The latter is a more stable reagent and stains nuclei scarlet. This colour contrasts well with blues and greens due to other dyes or to the end-products of histochemical reactions. Carmalum gives a crimson nuclear stain. Gabe (1976) highly recommends the aluminium lake of kernechtrot as a red nuclear stain. Red nuclear stains are most often employed as counterstains to histochemical or other specialized methods that impart other colours to the objects in which the microscopist is primarily interested.

6.2. ANIONIC COUNTERSTAINS

In most of the general "oversight" methods used in histology a blue, purple, or black nuclear stain is followed by a paler, usually pink, counterstain, which colours all the other components of the tissue. Eosin is suitable for this purpose. It is an anionic dye, so it is bound principally by ionized cationic groups of protein molecules. The most numerous of these are the ε-amino group of lysine and the guanidino group of arginine. Nearly all proteins contain these two amino acids, so eosin and other anionic dyes are bound by almost all the structures present in any tissue. A few objects are, however, missed by anionic counterstains. Thus, the extracellular matrix surrounding the collagen fibres of connective tissue is composed largely of proteoglycans (see Chapter 11), which have negatively charged molecules and therefore cannot bind anionic dyes. The same is true of many glycoprotein secretory products and of the matrix of cartilage. Glycogen, a neutral polysaccharide, is unable to bind anionic or cationic dyes. The granules of mast cells contain a basic protein, but this is already neutralized by heparin, a strongly acid proteoglycan also present in the granules. These are therefore stained by cationic dyes but not by mordant dyes or the usual anionic counterstains. Usually it is not possible to see very fine fibres (e.g. reticulin, nerve fibres) or cytoplasmic organelles in sections coloured by a single anionic counterstain. Although these delicate structures bind the dye, the degree of contrast is insufficient to permit their resolution in sections more than 0.5–1.0 μm thick.

Anionic dyes other than eosin may also be used as counterstains, but those with strong tinctorial power (e.g. methyl blue, acid fuchsine) are usually avoided because they might obscure the primary staining of the nuclei. When a red nuclear stain has been used, a green counterstain (e.g. fast green FCF) is suitable.

6.3. SINGLE SOLUTION METHODS

It has already been pointed out that a cationic dye applied from a solution whose pH is higher than about 5 will stain nearly all the components of a tissue (see p. 78). Neutral or alkaline solutions of cationic dyes can therefore be used as single reagents for general purpose oversight staining. The contrast is less than when two dyes are used, but the speed and simplicity of a one-step method are sometimes advantageous. Although most anionic dyes are of no value when used alone, chlorazol black E is exceptional, for reasons that are obscure. The blackness of this dye may possibly be more important than its physical and chemical properties in producing optical contrast in stained sections.

Mordant dyes can also be used by themselves: a moderate amount of detail can be seen in addition to the nuclei in a section slightly overstained by alum–haematein, alum–brazilein, or alum–carminic acid, but these dye–metal complexes are hardly ever used alone in this way. The Heidenhain haematoxylin method (see Exercise 3, p. 75) is a more complicated procedure, requiring critical differentiation but capable of revealing considerable cytoplasmic detail. Instructions for this technique are given by Gabe (1976) and Lillie & Fullmer (1976). The dye chromoxane cyanine R has been used without a mordant to stain tissues in two colours (blue and red), but it is difficult to obtain consistent results.

Mann's eosin–methyl blue is a mixture of anionic dyes used as a one-step oversight stain. Its possible mechanism of action has already been mentioned (p. 78). Mann's mixture is also used in pathology for the detection of viral inclusion-bodies in diseased cells.

The azure–eosin technique is a valuable one-step staining method in which an anionic and a cationic dye are applied simultaneously from a single solution at a carefully controlled pH. Results are similar to those obtained using the same dyes in sequence, but a greater variety of shades of colour is seen when the single solution is employed. This technique was developed from those used for staining blood. The properties of mixtures containing eosin and cationic thiazine dyes will be discussed in Chapter 7. The only disadvantage of the azure–eosin technique is the instability of the staining mixture. Instructions are given on p. 86, and a full account of the method and the results obtained with it is given by Lillie & Fullmer (1976).

6.4. INDIVIDUAL METHODS

6.4.1. Haematoxylin and eosin

The method described here is typical of the many H. & E. procedures. The traditional recipe for Mayer's haemalum has been modified by reducing the amount of sodium iodate for the reasons given in Chapter 5. The citric acid in this solution probably serves only to maintain appropriate acidity. The function of the chloral hydrate is obscure. Do not expect a perfect result by following the numbered steps of this method uncritically: careful attention to the appended notes is essential.

Solutions required

A. Mayer's haemalum

Dissolve the following, in the order given, in 750 ml of water:

Aluminium potassium sulphate ($KAl(SO_4)_2 . 12H_2O$):	50 g
Haematoxylin (C.I. 75290):	1.0 g
Sodium iodate ($NaIO_3$):	0.1 g
Citric acid (monohydrate):	1.0 g
Chloral hydrate:	50 g

Make up with water to 1000 ml. Often keeps for a year but some batches lose their potency after only a few months. The solution may be reused many times. If the solution fails to stain nuclei properly, it may be over-oxidized. Try making another batch with only 50 mg of sodium iodate.

See *Note 5* below for an alternative alum–haematein solution.

B. Eosin

Eosin (C.I. 45380):	2.5 g
Water:	500 ml

This solution keeps indefinitely and may be used repeatedly. Moulds often grow in it and need to be removed by filtration. **Alternatively**, use 0.2% ethyl eosin (C.I. 45386) in 95% alcohol.

Procedure

1. De-wax and hydrate paraffin sections. Frozen sections should be dried onto slides.
2. Stain in Mayer's haemalum (solution A) for 1–15 min (usually 2–5 min, but this should be tested before staining a large batch of slides). Overstained sections can easily be differentiated by agitating for a few seconds in 1% (v/v) concentrated hydrochloric acid in 95% alcohol, then washing thoroughly in tap water.
3. Wash in running tap water for 2 or 3 min or until the sections turn blue. If the tap water is

not sufficiently alkaline to blue the sections, add a few drops of ammonium hydroxide (S.G. 0.9) **or** of saturated aqueous lithium carbonate **or** a small pinch of calcium hydroxide to about 500 ml of water and leave the washed sections in this for 30–60 s, then rinse in tap water again. Examine the wet slide under a microscope to check that selective nuclear staining has been achieved. Any blue coloration of cytoplasm and connective tissue should be extremely faint. (See also *Note 3* below.)

4. Immerse slides in eosin (solution B) for 30 s with agitation. (See also *Note 4* below.)
5. Wash (and differentiate) in running tap water for about 30 s. (See *Note 1* below.)
6. Dehydrate in 70%, 95%, and two changes of absolute ethanol (with agitation, about 30 s in each change; without agitation, 2–3 min in each change). (See also *Note 2* below.)
7. Clear in xylene and cover, using a resinous medium.

Result

Nuclear chromatin—blue to purple, cytoplasm, collagen, keratin, erythrocytes—pink.

Notes

1. The ideal balance between the two components of the H. & E. stain is a matter of personal taste and is determined by the intensity of coloration due to the eosin. For a weaker counterstain, use 0.2% eosin or prolong the differentiation (stage 5). Differentiation also occurs in the 70% alcohol used for dehydration and, to a lesser extent, in the higher alcohols. In some objects stained by eosin, a yellow to orange cast can be discerned. This is most easily seen in erythrocytes. Staining by eosin should never be so strong that the nuclei are obscured.
2. For celloidin sections, avoid absolute alcohol and complete the dehydration in two changes (each 10 min) of *n*-butanol. Frozen sections usually require shorter times in the dyes and longer times for differentiation, washing, dehydration, and clearing.
3. Poor nuclear staining may be due to prior excessive exposure of the tissue to acidic reagents (e.g. unneutralized formalin or decalcifying fluids). To restore the chromophilia, Luna (1968) recommends treatment of the hydrated sections with **either** 5% aqueous sodium bicarbonate ($NaHCO_3$) **or** 5% aqueous periodic acid, overnight, followed by a 5-min wash in water before staining.
4. A satisfactory alternative to eosin is **fast green FCF** (0.5% aqueous), which is differentiated by water more readily than by alcohol. This dye stains acidophilic elements a bluish-green colour. It is valuable if some components of the section have already been stained pink or red with, for example, periodic acid–Schiff.
5. An alternative aluminium–haematein solution is Baker's (1962) "haematal-16". This contains 16 ions of Al^{3+} for every one molecule of haematein. Ethylene glycol is present to dissolve the haematein, which is only sparingly soluble in water. Haematal-16 is a slow, progressive nuclear stain and should be applied to the sections for 10–30 min. Dissolve 0.94 g of haematoxylin in a mixture of water (250 ml) and ethylene glycol (250 ml). Bubble air through the solution for about 4 weeks. Every few days make up again to 500 ml with water to compensate for evaporation. Add to the above solution 500 ml of ethylene glycol followed by 1000 ml of an aqueous solution of aluminium sulphate ($Al_2(SO_4)_3 . 12H_2O$—15.8 g; water to 1000 ml). Final volume is 2000 ml. When the haematein is made by atmospheric oxidation, as described here, this haematal-16 solution is stable for at least 2 years, possibly longer.

6.4.2. Other nuclear stains

6.4.2.1. Chromoxane cyanine R

The iron(III) lake of this dye is a satisfactory alternative to haematoxylin, though nuclei are usually less deeply coloured.

Staining mixture (Hogg & Simpson, 1975)

Chromoxane cyanine R (C.I. 43820):	1 g
Concentrated sulphuric acid:	2.5 ml

Dissolve the dye in the acid in a flask by mixing with a glass rod, then add:

Water:	450 ml
4% aqueous iron alum $NH_4Fe(SO_4)_2 . 12H_2O$:	50 ml

Mix thoroughly and filter. Stable for several months. Deterioration is very gradual and is detected by lessening intensity of the colour imparted to nuclei. The mixture is just as effective if the dye is first dissolved in the water and the sulphuric acid is then added, followed by the iron alum. A 2.8% solution of ferric nitrate ($Fe(NO_3)_3 . 7H_2O$) may be substituted for the 4% iron alum. Alternative mixtures are described by Llewellyn (1974, 1978).

Procedure

1. Stain hydrated sections for 5 min.
2. Wash in tap water to remove excess dye.
3. Differentiate by immersion and agitation in 70% ethanol to each 100 ml of which has been added 0.5 ml of concentrated hydrochloric acid. Hogg & Simpson (1975) recommend differentiation for 10 s, but in my experience 30 s are usually necessary.
4. Wash in running tap water (a trace of alkali may have to be added) until colour of sections changes from pinkish brown to blue.
5. Counterstain as in the H. & E. method (see above) or with the van Gieson mixture (p. 99).
6. Wash in tap water, dehydrate, clear, and cover.

Result

Nuclei blue; cytoplasm and collagen pink (with eosin counterstain). When a van Gieson counterstain is used, the nuclei are green–blue, cytoplasm yellow, collagen red.

6.4.2.2. Iron-haematoxylin

Weigert's iron–haematoxylin (as described in conjunction with the van Gieson method, p. 99) is recommended. Nuclei are coloured black and the stain is usually unaffected by subsequent reagents. If necessary, differentiate in acid alcohol as for alum–haematein or chromoxane cyanine R.

6.4.2.3. Brazilin

This dye (see p. 61) is closely related to haematoxylin. Its oxidation product (brazilein) forms with aluminium ions a lake which stains nuclei red. Since the colour (like that imparted by alum–haematein) is extracted by acids, alum–brazilein is best applied as a counterstain to follow a method that imparts green or blue colours to other components of the tissue.

A suitable mixture is **Mayer's brazalum**. This is made in exactly the same way as Mayer's haemalum (p. 81), but substituting brazilin (C.I. 75280) for haematoxylin. It is more stable than haemalum, probably because brazilin has only one pair of hydroxyl groups in the easily oxidized catechol configuration. The staining time is usually about 5 min, with differentiation, if necessary, in acid alcohol. Nuclei are stained red.

6.4.2.4. Mayer's carmalum

Staining mixture (after Gatenby & Beams, 1950)

Carminic acid (C.I. 75470; not to be confused with carmine):	1.0 g
Aluminium potassium sulphate, $KAl(SO_4)_2 . 12H_2O$:	10 g
Water:	200 ml

Heat until it boils, then allow to cool to room temperature. Filter. Add 1.0 ml formalin (37–40% HCHO) as an antibacterial preservative. The solution loses most of its potency after 2–3 weeks.

Procedure

1. Stain hydrated sections for 10–30 min. Overstaining does not occur.
2. Wash in running tap water for about 1 min.
3. Dehydrate, clear, and cover.

Result

Nuclei crimson. Like alum–brazilein, carmalum is valuable as a counterstain. It gives pleasing appearances following indigogenic methods for esterases (see Chapter 15) or silver methods for nervous tissue (Chapter 18).

6.4.2.5. Alum–tetracycline

Tetracycline is a yellow antibiotic with the structure

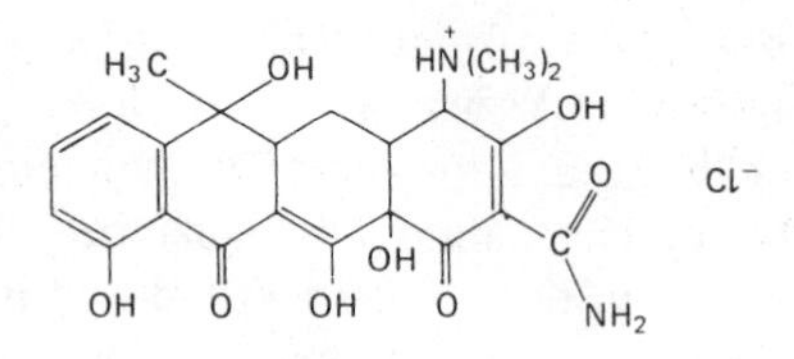

As a dye it could be classified as a hydroxyketone or aminoketone cation. It is strongly fluorescent and is able to form chelates with many metals.

The simple procedure described below (previously unpublished) is essentially an alternative, for fluorescence microscopy, to alum–haematein. Similar effects cannot be obtained with tetracycline alone, but the mordanting action of aluminium in this method has not been studied.

Staining solution

Tetracycline hydrochloride:	0.5 g
Water:	100 ml
Aluminium potassium sulphate ($KAl(SO_4)_2 . 12H_2O$):	5.0 g

The solution is stable for about 1 month, becoming decreasingly efficacious as its colour changes from yellow to orange to brown.

Procedure

1. Stain hydrated sections for 5 min.
2. Wash in three changes of water.
3. Dehydrated through graded alcohols, clear in xylene, and mount in a non-fluorescent resinous medium (e.g. DPX).
4. Examine by fluorescence microscopy with excitation by blue or ultraviolet light.

Result

Nuclei emit a yellow to orange fluorescence.

6.4.2.6. CATIONIC DYES AS NUCLEAR STAINS

As explained above, cationic dyes will not stain nuclei specifically. The following solutions are suitable for use as counterstains to other methods.

(a) Toluidine blue O (C.I. 52040)

A 0.5% aqueous solution, acidified by adding approximately 1.0 ml of glacial acetic acid to each 100 ml of dye solution. Alternatively, make up in 0.1 M acetate buffer, pH 4.0. The optimum pH may vary between 3.0 and 5.5 according to the fixative and the tissue to be stained. Stain in this solution for 3 min, rinse in water, dehydrate in 95% followed by two changes of absolute alcohol, clear in xylene, and mount in a resinous medium.

If too much colour is lost during washing or alcoholic dehydration, move the slides directly onto filter paper and blot firmly with two or three layers of filter paper. Transfer to a clean, dry staining rack or coplin jar and dehydrate in two changes (each 4–5 min) of *n*-butanol. Clear and mount as above. If detachment of the sections seems imminent, the clearing in xylene may be omitted, but the resinous medium will then take longer to become fully transparent.

The dye may also be immobilized in stained structures by converting it to an insoluble molybdate. After staining, wash in water and immerse for 5 min in 5% aqueous ammonium molybdate (= ammonium paramolybdate, $(NH_4)_6Mo_7O_{24} . 4H_2O$) solution (which may be used repeatedly until it becomes cloudy; usually stable for 1 or 2 weeks). Wash in running tap water for 2–3 min. The molybdate of toluidine blue is not extracted by water or alcohols and even resists counterstaining by acidic mixtures of dyes, such as that of van Gieson.

Nuclei and acid carbohydrate components are stained blue. Some carbohydrate-containing structures are metachromatically stained (red) as explained in Chapter 11. Cytoplasmic RNA (e.g. Nissl bodies of neurons) is also blue. Other thiazine dyes (e.g. Borrel's methylene blue, azure A) may be substituted for toluidine blue.

(b) Neutral red (C.I. 50040)

Use a 0.5% aqueous solution for 1–5 min. Acidification with acetic acid (see above) is usually desirable. Wash, dehydrate, and clear as described for toluidine blue O.

Nuclei and other basophilic structures are stained red. Safranine O (C.I. 50240) gives a similar result, but longer staining times (often 30–60 min) are usually required.

(c) Methyl green (C.I. 42585)

Instructions for preparing 2% aqueous methyl green and extracting chloroform-soluble contaminants are given with the methyl green–pyronine method (Chapter 9, p. 109). For staining with methyl green alone, dilute with water to give a 0.25% solution (which is stable for a few months) and stain for 15 min. Rinse quickly in water, blot the slides dry, and dehydrate in two changes of *n*-butanol (as described above for toluidine blue O). Clear in xylene and cover, using a resinous mounting medium.

Nuclei and other basophilic structures are stained bluish green.

(d) Polychrome methylene blue

The staining solution is a mixture of thiazine dyes produced by oxidizing methylene blue. The oxidation (polychroming—see Chapters 5 and 7) may be accomplished in various ways. The following method, in which silver oxide is used as an oxidizing agent, is recommended for its simplicity and speed.

Borrel's methylene blue

Add 25 ml of 4% sodium hydroxide (NaOH) to 100 ml of 1% silver nitrate ($AgNO_3$). Allow the precipitate of silver oxide (Ag_2O) to settle, decant off the supernatant, and wash the precipitate by shaking and decantation, with two changes of 100 ml of water. Dissolve 1.0 g of methylene blue (C.I. 52015) in 100 ml of water, boil, and add to the washed Ag_2O. Boil for 5 min, leave to cool, then filter. This stock solution is diluted to five times its volume with water to give the working solution. Both the stock and the diluted solutions are stable for at least 3 years.

Method

Stain frozen or hydrated paraffin sections for 2 min in the diluted solution of Borrel's methylene blue. Rinse rapidly in running tap water, blot slides dry, and dehydrate in two changes (each 30 s, with agitation) of absolute ethanol. Clear in xylene and cover, using a resinous medium.

Result

All components of the tissue are stained in various shades of blue. Strongly acid carbohydrates stain metachromatically (red), especially if some differentiation is allowed to occur in the water wash. With deliberate overstaining, as described, the objects that are normally chromotropic will be dark blue or purple.

6.4.3. Mann's eosin–methyl blue

The procedure given below is the "short method" of Mann (1902). The "long method" (Mann, 1902; also discussed at length by Gabe, 1976) is more controllable and is used for demonstrating intracellular objects such as secretory granules and viral inclusion bodies. Fixation is not critical, but superior contrast is obtained if mixtures containing mercuric chloride, potassium dichromate, or picric acid have been used.

Preparation of stain

Mix:	
1% aqueous eosin (C.I. 45380):	45 ml
1% aqueous methyl blue (C.I. 42780):	35 ml
Water:	100 ml

The solution keeps for several months. Precipitates which may form in it do not matter but should be removed by filtration. Aniline blue WS (C.I. 42755) or any water-soluble "aniline blue" (see Chapter 5) may be used in place of methyl blue.

Procedure

1. De-wax and hydrate paraffin sections.
2. Stain in eosin–methyl blue for 10 min.
3. Wash in running tap water for a few seconds to remove excess dyes from slides.
4. Dehydrate in 95% and two changes of absolute alcohol.
5. Clear in xylene and cover, using a resinous mounting medium.

Result

Nuclei and collagen—blue; erythrocytes, cytoplasm, nucleoli—red. With this "short" method, the colours obtained are rather variable. Like Borrel's methylene blue, this is a "quick look" method for determining which of a series of sections or specimens are worthy of more critical examination.

6.4.4. Chlorazol black E

Staining solution

Chlorazol black E (C.I. 30235):	3.0 g
70% aqueous ethanol:	300 ml

Dissolve the dye (magnetic stirrer; 30 min at room temperature), then filter the solution. It keeps for 1 year.

Procedure

1. De-wax paraffin sections and take to 70% alcohol.

2. Stain in the chlorazol black E solution for about 10 min. The time is not critical.
3. Rinse in 95% ethanol, two changes, 1 min in each. (See *Note 1* below.)
4. Complete the dehydration in two changes of absolute alcohol.
5. Clear in xylene and cover, using a resinous mounting medium.

Result

Nuclei black. Elastin black. Other components of tissue in various shades of grey. Cytoplasm often has a greenish tinge and cartilage matrix is usually pinkish grey. Collagen is rather lightly stained. Cytoplasmic organelles such as mitochondria and secretory granules are well displayed in suitably fixed material. The colour does not fade. (See also *Note 3* below.)

Notes

1. The method will work after any fixation but, as with most other dye staining methods, greater contrast is obtained after fixation in a mixture containing picric acid, mercuric chloride, or potassium dichromate than after fixation in formaldehyde alone.
2. Over-staining rarely occurs if the time at stage 2 is less than 20 min. Differentiation occurs slowly and controllably in 95% ethanol. Cannon (1937), the originator of the method, recommended terpineol as a differentiator. He also commented on the similarity of the end result to that obtained with Heidenhain's iron-haematoxylin, a traditional but time-consuming technique involving a critical differentiation.
3. The most informative preparations are thin sections (5 μm or less) examined at high magnification. Other methods of staining with chlorazol black E are given by Clark (1973).

6.4.5. Azure–eosin method (Lillie's technique)

Staining solution

This is made from stock solutions, which all keep for several months.

Azure A (C.I. 52005),	
0.1% aqueous stock solution:	16 ml
Eosin B (C.I. 45400),	
0.1% aqueous stock solution:	16 ml
0.2 M acetic acid:	6.8 ml
0.2 M sodium acetate:	1.2 ml
Acetone:	20 ml
Water:	100 ml

This working solution should be mixed immediately before using. The quantity above suffices for one batch of 8–10 slides in a 100 or 125 ml staining vessel. Its pH is 4.0, which is usually optimum for formaldehyde-fixed tissues. (See also *Note 2*, below.)

Procedure

1. De-wax and hydrate paraffin sections.
2. Stain in the working solution for 1 h.
3. Pour off the staining solution and replace it with acetone, three changes, each 45–60 s, with agitation. (See also *Note 1*, below.)
4. Clear in two changes of xylene.
5. Apply coverslips, using a synthetic resinous mounting medium.

Result

Nuclei and cytoplasmic RNA blue. Cartilage matrix and other metachromatic materials red to purple. Muscle cells pink. Cytoplasm of other cells pale blue to pink, according to type. Collagen and erythrocytes pink. (See *Notes 2* and *3* below.)

Notes

1. As an alternative to dehydration in acetone, the sections may be blotted dry and passed through two changes of *n*-butanol, each 3–5 min with occasional agitation.
2. If satisfactory results are not obtained, the pH of the staining solution should be changed. The acetic acid and sodium acetate may be replaced by 8 ml of 0.2M acetate buffer (see Chapter 20) of any pH from 3.5 to 5.5. Phosphate and other buffers may be used outside this range. Before applying this method to large numbers of slides, test it from pH 3.5 to 5.5 at intervals of 0.5 pH unit. Staining solutions buffered beyond this range will rarely be needed. Solutions of pH > 4.0 are often needed after fixation in agents other than formaldehyde.

3. A full account of this technique, with descriptions of the tinctorial effects in normal and pathological tissues, is given by Lillie & Fullmer (1976).
4. Similar results can be obtained with Giemsa's stain, which is described in Chapter 7.

6.5. EXERCISES

Theoretical

1. What differences would you expect to see between two similar paraffin sections of an autonomic ganglion: one stained with H. & E., the other by the azure–eosin method?

2. Why is Mayer's haemalum stable for several weeks while Weigert's iron–haematoxylin deteriorates after a few days?

3. Which of the following dyes would be suitable counterstains for paraffin sections in which the nuclei have been stained with chromoxane cyanine R? Give reasons: (a) neutral red; (b) erythrosin; (c) methyl green; (d) methyl blue; (e) orange G; (f) Bismarck brown Y.

4. When a specimen has stood for several months in an aqueous formaldehyde solution, it is often very difficult to obtain sufficiently intense staining by eosin when paraffin sections are subjected to the H. & E. procedure. Suggest a reason for this.

5. What result would you expect if you stained a paraffin section of a specimen fixed in Altmann's fluid (p. 23) with toluidine blue O at pH 4.0?

6. Making use of the dyes discussed in this chapter, devise a staining method which would colour nuclei but not cytoplasmic RNA red and cytoplasm and collagen in some shade of green.

7. Discuss the chemical mechanisms involved in the staining of: (a) nuclei by mordant dyes; (b) nuclei by cationic dyes; (c) cytoplasm by anionic dyes.

Practical

8. Take paraffin sections of several of the variously fixed tissues you have prepared and stain them by the following methods. Use thin (4–5 μm) and thicker (7–12 μm) sections of each tissue if possible.

(a) Mayer's haemalum and eosin (H. & E.) **or** chromoxane cyanine R and eosin.
(b) Mann's short eosin–methyl blue method.
(c) Chlorazol black E.

When you examine the stained slides, pay special attention to:

(1) The disposition of smooth and striated muscle and of collagen.
(2) The different types of cells, as distinguished by their nuclei.
(3) The layers making up the walls of tubular structures such as blood-vessels, ducts of glands, and the alimentary tract.
(4) The resolution of detail within the cytoplasms of epithelial cells, secretory cells, neurons, and muscle fibres.

Decide which techniques are best suited for demonstrating features (1)–(4) above and note the effects of different fixatives on the appearances of nuclei, cytoplasm, and general organization of the tissues.

9. Take paraffin sections of any convenient tissue and subject half of them to methylation for 24 h as described in Chapter 10, p. 141. This procedure causes esterification of carboxyl groups and phosphate ester groups and removes sulphate ester groups. Stain methylated and control sections with: (a) an alum–haematein, (b) toluidine blue O or neutral red.

Observe and explain the effects of methylation on the staining of nuclei and other structures by the two dyes.

7

Blood Stains

7.1. Neutral stains 88
7.2. Preparation of blood-films 89
7.3. Leishman's and Wright's stains 89
7.4. Giemsa's stain 90
7.5. Exercises 91

Special staining methods are used for the identification of the various types of cells occurring in blood and haemopoietic tissue. The same techniques also reveal abnormalities in the cells and the presence of pathogenic protozoa such as malaria parasites. The reader who is unfamiliar with haematology is advised to consult the relevant parts of a textbook of histology or clinical pathology before attempting to identify cell types in the blood.

7.1. NEUTRAL STAINS

A useful type of staining reagent may be made by mixing aqueous solutions of suitable cationic and anionic dyes (e.g. methylene blue and eosin; neutral red and fast green FCF). The precipitate which is formed contains ionic species of both dyes and is almost insoluble in water but freely soluble in alcohol. It is also soluble in water containing an excess of either the anionic or the cationic dye. Mixed dyes (often called "neutral stains") of this type find their most important applications in haematology, but may also be used for general staining of any tissue. When the anionic component is eosin, the neutral stain is called an "eosinate".

The alcoholic solution of the combined stain must be diluted with water immediately before use in order to liberate the colorant ions in an active form. The neutral stains in common use for blood are carefully balanced mixtures containing two or three thiazine dyes (methylene blue and its oxidation products) and eosin, dissolved usually in methanol and glycerol. They are known generically as "Romanowsky stains" after the Russian haematologist who first described their properties in 1891. The commercially available ready-mixed preparations are more reliable than those made in the laboratory from the separate ingredients. However, for non-haematological purposes extempore mixtures are quite satisfactory (see Lillie & Fullmer, 1976). The pH of the stain and of the water used for rinsing is a critical factor, especially in applying these mixtures of dyes to sections of tissues.

The different kinds of leukocyte in a blood-film are identified by virtue of the colours imparted by the staining mixture to the nuclei and cytoplasmic granules of the cells. Both the cationic and the anionic components of the mixture contribute to these colours, and the desired result cannot be obtained by using the dyes sequentially. Erythrocytes are coloured by the eosin anions, with intensity approximately proportional to their intracellular concentrations of haemoglobin. Immature mammalian red cells (reticulocytes) also contain material stained by the cationic dyes, this representing the remnants of the nuclei of the cells. Abnormal objects such as malaria parasites and leukaemic white cells also have their characteristic tinctorial properties. The nuclei of protozoa should be coloured bright red by a satisfactory haematological-staining solution. Since abnormalities are seen only from time to time, it is essential for the pathologist that mixtures used for staining blood give consistent results, even in the hands of inexperienced technicians.

The main source of variation given by different batches of blood stains is the rather unpredictable assortment of dyes produced by the oxidation of methylene blue (see Chapter 5, pp. 70–71). It is now known that the properties of Romanowsky stains are attributable to methylene blue, azure B, and eosin. Other thiazine dyes, notably methylene violet (Bernthsen) (see p. 71), are always present in the mixtures, but at low concentrations they do not have adverse effects. Although eosin is usually impure, its contamination with other xanthine dyes is unimportant.

Haematological stains have been developed empirically, and the reasons for their valuable properties are still largely unknown. The subject has been reviewed by Lillie (1977) and Marshall (1978).

7.2. PREPARATION OF BLOOD-FILMS

The glass slides must be clean and have been degreased in alcohol or acetone. A small drop of blood is spread over a slide as shown (Fig. 7.1). The two slides meet at an angle of 45°. The film is then allowed to dry. In the technique to be described below the stock solution of mixed dyes dissolved in methanol and glycerol serves as a fixative and then as a stain.

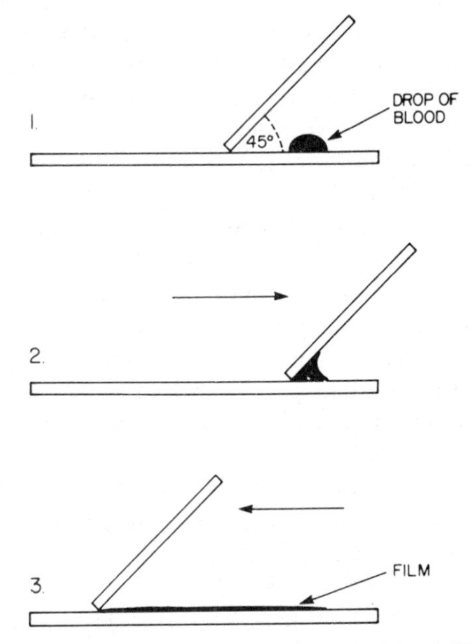

FIG. 7.1. Technique for preparation of a blood-film.

7.3. LEISHMAN'S AND WRIGHT'S STAINS

These are the methods of choice for blood-films.

Leishman's stain and Wright's stain can be made up from their ingredients (see *Note 1* below), but the commercially available ready-made solutions are usually to be preferred.

A buffer solution, pH 6.4–6.8, is also required, but tap water, if not too alkaline, will often serve quite adequately in its place. Distilled water is usually rather too acid (pH around 5.5), owing to dissolved carbon dioxide derived from the atmosphere. A suitable solution for use with blood stains may be made by diluting a phosphate buffer (Chapter 20) with 5–10 times its volume of distilled water.

Procedure

1. Place slide with film on a horizontal staining rack over a sink.
2. Flood the slide with the stock solution of Leishman's or Wright's stain. Wait for 1 min. The solvent fixes the cells and plasma of the film. (See *Note 4*, below.)
3. Add buffer solution in sufficient quantity to dilute the stain two or three times. (The excess liquid will spill over into the sink.) Leave for 3 min.
4. Wash off the stain with copious buffer solution (or tap water if satisfactory).
5. Stand the slide on end to drain and dry.
6. Clear and cover (see *Note 2* below) if permanent mount required.

Result

Erythrocytes—pink; nuclei of leukocytes—deep purple; cytoplasm of agranular leukocytes—pale blue or lilac; basophil granules—dark blue; neutrophil granules—purple; eosinophil granules—red to orange. Platelets—blue to purple. Nuclei of protozoa, including malaria parasites, are coloured bright red.

Notes

1. Both Leishman's and Wright's stains are made by dissolving in methanol the precipitated eosinates of polychromed methylene blue. They differ only in the methods of polychroming. If the stain is supplied as a powder, make a 0.15% solution in methanol. Shake well and leave to stand for 2 days or longer. Filter. The caps of the bottles in which the solution is kept should not have metal liners, since these have been shown to cause deterioration of the stains.
2. The preparations are commonly examined by placing immersion-oil directly onto the dried films and examining them with an oil-immersion objective, which is necessary for resolution of the cytological detail. A permanent preparation is made by clearing in xylene for about 2 min (**without** prior exposure to alcohols) and applying a resinous mounting medium and a coverslip.
3. Unsatisfactory staining can result from deterioration of the stain due to evaporation of the solvent or from using water at the wrong

pH. It is occasionally necessary to vary the pH of the buffer between 6.0 and 7.0 in order to secure optimum coloration of the film.

4. If blood-films cannot be stained within about 2 hr of preparation, they should be fixed by immersion for a few minutes in methanol. Stage 2 of the method described above is then unnecessary, and the films may be stained in freshly diluted Leishman's or Wright's solution.

7.4. GIEMSA'S STAIN

Giemsa's stain is sometimes used for blood-films, but is more suitable for sections with which more consistent results are obtained than with Leishman's or Wright's staining mixtures. Giemsa powder is a mixture of "Azure II" (21% w/w) with its eosinate (79% w/w). Azure II is believed to be a mixture of equal weights of methylene blue and azure B, originally marketed by the German firm of K. Hollborn, which bought Dr. G. Grübler's business (see Chapter 5, p. 47) in 1897 (Lillie, 1977). According to Lillie & Fullmer (1976), a Giemsa powder may be made by mixing methylene blue eosinate (4 g), azure B eosinate (5 g), azure A eosinate (1 g) and methylene blue (2 g if dye content is 85–88%, as the chloride). Giemsa powder obtained ready-made, from a reputable supply house and if possible certified by the Biological Stains Commission, is recommended for all purposes. The glycerol added to the methanol used as solvent has the effect of increasing the stability of the diluted Giemsa stain, so that it is possible to expose sections to the aqueous working solution for several hours if necessary.

Helly's fixative is recommended for haemopoietic tissue. Decalcification may be necessary.

Preparation of stain

A. Giemsa stock solution

Giemsa powder:	2.0 g
Glycerol:	132 ml

Mix thoroughly in a 300 ml bottle with an air-tight screw cap. Put the bottle containing the mixture in an oven at 60°C for 2 h. Then add:

Absolute methanol: 132 ml

Mix gently. When cool, tighten the screw cap and shake. Stable for about 2 years.

B. Buffer solutions

Reagents should be available for the preparation of acetate or phosphate buffers from pH 4.0 to pH 7 (see Chapter 20). Buffered water is made by diluting the buffer solutions 5–10 times with distilled water. See *Note 1* below.

Procedure

1. De-wax and hydrate paraffin sections.
2. Immerse slides for 5 min in buffer solution at the optimum pH, as determined by experiment (see *Note 1* below). For Helly-fixed cancellous bone this is often pH 6.8. With formaldehyde- or Carnoy-fixed tissue, a pH as low as 4.0 is sometimes needed.
3. The stock Giemsa (solution A) is diluted with 50 volumes of buffered water at the same pH as that used in step 2. The slides are immersed in the diluted stain for 2 h. Alternatively, preheat the diluted Giemsa solution in a closed coplin jar in the wax oven (60–65°C) for about 20 min. Insert the slides and leave them for a further 15 min. If slides are placed in cold, diluted Giemsa and then put into the oven, allow about 30 min for staining.
4. Rinse the slides rapidly in buffered water and differentiate if necessary as described in *Note 2* below.
5. Blot sections dry with filter paper.
6. Dehydrate in two changes (each 3–5 min) of *n*-butanol. Alternatively, dehydrate rapidly in three changes of absolute ethanol.
7. Clear in xylene and cover.

Result

Nuclei—blue to purple; erythrocytes, collagen, and keratin—pink; leukocytes—as described above (Section 7.3) for blood-films except that basophil granules (and also mast cell granules) are deep purple rather than dark blue; cartilage matrix—purple. Differentiation (see *Note 2*) accentuates the metachromasia (red–purple colour) of cartilage and of mast cell and basophil granules.

Notes

1. This and related methods are influenced greatly by the pH with which the section is

equilibrated before staining, which should also be the pH of the staining solution itself. Distilled water is often satisfactory as a diluent for the stain and for washing. Check the pH of buffered water with a meter after dilution. Adjustment is sometimes necessary, especially with acetate buffers.

2. After rinsing with buffer, the section should be examined with a microscope. If the blue component of the stain is too strong, the slide should be differentiated by dipping it quickly into a 0.01% aqueous solution of glacial acetic acid and then returning it to the buffer. Several dips in the very weak acetic acid may be needed. If the pink component predominates, the pH of the stain is probably inappropriate.
3. Lillie & Fullmer (1976) strongly advocate this type of technique as a general oversight method, since more cytoplasmic details (granules, RNA, etc.) can be seen than with haematoxylin and eosin. The azure–eosin method (Chapter 6) is similar in principle, but does not provide the same differential coloration of leukocytes as the Giemsa stain.
4. If blood-films are to be stained with Giemsa, fix them first in methanol. Dilute the stock Giemsa solution (A) with 40–50 volumes of distilled water and stain by immersing the slides for 15–40 min. The time will vary with different batches of Giemsa powder. Rinse in water, dry, and, if desired, clear and mount as described for Leishman's and Wright's stains (Section 7.3).

7.5. EXERCISES

Theoretical

1. How much information could you obtain about the cells contained in a blood-film stained with alum–haematoxylin and eosin?

2. Why are "neutral stains" dissolved in non-aqueous solvents and then considerably diluted with water immediately before using?

3. Why is it necessary to stain sections of formaldehyde-fixed tissues with an azure–eosin combination at a lower pH than that required for blood-films?

Practical

4. Prepare some blood films and stain them with Leishman's or Wright's stain. Identify the various types of leukocyte. Stain a film with toluidine blue O at pH 4 (see Chapter 6, p. 84) after preliminary fixation in methanol. Does the result suggest any special purpose for which blood-films might be stained with a thiazine dye alone?

5. Stain sections of decalcified bone in order to demonstrate the cell types in the marrow.

8

Methods for Connective Tissue

8.1. Collagen, reticulin, and elastin 92
8.2. Methods using anionic dyes 93
8.2.1. van Gieson's method 93
8.2.2. Trichrome methods 94
8.2.2.1. The heteropolyacids 95
8.2.2.2. How do trichrome methods work? 96
8.3. Methods for reticulin 97
8.4. Methods for elastin 98
8.5. Individual methods 99
8.5.1. Iron–haematoxylin and van Gieson 99
8.5.2. Cason's trichrome stain 100
8.5.3. Method for reticulin 100
8.5.4. Aldehyde–fuchsine for elastic tissue 102
8.6. Exercises 102

A connective tissue stain is employed when it is desired to identify and study the various extracellular fibrous elements of animal tissues. Smooth and striated muscle fibres, erythrocytes, and the cytoplasms of other cell types are also clearly shown by the same techniques. Nuclei are commonly stained in the same section in order to facilitate correlation of the multi-coloured histological picture with the more familiar appearance obtained with an oversight method such as haematoxylin and eosin. The amorphous ground substance of connective tissue, however, is most easily demonstrated by histochemical methods for proteoglycans (see Chapter 11).

8.1. COLLAGEN, RETICULIN, AND ELASTIN

The three principal types of fibre found in connective tissue are **collagen, reticulin,** and **elastin**. Apart from their morphological characteristics (for which consult a textbook of histology), these fibres have certain physical and chemical properties which enable them to be stained selectively.

Collagen is a basic glycoprotein containing high proportions of glycine and proline. Another amino acid, hydroxylysine, though present in smaller quantities, is found only in collagen. The conformation of the collagen molecule is maintained by hydrogen bonding between peptide groups (as in all proteins) and the peptide chains are joined to one another by various types of covalent linkage formed by condensation of hydroxylysine with lysine. This results in the formation of cables, each consisting of a triple helix of polypeptide strands. The carbohydrate components are mostly α-D-glucosyl and β-D-galactosyl residues, attached mainly to hydroxylysine. The triple helices are joined end to end and side to side by hydrogen bonds, electrostatic attractions, and covalent bridges to form structures visible in the electron microscope as collagen **fibrils**. The collagen **fibres** seen with the light microscope are bundles of these fibrils.

Since it is a glycoprotein, collagen is stained, though not strongly, by the periodic acid–Schiff method (see Chapter 11). It is possible to detect free aldehyde groups in collagen by means of reagents more sensitive than Schiff's (Davis & Janis, 1966). These may arise from the oxidative deamination of the side-chains of lysine, as in immature elastin (see below).

The term "reticulin" includes the basement membranes of epithelia and blood capillaries and very fine (including immature) collagen fibres. Both types of reticulin contain more carbohydrate than does ordinary collagen. In the electron microscope, the reticular fibril shows the characteristic banding pattern of collagen. The staining methods for reticulin probably demonstrate a carbohydrate-containing matrix in which the collagenous fibrils are embedded, rather than the fibrils themselves (Puchtler & Waldrop, 1978).

Elastic fibres and elastic laminae are made of elastin, a protein different from that of collagen and reticulin. Elastin is a hydrophobic protein, rich in glycine, alanine, and valine. It contains remarkably few amino acids with ionizable side-chains. The peptide strands of elastin do not form fibrils but are united by desmosine and isodesmosine linkages to form a material with a predominantly amorphous appearance in the electron microscope. A desmosine bridge is formed from four lysine side-chains. Three of these are oxidatively deaminated to give aldehydes, which then unite, together with one unoxidized lysine side-chain, to form an aromatic ring similar to that of pyridine:

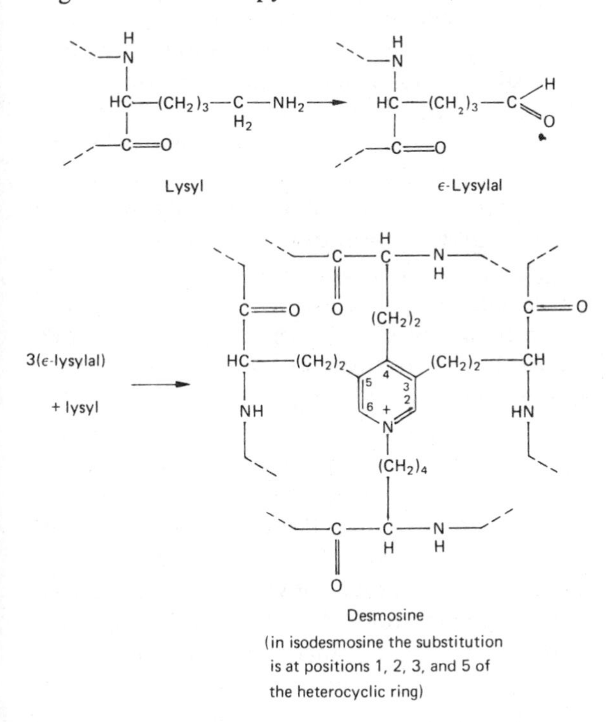

Desmosine
(in isodesmosine the substitution is at positions 1, 2, 3, and 5 of the heterocyclic ring)

In immature elastin, the formation of desmosine is incomplete, and free aldehydes can be detected histochemically (Nakao & Angrist, 1968; MacCallum, 1973). Elastin is resistant to digestion by most proteolytic enzymes. With ordinary staining methods it is acidophilic but contrasts poorly with other components of the tissue.

The histological techniques for connective tissues fall into three categories: those based on the use of mixtures of anionic dyes to impart different colours to collagen and cytoplasm, the methods for reticulin (which are essentially histochemical in nature) and the largely empirical methods for elastin.

8.2. METHODS USING ANIONIC DYES

Numerous techniques are available for staining with mixtures of dyes, but most are variants of one or the other of two main groups. In the first group, typified by van Gieson's method, a mixture of two anionic dyes imparts one colour to collagen and another to cytoplasm, including that of muscle fibres and erythrocytes. The second group embraces the "trichrome" procedures in which two, three, or rarely four anionic dyes are used in conjunction with phosphotungstic or phosphomolybdic acid. Collagen, cytoplasm (of muscle and most other cells), and erythrocytes are all coloured differently, and other elements such as cartilage, fibrin, and secretory granules also acquire characteristic colours.

8.2.1. van Gieson's method

Following a nuclear stain with an iron–haematoxylin, the sections are immersed in a solution of acid fuchsine in near-saturated aqueous picric acid. Collagen is stained red, cytoplasm is yellow, and nuclei are brown or black. In related techniques other dyes (most frequently methyl blue) are substituted for acid fuchsine and occasionally a third dye is used, either by itself or mixed with the other two. In all the methods, the principal staining mixture is quite strongly acid, with pH of 1.0–2.0 (see Clark, 1973; Lillie & Fullmer, 1976, for details).

Baker (1958) explained differential staining of this type on the basis of the different sizes of the molecules of the dyes used. He suggested that when two anionic dyes were applied simultaneously to a section there would be competition between them for cationic binding sites on the protein molecules of the tissue. The dye with smaller, more rapidly diffusing ions would penetrate quickly into a tightly woven proteinaceous matrix. There it would occupy the binding sites and exclude the dye whose particles were larger. This latter dye would, however, enter the loosely textured, more easily permeated parts of the tissue. It might not immediately attach to cationic groups there, however, since these would already have been occupied by the anions of the first, more diffusible dye. When the section is washed in water or alcohol, the small dye ions would be the first to leave the tissue and would come out of the more permeable regions first. The large dye ions could then attach themselves to the vacated cationic binding sites. Evidence in support of this explanation was derived from experiments in which gelatine gels were stained by mixtures of eosin (fairly small molecules) and methyl blue (aggregates of large molecules). Concentrated gels were coloured red by eosin and dilute ones blue, by methyl blue. Further support came from studies of the rates of diffusion of several dyes in gelatinous matrices. In general, the rates of diffusion varied inversely with the molecular weights.

It is assumed in this hypothesis that fixed cytoplasm is in some way less permeable than fixed collagen, though there is no evidence other than that derived from the behaviour of dyes to support or refute such an assertion. In one variant of the van Gieson method (see Lillie & Fullmer, 1976) the sections are stained with fast green FCF, a dye of higher molecular weight than either picric acid or acid fuchsine, prior to immersion in the van Gieson mixture. Collagen is coloured red, erythrocytes green, and the cytoplasm of muscle yellowish green. From Baker's hypothesis, however, one would have expected green collagen, yellow erythrocytes, and probably red muscle. Another paradoxical observation is even more easily made. When sections are stained only with the van Gieson mixture (with no prior nuclear stain), nuclei are coloured yellow by the picric acid. However, nuclei are stained blue by Mann's eosin–methyl blue method (see Chapter 6, p. 78). Thus, with one pair of anionic dyes the nuclei of cells show affinity for the smaller molecules, while with another pair they take up the larger molecules.

Lillie (1964) sought a chemical explanation for the differential dyeing of collagen and cytoplasm. He subjected sections to various pre-treatments before staining and found that:

(a) Treatment with nitrous acid abolished all staining by ordinary anionic dyes such as eosin and biebrich scarlet but did not prevent the red coloration of collagen by van Gieson's method. Nitrous acid removes amino groups: —NH_2 is probably usually replaced by —OH (see Chapter 10).

(b) Treatment with a mixture of acetic anhydride, acetic acid, and sulphuric acid abolished all acidophilia, including the affinity of collagen for acid fuchsine in the van Gieson mixture. The formerly acidophilic structures became basophilic, a property attributable to the formation of —$NHSO_3^-$ groups with primary amines and of —OSO_3^- groups with hydroxyls. These changes were reversed by hydrolysis of the sulphate esters in a methanol–sulphuric acid mixture.

The reagents used for deamination and sulphation have small molecules and may reasonably be assumed to attack any parts of the tissue that are accessible to dye molecules. Binding of acid fuchsine to hydroxyl groups of serine, threonine, and hydroxylysine, presumably by hydrogen bonding to the nitrogen atoms of the dye, was tentatively suggested by Lillie (1964) as the mechanism of staining of collagen. His contention was supported by Puchtler & Sweat (1964a), who found that direct cotton dyes were effective substitutes for acid fuchsine in mixtures of the van Gieson type. Such dyes are capable of binding to substrates by non-ionic mechanisms, though the occurrence of hydrogen bonding is questionable (see Chapter 5). The staining of cytoplasm and erythrocytes by picric acid has not been studied critically. The molecular size may have some significance, however, since other dyes with small molecules, such as naphthol yellow S, martius yellow, and orange G, are used for the same purpose in other techniques. The chemistry and physics of staining by van Gieson's method are clearly in need of further investigation.

8.2.2. Trichrome methods

The name "trichrome" identifies staining techniques in which two or more anionic dyes are used in conjunction with either phosphomolybdic or phosphotungstic acid. These acids may be mixed with the dyes, or solutions of the individual reagents may be applied sequentially to the sections in various ways. Whatever technique is employed, the result is a selective colouring of collagen by one of the dyes. Cartilage and some mucous secretions acquire the same colour as collagen, but their intensity of staining is usually less. If one other dye is applied, it stains cytoplasm and erythrocytes. If two other dyes are used, they impart their different colours to erythrocytes and to the cytoplasms of other types of cell. Secretory granules are variously stained: sometimes the same colour as collagen, sometimes the same colour as erythrocytes.

Nuclear staining can also be obtained with some of these methods, but it is more usual to stain the nuclei black with an iron–haematoxylin before carrying out the trichrome procedure. The trichrome techniques reveal fine collagenous and reticular fibres and basement membranes, which are not demonstrated by simpler methods of the van Gieson type.

8.2.2.1. THE HETEROPOLYACIDS

Phosphomolybdic acid (PMA) and phosphotungstic acid (PTA) are known as heteropolyacids. They are formed by coordination of molybdate or tungstate ions with phosphoric acid. They are sold as hydrated crystals, which are freely soluble in water to give strongly acid solutions:

$$H_3PO_4.12MoO_3.24H_2O \longrightarrow 3H^+ + [PMo_{12}O_{40}]^{3-} + 24\,H_2O$$

Phosphomolybdic acid (=*dodeca*molybdophosphoric acid) — Phosphomolybdate anion

$$H_3PO_4.12WO_3.24H_2O \longrightarrow 3H^+ + [PW_{12}O_{40}]^{3-} + 24\,H_2O$$

Phosphotungstic acid (= *dodeca*tungstophosphoric acid) — Phosphotungstate anion

The oxidation number of Mo and of W in these compounds is +6. The heteropolyacids are decomposed by alkalis to give molybdate (MoO_4^{2-}) or tungstate (WO_4^{2-}) and dibasic phosphate ions (see Cotton & Wilkinson, 1972). In aqueous solutions, PTA forms complex anions, $[PO_4(WO_3)_{12}]^{3-}$ as shown above, but these decompose if the pH rises above 2.0, to give $[(PO_5)_2(WO_3)_{17}]^{10-}$ and $[PO_6(WO_3)_{11}]^{7-}$ (Rieck, 1967).

The heteropolyacids are able to bind to tissues from aqueous or alcoholic solutions. Baker (1958) called them "colourless anionic dyes". Sites of attachment of PMA are easily demonstrated by subsequent treatment of the sections with either ultraviolet radiation or a chemical reducing agent such as stannous chloride. A blue mixture of insoluble oxides of Mo(V) and Mo(VI), with compositions such as $MoO_2(OH)$ and $MoO_{2.5}(OH)_{0.5}$, is formed. It is known as molybdenum blue. A corresponding but less intensely coloured tungsten blue, formulated as $WO_{2.7}$, can also be produced, but sites of binding of PTA to tissues have been more thoroughly studied under the electron microscope, with which the electron-dense tungsten-containing deposits can be accurately localized.

The various studies of the binding of heteropolyacids to tissues (Baker, 1958; Puchtler & Isler, 1958; Bulmer, 1962; Puchtler & Sweat, 1964b; Everett & Miller, 1974; Hayat, 1975) are not all in agreement with one another, but the following facts appear to be undisputed:

1. Chemical studies indicate that PTA binds to proteins and amino acids but not to carbohydrates. Both the heteropolyacids are used as precipitants for proteins, amino acids, and alkaloids.
2. Applied at pH < 1.5, PTA imparts electron density to carbohydrate-containing structures, but at pH > 1.5 it binds to proteins. PTA can oxidase carbohydrate hydroxyl groups to aldehydes, so the electron-dense deposits resulting from staining at pH < 1.5 may be of insoluble compounds in which the oxidation state of the tungsten is less than +6.
3. Collagen fibres bind large amounts of PMA. Cytoplasm binds smaller amounts. Nuclei of cells have very little affinity for PMA. PTA behaves similarly.
4. Affinity for PMA and PTA is depressed or abolished if amino groups in the tissue are first removed by treatment with nitrous acid or esterified by reaction with acetic anhydride or benzoyl chloride. In light microscopy there is no evidence for binding of PMA or PTA to carbohydrates or to hydroxyl groups of amino acids.
5. Methylation of sections results in increased attachment of PMA to all parts of the tissue, including erythrocytes. Methylating agents add methyl groups to amine nitrogen atoms (increasing their basicity) and to hydroxyl oxygen atoms (forming ethers or glycosides).
6. Structures that have bound PMA become stainable by cationic dyes.
7. Treatment with PMA or PTA affects stainability by anionic dyes. The effects are variable:

   (a) There is considerable suppression of the staining of all parts of the tissue by some anionic dyes, including picric acid, Martius yellow, eosin, orange G, and biebrich scarlet. The amount of suppression is greater in collagen than in cytoplasm.

(b) There is similar suppression of cytoplasmic staining by other dyes, including aniline blue WS, light green SF, fast green FCF, and acid fuchsine, but collagen is stained with only slightly reduced intensity by these dyes after treatment of the sections with a heteropolyacid.

(c) Treatment of sections with PMA or PTA either before or at the same time as staining with aniline blue WS, light green SF, or fast green FCF has the effect of preventing the attachment of these dyes to materials other than collagen, cartilage matrix, and certain carbohydrate-containing secretory products.

(d) If sections are treated with PTA, stained with aniline blue WS, and then exposed to 6 M urea, a reagent which disrupts hydrogen bonds, the dye is removed, but the PTA remains attached to the collagen in the tissue. The treatment with urea may not, however, be a specific test for hydrogen bonding.

8.2.2.2. HOW DO TRICHROME METHODS WORK?

Three hypotheses have been advanced to account for the differential colouring of tissues by anionic dyes used in association with heteropolyacids.

In the first theory, championed by Baker (1958), it is held that the anions of the dyes and of the heteropolyacids compete with one another for cationic binding sites and that the textures of the various structural components of a tissue determine their penetration by molecules of different size. PMA and PTA form anions intermediate in size between those of the dyes generally used as cytoplasmic stains and those which stain collagen. The smallest molecules would therefore enter and bind to the supposedly dense network of proteins forming the stroma of the erythrocyte, which would not be penetrated by the heteropolyacid. The much less "dense" matrix of the collagen fibre would accommodate both the ions of PMA or PTA and those of a dye with large molecules such as aniline blue WS or light green SF. Small dye molcules could also penetrate the collagen fibres but, since they diffuse rapidly, they would be able to enter and leave freely. The larger, more slowly diffusing molecules of the heteropolyacids and of dyes such as aniline blue WS would remain in the collagen and become attached to cationic binding sites there. Dyes with molecules of intermediate size (e.g. acid fuchsine) would compete with PMA or PTA for binding sites in the cytoplasm of muscle fibres and other cells. These dye molecules would be too big to enter erythrocytes in the presence of dyes with small molecules, but they would be small enough to escape from collagen fibres more quickly than the largest dye molecules. Bulmer (1962) demonstrated that heteropolyacids were bound to tissues only when the latter contained ionizable amino groups, but agreed with Baker (1958) in attributing the trichrome staining effects to differential permeability of cytoplasm and collagen to large and small molecules.

A similar postulated mechanism was discussed earlier in relation to the differential staining of cytoplasm and collagen by the van Gieson method. The objections raised there also apply to the application of this hypothesis to the trichrome techniques. Furthermore, it is difficult, in terms of this theory, to account for the fact that treatment with heteropolyacids induces basophilia, as well as selective though depressed affinity for certain anionic dyes, in collagen.

The second hypothesis accounts adequately for the basophilia produced by treatment of collagen with PMA or PTA. Puchtler & Isler (1958) proposed that cationic dyes were attracted by the free negatively charged groups of the collagen-bound ions of heteropolyacid. For example:

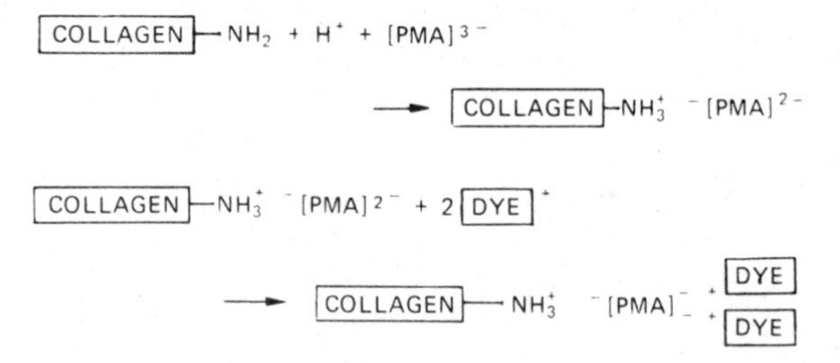

Thus a function similar to that of a mordant is attributed to the heteropolyacid. Puchtler & Isler noted that the dyes which stain collagen in the trichrome techniques are all amphoteric. They suggested that these dyes were bound by ionic forces to the PMA or PTA, which was itself attached electrovalently to the tissue. The cytoplasmic stains used in trichrome techniques are wholly anionic dyes, so they would not attach to the free negatively charged sites of the bound heteropolyacid molecules.

Several objections to this hypothesis can be made. The staining of collagen is attributed to its content of amino acids with basic side-chains: lysine, arginine, and histidine. Haemoglobin, the principal protein of erythrocytes, is more basic than collagen but it is not similarly stained in the trichrome procedures. If collagen did contain more cationic binding sites than cytoplasm, mordanting by PMA or PTA would be expected to result in intensification of staining by amphoteric dyes, since two molecules of dye could attach to each of the bound trivalent anions of the heteropolyacid. However, all investigators agree that pre-treatment with PMA or PTA reduces the intensity of staining by aniline blue WS and similar dyes, even in collagen. The matrix of cartilage and the granules of mast cells, although they are composed of strongly acid proteoglycans (see Chapter 11), are not ordinarily stained strongly by amphoteric dyes such as acid fuchsine and aniline blue WS, so these dyes should not be expected to behave as if they were cationic. Finally, the effect of 6 M urea on trichrome-stained sections (see above, p. 96) indicates that the heteropolyacid and the dye are bound to collagen by different mechanisms.

A third explanation for the actions of PMA and PTA in trichrome procedures was offered by Everett & Miller (1974), who provided evidence of two different modes of binding of these acids to tissues. Attachment by electrovalent forces was thought to occur in cytoplasm and to inhibit there the binding of anionic dyes. The binding of the heteropolyacids to collagen was believed to be non-ionic, so that staining by anionic dyes was not prevented. The principal objection to this hypothesis is that it does not account for the fact that the staining of collagen after treatment with PMA or PTA can be effected by some anionic dyes but not by others.

From the foregoing discussion it will be seen that there is, as yet, no satisfactory explanation of the effects of heteropolyacids on the stainability of tissues by different anionic dyes. Further research, based on innovative hypotheses, is clearly needed. The relationships between the applications of PTA in light and electron microscopy also require elucidation.

8.3. METHODS FOR RETICULIN

Reticulin is stained by virtue of its content of hexose sugars. The simplest method, therefore, is the periodic acid–Schiff (PAS) procedure, which will be described, along with other techniques of carbohydrate histochemistry, in Chapter 11. There are, however, older methods for reticulin in which the fibres and basement membranes are rendered visible by deposition upon them of finely divided metallic silver. These older methods show reticulin in more striking contrast to their background than does PAS, so they will be discussed briefly alongside other methods for connective tissue.

The histochemical basis of the methods for reticulin has been shown by Lhotka (1956) and by Velican & Velican (1970) to be essentially similar to that of the PAS procedure. Adjacent hydroxyl groups of the hexose sugars of the glycoprotein are oxidized to aldehydes by potassium permanganate or periodic acid. The oxidation is sometimes followed by an empirically discovered "sensitizing" treatment with a ferric or uranyl salt. The chemical rationale of the "sensitization" is unknown, but the result is enhancement of contrast in the final preparation. The aldehydes then reduce the silver diammine ion, $[Ag(NH_3)_2]^+$, to the metal. Silver diammine is produced by adding ammonium hydroxide to an aqueous solution of silver nitrate. A precipitate of silver oxide is formed, but dissolves as the addition is continued:

$$2NH_4OH + 2AgNO_3 \longrightarrow 2NH_4NO_3 + Ag_2O\downarrow + H_2O$$

$$Ag_2O + 4NH_4OH \longrightarrow 2[Ag(NH_3)_2]^+ + 2OH^- + 3H_2O$$

The sequence of reactions involved in the silver staining of reticulin is as follows:

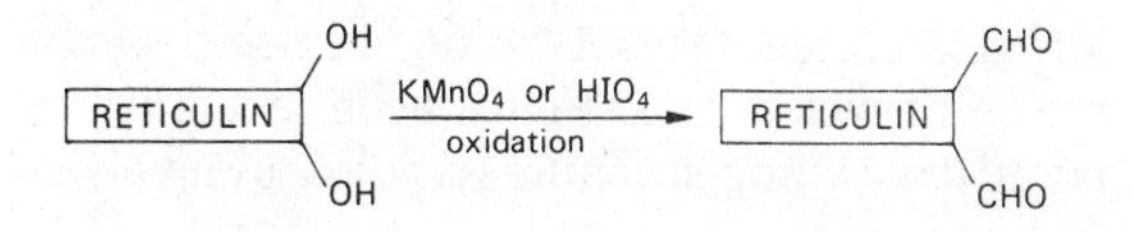

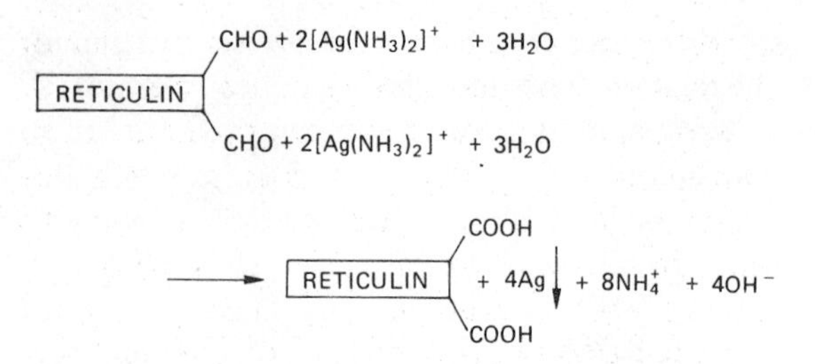

It is to be expected that four atoms of silver will be deposited at the site of each glucosyl, galactosyl, or other reactive sugar residue. This is not enough to provide adequate visibility, but fortunately it is possible to precipitate more silver upon that initially deposited by a process similar to photographic development. This occurs when the incompletely washed sections are transferred to an aqueous formaldehyde solution. Residual silver diammine ions are reduced to silver by the formaldehyde. Metallic silver catalyses the reaction, so precipitation of the metal due to the action of formaldehyde occurs mainly at the original sites of the sugar molecules of the reticulin. Yet further contrast may be obtained by the process known as gold toning. The chemistry of development and toning will be described in connection with neurological staining (Chapter 18) in which the two processes are used extensively.

In some techniques, the aldehyde-detecting reagent is not silver diammine but a complex of silver with hexamethylene tetramine. This substance, $(CH_2)_6N_4$, is a solid cyclic condensation product of ammonia and formaldehyde. It forms with silver a complex cation similar to the one formed with ammonia. Hexamethylene tetramine (also called hexamine or methenamine) can be weighed accurately, whereas an aqueous solution of ammonia loses potency with time after a new bottle has been opened.

8.4. METHODS FOR ELASTIN

The available methods for staining elastin are many and varied and there is no completely satisfactory explanation of their modes of action. Dyes are certainly not attracted to elastin by electrostatic ionic forces (see Baker, 1958, and Pearse, 1968, for review of evidence), and it has been generally supposed that hydrogen bonding is primarily involved. However, Horobin and James (1970) have shown that elastin can be coloured by dyes incapable of forming hydrogen bonds. These investigators found that the only feature held in common by all of a large number of dyes which stained elastin was the presence of at least five aromatic rings in the molecule. Chemical blocking methods for many reactive groups did not appreciably reduce the staining of elastin by such dyes, and the prevention of hydrogen bonding caused only slight inhibition. Horobin and James proposed that large dye molecules were bound to elastic tissue in the same way that they are thought to be bound to some textile fibres, namely by van der Waals forces. The nature of this type of intermolecular attraction has been explained in Chapter 5. It should be remembered that direct dyes are anionic and also colour all acidophilic components of a tissue, though not always as intensely as elastin, when applied from aqueous solutions. Ordinary anionic dyes, such as eosin, stain elastin lightly. Vidal (1978) has provided evidence for the binding of a fluorochrome, sodium 8-anilino-l-naphthalene sulphonate, to elastin by hydrophobic interaction.

One of the simplest, though not the most specific, of techniques is Gomori's aldehyde–fuchsine method, which is presented below. Aldehyde–fuchsine is prepared by treating pararosaniline with acetaldehyde (formed by depolymerization of paraldehyde). Bangle (1954) considered it likely that Schiff's bases (= imines, anils, or azomethines: substances containing the grouping —N=C⟨) were formed with one or two of the free amino groups of pararosaniline:

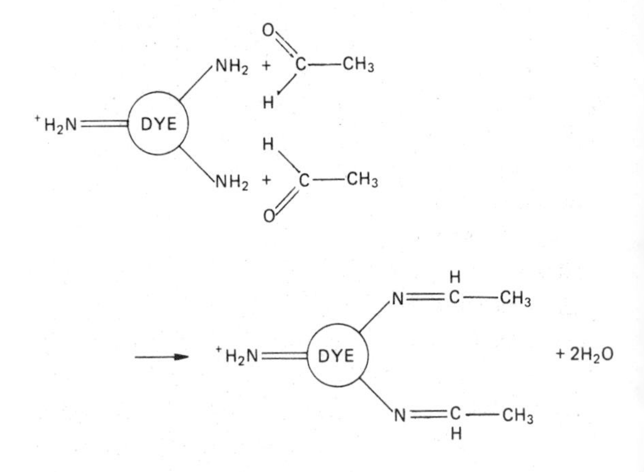

If all three nitrogen atoms of pararosaniline react with acetaldehyde, the characteristic staining properties are lost. It has been suggested by Buehner *et al.* (1979) that the active ingredient of aldehyde–fuchsine solutions is a tautomer of Bangle's proposed product, with the structure:

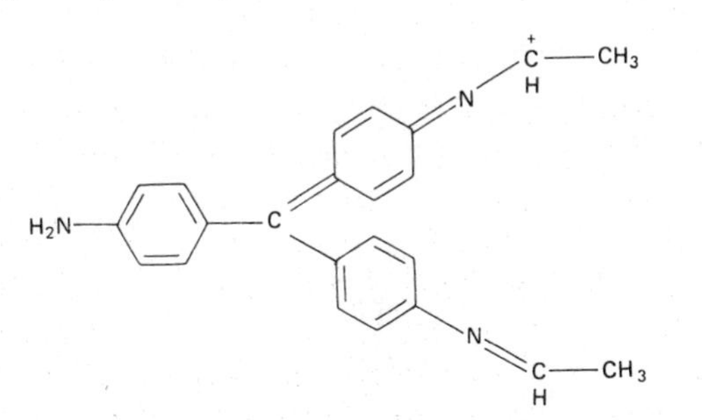

On theoretical grounds, this carbonium ion could be expected to be stable and to be capable of reacting with and forming covalent bonds. Horobin and James (1970) suggest that heterocyclic compounds of higher molecular weight may also be formed and be responsible for staining elastin by the same mechanism as that proposed for direct cotton dyes. Goldstein (1962) observed that aldehyde–fuchsine failed to stain elastic fibres in the presence of a high concentration of urea, indicating that this dye might be bound by hydrogen bonding.

Aldehyde–fuchsine has several other uses in histology and in at least some of these, it probably combines with carboxyl groups and half-sulphate esters (Sumner, 1965; Gabe, 1976), acting therefore in a manner quite different from that involved in the staining of connective tissue. Aldehyde–fuchsine also stains the sites of aldehyde and sulphonic acid groups, when these are produced in tissues by appropriate chemical treatments. Structures stained by aldehyde–fuchsine are unusual in that they are not decolorized by prolonged treatment with acidified alcohol. This property may indicate that stable covalent bonds are formed between the dye and the tissue. Staining methods employing aldehyde fuchsine have been reviewed at length by Mowry (1978).

Elastic fibres and laminae are easily recognized by their morphology once they are rendered visible. The elastic laminae of arteries, especially in young rodents, give a positive reaction with Schiff's reagent, owing to the presence of free aldehyde groups (see p. 93) and elastic laminae are usually strongly autofluorescent when excited by blue or ultraviolet light in the fluorescence microscope.

Oxytalan fibres, found in the periodontal membranes of some animals, are stained by aldehyde–fuchsine only after prior treatment of the sections with strong oxidizing agents. They cannot be demonstrated by conventional staining methods for elastin, collagen, or reticulin (see Lillie & Fullmer, 1976, for more information).

8.5. INDIVIDUAL METHODS

8.5.1. Iron–haematoxylin and van Gieson

Solutions required

1. Weigert's iron–haematoxylin

Solution A. Haematoxylin (C.I. 75290): 5 g
95% ethanol: 500 ml

Keeps for a few years

Solution B. Ferric chloride ($FeCl_3.6H_2O$): 5.8 g
Water: 495 ml
Concentrated hydrochloric acid: 5 ml

Keeps indefinitely

Working solution: Mix equal volumes of A and B. Put A in the staining jar or tank first for more rapid mixing. The mixture should be made just before using, but can be kept for about 2 weeks at 4°C.

2. van Gieson's solution

Acid fuchsine (C.I. 42685): 0.5 g
Saturated aqueous picric acid: 500 ml
(Keeps indefinitely)

See *Note 2* below.

Procedure

1. De-wax and hydrate paraffin sections.
2. Stain in working solution of Weigert's haematoxylin for 5 min (10 min if the solution is more than a few days old).
3. Wash in running tap water.
4. Stain in van Gieson's solution, 2–5 min. The time is not critical.

5. Wash briefly in running tap water. This also differentiates the acid fuchsine.
6. Dehydrate rapidly three changes of 100% ethanol. This step also differentiates the picric acid.
7. Clear in xylene and mount in a resinous medium.

Result

Nuclei—black or brown; collagen—red; cytoplasm (especially smooth and striated muscle), keratin and erythrocytes—yellow.

Notes

1. An alternative nuclear stain is iron–alum–chromoxane cyanine R (see p. 82). If this is used, the nuclei will be blue or greenish blue. Stage 2 may also be replaced by a 5-min staining in toluidine blue O (p. 84), followed by washing in water and insolubilization of the dye by immersion of the slides for 5 min in 5% aqueous ammonium molybdate. The toluidine blue O colours nuclei and cytoplasmic RNA blue, cartilage, mast cell granules, and other metachromatic materials red–purple. A green colour is formed in the cytoplasms of many epithelial and secretory cells, presumably as a consequence of staining by both the blue and the yellow dyes.
2. Owing to the variable dye contents of samples of acid fuchsine, a van Gieson mixture may be found to perform unsatisfactorily. If stained sections are too red (e.g. orange muscle fibres), pour out about one-quarter of the mixture and top up the remaining three-quarters to the original volume with saturated aqueous picric acid. If red is too weak (inadequate staining of collagen), add another 0.1 g of acid fuchsine. Once the mixture has been adjusted to give correct staining, it will retain its properties for several years.

8.5.2. Cason's trichrome stain

This method (Cason, 1950) is technically very simple, but does not always work as well as the longer trichrome methods (Mallory, Masson, Heidenhain's Azan—see Luna, 1968; Clark, 1973; Culling, 1974; Gabe, 1976) in which differentiations can be more precisely controlled. An iron–haematoxylin nuclear stain, not included in the original procedure, has been inserted in the technique given below.

Solutions required

A. Weigert's iron–haematoxylin (working solution)
See above (p. 99).

B. Cason's trichrome solution

Water:	200 ml	Keeps for 1–2 years
Dissolve in order stated:		
Phosphotungstic acid:	1 g	
Orange G (C.I. 16230):	2 g	
Aniline blue WS (C.I. 42755):	1 g	
Acid fuchsine (C.I. 42685):	3 g	

Procedure

1. De-wax and hydrate sections.
2. Stain in Weigert's haematoxylin, 5 min.
3. Wash in running tap water, 2 min.
4. Immerse in Cason's trichrome solution, 5 min.
5. Wash in running water, 3–5 s.
6. Blot slides dry with filter paper.
7. Dehydrate rapidly in three changes of 100% ethanol.
8. Clear in xylene and cover, using a resinous medium.

Results

Collagen—blue; cytoplasm, muscle—red; keratin, erythrocytes—orange; nuclei—brown (but sometimes blue). The pre-staining with iron–haematoxylin makes the trichrome colours a little unpredictable with some material. If the nuclear stain with iron–haematoxylin is omitted, most nuclei are coloured red; others may be blue or unstained.

8.5.3. Method for reticulin

This method (Gordon & Sweets, 1936) is just one of many available silver methods for reticulin. The oxidizing agent is $KMnO_4$ rather than HIO_4, probably because the latter was not used in histochemistry when methods of this type were invented. The function of the iron alum is obscure, but it has been shown to improve the histological picture (Velican & Velican, 1972). Any fixative may be used. The sections must be firmly attached to the slides with coagulated albumen or chrome–gelatine if they are to remain in place after treatment with alkaline reagents.

Nearly all fixatives are suitable. After mixtures containing potassium dichromate or chromic acid, it is sometimes necessary to omit the oxidation with potassium permanganate.

Solutions required

A. Acid permanganate

Potassium permanganate ($KMnO_4$):	1.0 g
Water:	95 ml
3% aqueous H_2SO_4:	5 ml

Prepare just before use from a stock 6% aqueous $KMnO_4$. The addition of H_2SO_4 is not necessary

B. 1% oxalic acid

Oxalic acid (HOOC.COOH.$2H_2O$):	5 g
Water:	to 500 ml

Keeps indefinitely

C. Iron alum

Iron alum ($NH_4Fe(SO_4)_2.12H_2O$):	10 g
Water:	to 500 ml

Prepare on the day it is to be used. Alternatively, use 4% aqueous ferric chloride ($FeCl_3.6H_2O$), which keeps indefinitely

D. Ammoniacal silver solution

Stock solutions: 1. 10% aqueous silver nitrate.
2. Ammonium hydroxide (28% NH_3).
3. 4% aqueous sodium hydroxide.

Working solution: Add ammonium hydroxide drop by drop to 10 ml of 10% $AgNO_3$ until the brown precipitate of Ag_2O is almost (not quite) re-dissolved. Add 7.5 ml of 4% NaOH, followed by a few more drops of ammonium hydroxide, until the newly formed precipitate is just dissolved. Be careful not to add too much ammonia. Swirl the solution for a few seconds after adding each drop, since dissolution of the precipitate is not quite instantaneous. Make up to 100 ml with water. **This solution should be made just before use and discarded afterwards by washing it down the sink with plenty of water.** Ammoniacal silver solutions decompose on evaporation to form explosive "fulminating silver", a mixture of silver amide and silver nitride.

E. Reducer

Neutralized formalin (40% HCHO which has stood over marble):	10 ml	Prepare just before using
Water:	90 ml	

F. Yellow gold chloride

Sodium tetrachloroaurate ($NaAuCl_4.2H_2O$):	1 g	Keeps for several months
Water:	500 ml	

This solution may be re-used repeatedly

G. Sodium thiosulphate

Sodium thiosulphate ($Na_2S_2O_3.5H_2O$):	25 g	Keeps for several months
Water:	to 500 ml	

Procedure

1. De-wax and hydrate paraffin sections.
2. Oxidize for 1 min in acid permanganate (solution A).
3. Wash in water.
4. Immerse in 1% oxalic acid (solution B) until the sections are white. Usually about 30 s.
5. Wash in water (three changes).
6. Treat with iron alum (solution C), 10 min.
7. Wash in water (three changes).
8. Immerse slides in the ammoniacal silver solution (solution D) 5–10 s.
9. Rinse in water (once only). See *Note 1* below.
10. Place in formaldehyde reducer (solution E), 30 s.
11. Wash in water (three changes). See *Note 2* below.
12. Tone in 0.2% yellow gold chloride (solution F), 2 m.
13. Wash in water (two changes).
14. Immerse in sodium thiosulphate (solution G), 3 min.
15. Wash in water (three changes).
16. Dehydrate through graded alcohols, clear in xylene and cover.

Result

Reticulin—black. Other elements in shades of grey.

Notes

1. This rinse is a critical step in the method. If it is excessive, staining of reticulin will be inadequate. If it is insufficient there will be non-specific precipitation of silver on the sections. If working with slides in a coplin jar, pour out the ammoniacal silver solution, fill up with water, agitate for about 5 s, pour out the water, and immediately pour in the reducer.
2. The sections should be examined after stage 11. If staining is excessive, the method should be repeated from stage 6. This will produce an effect equivalent to differentiation. Probably some of the finely divided metallic silver in the specimen is oxidized to silver ions, which are then precipitated as silver chloride:

$$Ag + Fe^{3+} \rightarrow Ag^{+} + Fe^{2+}$$
$$Ag^{+} + Cl^{-} \rightarrow AgCl \downarrow$$

This salt would dissolve in the slight excess of ammonia present in the ammoniacal silver solution:

$$\underset{(solid)}{AgCl} + 2NH_3 \rightarrow [Ag(NH_3)_2]^{+}_{+}$$

Any undissolved silver chloride would be removed at stage 14 of the method:

$$\underset{(solid)}{AgCl} + 2S_2O_3^{2-} \rightarrow [Ag(S_2O_3)_2]^{3-} + Cl^{-}$$

8.5.4. Aldehyde–fuchsine for elastic tissue

Preparation of stain

There are many ways of preparing aldehyde–fuchsine. The following (from Gabe, 1976) has the advantage of yielding a solid product which is stable for at least 2 years. It is important to use pararosaniline (C.I. 42500) or a sample of basic fuchsine consisting only of this dye (Mowry & Emmel, 1977).

Dissolve 1.0 g of pararosaniline in 200 ml of water (heat to boiling, then allow to cool to room temperature). Add 2.0 ml of concentrated hydrochloric acid and 1.0 ml of paraldehyde. Leave for 24 h, or longer if a pink ring appears when the deep purple mixture is spotted onto filter paper. Filter. Discard the filtrate. Wash the residue with 50 ml of water, then dry the filter paper and its contents in an oven at 60°C. Collect and keep the aldehyde fuchsine powder.

Working solution:

Aldehyde–fuchsine powder:	0.25 g
70% ethanol:	200 ml
Acetic acid (glacial):	2.0 ml

Stable for at least 2 years

Leave overnight to dissolve. Filtration is not needed.

Procedure

1. De-wax and hydrate paraffin sections.
2. Stain in aldehyde–fuchsine for 5 min.
3. Rinse in running tap water to remove most of excess dye.
4. Rinse in 95% ethanol containing 0.5% (v/v) concentrated hydrochloric acid until any remaining excess of dye is removed (usually 20 s. Longer times do no harm.)
5. Apply counterstain if desired (alum–haematoxylin and fast green FCF is a suitable combination).
6. Wash, dehydrate in graded alcohols, clear, and cover.

Result

Elastic fibres and laminae—red to purple.

Note

Aldehyde–fuchsine is also used to stain the endocrine pancreas, the adenohypophysis, mast cells, and neurosecretory material (see Chapter 18). The different techniques vary somewhat and are described by Luna (1968), Pearse (1968), and Mowry (1978).

8.6. EXERCISES

Theoretical

1. What are the chemical differences between collagen and elastin? Do these differences account for the stainability of the two substances by different techniques?

2. What is reticulin? Suggest a reason why the silver methods for reticulin stain this substance more intensely than collagen.

3. Discuss the role of phosphotungstic or phosphomolybdic acid in histological techniques for the differential staining of cytoplasm and collagen.

4. Devise a staining procedure whereby nuclei, muscle, collagen, and elastin would all be differently coloured.

Practical

5. De-wax and hydrate some paraffin sections of any convenient tissue. Immerse in 2% aqueous phosphomolybdic acid for 15 min. Wash in water and then immerse in a freshly prepared 0.5% aqueous solution of stannous chloride ($SnCl_2.2H_2O$) for 5 min. Wash, dehydrate, and mount in a resinous medium. What structures in the tissue are stained blue? What chemical reactions are responsible for the production of the colour?

6. Stain sections of various tissues (skin, kidney, and intestine are suitable) by the van Gieson method and by Cason's trichrome method. Note the influence of the fixative on the brightness of the colours seen in the stained preparations.

7. Stain sections of any convenient tissue for elastin using the aldehyde–fuchsine method. Positive staining will be observed in the elastic laminae of arteries. Elastic fibres occur in connective tissue: a few may be seen in sections of skin or mesentery. If tissues from a young rat are available, demonstrate incompletely formed desmosine and isodesmosine linkages, by a suitable histochemical method from Chapter 10, in arterial elastic laminae.

8. Stain a section of external ear with aldehyde–fuchsine. What, other than elastin, is stained? Use other staining methods (Chapter 6 and this chapter) to show that aldehyde–fuchsine positive material other than that in elastic laminae of arteries and occasional fibres in the dermis is not composed of elastin.

9. Stain some sections of liver or kidney with (a) van Gieson's method, (b) silver method for reticulin, (c) the PAS method (see Chapter 11). Compare the distribution of reticulin with that of collagen. Other suitable tissues for this exercise are lymph node, spleen, and decalcified bone.

9

Methods for Nucleic Acids

9.1. Chemistry and distribution of nucleic acids 104
9.2. Demonstration of nucleic acids with cationic dyes 105
9.3. The Feulgen method 106
9.4. Enzymatic extraction of nucleic acids 107
9.4.1. Enzymes as reagents in histochemistry 107
9.4.2. Ribonuclease and deoxyribonuclease 107
9.5. Chemical extraction of nucleic acids 108
9.6. Individual methods 108
9.6.1. Single cationic dyes 108
9.6.2. Methyl green–pyronine 109
9.6.3. The Feulgen method 110
9.6.4. Enzymatic extraction of nucleic acids 112
9.7. Exercises 112

9.1. CHEMISTRY AND DISTRIBUTION OF NUCLEIC ACIDS

The nucleic acids are macromolecular compounds with the general structure:

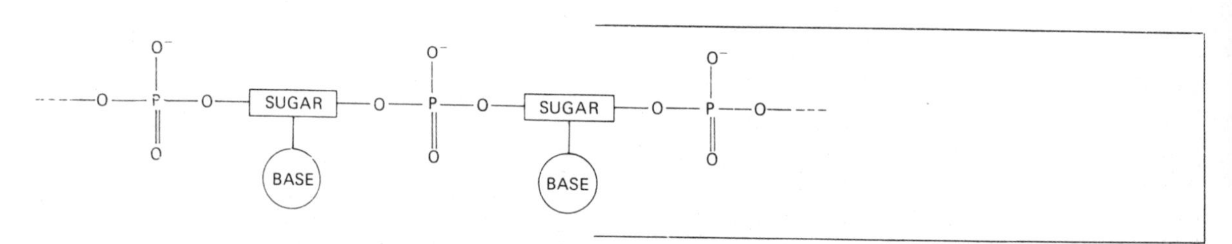

The phosphate and sugar moieties are shown in detail in Fig. 9.1. The bases are purines and pyrimidines. They are of little importance in the histochemical study of the nucleic acids by light microscopy, though they may be responsible for the binding of certain heavy metal ions and complexes used as electron-dense "stains" at the ultrastructural level. Their structural formulae can be found in textbooks of biochemistry. In deoxyribonucleic acid (DNA), the sugar is α-D-2-deoxyribofuranose. The molecules of DNA in chromosomes form paired strands, united to one another by hydrogen bonding between the bases and maintained in the double helical configuration discovered by Watson and Crick (1953). In eukaryotic cells the negative charges on the ionized free hydroxyl groups of the phosphoric acid residues are neutralized by the basic side-chains of the accompanying molecules of the **nucleoproteins**. The latter include strongly basic proteins of the type known as **histones**, which are rich in lysine and arginine and have molecular weights in the range 12,000–20,000. The nucleoproteins also include species with very large molecules, of molecular weight 10^7–10^9. Ribonucleic acid (RNA) has a more open structure than that of DNA and the sugar is α-D-ribofuranose. Most of the DNA in a eukaryotic cell is located in the chromosomes of the nucleus. Most of the RNA is in the ribosomes of the endoplasmic reticulum and in the nucleolus. The ribosomal and nucleolar RNA, like the chromosomal DNA, is associated with nucleoprotein. The nucleic acids of prokaryotic cells (organisms without nuclei in their protoplasm) have their phosphoric acid groups balanced by inorganic ions (Na^+, Mg^{2+}) and by certain amines.

Histochemical procedures demonstrate DNA only in the nucleus and RNA only in the nucleolus and endoplasmic reticulum, although it is known that the nucleic acids are by no means confined to these situations. The nucleic acids of prokaryotes, such as bacteria, can also be stained, but their localizations cannot be accurately resolved by light microscopy.

Both the nucleic acids can be stained with cationic dyes, and DNA can be demonstrated selectively by the Feulgen reaction. No technique is available for the staining only of RNA. The specificity of a histochemical method for DNA or RNA can be confirmed by specific removal of either of these substances by enzymatic hydrolysis or chemical extraction.

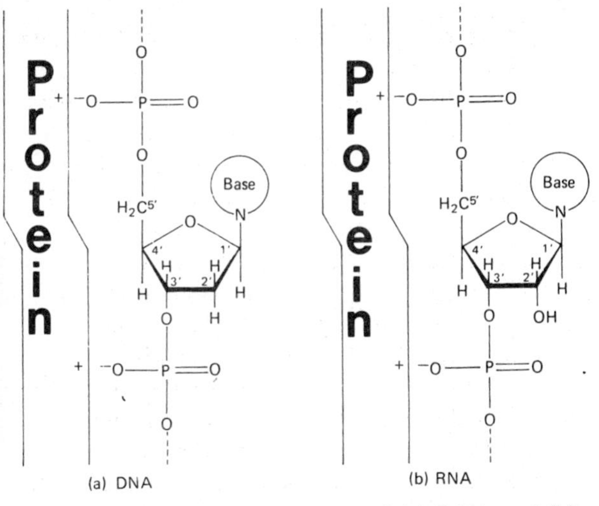

Fig. 9.1. Structures of fragments of (a) DNA and (b) RNA showing linkages of deoxyribose and ribose with phosphoric acid and with purine or pyrimidine bases. The negative charges on the phosphoric acid groups are neutralized by nucleoprotein cations. The numbers 1′ to 5′ identify the carbon atoms of the pentose sugars. Thickened lines indicate that the furanose rings lie perpendicular to the plane of the paper, with carbons 2′ and 3′ nearest to the reader (see also Chapter 11 for structural formulae of sugars).

9.2. DEMONSTRATION OF NUCLEIC ACIDS WITH CATIONIC DYES

The chemistry of the staining of nucleic acids by cationic dyes has already been discussed in Chapter 6. The process may be considered to be histochemical when conditions are carefully standardized and suitable control procedures are carried out. Any basic dye may be used for the purpose, but the specificity of staining of nucleic acids in any structure must be confirmed by chemical or enzymatic extraction, since nucleic acids are not the only substances capable of binding coloured cations. Controls are particularly important in the case of cytoplasmic RNA, which cannot easily be distinguished from the basophilic secretory products of some cells. The dye–metal complex formed from gallocyanine and chromic ions has the properties of a cationic dye and is also used to stain DNA and RNA. The chrome alum–gallocyanine method of Einarson (1951) is used in quantitative micro-photometric studies of the nucleic acid contents of cells because the bound dye–metal complex is more stable towards light, water, and organic solvents than are ordinary cationic dyes.

In the **methyl green–pyronine** procedure, two basic dyes are applied simultaneously under carefully controlled conditions of concentration and pH. This results in differential colouring of DNA (bluish green) and RNA (pink to red) in the same section. The mechanism by which this effect is achieved is not fully understood. An early suggestion was that methyl green, with its two tertiary amine groups (see p. 64), might bind selectively to suitably spaced phosphate ions on adjacent turns of the DNA helix. This was disproved by Baker & Williams (1965) who showed that methyl green used alone was not a selective stain for DNA and that it could be replaced in the methyl green–pyronine method by malachite green (C.I. 42000; Basic green 4), whose molecules each bear only a single positive charge. They speculated that nucleoli and ribosomes were more densely textured than chromosomes. Thus the former would admit pyronine Y, which has a compact coplanar system of three fused rings (see p. 66). The methyl green molecule, on the other hand, is more cumbersome: at any instant in time, two of its three rings are free to rotate about single bonds. This bulkiness, it was suggested, would make the methyl green molecules unsuccessful in the competition with pyronine Y for binding sites in nucleoli and ribosomes.

A rather similar explanation was proposed by Scott (1967), but with reversal of the penetrating powers of the two dyes. Scott suggested that the rigid helical configuration of DNA rendered its phosphoric acid groups more accessible to the large, non-planar molecules of methyl green, while RNA, with its more open conformation, was more readily penetrated by the smaller, planar molecule of pyronine Y. Scott's conclusions were derived from a study of the interactions of dyes with nucleic acids in solution. Although they appear to be contradictory, the two hypotheses may yet prove to be compatible, though further investigation is needed. The "textures" envisaged by Baker & Williams (1965) would involve the spaces amongst the molecules of nucleic acids, nucleoproteins, and other substances in tissues modified by fixation. The interactions considered by Scott (1967), however, were more intimate, being between the dyes and the nucleic acids themselves. Under histochemical conditions, the binding of dyes to nucleic acids could occur only after penetration by the dyes of the structural matrices of the organelles containing the nucleic acids.

These explanations of the differential staining of DNA and RNA by mixtures of methyl green and pyronine Y are both based on the supposed competition between the two dyes for anionic sites of different accessibility. Anyone who has had to adjust the compositions of mixtures of these two dyes will agree that the concentrations have to be nicely balanced if the desired tinctorial effects are to be obtained. A competitive staining mechanism is strongly suggested. There is, however, no direct evidence to relate the sizes and shapes of the dye molecules to their selectivities for DNA and RNA. In terms of the hypotheses outlined above, it may be possible to account for the well-known observation that pyronine Y is greatly superior to pyronine B in this technique. The latter dye differs from the former only in the possession of ethyl rather than methyl groups attached to the nitrogen atoms. Pyronine Y therefore has a less bulky auxochromic group than pyronine B.

In critical histochemical work it is always necessary to use extracted or enzyme-treated control sections in order to check the specificity of staining with methyl green and pyronine. Basophilic materials other than nucleic acids are stained mainly by the pyronine, sometimes metachromatically (orange rather than red or pink).

9.3. THE FEULGEN METHOD

When sections of tissue are treated with 1.0 N hydrochloric acid at 60°C for 5–20 min (according to the fixative used, see below), most of the RNA is broken down to soluble substances and lost from the section, but DNA is only partially hydrolysed. The purine and pyrimidine bases of the DNA are removed from the deoxyribose residues, which remain in their original positions and are capable of reacting as aldehydes (Overend & Stacey, 1949). The Feulgen hydrolysis does not break the ester linkages between the phosphoric acid and sugar units of the DNA.

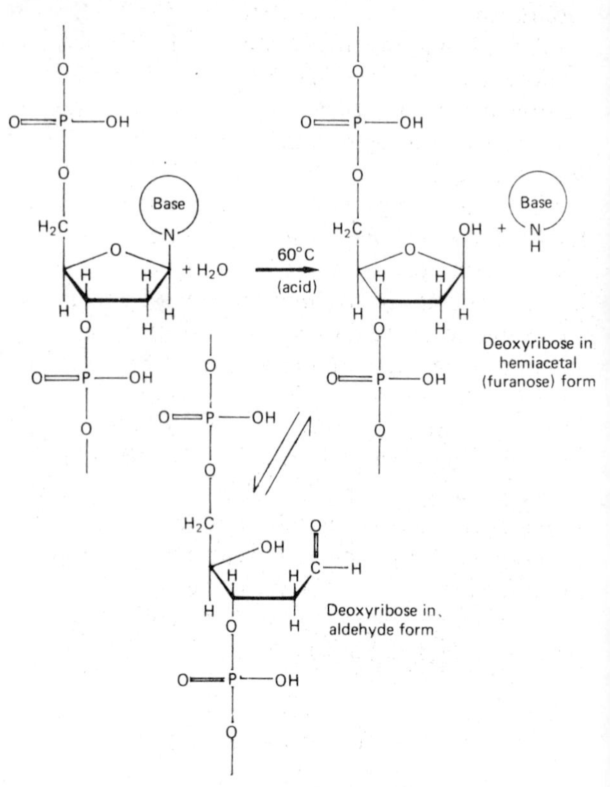

The exact reasons for these changes and for the fact that RNA does not behave similarly are imperfectly understood. Several theories are discussed at some length by Pearse (1968). The ribosyl residues of any RNA that is not removed during the hydrolysis do not react as aldehydes. This property is probably due to the presence of a hydroxyl group at position 2′ of ribose.

The aldehyde groups liberated by the Feulgen hydrolysis are stained with Schiff's reagent, and the chemistry of this process will be considered in Chapter 10. Other reagents, notably silver diammine and naphthoic acid hydrazide, can also be used to detect aldehydes formed from partially hydrolysed DNA.

The Feulgen reaction is, in practice, a highly specific method for DNA, though its specificity may be checked by incubating control sections with deoxyribonuclease. Any free aldehydes initially present in the tissue (i.e. introduced by fixative or present in immature elastic tissue and some lipids) will give false positive staining, but this will be obtained when the acid hydrolysis is omitted (except, rarely, in the case of some lipids), so is unlikely to cause confusion.

Pre-existing aldehydes are easily destroyed before staining by treating the sections with sodium borohydride. This manoeuvre (see Chapter 10) is always necessary with tissue fixed in glutaraldehyde (Kasten & Lala, 1975). It should be noted that Bouin's fluid and acid-decalcifying agents can attack DNA and may hydrolyse it excessively with consequent loss of stainability by the Feulgen method. For the histochemistry of nucleic acids it is desirable to avoid the use of fixatives containing picric acid, and chelating agents are to be preferred to acids when decalcification is necessary.

The intensity of staining by the Feulgen method is proportional to the concentration of DNA. Thus it is possible to make micro-photometric and morphometric observations and determine the quantities of DNA in the nuclei of cells. Important studies of the cell cycle have been carried out in this way. The quantitative methodology is described and discussed in detail by Swift (1950).

9.4. ENZYMATIC EXTRACTION OF NUCLEIC ACIDS

9.4.1. Enzymes as reagents in histochemistry

Sections of tissue can be treated with the pancreatic enzymes ribonuclease (RNase) and deoxyribonuclease (DNase) in order to remove, selectively, the two nucleic acids. Other enzymes are also used in histochemistry to remove (or prevent the staining of) such substances as collagen, glycogen, and various carbohydrate constituents of mucosubstances. Some caution is required in interpreting the results of enzymatic treatments, and the following conditions must always be satisfied in order to draw valid conclusions:

1. The substrate must be present in a suitable state for being attacked by the enzyme. Fixatives containing picric acid, mercury, or chromium should be avoided since these substances are inhibitory to many enzymes. Fixation in formaldehyde prevents the subsequent digestion of collagen and reticulin by collagenase, but for most other enzymes formaldehyde (aqueous or alcoholic) and mixtures containing alcohol and acetic acid as the main active ingredients are the least objectionable fixatives.
2. The preparation of the enzyme used must be free from contamination by other enzymes which might produce confusing effects. The presence of proteolytic enzymes in a specimen of RNase, for example, would be undesirable. Similarly, DNase must be free of RNase.
3. The conditions of incubation must be adequate for complete action of the enzyme upon all of the substrate present in the section. The use of an identically fixed and processed control section known to contain the substrate is recommended.
4. Non-specific extraction of the substrate should not occur and should be sought in control sections incubated only in the buffer or other solvent used for the enzyme. Buffer solutions alone sometimes extract appreciable quantities of RNA and glycogen from sections. In some circumstances it is preferable to dissolve the enzyme in water, even if this does not provide the optimum pH for the enzyme.
5. Any necessary co-factors or co-enzymes must be provided. Neuraminidase, for example, will function only in the presence of calcium ions, and DNase requires magnesium.

9.4.2. Ribonuclease and deoxyribonuclease

Enzymes from various animal and plant sources catalyse the decomposition of nucleic acids. In histochemical practice, enzymes extracted from the bovine or porcine pancreas are usually used. The enzymes should be of the highest available purity, regardless of expense, for the reasons given in the preceding section. Cruder preparations are cheaper and will work, but when they are used there can be no certainty that the observed alterations in stainability of the tissue are due solely to removal of RNA or DNA. The reason for using an enzyme is that it acts with high specificity upon its substrate. With a mixture of enzymes, this specificity will be lost.

Ribonuclease (systematic name: polyribonucleotide 2-oligonucleotido-transferase (cyclizing); E.C. 2.7.7.16†) acts on the phosphate group attached to position 3′ of those ribose units of RNA that bear pyrimidine bases at position 1′. The phosphate-ester linkage to position 5′ of the adjacent sugar residue is transferred to position 2′ of the first ribose unit—the one bearing the pyrimidine base.

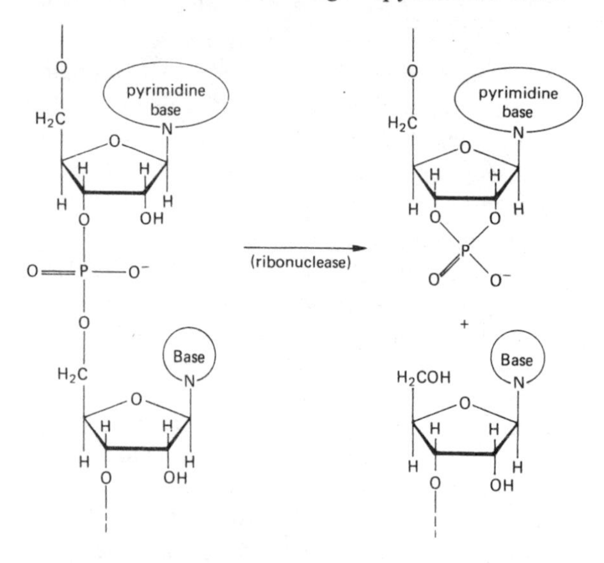

The long molecule of RNA is thus broken into small, soluble oligonucleotides. These products are lost from the tissue by diffusion into the RNase solution and the water used for subsequent washing.

Deoxyribonuclease (systematic name: deoxyribonucleate oligonucleotido-hydrolase; E.C. 3.1.4.5) of pancreatic origin catalyses the attack of water upon multiple phosphoric acid ester linkages to position 5′ of deoxyribose. The DNA is thus hydrolysed to yield soluble oligonucleotides. A related enzyme is deoxyribonuclease II (systematic name: deoxyribonucleate 3′-nucleotidohydrolase; E.C. 3.1.4.6), extracted from spleen and from various micro-organisms, which brings about hydrolysis at position 3′ of deoxyribose. The latter enzyme has been used less often in histochemistry than pancreatic DNase.

† Systematic names and E.C. numbers of enzymes are explained in Chapter 14.

The control procedures required in conjunction with the histochemical use of RNase and DNase are explained alongside the practical instructions given on p. 112.

9.5. CHEMICAL EXTRACTION OF NUCLEIC ACIDS

It is also possible to extract nucleic acids non-enzymatically from sections of tissue. Trichloroacetic acid (4% aqueous, 90°C, 15 min) removes both DNA and RNA. Perchloric acid (10% aqueous $HClO_4$, 4°C, overnight) removes RNA but not DNA, while a more drastic treatment with perchloric acid (5% aqueous $HClO_4$, 30 min, 60°C) extracts both nucleic acids. These chemical procedures are based on the catalysis by acids of the hydrolysis of the bonds between the sugars and the bases and between the sugars and phosphoric acid. The phosphate-ester bonds of RNA are more labile to hydrolysis than those of DNA. It is also possible to remove RNA by treatment with hot 1.0 N hydrochloric acid, as in the Feulgen hydrolysis. The extraction of RNA by perchloric acid also removes the purine and pyrimidine bases from DNA, thus engendering stainability of this nucleic acid by Schiff's reagent. The reagents used for chemical extraction are less expensive than RNase and DNase, but they are also less specific for the nucleic acids and their actions are more difficult to control.

9.6. INDIVIDUAL METHODS

9.6.1. Single cationic dyes

The use of cationic dyes is described in Chapter 6. Nucleic acids are easily stained by dilute solutions of thiazine dyes. Sulphated mucosubstances are also stained, often metachromatically, but proteins do not interfere at $pH < 4$.

It must be remembered that the aluminium– and iron–haematoxylins do not stain nuclei by virtue of the presence of DNA, and the term "basophilic" should not be applied to objects stained by these metal–haematein complexes.

With some conditions of staining, especially if an aqueous mounting medium is used, cationic dyes may stain nucleic acids, especially RNA, metachromatically, but this phenomenon is not sufficiently reproducible to form the basis of a histochemical method. Some fluorochromes, notably acridine orange (Bertalanffy & Bickis, 1956), also give different colours with the two nucleic acids.

9.6.2. Methyl green–pyronine

In the technique presented below, the staining mixture is based on that of Brachet (1953) as modified by Pearse (1968) but with allowance for variations among different batches of dyes. The stained slides are dehydrated with *n*-butanol as described by Kurnick (1955). Other methods of dehydration (e.g. in alcohol, acetone, or *t*-butanol) commonly used with this procedure are liable to extract too much of one or other of the dyes. For the same reason, the slides are not washed in water after staining.

Solutions required

Dyes must be of grades designated for nucleic acids or for the Unna–Pappenheim technique. Batches certified by the Biological Stains Commission should be used if possible.

A. Methyl green

Dissolve 2.0 g of methyl green (C.I. 42585) in 100 ml of water. Shake with 100 ml of chloroform in a separating funnel. Discard the chloroform (heavier) layer. Repeat the extractions until no more violet colour passes into the chloroform. Up to ten changes of chloroform may be needed. The extracted aqueous solution of methyl green is stable for several months.

B. Pyronine Y

Dissolve 8.7 g of pyronine Y (C.I. 45005) in 175 ml of water. **Do not** use pyronine B (C.I. 45010), which is unsuitable.

C. Buffer solution

(This is a 0.2 M acetate buffer, pH 4.8.)

Glacial acetic acid: 4.6 ml
Sodium acetate
(CH_3COONa): 9.85 g
Water: to 1000 ml

D. Methyl green–pyronine working solution

Because the dye contents of the two components of the stain are variable, it is necessary to determine the optimum mixture by experiment. As test objects use sections of central nervous tissue (DNA in all nuclei and RNA in Nissl substance of neuronal cytoplasm) or of lymph node (preferably a reactive one; DNA in all nuclei; RNA in cytoplasm of plasma cells). The test tissues should, ideally, be fixed and prepared in the same way as any other tissues that are to be investigated.

As a first attempt, mix:

Solution A: 5 ml
Solution B: 8 ml
Solution C: 13 ml

Stain a slide in a small coplin jar, by the **complete** procedure described below. Examine the mounted preparation. If staining is perfect, the nuclear DNA should be bluish green and the nucleoli and cytoplasmic RNA red or deep, bright pink. Collagen and cytoplasm of most cells, including muscle and the neuropil of the central nervous system, should be unstained. If the mixture is not right first time, modify it according to the following scheme. Notice that it is necessary to vary the **absolute** concentrations of the two dyes in the working solution. Attempts to correct unbalanced staining by adding more of one dye to an already mixed solution are nearly always unsuccessful.

1. If the pyronine Y predominates (red or purple nuclei; pink colour in areas that should be unstained), the concentration of this dye must be reduced. This is the commonest fault with the initial mixture.

Try: Solution A: 5 ml
Solution B: 5 ml
Solution C: 16 ml

2. If the methyl green predominates (blue–green nuclei, purple staining of cytoplasmic RNA), reduce the concentration of this dye.

Try: Solution A: 3 ml
Solution B: 8 ml
Solution C: 15 ml

3. If staining by methyl green is too weak (nuclei unstained, pale pink, or very pale greenish blue) despite reduction if necessary of the concentration

of pyronine Y, the concentration of methyl green must be increased.

Try: Solution A: 7 ml
Solution B: 8 ml (**or** revised amount from modification No. 1, above)
Solution C: 11 ml (**or** whatever amount gives final volume 26 ml)

4. If staining by pyronine Y is too weak (nucleoli and cytoplasmic RNA pale pink or pale greenish blue) despite reduction if necessary of the concentration of methyl green, the concentration of pyronine Y must be increased.

Try: Solution A: 5 ml (**or** revised amount from modification No. 2 or 3 above)
Solution B: 10 ml
Solution C: 11 ml (**or** whatever amount gives final volume 26 ml)

By proceeding in the manner described above, it will be possible to obtain a satisfactory staining solution in two or three attempts. A larger volume of this working solution may now be prepared. It may be used repeatedly and will retain its efficacy for at least one month. It will sometimes last for 6–12 months.

The stock solutions (A, B, and C) are all stable for about 2 years, but the buffer (solution C) should be renewed if fungi grow in it.

Procedure

1. De-wax and hydrate paraffin sections or use frozen sections which have been dried onto slides.
2. Stain in working solution of methyl green–pyronine for 30 min.
3. Transfer slides (without washing) to filter paper and blot dry.
4. Place slides (in a clean, dry staining rack) in *n*-butanol, two changes, each 5 min (see *Note 2*).
5. Clear in xylene, 5 min (see *Note 3*).
6. Cover, using a resinous mounting medium.

Result

DNA—bluish green; RNA—pink to red; acid mucosubstances (cartilage, mast cell granules, etc.)—orange (see *Note 4*).

Notes

1. It is possible to correct faulty staining by washing in water (removes methyl green) and by dehydrating in acetone (removes pyronine) or alcohol (removes both dyes), but greater control of the method is obtained if adjustments are systematically made to the composition of the working solution (D) containing the dyes, as described above, and not varying steps 2 to 4 of the procedure.
2. The *n*-butanol slowly differentiates the pyronine in addition to dehydrating the sections.
3. Clearing in xylene is optional. Sections which look as if they might come off the slides can be mounted directly from *n*-butanol, but the preparations will then take a few hours to become fully transparent.
4. The colours fade on prolonged exposure to daylight. The histochemical specificity of the method (especially for RNA) should always be confirmed by enzymatic treatment of control sections.
5. A methyl green–pyronine method for application to sections of plastic-embedded tissue has been described by Jurand & Goel (1976).

9.6.3. The Feulgen method

The optimum time of hydrolysis by 1.0 N HCl at 60°C varies with the fixative:

Carnoy:	8 min
Formaldehyde:	8 min
Helly:	5 min
Susa:	18 min

(Data summarized from Pearse, 1968.) See also *Note 4* below.

Preparation of Schiff's reagent

This is not the only way to make Schiff's reagent. For other methods, see Pearse (1968) and Lillie & Fullmer (1976).

It is important to use either basic fuchsine designated "special for DNA" or pure pararosaniline.

Boil 400 ml of water, add 1.0 g of basic fuchsine, cool to room temperature, and filter. Add 1.0 ml of thionyl chloride (**cautiously:** the reaction $SOCl_2 + H_2O \rightarrow SO_2 + 2HCl$ is very rapid. Do not pipette thionyl chloride by mouth. Use fume hood.) Allow to stand in a stoppered bottle overnight. Shake the solution with about 2 g of activated charcoal, which should remove all residual colour. A slight yellowish-brown tinge does not matter, but the solution must not be pink. Filter to remove charcoal and store in a tightly closed bottle at 4°C.

Schiff's reagent will keep for about 6 months, but should be discarded if a pink colour develops or if it no longer smells strongly of sulphur dioxide. See *Note 3* below for an alternative to Schiff's reagent.

Other solutions required

A. 1.0 N hydrochloric acid

B. Bisulphite water

Approximately 5 g of sodium or potassium metabisulphite (anhydrous) dissolved in 1 litre of water and 3–5 ml of concentrated hydrochloric acid added. This solution must be freshly made and put into three staining tanks.

C. A counterstain, if desired

Fast green FCF (0.5%, aqueous) is suitable.

Procedure

1. Pre-heat the 1.0 N hydrochloric acid to 60°C in a covered container (see *Note 1*) in the wax oven. Take Schiff's reagent out of refrigerator and allow it to warm to room temperature.
2. De-wax and hydrate paraffin sections.
3. Immerse in 1.0 N HCl at 60°C for the optimum time.
4. Rinse in water (two changes).
5. Immerse in Schiff's reagent, for 15–30 min (until nuclei are stained; the sections should be pale pink).
6. Transfer slides directly to bisulphite water; three changes, each 10–15 s with agitation (see *Note 2*).
7. Wash in running tap water for 2 min.
8. Counterstain (if desired). Wash briefly in water (which also differentiates fast green FCF).
9. Dehydrate, clear, and mount in a resinous medium.

Result

DNA (nuclear chromatin; chromosomes)—pink to purple.

Notes

1. Do not allow water vapour to contaminate the melted wax in the oven.
2. The purpose of the bisulphite rinse is to prevent the re-colorization of Schiff's reagent which occurs on dilution with water alone and can cause artifactual background staining. Demalsy & Callebaut (1967) recommend omission of this rinse and prescribe washing in **copious** running tap water instead.
3. Horobin & Kevill-Davies (1971a) have shown that an acidified alcoholic solution of basic fuchsine (not decolorized with SO_2) will stain aldehyde groups. This solution is made as follows:

Basic fuchsine:	1 g
Ethanol:	160 ml
Water:	40 ml
Concentrated hydrochloric acid:	2 ml

Keeps for a few weeks. Without the hydrochloric acid, it is indefinitely stable. The HCl can be added before use. Store in a tightly closed bottle

After staining for 20 min in this solution, the slides are washed in absolute or 95% alcohol to remove all excess dye, cleared, and covered. (See Chapter 10 for discussion of histochemical detection of aldehydes.)

4. Controls.

(a) Omit the acid hydrolysis (step 3). Nothing should be stained. A positive result may indicate preexisting aldehyde groups derived from the fixative (especially glutaraldehyde) or hydrolysis of DNA during fixation (picric acid does this) or decalcification. Elastin of immature animals normally has free aldehyde groups (see Chapter 8), but is unlikely to be mistaken for DNA. The pseudo-plasmal reaction of some lipids (see Chapter 12) can be a cause of direct reactivity with Schiff's reagent in frozen sections, especially after fixation and storage in formaldehyde solutions.

(b) The times for hydrolysis given above are only approximate. In critical work the optimum time must be determined experimentally for each batch of sections. The optimum hydrolysis is that which yields the most darkly stained nuclei. If an unexpected negative result is obtained, stain a hydrolysed section with a basic dye: the DNA may have been completely depolymerized and extracted.

(c) From the results of a chemical study of the acid-catalysed hydrolysis of DNA (Kjellstrand, 1977), it has been suggested that the highest yields of aldehydes might be attained by using concentrations of HCl considerably higher than 1.0 N, at or only slightly above room temperature.

5. If staining seems to be too weak, even with an optimum hydrolysis, the Schiff's reagent is probably unsatisfactory.

9.6.4. Enzymatic extraction of nucleic acids

Since DNase and RNase are expensive they can be used only in small quantities, applied as drops to individual sections. This is conveniently done in a petri dish containing damp gauze or filter paper to saturate the atmosphere inside with water vapour and prevent evaporation of the drop of enzyme solution. The dish containing the slide is placed in an incubator at 37°C. One section on a slide can be treated with enzyme while another is treated only with the solvent. Any other sections on the slide remain dry. In this way the effects of both the enzyme and its solvent can be examined on adjacent serial sections.

Enzyme solutions

The enzymes should be free of other interfering enzymes. Very small quantities may be weighed with a torsion balance. The solutions of the enzymes can be recovered after use (by sucking off the slide with a syringe and needle) and stored frozen at −20°C, but the potency declines with such storage.

Ribonuclease is used as a solution containing 0.2–0.5 mg of enzyme per ml of water. Incubate for 1 h at 37°C.

Deoxyribonuclease is used as a solution containing 0.05 mg of enzyme per ml in TRIS buffer (pH 7.5) containing 0.2 M $MgSO_4$. The usual 0.2 M TRIS buffer should be diluted five times with water. Incubate for 3–6 h at 37°C.

Sections of central nervous tissue are valuable for determining the adequacy of extraction of nucleic acids. The Nissl substance of neurons (stainable by cationic dyes) should be completely removed by an adequate RNase treatment. The nuclear chromatin (but not the nucleoli) of all cells should fail to stain with cationic dyes or by the Feulgen method after adequate treatment with DNase. The control sections of central nervous tissue should have been fixed and processed in parallel with other specimens being studied.

Chemical extraction of nucleic acids may be carried out as described in Section 9.5 of this chapter. Perchloric and trichloroacetic acids are highly caustic and must be handled carefully. They should be discarded by flushing down the sink with copious tap water.

9.7. EXERCISES

Theoretical

1. How would you proceed to carry out the Feulgen reaction on sections of a tissue which had been fixed in glutaraldehyde?

2. Using sections of Carnoy-fixed central nervous tissue as test objects, how would you determine (a) whether a sample of DNase is substantially free of RNase, (b) whether a sample of RNase is substantially free of DNase?

3. What types of dyes are suitable for staining the nucleoprotein of the nuclei of mammalian cells? What would be the effect of prior extraction of DNA upon the results of such staining?

4. Why do enzymes such as RNase and DNase work more rapidly and more completely on sections of tissue after fixation in alcohol–acetic acid mixtures than after fixation in aqueous formaldehyde or glutaraldehyde?

Practical

5. Stain paraffin sections of liver, intestine, salivary gland, kidney, sternum, and formaldehyde-fixed brain by the methyl green–pyronine method. Apply control procedures to identify cytoplasmic RNA in the presence of other basophilic materials.

6. Stain sections of the same tissues by the Feulgen method. Why do some nuclei become more intensely coloured than others when nearly all of them contain the same amount of DNA?

7. Bearing in mind the limitations of enzymes as histochemical reagents, attempt to prove that:

(a) The Nissl substance of the neuron is stained by pyronine Y on account of its content of RNA.
(b) Mast cell granules (skin and tongue are suitable tissues) must owe their stainability with pyronine Y to a substance other than RNA.
(c) The Feulgen reaction in the nuclei of cells of the adrenal cortex is due to the presence of DNA.
(d) The stainability of nuclei (in any tissue) by alum–haematoxylin is not dependent upon the presence of either type of nucleic acid.

10

Organic Functional Groups and Protein Histochemistry

10.1. General considerations and methodology 114
10.2. Alcohols 117
10.2.1. Histochemical detection 117
10.2.2. Blocking procedures 117
10.3. Carboxylic acids 118
10.3.1. Histochemical detection 118
10.3.2. Blocking procedures 118
10.4. Amino groups 119
10.4.1. Histochemical detection 119
10.4.2. Blocking reactions 119
10.5. Arginine 120
10.5.1. The Sakaguchi reaction 121
10.5.2. Blocking reactions 121
10.6. Tryptophan 122
10.6.1. The rosindole reaction 122
10.6.2. Blocking reactions 123
10.7. Tyrosine 123
10.7.1. The Millon reaction 123
10.7.2. Blocking reactions 124
10.8. Cysteine and cystine 124
10.8.1. Some properties of cysteine and cystine 124
10.8.2. Histochemical methods for cysteine 125
10.8.3. Cysteic acid methods for cystine 125
10.8.4. Blocking procedures 126
10.9. Methods for proteins in general 126
10.9.1. Demonstration of proteins with dyes 126
10.9.2. The coupled tetrazonium reaction and related methods 127
10.10. Aldehydes and ketones 128
10.10.1. Schiff's reagent 128
10.10.2. Hydrazines and hydrazides 129
10.10.3. Aromatic amines 130
10.10.4. Ammoniacal silver nitrate 130
10.10.5. Other methods for aldehydes 131
10.10.6. Blocking procedures 131
10.11. Individual methods 132
10.11.1. Sulphation-induced basophilia 132
10.11.2. Acid anhydride method for carboxyl groups 133
10.11.3. Hydroxynaphthaldehyde method for amino groups 134
10.11.4. Sakaguchi reaction for arginine 134
10.11.5. NQS method for arginine 135
10.11.6. Rosindole reaction for tryptophan 135
10.11.7. Millon reaction for tyrosine 136
10.11.8. Ferric ferricyanide reaction 136
10.11.9. Mercurochrome method for cysteine 137
10.11.10. Cysteic acid method for cystine 137
10.11.11. Brilliant indocyanine 6B for proteins 138
10.11.12. The coupled tetrazonium reaction 139
10.11.13. Some methods for aldehydes and ketones 140
10.11.14. Blocking and unblocking procedures 140
10.11.14.1. Acetylation 140
10.11.14.2. Benzoylation 140
10.11.14.3. Sulphation–acetylation 140
10.11.14.4. Methylation and desulphation 141
10.11.14.5. Saponification 141
10.11.14.6. Deamination with nitrous acid 141
10.11.14.7. Benzil blockade of arginine 142
10.11.14.8. Iodination of tyrosine 142
10.11.14.9. Reduction of cystine to cysteine 142
10.11.14.10. Sulphydryl blocking reagents 142
10.11.14.11. Blocking reactions for aldehydes and ketones 143
10.12.1. Exercises 143

This chapter is devoted to the histochemistry of the side-chains of some amino acids and to the reactions of a few functional groups artificially produced by the action of reagents upon tissues. The histochemical reactions of other organic compounds such as carbohydrates, lipids, nucleic acids, and amines are described in other chapters.

As elsewhere in this book, no attempt is made to describe more than a small selection of the methods available for the various organic functional groups. The subject is treated at length by Lillie & Fullmer (1976) and much information can also be found in the texts of Pearse (1968) and Gabe (1976).

10.1. GENERAL CONSIDERATIONS AND METHODOLOGY

Functional groups are detected histochemically by making use of their characteristic chemical reactions to produce coloured compounds. The techniques can be properly understood and intelligently used only when the underlying organic chemistry is constantly kept in mind. For more information the reader should consult a textbook of organic chemistry. Those of Noller (1965), Morrison & Boyd (1973), and March (1977) are recommended. Some histochemically valuable reactions are not included in general texts, but many of them are described by Feigl (1960) and Glazer (1976).

The methods discussed in this chapter are for functional groups which form part of the macromolecular structure of the tissue. Diffusion and extraction by solvents prior to staining cannot, therefore, give rise to false localizations. The reactions can all be carried out on frozen or paraffin sections of suitably fixed material. It is necessary, however, to take into account the chemical reactions of fixation. The non-additive coagulant fixatives do not cause chemical changes in tissues, so alcohol-based mixtures such as Carnoy's fluid are ideal for the histochemical study of proteins (except for a few secretory products which are not fixed by alcohol). Formaldehyde is also acceptable, since most of its reactions with the side-chains of amino acids are reversed when the fixative is washed out (see Chapter 2). The same is true of picric acid and, except when sulphydryl groups are to be demonstrated, of

TABLE 10.1. *Histochemically demonstrable functional groups*

Functional group	Structure	Occurrence
Hydroxyl	—OH	Proteins: serine, threonine, hydroxyproline, hydroxylysine All carbohydrates (see Chapter 11) Some lipids (see Chapter 12)
Phenol	Ar—OH	Proteins: tyrosine In plants: tannins Small molecules: serotonin, catecholamines (see Chapter 17)
Carboxyl	—COOH	Proteins: C-terminus of all peptide chains; side-chains of glutamic and aspartic acids Carbohydrates: most glycoproteins and proteoglycans (see Chapter 11) Lipids: free fatty acids (see Chapter 12)
Amino	$—NH_2$	Proteins: N-terminus of all peptide chains; side-chain of lysine Some carbohydrates (see Chapter 11) Some lipids (see Chapter 12) Small molecules (see Chapter 17)
Guanidino	$—NH—C(=NH)—NH_2$	Proteins: side-chains of arginine
Indolyl	(indole ring structure, H N)	Proteins: side-chains of tryptophan Small molecule: serotonin (see Chapter 17)

TABLE 10.1 *(cont.)*

Functional group	Structure	Occurrence
Sulphydryl (=thiol)	—SH	Proteins: side-chain of cysteine
Disulphide	—S—S—	Proteins: cystine bridges between peptide chains
Sulphate ester	$—OSO_3H$ $(—OSO_3^- + H^+)$	Carbohydrates: some proteoglycans (see Chapter 11) Lipids: sulphatides (see Chapter 12) Also introduced artificially in histochemical sulphation procedures
Sulphonic acid	$—SO_3H$ $(—SO_3^- + H^+)$	Does not occur naturally Produced or introduced artificially in several histochemical methods
Phosphate-esters	$—\overset{O}{\overset{\|}{P}}(OH)_2$; $—\underset{O}{\underset{\|}{P}}(OH)O—$	Nucleic acids (see Chapter 9) Some lipids (see Chapter 12) Some proteins, notably casein (in mammary gland and in milk), with —OH of serine esterified by H_3PO_4
Aldehyde	—CHO	Proteins: elastin of immature animals (see Chapter 8) Lipids, when oxidized by air (see Chapter 12) Produced artificially in many histochemical methods
Ketone	>C=O	Lipids: some steroids (see Chapter 12) Produced artificially in a few histochemical procedures
Olefinic double bone	—C=C—	Lipids (see Chapter 12)

mercuric chloride. Chromium trioxide, potassium dichromate, osmium tetroxide, and glutaraldehyde are best avoided, since they all produce imperfectly understood irreversible chemical changes in proteins.

Often a chromogenic reaction for a functional group is not as specific as we would like it to be. It is then necessary to examine control sections in which the group has been chemically altered (i.e. "blocked") so that it can no longer take part in the reaction that yields a visible deposit. Many of the blocking reactions are also of low specificity, however, so it is frequently necessary to use more than one of them in order to discover, by elimination, the sites at which a single functional group is present in the tissue. Sometimes it is possible to undo the effect of blockade of a functional group by judicious application of a second reagent. This strategy is valuable when the blockage of one functional group is reversible but that of another is not.

Despite the intrinsic interest and great instructional value of functional group histochemistry, the methods derived from this branch of the science are used only on rare occasions in routine histological and pathological practice and in the investigation of biological problems. This is unfortunate, since the techniques are often valuable for displaying structural features of tissues and have the added advantage of providing information about the chemical natures of the objects that are stained.

The major histochemically demonstrable organic functional groups present in animal tissues are listed in Table 10.1 (pp. 115–116).

10.2. ALCOHOLS

Aliphatic hydroxyl groups occur mainly in carbohydrates, but also in the widely distributed amino acids serine, a primary alcohol, and threonine, a secondary alcohol. Hydroxylysine and hydroxyproline, which occur principally in collagen, are also secondary alcohols. Some lipids are alcohols of high molecular weight, but in tissues they are usually esterified.

10.2.1. Histochemical detection

Hydroxyl groups in tissues can be detected by converting them to sulphate esters and then staining with a cationic dye at pH 1.0 or lower (see p. 78). This procedure demonstrates all hydroxyl groups. Those of proteins cannot be distinguished from those of carbohydrates, which are generally more abundant. The dye will also bind to preexisting sulphate-ester groups in the tissue, such as those of some proteoglycans and of sulphatide lipids. The native sulphate esters are recognized by staining control sections that have not been subjected to the sulphation treatment. Both natural and artificially produced sulphate esters are hydrolysed, with loss of the sulphate group, by exposure to hot acidified methanol (see Section 10.11.14, p. 141).

Various reagents are available for sulphation of hydroxyl groups in sections. Concentrated sulphuric acid (1 min at room temperature) is the simplest, but is physically injurious. It works well, however, with semi-thin sections (0.5–1.0 μm) of plastic-embedded material. A mixture of sulphuric acid and diethyl ether is milder. An even more gentle reagent is 0.25% sulphuric acid in a mixture of acetic acid and acetic anhydride (Lillie, 1964), but this also brings about N-sulphation of amino groups.

10.2.2. Blocking procedures

The reactivity of hydroxyl groups is blocked by base-catalysed acylation with either acetic anhydride or benzoyl chloride:

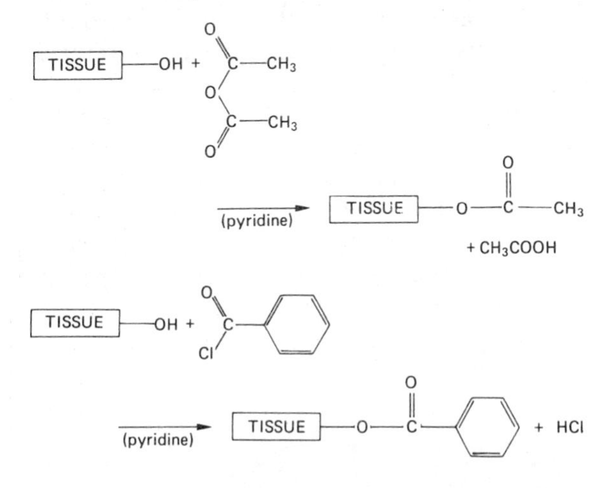

These reactions generate esters at the sites of hydroxyl groups. Acetic anhydride also produces N-acetylation of some amino groups. Benzoyl chloride causes benzoylation of almost all amino groups. The acylation reactions also block the reactivity of tyrosine, tryptophan, and histidine residues. The histochemical reactivity of arginine, however, is only slightly depressed by acylation. Esters formed with hydroxyl groups and amides formed by N-acylation of amines are hydrolysed, with reversal of the blockade, by exposure of the sections to alkaline ethanol, a procedure known in histochemical parlance as "saponification".

Diisopropylfluorophosphate (DFP) and related organophosphorus compounds combine irreversibly with the hydroxyl group of serine, but not in all proteins. The value of these reagents in histochemistry is therefore limited to their uses as inhibitors of certain serine-containing enzymes (see Chapter 15).

10.3. CARBOXYLIC ACIDS

Free carboxyl groups occur in glycoproteins and proteoglycans (Chapter 11), in fatty acids (Chapter 12), and in proteins. Every protein molecule has a terminal carboxyl group, but this is usually outnumbered by the side-chain carboxyls of glutamic and aspartic acids.

10.3.1. Histochemical detection

Cationic dyes will bind to carboxylic acids when the latter are ionized. For this to be so, the pH of the dye solution must usually be 4.0 or higher, so sulphate esters of proteoglycans and phosphoric acid groups of nucleic acids are also stained (see Chapter 6). Dyes are therefore of only limited use for the identification of sites of carboxyl groups in tissues.

The acid anhydride method is one which supposedly demonstrates only the carboxyl groups of proteins. Sections are first treated with a solution of acetic anhydride in pyridine at 60°C. This reagent is thought to react in different ways with C-terminal and with side-chain carboxyl groups:

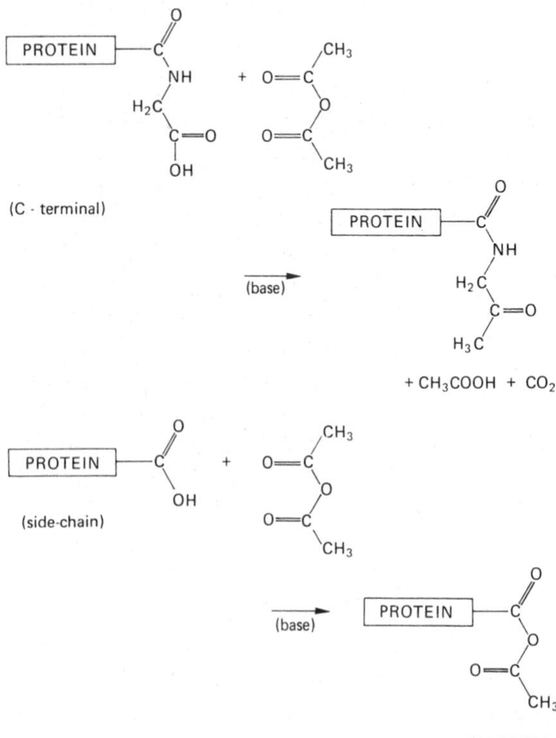

The product of the first reaction (Barrnett & Seligman, 1958) is a ketone and combines with the next reagent to be applied to the sections, 2-hydroxy-3-naphthoic acid hydrazide (HNAH) as described in Section 10.10.2. The reaction of acetic anhydride with side-chain carboxyls yields a mixed anhydride (Karnovsky & Fasman, 1960). This also combines with HNAH. The reaction is analogous to the acylation of an amine:

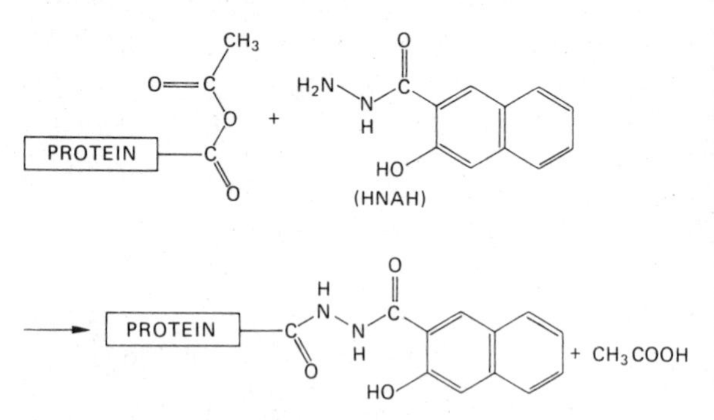

Both reactions result in the covalent binding of a naphthol group at the site of the carboxyl of the protein molecule. The naphthol can easily be made to couple with a diazonium salt to yield a coloured product (see Chapter 5, p. 52).

The carboxyl groups of carbohydrates do not react with acetic anhydride to give mixed anhydrides (Karnovsky & Mann, 1961), though there is no obvious reason why they should not behave in the same way as those of protein side-chains. A chemical study by Stoward & Burns (1971) indicates that the principal effect of hot acetic anhydride in pyridine is the formation of ketones at the sites of C-terminal carboxyl groups and that the side-chains of glutamic and aspartic acids form mixed anhydrides only to a very slight extent. This method is therefore considered to be specific for protein-bound carboxyl groups. The histochemistry of carbohydrates with free carboxyl groups is discussed in Chapter 11.

10.3.2. Blocking procedures

Carboxyl groups are easily blocked by converting them to their methyl esters. This may be accomplished with a variety of reagents. The simplest is 0.1 N HCl in methanol:

$$\boxed{\text{TISSUE}}\text{—COOH} + \text{HOHC}_3 \xrightarrow[\text{(acid)}]{} \boxed{\text{TISSUE}}\text{—COOCH}_3 + H_2O$$

(The active component of the reagent may be methyl chloride or methylene chloride rather than methanol itself.)

Other methylating agents occasionally used by histochemists are methyl iodide, diazomethane, and a solution of thionyl chloride in methanol.

Methylation also blocks amino groups, sulphydryl groups, and the phosphoric acid moieties of nucleic acids. The reagent also brings about hydrolysis of sulphate esters. It is a blocking reaction of greater importance in carbohydrate histochemistry than in the examination of carboxyl groups of proteins. Methyl esters of carboxylic acids are hydrolysed by saponification with alkaline ethanol.

10.4. AMINO GROUPS

This section is concerned with primary amino (—NH_2) groups of proteins. Some important soluble amines are also detectable histochemically (see Chapter 17), but these are no longer present in sections of tissues prepared by ordinary techniques. The most numerous amino groups of proteins are those in the ε-position on the side-chain of lysine. N-terminal amino groups are less abundant since there can be only one for each peptide chain.

Fixatives which combine covalently with amines, such as aldehydes, should be avoided if possible. They always impair staining of amino groups and sometimes prevent it completely. Strong oxidizing agents are also unsuitable for fixation, since they can cause oxidative deamination of proteins.

10.4.1. Histochemical detection

There are two main groups of methods for the demonstration of the amino groups of proteins. The first group consists simply of staining the tissue with a suitable anionic dye. This must be a dye that binds to its substrate exclusively by ionic attraction: usually one with fairly small molecules. Anionic dyes of high molecular weight, which are bound to tissues by forces other than electrostatic ones, are unsuitable. The mechanisms of staining by anionic dyes have already been discussed in Chapters 5, 6, and 8. Staining is to be expected at the sites of all tissue-bound cations, including amino and guanidino groups of proteins and organic bases such as choline that occur in some lipids. Blocking reactions (see Section 10.4.2) and extraction of lipids (see Chapter 12) must be used in order to establish that the acidophilia of a structure is due to the presence of amino groups.

The methods of the second group depend upon the formation of coloured deposits as the result of covalent combination of reagents with either intact or chemically modified amino groups. Many techniques are available. A typical one is the hydroxynaphthaldehyde method. The reagent, 2-hydroxy-3-naphthaldehyde, condenses with the protein-bound amino group to form an imine (also known as an azomethine or Schiff's base):

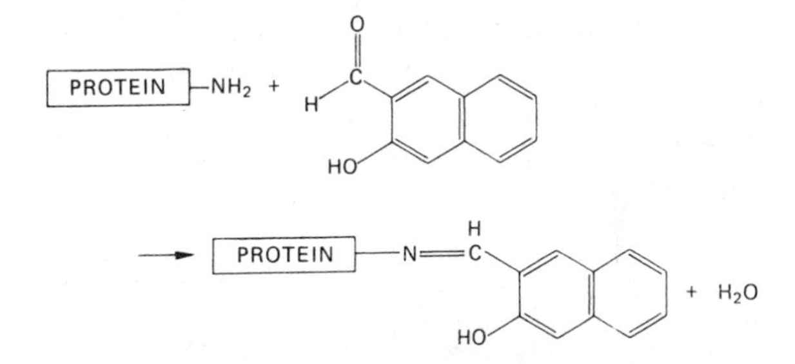

Thus, a naphthol becomes covalently bound at the site of the original NH_2 group. In the second stage of the method, the naphthol is made to couple with a diazonium salt to form an azo dye. The diazonium salt will, of course, also couple with the aromatic ring of tyrosine and with some other amino acids (see Section 10.9.2). However, the colours produced with these amino acids are different from and paler than that of the naphtholic azo dye. In any case they are easily allowed for in control sections which have not been treated with hydroxynaphthaldehyde. Blocking reactions may be used to confirm the specificity of the staining of amino groups.

10.4.2. Blocking reactions

Primary amines may be blocked either by removing them or by converting them to amides or sulphoamino compounds.

Deamination is most easily accomplished by treating the sections with nitrous acid, which is usually applied as a solution of sodium nitrite in aqueous acetic acid. The reaction is most simply expressed as:

$$\boxed{\text{PROTEIN}}\text{—}NH_2 + HONO \longrightarrow \boxed{\text{PROTEIN}}\text{—}OH + N_2\uparrow + H_2O$$

but is much more complicated in reality (White & Woodcock, 1968). Nitrous acid also reacts with tyrosine (see Section 10.7.1) and with sulphur-containing amino acids, but these reactions are unlikely to cause confusion in the testing of histochemical methods for amino groups. The guanidino group of arginine is not removed by treatment with nitrous acid (Lillie *et al.*, 1971). Strong oxidizing agents such as osmium tetroxide and potassium permanganate also bring about deamination of lysine side-chains, but their actions are rather unpredictable.

Acylation of amines is brought about by reaction with either acetic anhydride or benzoyl chloride:

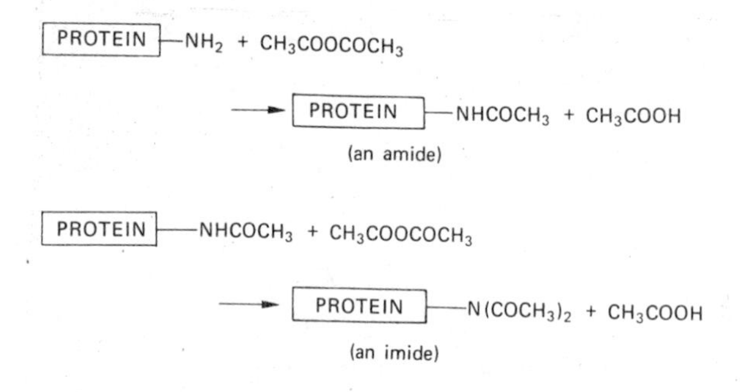

Equivalent reactions occur with benzoyl chloride. The reaction of acylation is catalysed either by a weak base such as pyridine or by a strong acid such as perchloric acid. Longer times and higher temperatures are needed for the base-catalysed acetylation and benzoylation of amines than for the esterification of hydroxyl groups by the same reagents. The amides and imides produced by these reactions are hydrolysed, with regeneration of the amines by "saponification" with an alcoholic solution of potassium hydroxide. For example:

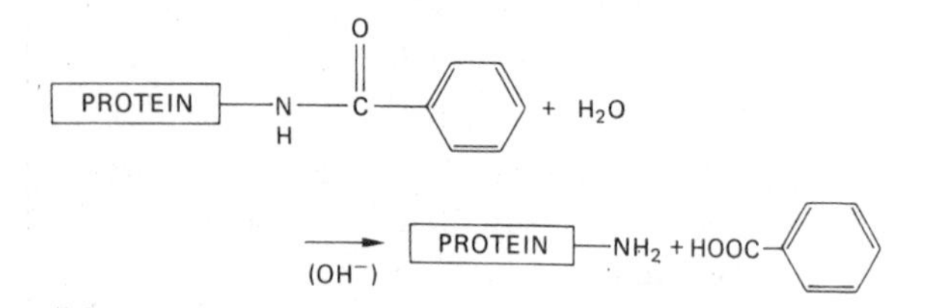

When sections are treated with acetic anhydride in the presence of a little sulphuric acid, amino groups are blocked in a few minutes and their sites become strongly basophilic. Although some acetylation may occur, the principal reaction is probably N-sulphation:

$$\boxed{\text{PROTEIN}}\text{—}NH_2 + H_2SO_4 \longrightarrow \boxed{\text{PROTEIN}}\text{—}N(H)\text{—}SO_3H + H_2O$$

(a sulphoamino compound)

The product ionizes as a strong acid, thus accounting for the induced basophilia. This blockade is not reversible by saponification, but it is reversed by treatment with hot acidified methanol, which presumably causes breaking of the nitrogen–sulphur bond. It should be noted that the chemistry of this sulphation–acetylation procedure (Lillie, 1964) is still poorly understood, though the technique is useful on account of the rapidity and completeness of the reaction. The reagent also brings about O-sulphation of hydroxyl groups, as has already been described (p. 117).

Primary amines may be converted to secondary and tertiary amines by alkylation with reagents such as methanolic hydrogen chloride:

$$\boxed{\text{PROTEIN}}\text{—}NH_2 + CH_3OH \xrightarrow{(HCl)} \boxed{\text{PROTEIN}}\text{—}N(H)\text{—}CH_3 + H_2O$$

$$\boxed{\text{PROTEIN}}\text{—}N(H)\text{—}CH_3 + CH_3OH \xrightarrow{(HCl)} \boxed{\text{PROTEIN}}\text{—}N(CH_3)_2 + H_2O$$

This reaction, which is slower than the methylation of carboxyl groups, does not prevent the binding of anionic dyes because the products are even more strongly basic than primary amines. Arylation of amino groups can be achieved by reaction with 2,4-dinitrofluorobenzene, but this reagent also combines with the side-chains of cysteine, histidine, and tyrosine.

10.5. ARGININE

Most proteins contain some arginine, but the highest concentrations are found in the histones associated with DNA and in certain cytoplasmic granules. The side-chain of this amino acid is strongly basic, owing to its guanidino group

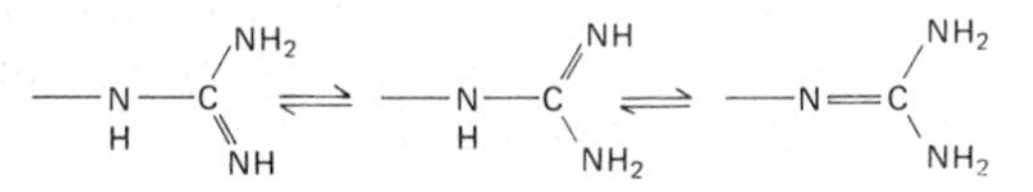

This is an amino acid for which highly specific histochemical tests are available.

10.5.1. The Sakaguchi reaction

Sections are treated with a freshly prepared solution containing sodium hypochlorite and α-naphthol, made strongly alkaline by the addition of sodium hydroxide. A pink or red colour develops at the sites of arginine residues. It is unstable, so permanent preparations cannot be made. Baker (1947) tested the Sakaguchi reaction with several pure compounds of known structure and concluded that coloured products were formed by reaction of the reagent with substances containing the arrangement

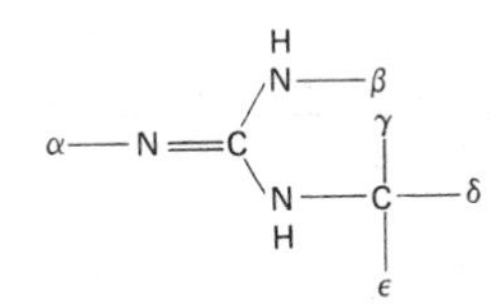

in which α and β were —H or —CH_3 and γ, δ and ε could be any radicals. Arginine is the only known component of fixed tissues to conform to this structural requirement.

The chemistry of the Sakaguchi reaction has been a mystery for many years, but a plausible mechanism has been proposed by Lillie *et al.* (1971). They discovered that an identical pattern of staining could be obtained with an alkaline solution of the sodium salt of 1,2-naphthoquinone-4-sulphonic acid (NQS)

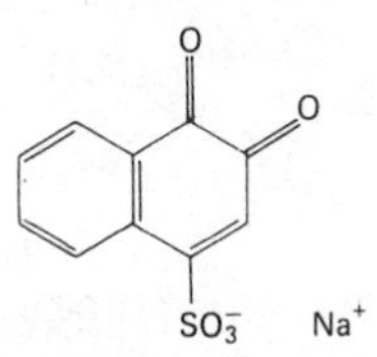

The alkalinity was produced by barium or strontium hydroxide. The hydroxides of sodium, calcium, and magnesium did not permit full development of the colour. The product of the reaction with NQS, unlike that formed in the Sakaguchi reaction, resisted extraction by dehydrating and clearing agents and was permanently stable in resinous mounting media. Studies with blocking reactions indicated that NQS stained arginine specifically. Lillie *et al.* (1971) speculated that the coloured product of the reaction of the guanidino group with NQS was

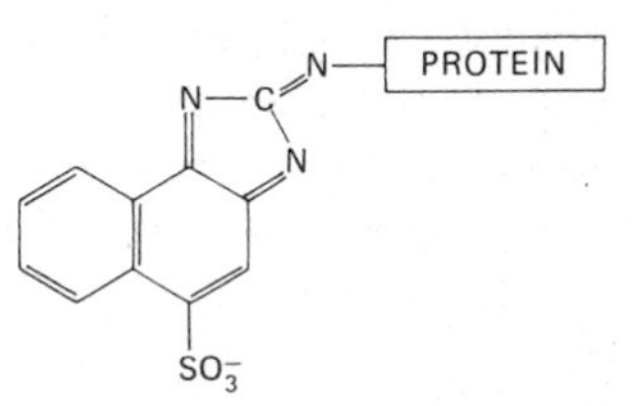

These authors also suggested that in the Sakaguchi reaction α-naphthol was oxidized by hypochlorite to a mixture of 1,2- and 1,4-naphthoquinones and that the former condensed with guanidino groups to give a coloured product. It was later suggested (Lillie & Fullmer, 1976) that cleavage of the chromogenic part of the structure from the remainder of the arginine residue might occur during the course of staining to give

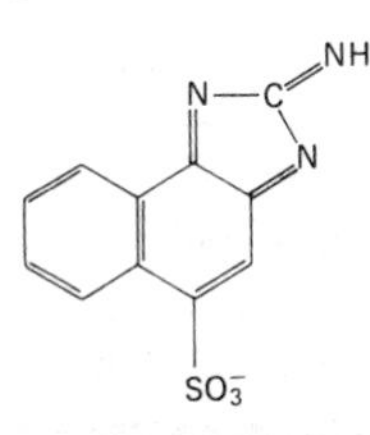

The sodium salt of this anion is soluble, but the barium salt is insoluble. Formation of such a product would account for the requirement for barium ions in the NQS reagent and for the observation that only weak histochemical reactions could be obtained when 1,2-naphthoquinone was substituted for its sulphonic acid derivative. These speculations merit further investigation.

10.5.2. Blocking reactions

The histochemical specificity of the Sakaguchi and related reactions is so high that blocking reactions are rarely considered necessary, though several are available. Lillie *et al.* (1971) used an alkaline alcoholic solution of benzil to convert the guanidino group to an unreactive imidazole:

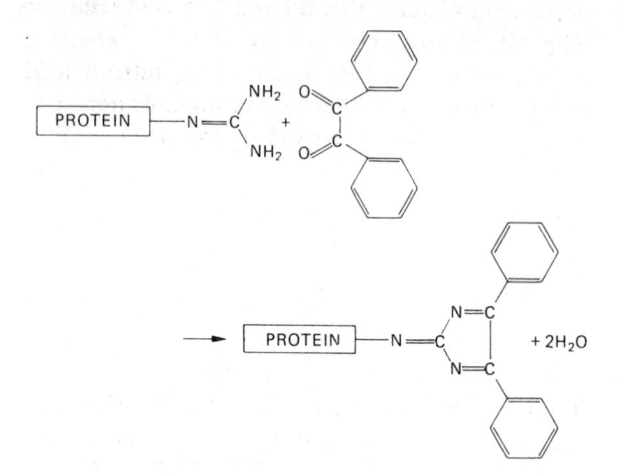

Another reagent, used by biochemists, is the trimer of 2,3-butanedione (also called diacetyl trimer), which reacts with arginine in neutral solution. It might have value as a histochemical blocking agent.

10.6. TRYPTOPHAN

Since tryptophan is a ubiquitous component of proteins, histochemical methods are employed only for the study of sites of high concentration of this amino acid, such as the Paneth cells of the intestine, the zymogen granules of the exocrine pancreas and several other secretory products. Amyloid (see Chapter 11, p. 149) also contains much tryptophan.

Histochemical techniques are based on the reactions of the indole ring, which remains free to react when the carboxyl and amino groups of tryptophan are incorporated into peptide linkages. Other naturally occurring indole-containing compounds are soluble and are therefore unlikely to be confused with tryptophan in paraffin sections. A possible exception is the serotonin of argentaffin cells and mast cells of rodents (see Chapter 17). Reactions for tryptophan can be obtained after almost any fixation but are strongest following formaldehyde or an alcoholic fixative. Picric acid, mercuric chloride, and potassium dichromate partially inhibit staining. Only one histochemical method, the rosindole reaction of Glenner (1957), will be described here, but several others are available.

10.6.1. The rosindole reaction

This method is based on the addition of *p*-dimethylaminobenzaldehyde (DMAB) to the indole ring, which occurs at the position adjacent to the nitrogen atom. Sections are first treated with a solution of DMAB in acetic acid to which a strong acid (perchloric, hydrochloric, or a mixture of the two) has been added. The DMAB combines with the indole ring of tryptophan in an acid-catalysed hydroxyalkylation reaction:

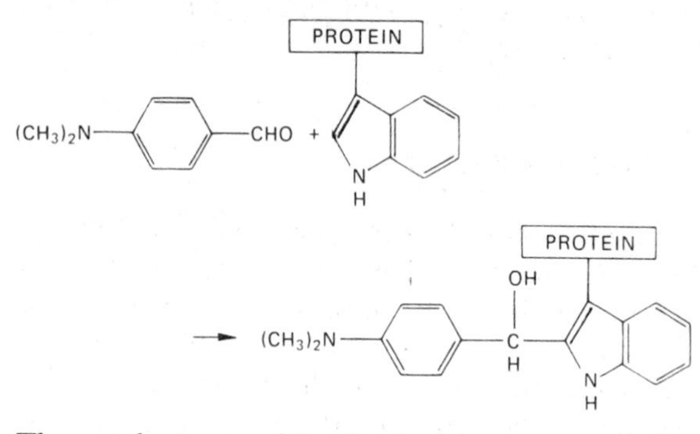

The product resembles the leuko-compound of a diarylmethane dye. In the second stage of the method an oxidizing agent, usually a strongly acidified solution of sodium nitrite, is applied. A blue colour develops due to the formation of a "rosindole" which may have a structure such as

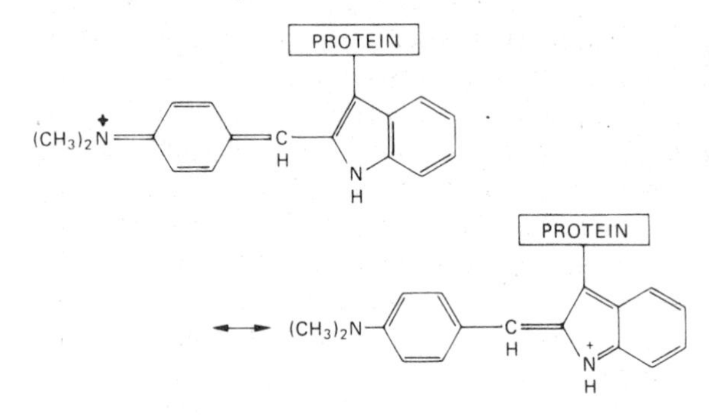

The resonance structure of such a compound shows similarity to those of diphenylmethane and cyanine dyes (see Chapter 5).

Glenner (1957) suggested that the product of the rosindole reaction was a triarylmethane dye:

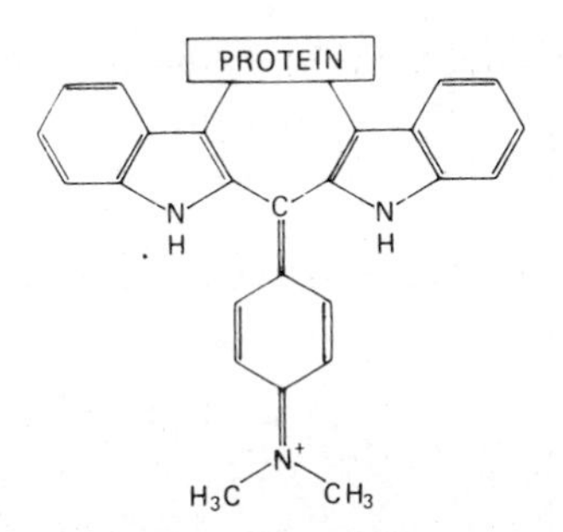

Such a compound could only be formed with fixed proteins if the two indole rings were exactly the right distance apart to be able to combine with one molecule of DMAB. A triarylmethane dye is a likely product of the reaction of DMAB with solutions of indole-containing compounds but an unlikely one under the conditions of the histochemical procedure.

10.6.2. Blocking reactions

There is no reaction of high specificity for preventing the characteristic reactions of the indole group of tryptophan. Strong oxidizing agents cause opening of the heterocyclic ring: an alkaline solution of potassium persulphate is suitable for this purpose. The products of the reaction have not been identified and the blockade is irreversible. Performic and peracetic acids (see pp. 137–138) may also be used.

10.7. TYROSINE

Tyrosine is another ubiquitous amino acid. Its histochemical demonstration may be considered indicative of the presence of proteins. Several techniques for the detection of the simple phenolic side-chain are available. Other phenols such as serotonin, the catecholamines, and tannins will give the same reactions, but protein-bound tyrosine and the serotonin of argentaffin cells are the only reactive compounds likely to be present in fixed animal tissues.

10.7.1. The Millon reaction

This is a modification for histochemical use of the classical Millon's test for protein. A red colour develops when tissues are treated with an acid solution containing mercuric and nitrite ions. All investigators agree that the reaction is specific for phenolic compounds.

In the first stage of the reaction, nitrous acid reacts with the phenolic ring causing C-nitrosation, principally *ortho* to the hydroxyl group

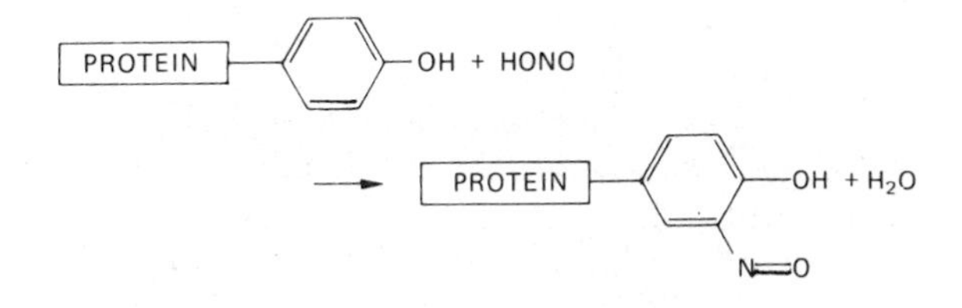

The nitroso compound forms with mercuric ions a stable chelate, which is red and may have a structure such as

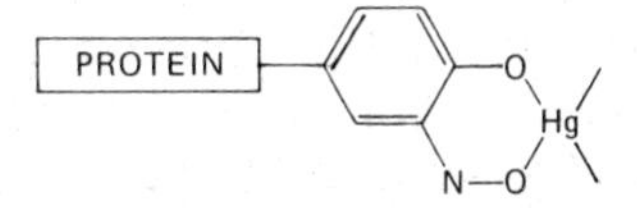

The other two coordination positions of the mercury atom may be satisfied by another nitrosophenol molecule (if one is suitably positioned in the fixed tissue) or by an inorganic ligand such as water.

In the absence of a metal such as mercury, further reaction of tyrosine with nitrous acid results in the formation of a diazonium salt:

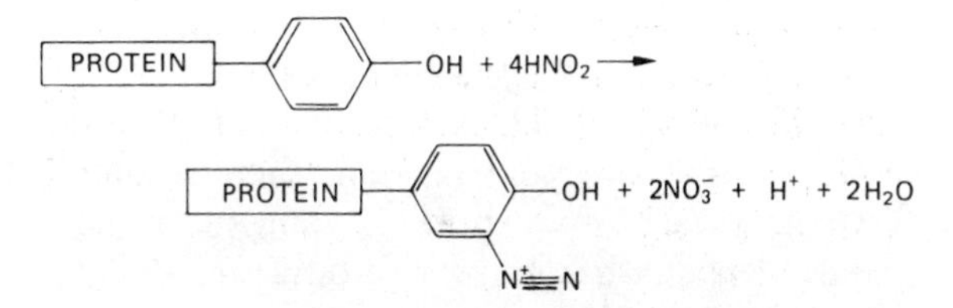

This product is rapidly decomposed by light, but in darkness it is stable enough to be coupled with a naphthol to form an azoic dye. These reactions form the basis of another histochemical technique for tyrosine.

As a phenol, tyrosine can couple with exogenous diazonium salts to form azo dyes, but a similar reaction occurs with the imidazole ring of histidine and probably also with the indole ring of tryptophan. Diazonium salts are useful as reagents for the detection of proteins (see Section 10.9.2) but lack specificity for any individual amino-acid side-chains.

10.7.2. Blocking reactions

The phenolic hydroxyl group of tyrosine is readily acylated by reaction with acetic anhydride or benzoyl chloride. The aromatic rings of the resultant esters no longer possess the reactivity of phenols towards nitrous acid and other reagents. Aliphatic hydroxyls and amino groups are also acylated (see p. 120).

The hydrogen atoms of phenolic rings are quite easily replaced by iodine atoms, so adequate treatment with a solution of iodine will prevent the histochemical reactions of tyrosine. Iodination can also cause oxidation of free sulphydryl groups of cysteine and partial inhibition of the reactions of tryptophan.

Blockade of tyrosine can also be achieved with tetranitromethane, which introduces a nitro group *ortho* to the hydroxyl group. However, this reagent also decomposes tryptophan and the sulphur-containing amino acids.

10.8. CYSTEINE AND CYSTINE

10.8.1. Some properties of cysteine and cystine

The free sulphydryl (also known as a mercaptan or thiol) group of cysteine is very reactive, mainly because it is a strong reducing agent. In ordinary paraffin sections, sulphydryl groups are demonstrable only if an unreactive fixative, such as a mixture based on alcohol and acetic acid, has been used and if exposure of the tissue to atmospheric oxygen has been kept to a minimum at all stages of processing. Fixatives that react with the —SH group, such as mercuric chloride, oxidizing agents, and, to a lesser extent, aldehydes, must be avoided.

Suitably positioned sulphydryl groups may be oxidized in pairs to yield disulphides:

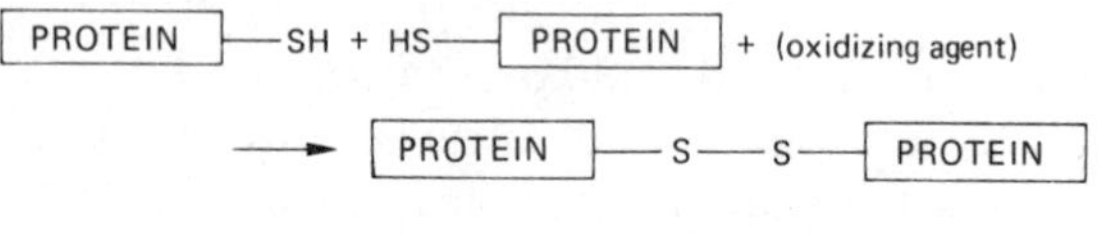

This is the normal mode of formation of cystine, the predominant sulphur-containing amino acid in both living and fixed tissues. Strong oxidizing agents, such as the permanganate ion and performic and peracetic acids, will cleave the disulphide bridge of cystine to yield two molecules of cysteic acid, which is an aliphatic sulphonic acid:

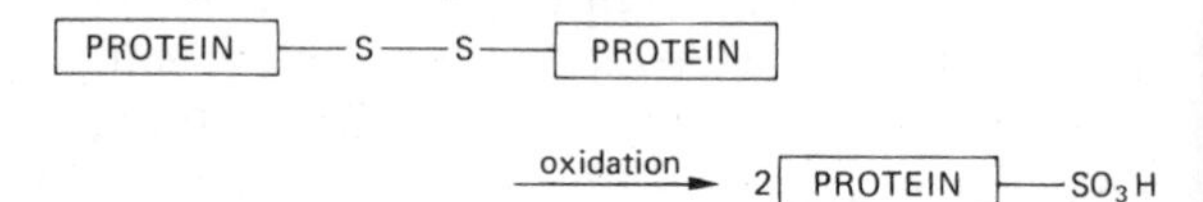

Cystine is easily reduced to cysteine. The reducing agents used in histochemistry include the alkali metal cyanides

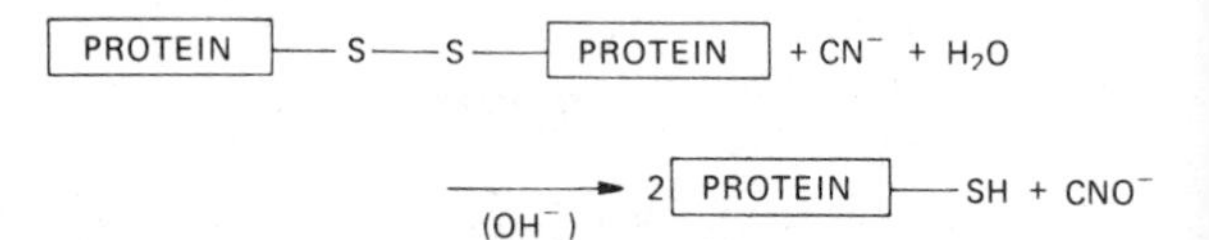

and a variety of compounds containing the sulphydryl group. For example:

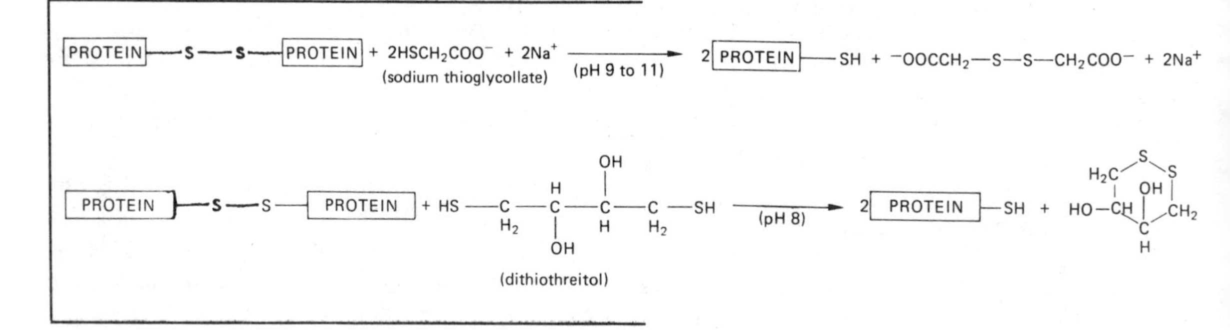

Dithiothreitol (Cleland, 1964) can be used in a less strongly alkaline solution than sodium thioglycollate, but is more expensive.

The other histochemically important property of the sulphydryl group is its ability to combine with both inorganic and organic compounds of mercury. For example:

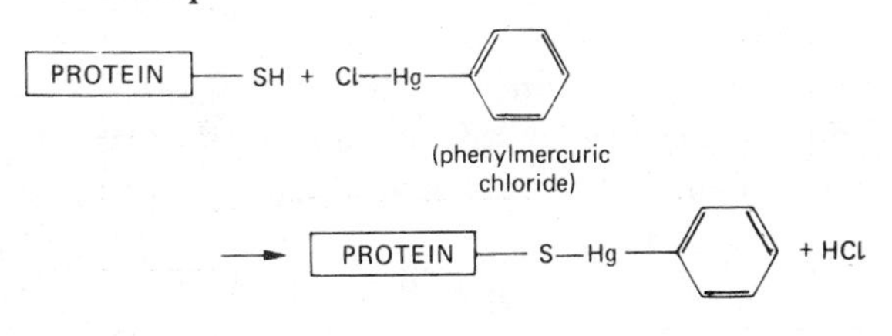

10.8.2. Histochemical methods for cysteine

It is possible to demonstrate sulphydryl groups either by virtue of their reducing properties or by the binding of chromogenic mercury-containing reagents. These methods may be applied directly to sections, to detect any cysteine that has survived fixation, or they may be applied to sections previously treated for the reduction of —S—S— to —SH. In the latter case, cysteine and cystine will both be demonstrated.

The simplest and probably the most sensitive technique based on the reducing properties of the sulphydryl group is the ferric ferricyanide method. A solution containing the ions Fe^{3+} and $[Fe(CN)_6]^{3-}$ yields an insoluble blue pigment when it is acted upon by a reducing agent. According to Lillie & Donaldson (1974) it is the ferric ion rather than the ferricyanide which is reduced by the —SH groups, so the pigment would be expected to be ferrous ferricyanide, also known as Turnbull's blue. However, it is known that this substance is identical to Prussian blue, the pigment precipitated when ferric and ferrocyanide ions meet one another. Prussian blue is ferric ferrocyanide, $Fe_4[Fe(CN)_6]_3$, but its crystals also include ions of sodium or potassium and molecules of water (see Cotton & Wilkinson, 1972).

The ferric ferricyanide reaction is positive with reducing agents other than cysteine, including serotonin, catecholamines, and some of the metabolic precursors of melanin. The reagent is also reduced by lipofuscin pigment and (sometimes) by elastin, for reasons that are incompletely understood. In the histochemical study of cysteine and cystine, blocking reactions must be used to confirm the specificity of staining.

Several organic compounds of mercury are used as histochemical reagents for the sulphydryl group (see Cowden & Curtis, 1970; Lillie, 1977). Examples are:

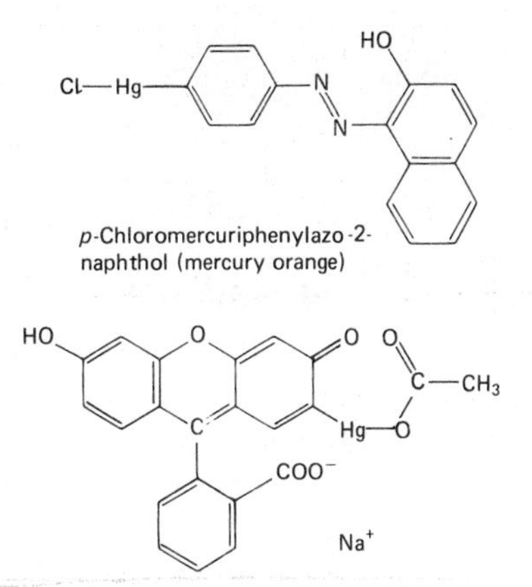

p-Chloromercuriphenylazo-2-naphthol (mercury orange)

Fluorescein mercuric acetate

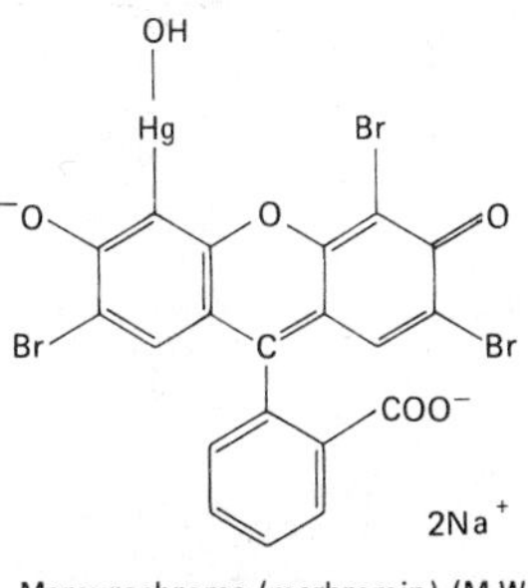

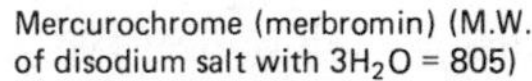
Mercurochrome (merbromin) (M.W. of disodium salt with $3H_2O$ = 805)

These and other mercurials react rather slowly with sulphydryl groups to give covalent coloured adducts. Reaction is favoured by an aprotic polar solvent such as *N*,*N*-dimethylformamide, in which the active anion of the reagent is not hindered by dipole–dipole attractions to the solvent molecules as would occur in water.

10.8.3. Cysteic acid methods for cystine

Cystine may be demonstrated without prior reduction to cysteine by oxidizing it to cysteic acid (see p. 124). This sulphonic acid is ionized even at very low pH and can therefore bind cationic dyes from strongly acid solutions. Alcian blue (at pH 0.2), thiazine dyes (at pH 1.0), and aldehyde–fuchsine are often used for the staining of cysteic acid. The other properties of these dyes, especially their attachment to sulphated carbohydrates, must, of course, be allowed for in suitable control sections.

Several other methods for cysteine and cystine are described in the larger textbooks of histochemistry. No histochemical method is yet available for methionine, the third sulphur-containing amino acid.

10.8.4. Blocking procedures

Two convenient blocking agents for the sulphydryl group are *N*-ethylmaleimide and iodoacetic acid:

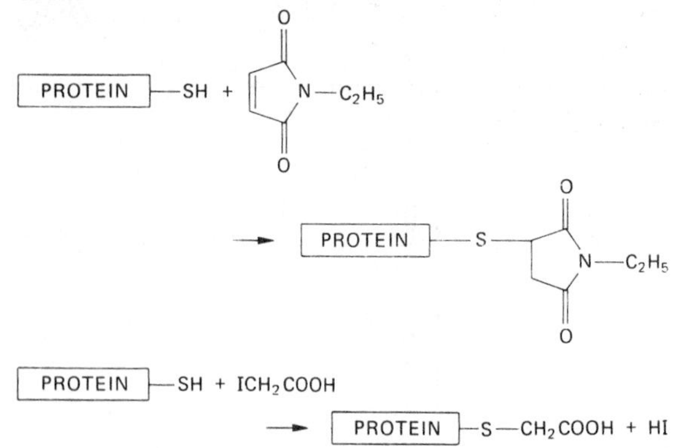

Both these reagents completely block the reactivity of SH groups. Other blocking agents, including iodine and various compounds of mercury, are likely to give incomplete or easily reversible blockades.

10.9. METHODS FOR PROTEINS IN GENERAL

The histologist or pathologist often wants to know whether or not an intracellular inclusion or an extracellular material is composed largely of protein. For the simple staining of protein, techniques are available which do not depend upon the presence of individual amino acids. Two will be considered here.

10.9.1. Demonstration of proteins with dyes

Staining by anionic dyes, especially those with small molecules, is due largely to electrostatic attachment to the basic side-chains of lysine and arginine. Such staining has already been discussed in Chapters 5 and 6 and in Section 10.4.1 of this chapter. Dyes with large molecules, especially those used industrially for the direct dyeing of cotton, are probably bound to tissues by hydrogen bonds, van der Waals forces, or charge-transfer bonds as well as by ionic attraction (see Chapter 5). Consequently such dyes are likely to stain nearly everything in a section.

An anionic dye is useful for the demonstration of protein only if the intensity of the colour produced at any site in the tissue is proportional to the local concentration of protein there. A triphenylmethane dye, which has been shown to bind stoichiometrically to proteins under histochemical conditions, is BRILLIANT INDOCYANINE 6B (C.I. 42660; Acid blue 83; M.W. 826):

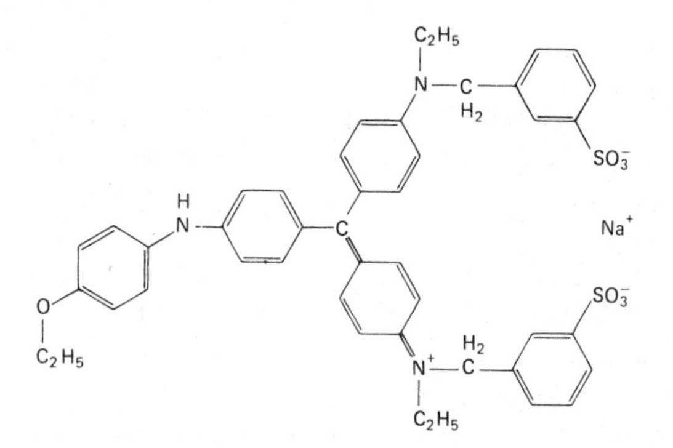

This dye has several synonyms, the best known of which are brilliant blue R and coomassie brilliant blue R-250. It is used by biochemists for staining proteins in acrylamide gels. Brilliant indocyanine 6B is almost insoluble in water, so sections are stained from a solution of the dye in a mixture of ethanol and acetic acid. The chemical specificity of the technique and the mode of binding of the dye are in need of investigation, but it has been shown that staining is prevented by prior treatment of plastic-embedded sections with proteolytic enzymes (Cawood *et al.*, 1978).

10.9.2. The coupled tetrazonium reaction and related methods

Alkaline solutions of diazonium salts combine with proteins in at least three different ways. Thus the expected coupling reaction occurs with the phenolic side-chains of tyrosine:

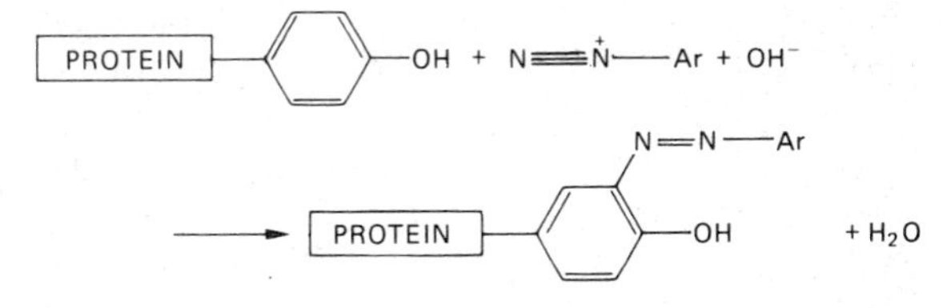

A similar coupling reaction occurs with the imidazole ring of histidine. The nitrogen atoms of this aromatic heterocycle confer some of the properties of an aromatic amine to the ring:

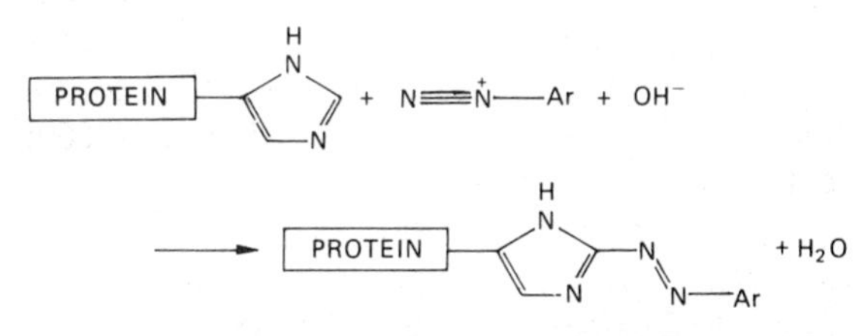

Coupling probably also occurs *ortho* to the nitrogen atom in the indole ring of tryptophan. A different reaction takes place with the ε-amino group of lysine:

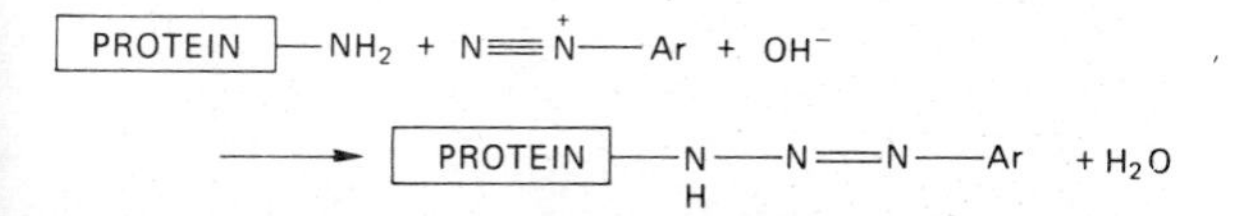

The product is a triazene. Similar reactions may occur with the secondary amine nitrogen atoms of proline and hydroxyproline and with the guanidino and sulphydryl groups of arginine and cysteine respectively. The nitrogen atoms of the purine and pyrimidine bases of the nucleic acids almost certainly do not couple with diazonium salts (see Lillie & Fullmer, 1976). Thus a diazonium salt applied to a section of tissue will be covalently bound to several organic functional groups, all of which are found only in proteins.†

In the coupled tetrazonium reaction (Danielli, 1947), the diazonium salt used is tetraazotized benzidine, which has to be freshly prepared in the laboratory from benzidine and nitrous acid:

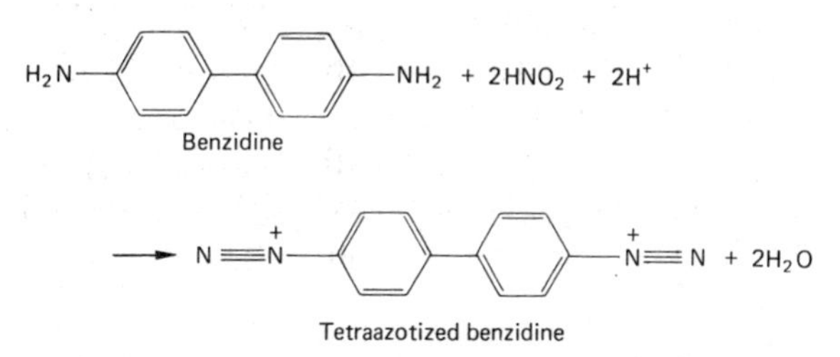

An alternative, more convenient reagent is the stabilized tetrazonium salt derived from *o*-dianisidine, known as fast blue B salt (see p. 57 for formula). It will be noticed that each of these compounds bears two diazonium groups per molecule. Usually only one of these will couple with a reactive site in a section, since not often will two coupling sites in the fixed tissue be exactly the right distance apart to react with diazonium groups of the same molecule of reagent.

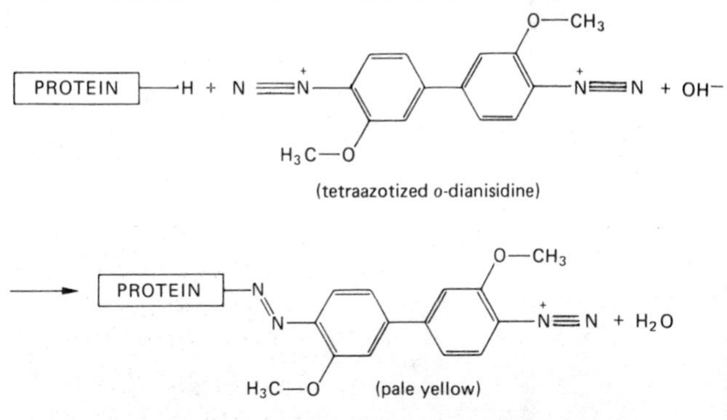

Azo compounds formed by coupling of simple diazonium salts with proteins are only feebly coloured. However, the free —$N^+{\equiv}N$ group of a bound tetrazonium salt is able to couple with any phenol or aromatic amine subsequently applied to the section, to give a strongly coloured dye containing two azo

†Exceptions are certain phenols: the catecholamines of chromaffin cells and serotonin in argentaffin cells (see Chapter 17), but these substances are absent from most tissues and their presence can be detected by other methods. Phenolic compounds of plants, such as tannins, may also be expected to couple with diazonium salts.

linkages. H-acid (see p. 58) is a suitable azoic coupling component for this purpose:

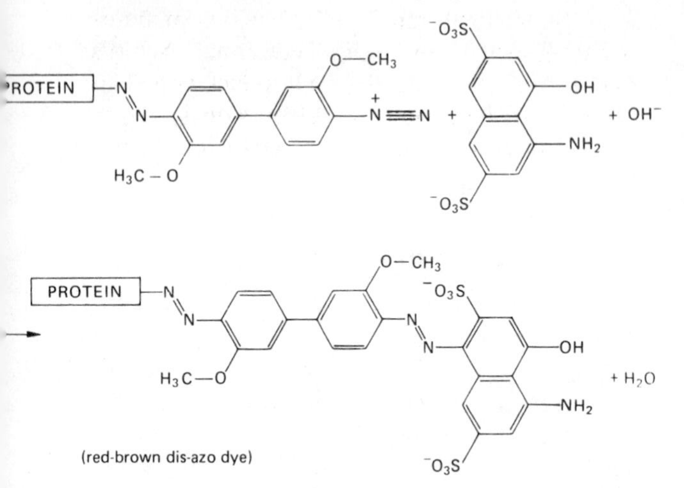

As an alternative to the coupled tetrazonium reaction, one may use as diazonium salt (with only one $N^+{\equiv}N$ group), which is itself an azo dye. Such a salt couples with proteins to give dis-azo dyes in a single-step reaction. A coloured stabilized diazonium salt suitable for staining of proteins is fast black K salt (see p. 57). Freshly diazotized safranine can also be used (Lillie *et al.*, 1968).

In attempts to make the coupled tetrazonium reaction specific for aromatic amino acids, sections are commonly treated with dilute hydrochloric acid after the first coupling, to destroy triazenes, but the efficacy of this manoeuvre is questionable.

10.10. ALDEHYDES AND KETONES

Free aldehyde groups are largely absent from tissues† but they are often produced in the course of histochemical manipulations. The Feulgen reaction (Chapter 9) and the PAS reaction (Chapter 11) are the most commonly used methods of this type. Ketones occur naturally as ketosteroids, for which histochemical techniques have been described, and as intermediate products in some staining methods (e.g. see p. 118), but are generally less important to the histochemist than are aldehydes.

†They occur in immature elastin (see Chapter 8), in lignin, in lipids that have been oxidized by air (see Chapter 12), and in specimens fixed in glutaraldehyde (see Chapter 2).

10.10.1. Schiff's reagent

Before reading this section the reader may wish to review the chemistry of pararosaniline and related triphenylmethane dyes (Chapter 5, pp. 63–64). In this account it is assumed that Schiff's reagent is prepared from pararosaniline, but equivalent chemical reactions occur with the other components of basic fuchsine.

Schiff's reagent is made by treating a solution of basic fuchsine with sulphurous acid. This weak acid is formed when sulphur dioxide combines with water:

$$H_2O + SO_2 \rightleftharpoons H_2SO_3 \rightleftharpoons H^+ + HSO_3^- \rightleftharpoons 2H^+ + SO_3^{2-}$$

These equilibria are displaced to the left when $[H^+]$ rises. Sulphurous acid, containing free sulphur dioxide, is prepared in the course of making Schiff's reagent, either from the reaction of a bisulphite (sodium metabisulphite, $Na_2S_2O_5$ is usually used) with a mineral acid, or of thionyl chloride with water (see p. 110), or by bubbling gaseous sulphur dioxide through an aqueous solution of the dye. Whatever method is used, a sulphonic acid group is added to the central carbon of the triphenylmethane structure

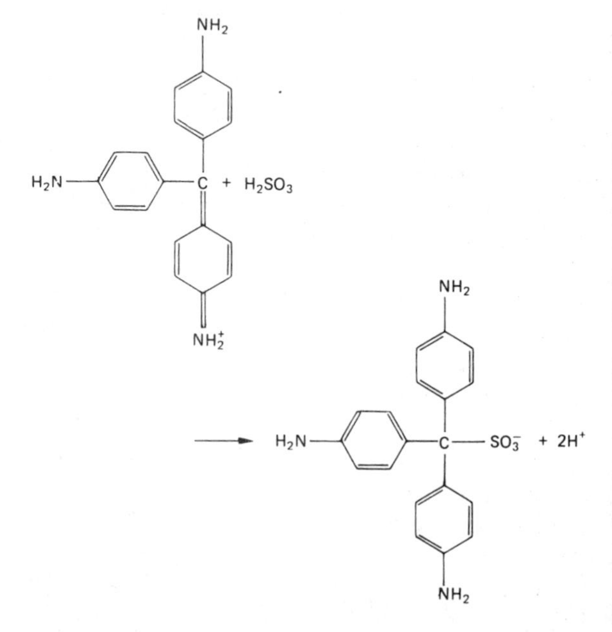

Schiff's reagent therefore consists of a solution of this colourless derivative of the dye in water, which contains an excess of sulphurous acid.

Schiff's reagent reacts with aldehydes, but not with ketones, to form brightly coloured products. This is not simply a matter of recolorizing the dye. New compounds are formed with colours somewhat different from that of basic fuchsine. The chemistry of the histochemical reaction is still not fully understood, but the results of three recent investigations (Dujindam & van Duijn, 1975; Gill & Jotz, 1976; Nettleton & Carpenter, 1977) concur in attributing the structure

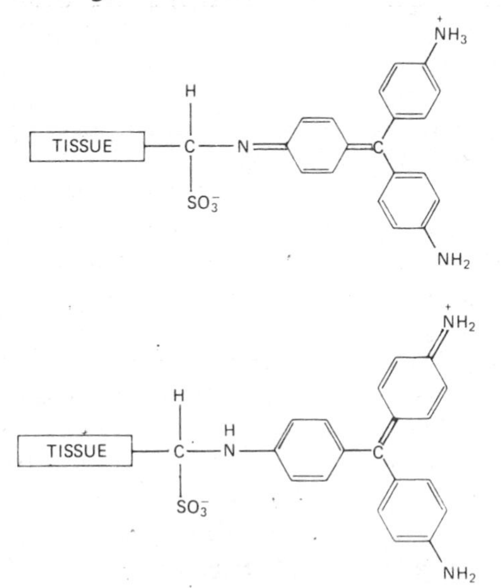

or

to the principal coloured compound formed when Schiff's reagent reacts with tissue-bound aldehyde groups. Both these compounds are alkylsulphonic acid derivatives of pararosaniline, and the two structures would be expected to exist in tautomeric equilibrium. There is still some uncertainty as to whether the aldehyde group first reacts with sulphurous acid to form a bisulphite compound, which then condenses with an amino group of Schiff's reagent, or whether the amino group of the reagent first reacts with the aldehyde to form an imine (= Schiff's base, anil, or azomethine), which is subsequently sulphonated. The second stage of either scheme must involve removal of the sulphonic acid group from the dye molecule and re-establishment of the triphenylmethane chromophore.

Several dyes and fluorochromes other than basic fuchsine will serve as substitutes for Schiff's reagent if their solutions are treated with sulphur dioxide, but usually these "pseudo-Schiff" reagents are coloured. The most widely used is one derived from acriflavine and thionyl chloride (Ornstein *et al.*, 1957), which is useful for the identification of aldehyde-containing components of tissues in fluorescence microscopy. The traditional Schiff's reagent also gives a fluorescent product with aldehydes, but its orange–brown emission is rather weak and contrasts poorly with a dark background.

10.10.2. Hydrazines and hydrazides

Organic hydrazines are compounds in which one hydrogen atom of hydrazine, $H_2N—NH_2$, is replaced by an organic radical. They have the general structure

$$\text{R(H)N—NH}_2 \qquad \text{(R = alkyl or aryl radical)}$$

Hydrazides are similar but with an acyl in place of R in the above general formula. Hydrazines and hydrazides condense readily with both aldehydes and ketones:

$$\text{R—NH—NH}_2 + \text{O=C}\langle\text{TISSUE} \rightarrow \text{R—NH—N=C}\langle\text{TISSUE} + H_2O$$

The reagent $RNHNH_2$ in this equation will be potentially useful for the demonstration of tissue-bound carbonyl groups if R is either coloured or able to form a coloured compound by reaction with another reagent. Several hydrazines and hydrazides serve as histochemical reagents in this way. They include the following:

$$H_2N—NH—C_6H_4—SO_3H$$

Phenylhydrazine-4-sulphonic acid (Shackleford, 1963): confers affinity for cationic dyes at low pH on sites of aldehydes

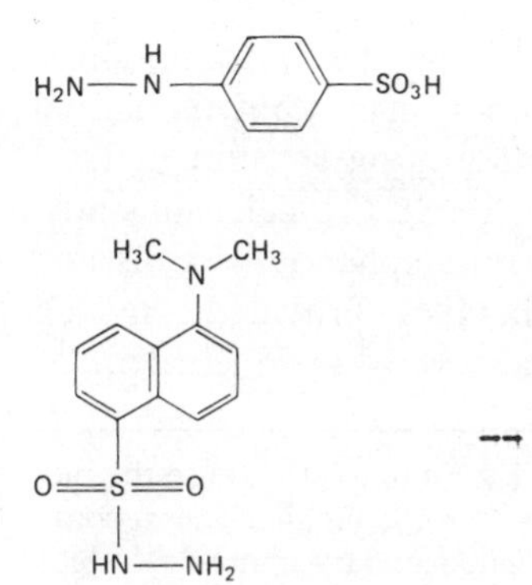

Dansylhydrazine (Weber *et al.*, 1975): forms fluorescent adducts, which can be made more chemically stable by reduction of the C=N bond to C—NH, to give a secondary amine

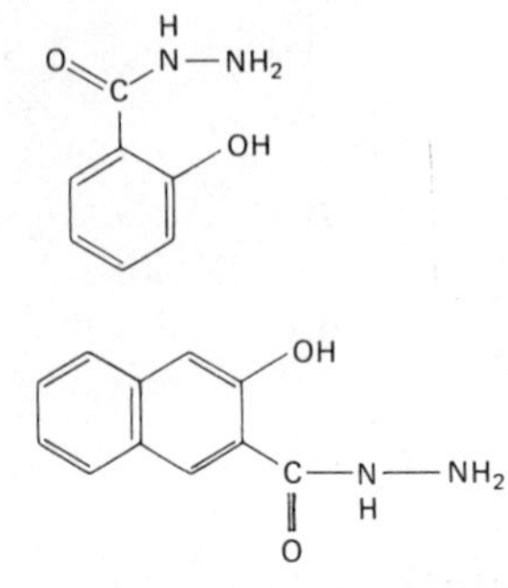

Salicylhydrazide (Stoward, 1968b): forms fluorescent adducts. Chelation with certain metals (Zn^{2+}, Al^{3+}) can be employed to distinguish between products of condensation with aldehydes or ketones

2-Hydroxy-3-naphthoic acid hydrazide (HNAH): this is the most popular reagent of its type. Sites of attachment are made visible by coupling with a diazonium salt

Phenylhydrazine ($C_6H_5NHNH_2$) is a useful reagent for blocking the reactivity of carbonyl compounds (see below).

10.10.3. Aromatic amines

The simplest reaction of a primary amine with an aldehyde is

$$R{-}NH_2 + R'{-}C(=O){-}H \longrightarrow R{-}N{=}C(H){-}R' + H_2O$$

The product, containing the C=N linkage, is variously known as an imine, an azomethine, a Schiff's base, or an anil. It is stable only when R or R′ is an aromatic ring. If the amine is present in excess, as it must be when it is a reagent applied to a section of a tissue, two of its molecules may react with one aldehyde group:

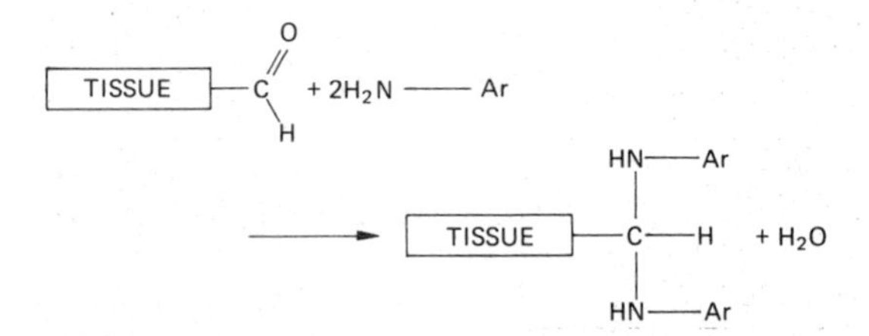

The product is known as an **aminal.** Its formation is favoured when the group Ar in the above equation carries an electron-withdrawing substituent (see Sollenberger & Martin, 1968). In an elegant study of the reaction of *m*-aminophenol with tissue-bound aldehydes, Lillie (1962) provided convincing evidence that aminals† were the principal products of the reaction, which was acid-catalysed and occurred in the absence of water. Sites of attachment of *m*-aminophenol could be demonstrated by coupling with a diazonium salt to give coloured products with structures such as

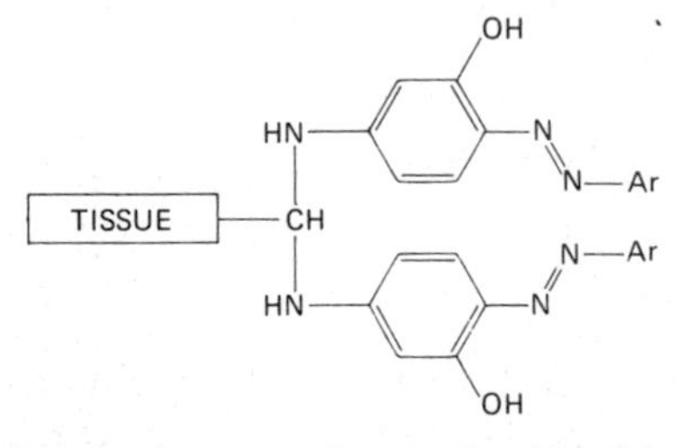

Aromatic amines are also used as blocking agents for aldehyde groups. Blockade is achieved more rapidly with *m*-aminophenol than with aniline. Both reagents are used as solutions in glacial acetic acid.

Many dyes are primary aromatic amines. An acidified alcoholic solution of basic fuchsine (Horobin & Kevill-Davies, 1971a) is a useful substitute for Schiff's reagent (see pp. 111, 140). The dye in this solution almost certainly condenses with aldehydes to form aminals in the same way as does *m*-aminophenol (Horobin & Kevill-Davies, 1971b). A solution of acriflavine in ethanol, used similarly, serves as a fluorescent reagent for aldehydes and has some advantages over the "pseudo-Schiff" reagent prepared by treatment of the same dye with sulphurous acid (Levinson *et al.*, 1977).

10.10.4. Ammoniacal silver nitrate

An aqueous solution of ammonia contains the dissolved gas in equilibrium with ammonium hydroxide:

$$NH_3 + H_2O \rightleftharpoons NH_4OH$$

Ammonium hydroxide ionizes as a weak base, so the solution is alkaline:

---

†Lillie (1962) and Lillie & Fullmer (1976) use the old term "diphenamine bases" for aminals. These compounds have also been called secondary amines in the histochemical literature, but "aminal" is the correct name for a compound containing the grouping $\overset{H}{C{-}N}{-}\underset{H}{\overset{|}{C}}{-}\overset{H}{N{-}C}$ just as the better-known acetals include the structure $C{-}O{-}\underset{H}{\overset{|}{C}}{-}O{-}C$.

$$NH_4OH \rightleftharpoons NH_4^+ + OH^-$$

When an ammonia solution is added in small aliquots to aqueous silver nitrate, a black or brown precipitate of silver oxide is first formed:

$$2Ag^+ + 2OH^- \longrightarrow Ag_2O\downarrow + H_2O$$

With further addition of ammonia the precipitate dissolves. The solution contains the silver diammine ion, $[Ag(NH_3)_2]^+$:

$$Ag_2O + 4NH_3 + H_2O \longrightarrow 2[Ag(NH_3)_2]^+ + 2OH^-$$

The silver diammine ion rapidly oxidizes aldehydes (but not ketones) and is itself reduced to metallic silver:

$$\boxed{\text{TISSUE}}\!-\!CHO + 2[Ag(NH_3)_2]^+ + H_2O$$

$$\longrightarrow \boxed{\text{TISSUE}}\!-\!COOH + 2Ag\downarrow + 4NH_3 + 2H^+$$

The precipitated silver is black and the deposits can be further intensified by gold toning or by physical development (see Chapter 18), so the test is a very sensitive one. Since the silver is electron opaque, ammoniacal silver nitrate solutions are sometimes used as substitutes for Schiff's reagent in ultrastructural histochemical studies. Other complexes of silver, notably that formed with hexamethylenetetramine (also known as methenamine or hexamine), may be used instead of silver diammine.

Unfortunately the histochemical specificity of this method is low. The silver complexes are also reduced by sulphydryl groups and by *o*-diphenols (see Chapter 17). The staining method for reticulin described in Chapter 8 is an example of a useful histological technique based on the detection of artificially produced aldehyde groups.

10.10.5. Other methods for aldehydes

Three other methods that have been proposed for the localization of aldehydes are worth mentioning, but none are much used.

(a) Aldehydes have a peroxidase-like property (see Feigl, 1960) in that they catalyse the oxidation of *p*-phenylenediamine by hydrogen peroxide. This reaction is also used as a spot-test in analytical chemistry.

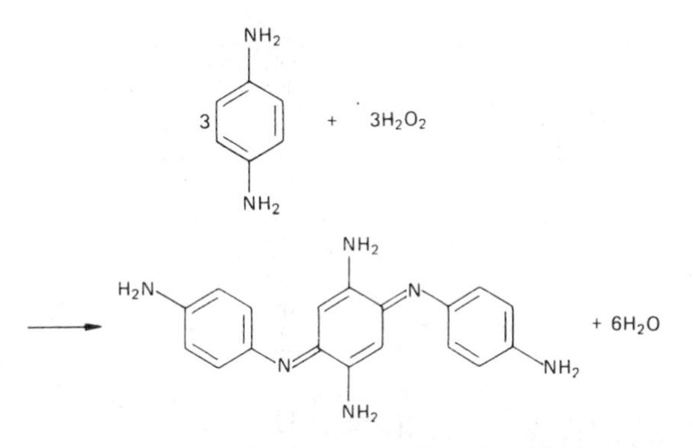

The product, which is known as Bandrowski's base, is darkly coloured and unstable, but is said to be stabilized by treatment with a solution of gold chloride (Scarselli, 1961). The structure of Bandrowski's base and the reactions involved in its formation are discussed by Corbett (1971).

(b) Another reagent used in spot-tests is 3-methyl-2-benzothiazolone hydrazone (MBTH) (Sanwicki *et al.*, 1961). It combines with aldehydes to form blue and green dyes and is a sensitive and specific histochemical reagent. The coloured product can be stabilized for a few days by treatment of the stained sections with ferric chloride or phosphomolybdic acid (Davis & Janis, 1966; Nakao & Angrist, 1968), though permanent preparations cannot be made.

(c) The aldehyde may be converted to its bisulphite addition compound:

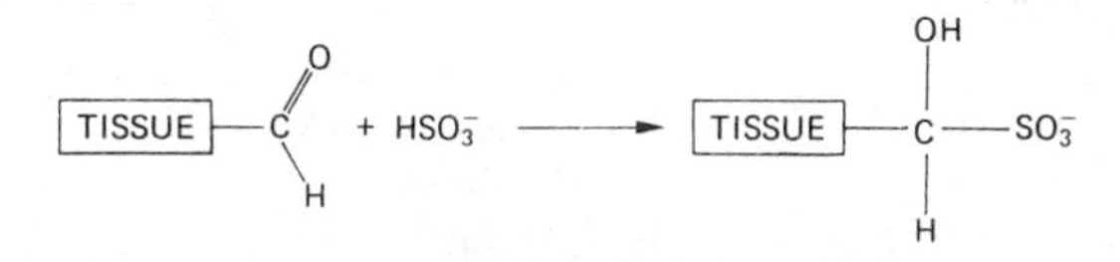

Since the addition compound is a sulphonic acid, it will bind cationic dyes such as toluidine blue (Rommanyi *et al.*, 1975) or alcian blue (Klessen, 1974) from solutions at pH 1.0 or lower.

10.10.6. Blocking procedures

Three blocking procedures for the carbonyl group are histochemically valuable. Most of the other available methods are unduly time-consuming, are damaging to the sections, or are too easily reversible. The three useful blocking agents are:

(a) Phenylhydrazine:

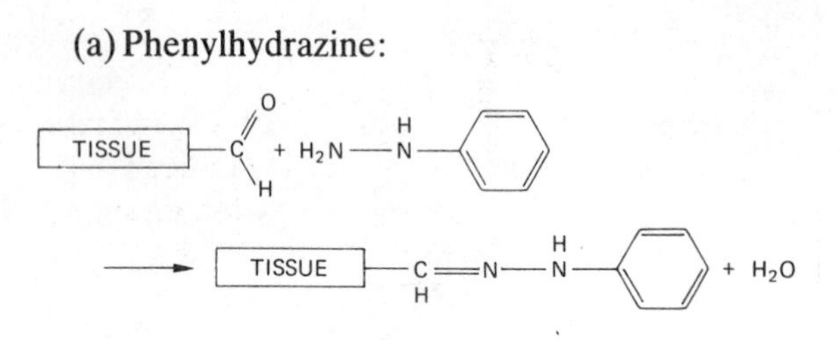

This is used as an aqueous solution of its hydrochloride or acetate.

(b) A solution of *m*-aminophenol in glacial acetic acid. The reaction is discussed in Section 10.10.3.

(c) Sodium borohydride:

$$4RCHO + NaBH_4 + 4H_2O \rightarrow 4RCH_2OH + B(OH)_3 + Na^+ + OH^-$$

This reducing agent is used as a mildly alkaline aqueous solution. It reduces aldehydes to primary alcohols and ketones to secondary alcohols. No other organic functional groups likely to be present in tissues are reduced, but it is worth noting that the C═N linkages of imines and hydrazones are reduced to C—N by this reagent (Billman & Diesing, 1957). Sodium borohydride is therefore the reagent of first choice for irreversible histochemical blocking of the carbonyl group. It is important that the pH of the solution be greater than 8.2. Neutral or less alkaline solutions will not reduce all the aldehyde groups present in a section (Bayliss & Adams, 1979).

10.11. INDIVIDUAL METHODS

The following procedures are all applicable to paraffin sections. Many fixatives are permissible, as explained in Section 10.1. Some of the methods are included for their instructional value rather than because they are practically useful. The chemical manipulations in Section 10.11.14 serve as controls when functional groups are being identified. The various blocking and unblocking reactions are summarized in Table 10.2.

10.11.1. Sulphation-induced basophilia

Solutions required

A. Sulphation reagent

Pack ice around a 100 ml conical flask containing 25 ml of diethyl ether. Slowly add 25 ml of concentrated sulphuric acid (which should have been pre-cooled by leaving at 4°C for 30–60 min). The mixture becomes hot. Use it when it has cooled to room

TABLE 10.2. *Some blocking and unblocking reactions*

Functional group	Blocking reaction	Product of blocking reaction	Unblocking reaction
—OH	Acetylation (3 h) Benzoylation Sulphation	$—OOCCH_3$ $—OOCC_6H_5$ $—OSO_3H$	Saponification Saponification Methylation (2 h)
—COOH	Methylation (2 h)	$—COOCH_3$	Saponification
$—OP(=O)(OH)_2$ $—OP(=O)(OH)O—$	Methylation (24 h)	$—OP(=O)(OCH_3)_2$ $—OP(=O)(OCH_3)O—$	Saponification
$—OSO_3H$ (of carbohydrates)	Methylation (24 h)	—OH	Irreversible
$—NH_2$	Acetylation (48 h) Benzoylation Sulphation Deamination (HNO_2)	$—NHCOCH_3$ $—NHCOC_6H_5$ $—NHSO_3H$ —OH, etc.	Saponification Saponification Methylation (2 h) Irreversible

temperature. After use, pour the mixture carefully into about 500 ml of tap water and then flush down the sink.

B. A cationic dye

A dilute solution of a thiazine dye (see Chapter 6) at pH 1.0 is suitable. Alcian blue at pH 1.0 (see Chapter 11) may also be used.

Procedure

1. Take sections to absolute ethanol and then into diethyl ether.
2. Place the slides in the sulphation reagent (A) for 5–10 min.
3. Rinse in 95% alcohol and then immerse in water, two changes, each 30 s.
4. Stain with a suitable cationic dye (solution B).
5. Wash in water, dehydrate appropriately for the dye used, clear in xylene, and mount in a resinous medium.

Result

Sites of sulphate esters are stained. When thiazine dyes are used, metachromatic effects are often seen.

Notes

1. A control section which has not been sulphated is necessary to demonstrate the distribution of sulphated mucosubstances.
2. Sulphation-induced basophilia is most prominent in carbohydrate-containing structures. Comparison with a section stained by the PAS method (see Chapter 11) is useful for distinguishing between carbohydrates and proteins.

10.11.2. Acid anhydride method for carboxyl groups

Solutions required

A. Acetic anhydride reagent

Acetic anhydride:	25 ml	Mix just before using
Pyridine:	25 ml	

This mixture is put into a screw-capped coplin jar in the wax oven and allowed to warm to about 60°C before adding the slides.

After using, allow the mixture to cool, then flush it down the sink, inside a fume cupboard, with copious tap water.

B. Hydroxynaphthoic acid hydrazide solution

2-hydroxy-3-naphthoic acid hydrazide:	50 mg	Keeps for several weeks
Glacial acetic acid:	2.5 ml	
Dissolve with aid of gentle heat, then add:		
Absolute ethanol:	to 50 ml	

This reagent is used at room temperature. It should be discarded after use.

C. 1% aqueous sodium bicarbonate (stable for several days)

D. 0.5 N hydrochloric acid (keeps indefinitely)

E. Coupling solution

Mix in order stated:

0.06 M phosphate buffer, pH 7.6:	15 ml	Mix immediately before using. Use once only
Absolute ethanol:	15 ml	
Fast blue B salt:	30 mg	

F. Glacial acetic acid, absolute ethanol and 50% ethanol should be at hand for some of the rinses

Procedure

1. De-wax paraffin sections and transfer the slides to glacial acetic acid.
2. Immerse in the pre-heated acetic anhydride reagent (solution A) for 1 h at about 60°C.
3. Rinse in glacial acetic acid.
4. Wash in absolute ethanol.
5. Immerse in the hydroxynaphthoic acid hydrazide solution (B) for 2 h.
6. Wash in three changes (each 10 min, with occasional agitation) of 50% ethanol.
7. Transfer slides to 0.5 N HCl (solution D) for 30 min.
8. Rinse in water and then in three changes of sodium bicarbonate (solution C).
9. Rinse in five changes of water. Make up solution E.
10. Immerse in the freshly prepared solution E for 2 min.
11. Wash in water, dehydrate through graded alcohols, clear in xylene, and cover, using a resinous mounting medium.

Result

Sites of protein-bound carboxyl groups red to purple.

Note

The specificity for carboxyl groups may be checked by appropriate blocking and unblocking procedures (Section 10.11.14).

10.11.3. Hydroxynaphthaldehyde method for amino groups

Solutions required

A. Hydroxynaphthaldehyde reagent

Dissolve 20 mg of 2,3-hydroxynaphthaldehyde in 20 ml of acetone. Add 30 ml of TRIS buffer, pH 8.5. Use immediately.

B. Coupling solution

TRIS buffer, pH 7.4:	50 ml	Make up immediately before using
Fast blue B salt:	25 mg	

Procedure

1. De-wax and hydrate paraffin sections. If a mercury-containing fixative has been used, **omit** the usual treatment with iodine and sodium thiosulphate.
2. Place slides in solution A for 1 h.
3. Wash in three changes of water, each 5 min with occasional agitation. Make up solution B.
4. Immerse in the coupling solution B for 3 min.
5. Wash in running tap water for 5 min.
6. Dehydrate through graded alcohols, clear in xylene, and mount in a resinous medium.

Result

Sites of high concentration of amino groups blue. Red or pink colours indicate lower density of amino groups.

Note

The specificity may be checked by means of suitable blocking procedures. See Weiss *et al.* (1954) for more information.

10.11.4. Sakaguchi reaction for arginine

This technique (Baker, 1947) is included mainly as an exercise in keeping sections on their slides in the presence of potentially damaging reagents. The slides are stained in a horizontal position, in the same way as blood-films (Chapter 7).

Solutions required

A. Sakaguchi reagent

Three stock solutions (all stable for several weeks) are needed:
(i) 1% α-naphthol in 70% ethanol, (ii) 1% aqueous sodium hydroxide, (iii) 1% aqueous sodium hypochlorite (Baker recommended Milton, a proprietary disinfectant containing 1% NaClO available in Britain. In North America, household bleaches such as Javex and Clorox contain approximately 5% NaClO. They work well after appropriate dilution with water. British bleaches are sometimes cloudy and may be unsuitable. Solutions of sodium hypochlorite are also available from chemical supply houses.)

Add two drops of the α-naphthol solution to 2 ml of the sodium hydroxide solution, then add four drops of the 1% sodium hypochlorite. This reagent must be mixed **immediately** before pouring it onto the sections.

B. Mounting medium

Pyridine:	3.0 ml
Chloroform:	1.0 ml

This is usually made on the day of use, but is probably quite stable.

Procedure

1. Paraffin sections are de-waxed, coated with a film of nitrocellulose (Baker recommended 1% celloidon, but the method using a 0.2% solution described in Chapter 4 is satisfactory), which is then hardened in 70% ethanol.
2. Transfer the slides to water. Make up solution A.
3. Shake off the water and pour on the freshly prepared Sakaguchi reagent (solution A). Leave for 15 min.
4. Drain the slides and blot carefully with filter paper.
5. Rinse in the pyridine–chloroform mixture (B).
6. Apply a coverslip, using the pyridine–chloroform mixture (B) as mounting medium. Examine immediately.

Result

Sites of arginine residues pink to red. High concentrations occur in nuclei, granules of eosinophils and Paneth cells, heads of spermatozoa, keratohyaline, and hairs. The colour fades in about 6 h.

Notes

1. Conventional clearing and mounting largely decolorize the sections, so permanent preparations cannot be made.
2. The next method is much more convenient but does not give such a bright colour as this one.
3. The specificity for arginine may be checked by blockade with benzil. Deamination, methylation, and acetylation do not prevent the reaction.

10.11.5. NQS method for arginine

This method (Lillie *et al.*, 1971) is technically easier than the preceding one, but still involves the use of an alkaline reagent, which is likely to cause detachment of the sections.

Solutions required

A. NQS reagent

Two stock solutions, both stable, are needed: (i) 0.4 N aqueous sodium hydroxide (1.6% NaOH), (ii) 1.0 M barium chloride (24.4% $BaCl_2 . 2H_2O$). Mix 10 ml of (i) with 10 ml of (ii) and 20 ml of water. Add 100 mg of 1,2-naphthoquinone-4-sulphonic acid sodium salt, which dissolves quickly. The solution is unstable and must be used at once.

B. Sodium acetate solution

1% CH_3COONa in water.

Procedure

1. De-wax and hydrate paraffin sections. It is advisable to coat with a film of nitrocellulose. Make up solution A.
2. Place slides in the NQS reagent (solution A) for 10 min.
3. Rinse in sodium acetate (solution B), 10 s.
4. Rinse in water, 10 s.
5. Dehydrate through graded alcohols, clear in xylene, and cover, using a resinous mounting medium.

Result

Sites of high concentration of arginine red to brown. Lower concentrations pink to orange. The colour does not fade.

Notes

See notes appended to the previous technique (Section 10.11.4).

10.11.6. Rosindole reaction for tryptophan

This technique (Glenner, 1957) works best with tissues fixed in alcohol–acetic acid mixtures or for less than 24 h in neutral buffered formaldehyde or formal–calcium.

Solutions required

A. Modified Ehrlich's reagent

p-dimethylaminobenzaldehyde:	1.0 g
Concentrated perchloric acid (60% $HClO_4$):	5.0 ml
Concentrated hydrochloric acid:	1.0 ml
Glacial acetic acid:	34 ml

This solution is stable for several days in a tightly closed container. The three acids used are all corrosive and must not be pipetted by mouth. The perchloric acid should not come into contact with heavy metals or their salts: explosive perchlorates could be formed. Discard the solution by pouring it into at least 1 litre of tap water and then flushing down the sink.

B. Oxidizing solution

Mix 35 ml glacial acetic acid with 5 ml concentrated hydrochloric acid. Weigh out 500 mg of sodium nitrite ($NaNO_2$) and put into a coplin jar. Immediately before use, pour the mixture of acids onto the sodium nitrite, which dissolves at once with evolution of pungent oxides of nitrogen. **This solution should be mixed and used in a fume cupboard.** After use, flush down sink with plenty of water.

C.

A mixture of equal parts of glacial acetic acid and xylene is needed for washes.

Procedure

1. De-wax paraffin sections and take to absolute ethanol. Wave the slides about in the air for a few seconds until the sections are almost dry.
2. Immerse slides in the modified Ehrlich's reagent (solution A) for 3–4 min.
3. Mix the oxidizing solution B and transfer the slides directly into it. Leave for 1 min. See *Note 1* below.
4. Wash in two changes of a mixture of equal parts acetic acid and xylene.

5. Clear in xylene (two changes) and mount in a resinous medium.

Result

Proteins containing tryptophan are stained blue.

Notes

1. Be careful not to leave the slides for too long at stage 3 since excessive oxidation causes bleaching of the colour.
2. The $HClO_4$ and HCl serve as strong acids for catalysis of the reaction. In a similar technique (Adams' "DMAB-nitrite" method), the DMAB is dissolved in concentrated HCl, which can be injurious to the sections.
3. This reaction is prevented by pre-treatment of the sections with a solution of 1.0 g potassium persulphate ($K_2S_2O_8$, also known as potassium peroxidisulphate) in 0.5 M aqueous potassium hydroxide, overnight. Coating with a film of celloidin is advisable.

10.11.7. Millon reaction for tyrosine

Solution required

Millon's reagent (Baker's (1956) modification)

Mercuric sulphate ($HgSO_4$):	10 g	Stable until the $NaNO_2$ has been added. Solution containing $NaNO_2$ should be used within one day
Water:	90 ml	
Concentrated sulphuric acid (S.G. 1.94):	10 ml	
Heat until the $HgSO_4$ is dissolved, then add:		
Water:	to 200 ml	
Sodium nitrite ($NaNO_2$):	50 mg	

Procedure

1. **Either** use free nitrocellulose sections **or** paraffin sections attached to slides, de-waxed, and hydrated.
2. Place the sections (or stand the slides) in a beaker containing Millon's reagent. Heat the beaker gently (electric hotplate or low flame of a Bunsen burner) until the solution boils.
3. Remove the beaker from the source of heat and leave it to cool to room temperature.
4. Wash the sections (or slides) carefully in three changes of water.
5. **Either** mount in an aqueous medium **or** dehydrate through alcohols, clear in xylene, and mount in a resinous medium.

Result

Proteins containing tyrosine appear in various shades of red, pink, and orange.

Note

Although the Millon reaction occurs specifically with tyrosine, virtually all proteins give a positive reaction since they all contain this amino acid.

10.11.8. Ferric ferricyanide reaction

This is a rather general reaction for reducing groups, as explained on p. 125. If the sections are treated for conversion of disulphide to sulphydryl groups (see p. 142) between stages 2 and 3, —SH generated from —S—S— will be demonstrated. Controls with sulphydryl blockade are necessary if any demonstrated site of reduction is to be attributed to cystine or cysteine. Free sulphydryl groups naturally present in tissues are seen to best advantage after fixation in Carnoy or a similar non-reactive coagulant fixative.

Solutions required

A. Ferric ferricyanide reagent

This is made from two stock solutions, both of which are indefinitely stable:

(i) Potassium ferricyanide, $K_3Fe(CN)_6$:	24 mg
Water:	60 ml
(ii) Ferric chloride ($FeCl_3.6H_2O$):	5.0 g
Water:	500 ml

The working solution, prepared immediately before use and used only once, is made by mixing 10 ml of (i) with 30 ml of (ii).

B. 1% acetic acid

Glacial acetic acid:	1.0 ml	Keeps indefinitely
Water:	to 100 ml	

Procedure

1. De-wax and hydrate paraffin sections. If desired reduce —S—S— to —SH (see Section 10.11.14.9). Blocking reactions for —SH (Section 10.11.14.10) may be carried out after or instead of the reduction, with control sections.

2. Immerse in solution A for 10 min.
3. Rinse in 1% acetic acid (solution B). See *Notes 1* and *2*.
4. Dehydrate through graded alcohols, clear in xylene, and mount in a resinous medium.

Results

Sites of strong reduction blue. Sites of weak reduction (most of the "background") yellow green. See Lillie & Burtner (1953) for a list of positively reacting substances.

Notes

1. The wash in dilute acetic acid is probably unnecessary. The idea is to preserve as much as possible of the Prussian blue, which is soluble in alkalis and might be slightly extracted by a neutral liquid such as tap water.
2. A counterstain may be applied after stage 3 if considered necessary. Mayer's carmalum (p. 83) is convenient for nuclei, but usually the background coloration is adequate for recognition of the architecture of the tissue.

10.11.9. Mercurochrome method for cysteine

Solutions required

A. Mercurochrome reagent

Mercurochrome (= merbromin):	20 mg	Stable for several weeks
Water:	0.5 ml	
Dissolve, then add: *N,N*-dimethylformamide:	100 ml	

B. *N,N*-dimethylformamide is required for washes

Procedure

1. De-wax and hydrate paraffin sections. If desired, apply a reaction for the conversion of cystine to cysteine. Wash in water, drain, and proceed to stage 2.
2. Immerse in mercurochrome reagent (solution A) for **either** 1 h **or** 48 h (for fluorescence or conventional microscopy, respectively).
3. Wash in two changes of *N,N*-dimethylformamide, each 3 min.
4. Dehydrate in absolute ethanol (two changes), clear in xylene, and cover, using a non-fluorescent resinous mounting medium.

Result

Bright green fluorescence (excitation by blue or ultraviolet) at sites of sulphydryl groups, after 1 h at stage 2.

Red to pink staining of sulphydryl sites after 48 h at stage 2.

Notes

1. The specificity of staining may be confirmed in control sections by applying a sulphydryl-blocking procedure before stage 2 of the technique. If non-specific background staining is troublesome (mercurochrome is an anionic dye, but has little staining power in the solvent used), try washing it out with water or 70% alcohol.
2. Some batches of mercurochrome are unsatisfactory for this method (Lillie, 1977) for reasons unknown.

10.11.10. Cysteic acid method for cystine

This is a slightly modified version of the technique of Adams & Sloper (1955). Fixation is not critical, but neurosecretory material is not preserved by alcoholic fixatives.

Solutions required

A. Performic acid

This reagent should be prepared in a 150–200 ml conical flask. A magnetic stirrer should be used, but if one is not available, mix the ingredients in a 200 ml beaker and stir with a glass rod, taking great care to avoid splashing and spillage. See *Note 4* below.

Formic acid (the 98% acid is recommended, but the 88% acid is suitable):	80 ml
Hydrogen peroxide (30%; "100 volumes available oxygen"):	8 ml
Concentrated sulphuric acid (96%):	1 ml

Stir vigorously at intervals for 1 h. This solution is highly corrosive. It should be free of bubbles when used. It keeps for 24 h. Dilute with at least 1 litre of tap water before discarding. See *Note 5* below for alternatives to performic acid.

B. Alcian blue

Alcian blue (see p. 74):	6 g	Keeps for several months
Water:	200 ml	
Concentrated sulphuric acid (96%):	5.5 ml	

Heat to 70°C, cool to room temperature and filter. With some batches of alcian blue it is impossible to obtain a 3% solution, but this does not matter.

Procedure

1. De-wax and hydrate paraffin sections. Coating with a film of nitrocellulose (see Chapter 4) is advisable, though not always necessary. Blot dry with filter paper.
2. Immerse in performic acid (solution A) or alternative (see *Note 5* below) for 5 min.
3. Rinse in four changes of tap water, without agitation, for a total of 10 min.
4. Transfer to 70% and then to absolute alcohol. Blot the sections to flatten creases and return to water. Place slides on a hotplate until just dry.
5. Stain in alcian blue (solution B) for 1 h.
6. Wash in three or four changes of tap water until all excess dye is removed.
7. Counterstain if desired (see *Note 1* below).
8. Wash in water, blot dry, dehydrate in two changes (each 5 min) of *n*-butanol, clear in xylene, and cover. See *Note 2* below.

Result

Sites of high cystine concentration (>4% of total amino-acid content of protein) blue. Sulphated carbohydrates are also blue (see *Note 3*).

Notes

1. Suitable counterstains are: 0.5% aqueous neutral red or safranine (for nuclei, etc.); 0.5% aqueous eosin (for general pink background). The alcian blue resists extraction by most of the commonly used histological reagents.
2. If the sections look as if they will come off the slides, mount into a resinous medium directly from the second change of *n*-butanol.
3. To control for the specificity for cystine, omit the oxidation (stage 2).
4. Performic and peracetic acids can react explosively with copper and its salts. Use clean glassware and remember to dilute the reagent before pouring it down the sink.
5. An alternative oxidizing agent is peracetic acid, $H_3C{-}C(=O){-}O{-}OH$. This is much more stable than performic acid and is available commercially as a 40% solution. Peracetic acid can also be made in the laboratory: thoroughly mix 40 ml of acetic anhydride with 10 ml of 30% hydrogen peroxide; leave to stand for 24 h and then add an equal volume of water. A less noxious but less specific oxidizing agent is an acidified permanganate solution (Pasteels & Herlant, 1962; see Adams, 1965, and Gabe, 1976, for discussion): sections are immersed for 5 min in a freshly prepared 0.5% solution of potassium permanganate ($KMnO_4$) in 2% sulphuric acid and then rinsed in 1% aqueous oxalic acid ($H_2C_2O_4 . 2H_2O$) to remove the brown stain of manganese dioxide from the sections. A further wash in water precedes staining with alcian blue. Oxidation with permanganate is acceptable when the cysteic acid method is being used for purely histological purposes such as the study of cell types in the hypophysis or endocrine pancreas. Performic or peracetic acid must be used for critical histochemical identification of cystine-containing proteins.
6. Other cationic dyes in acid solution (pH <1.0) may be substituted for alcian blue.

10.11.11. Brilliant indocyanine 6B for proteins

This technique, based on that of Cawood *et al.* (1978), is suitable for paraffin sections of tissues fixed in formaldehyde, Carnoy or Bouin. See *Note* below.

Solutions required

A. Acetic–ethanol

Glacial acetic acid:	100 ml	Keeps indefinitely
Absolute ethanol:	300 ml	

B. Dye solution

Brilliant indocyanine 6B (see p. 126 for synonyms):	40 mg
Acetic ethanol (solution A above):	200 ml

The dye dissolves completely and filtration is not needed. The solution is stable for at least a year.

Procedure

1. De-wax and hydrate paraffin sections and take to absolute ethanol.
2. Stain in the dye solution (B) for 30 min.
3. Drain slides and immerse in acetic–ethanol (solution A) for 5 min with occasional agitation.
4. Rinse in 95% ethanol, dehydrate in absolute ethanol, clear in xylene, and cover, using a resinous mounting medium.

Result

Proteins are stained a bright royal blue. The depth of colour is proportional to the local concentration of protein.

Note

In the original description of this technique (Cawood *et al.*, 1978), plastic-embedded sections were stained for 24 h and stage 3 lasted 20 min. When paraffin sections are used, there is no increase in intensity of staining after 15–20 min and subsequent washing in acetic–ethanol removes hardly any dye from the sections. The specificity for proteins is not fully established. Carbohydrate-rich components of tissues, such as the matrix of cartilage, are very weakly stained by comparison with cytoplasm and collagen. Nuclei are recognizable in the stained sections but are no more strongly coloured than the surrounding cytoplasm.

10.11.12. The coupled tetrazonium reaction

This is the technique of Danielli (1947) modified by the substitution of the tetrazonium salt of *o*-dianisidine, which is commercially available in stabilized form, for that of benzidine.

Solutions required

A. Tetrazonium solution

Fast blue B salt (tetraazotized *o*-dianisidine. **Caution:** handle with care since this compound may be carcinogenic):	200 mg	This solution is used immediately
TRIS buffer, pH 9.2, **cooled to 0–4°C:**	100 ml	

B. 0.01 N hydrochloric acid

Concentrated hydrochloric acid:	1.0 ml	Keeps indefinitely
Water:	to 1000 ml	

This must be **cooled to 0–4°C** for use.

C. Coupling solution

8-amino-1-naphthol-3,6-disulphonic acid, sodium salt (H-acid):	50 mg	Use within 30 min
TRIS buffer, pH 9.2, **cooled to 4°C:**	100 ml	

D. TRIS buffer, pH 9.2

About 500 ml, **cooled to 4°C**, will be needed for rinses.

Procedure

1. Use frozen sections, or de-wax and hydrate paraffin sections. The latter give clearer results. The alkaline reagents may cause detachment of sections from slides. Coating with a film of celloidin is often necessary.
2. Immerse in the freshly prepared tetrazonium solution (A) for 15 min **at 0–4°C.**
3. Pass sections through four changes of TRIS buffer, pH 9.2 **at 0–4°C** and into water, taking at least 5 min for the whole process.
4. Immerse in 0.01 N HCl (solution B) at 4°C for 2 min.
5. Rinse quickly in three changes of water.
6. Immerse in the coupling solution (C) for 15 min **at 0–4°C**.
7. Wash in three changes of water.
8. Dehydrate through graded alcohols, clear in xylene, and cover, using a resinous mounting medium.

Result

Proteins: red to brown. The colour is stronger if stages 4 and 5 are omitted.

Note

For a more rapid method, probably with higher specificity for proteins, substitute a solution of fast black K salt (approximately 1.0 mg per ml, in any convenient buffer, pH 8.5, 0–4°C) for solution A of the above method. Proceed as described, but omit stages 6 and 7. The colours obtained are similar to but a little lighter than those seen with the coupled tetrazonium reaction.

10.11.13. Some methods for aldehydes and ketones

Schiff's reagent. The preparation of Schiff's reagent is described on p. 110, in conjunction with its use in the Feulgen technique. It is used as described on pp. 111, 164, and 186.

Basic fuchsine in acid alcohol. This substitute for Schiff's reagent is described on p. 111. It may be used whenever an alcoholic reagent is acceptable. It is not suitable, of course, for the demonstration of lipid-bound aldehydes (see Chapter 12).

A related reagent is 0.01% acriflavine in 90% ethanol (Levinson *et al.*, 1977). This is applied to the sections for 10 min. Washing in three changes of 70% ethanol (total time 5 min) is followed by dehydration, clearing in xylene, and mounting in a non-fluorescent resinous medium. The preparations are examined by fluorescence microscopy with excitation by blue (490 nm) light. Green fluorescence is seen at aldehyde-containing sites.

Hydroxynaphthoic acid hydrazide. The reagent is made up as follows:

2-hydroxy-3-naphthoic acid hydrazide:	200 mg	Keeps for several weeks
Absolute ethanol:	100 ml	
Glacial acetic acid:	10 ml	
Water:	90 ml	

Dissolve with aid of gentle heat (up to 70°C) and vigorous stirring.

Sections of tissue in which aldehyde or ketone groups have been produced are immersed in the HNAH reagent for 1 h at 60°C or for 3 h at room temperature. They are then washed in three changes (each 10 min) of 50% ethanol, taken to water, and reacted at 0–4°C with fast blue B salt (1 mg per ml, pH 7.5, 3 min). The sections are then washed, dehydrated, cleared, and mounted in a resinous medium. Sites of aldehydes and ketones are shown in blue and purple. There is also a pink to light brown background staining, which has no histochemical significance.

Amine–aldehyde condensation with *m*-aminophenol, followed by coupling with fast black K salt, is described on p. 143.

10.11.14. Blocking and unblocking procedures

Some of these reactions affect more than one functional group. Their uses are summarized in Table 10.1 (p. 132). The remaining reactions are used for specific purposes.

10.11.14.1. Acetylation

Reagent

Pyridine (anhydrous):	24 ml	Use once only. Stable for a few days
Acetic anhydride:	16 ml	

Procedure

De-wax paraffin sections and take into absolute ethanol and then into pyridine. Transfer to the acetylating reagent and leave in a screw-capped coplin jar for **either** 3 h **or** 48 h at 37°C. Wash in absolute alcohol and take to water.

Effects

The shorter treatment blocks hydroxyl groups. The longer treatment blocks hydroxyl and amino groups. These blockades are largely reversed by saponification (see Section 10.11.14.5).

10.11.14.2. Benzoylation

Reagent

Benzoyl chloride:	2 ml	Use once only. Stable for about 1 week
Pyridine (anhydrous):	38 ml	

Procedure

De-wax paraffin sections and take into absolute ethanol and then into pyridine. Transfer to the reagent for 24 h at room temperature. Wash in absolute alcohol and take to water.

Effects

Hydroxyl and amino groups are blocked. The blockade can be largely reversed by saponification.

10.11.14.3. Sulphation–acetylation

This is the fastest way to block amino groups (Lillie, 1964). Sulphation of hydroxyl groups also occurs.

Reagent

Acetic anhydride:	10 ml	Stable for about 1 week. Use once only
Glacial acetic acid:	30 ml	
Concentrated sulphuric acid:	0.1 ml	

Keeps for several weeks.

Procedure

De-wax paraffin sections and take to absolute ethanol and then into glacial acetic acid. Transfer to the reagent for 10 min at room temperature. Wash in running tap water for 5 min.

Effects

Amino and hydroxyl groups are blocked, mainly by sulphation. This effect is reversed by treatment for 3 h at 60°C with the reagent used for methylation (see Section 10.11.4.4 below). Some acetylation may also occur. Histochemical reactions of arginine are unimpaired. It is possible, however, that N-sulphation of guanidino groups occurs (preventing the attachment of anionic dyes and inducing basophilia) but is reversed by hydrolysis in the strongly alkaline reagents used in specific methods for arginine.

10.11.14.4. Methylation and Desulphation

Reagent

Methanol (absolute):	60 ml	Use immediately
Concentrated hydrochloric acid:	0.5 ml	

Procedure

De-wax paraffin sections, take into absolute ethanol, and then into absolute methanol. Transfer to the reagent in a tightly screw-capped coplin jar, which should be no more than two-thirds full, at 60°C for the required time (see below). Rinse in 95% ethanol and bring to water.

Effects

The chemical changes produced depend on the length of time for which the reaction is allowed to proceed:

2 h: Carboxyl groups are esterified.
Sulphate groups introduced experimentally are released with re-establishment of original $—NH_2$ or $—OH$.
There is partial hydrolysis of naturally occurring sulphate esters (of carbohydrates) and partial esterification of phosphate-ester groups (of nucleic acids).

24 h: Carboxyl and phosphate groups are esterified.
All sulphate esters are hydrolysed. Primary amines are largely converted to tertiary amines. This increases their acidophilia but prevents chemical reactions characteristic of the $—NH_2$ group.

The effects of methylation, **except for removal of sulphate groups**, are reversible by saponification (see Section 10.11.14.5).

10.11.14.5. Saponification

This procedure is used for reversal of effects of acylation and methylation. The reagent is alkaline and may remove sections from slides. Coating with a film of celloidin is sometimes helpful in preventing losses of sections.

Reagent

Water:	25 ml	Use once only. Stable for a few days
Potassium hydroxide (KOH):	1.0 g	
Absolute ethanol:	75 ml	

Procedure

Take sections to 70% ethanol, then immerse in the reagent for 30 min at room temperature. Rinse with minimum agitation in 70% ethanol, blot to flatten sections if necessary. Rinse in water. Allow the sections to dry in air if they look as if they are only loosely adherent to their slides. Proceed with the staining method.

10.11.14.6. Deamination with Nitrous Acid

Reagent

Sodium nitrite ($NaNO_2$):	7.0 g	Use immediately
Water:	94 ml	
Dissolve, then add:		
Glacial acetic acid:	6.0 ml	

Procedure

De-wax and hydrate paraffin sections. Immerse in the reagent for 24 h. Wash in running tap water for 5 min.

Effects

Primary amine groups are removed. The guanidino group of arginine is only slightly affected. Most of the deamination occurs in the first 4 h, but it is occasionally necessary to apply the reagent for 48 h.

10.11.14.7. Benzil blockade of arginine

Reagent

Benzil:	1.6 g	Prepare just before using. Use only once
Ethanol (absolute):	32 ml	
Water:	8 ml	
Sodium hydroxide (NaOH):	0.8 g	

This is a saturated solution. Some benzil may remain undissolved.

Procedure

De-wax paraffin sections and take to 95% ethanol. Transfer to the reagent and leave for 1 h at room temperature. Rinse in 70% ethanol and then in three changes of water.

Effects

Histochemical reactions of arginine are prevented. Acidophilia generally is depressed, indicating that there may also be partial blockade of amino groups.

10.11.14.8. Iodination of tyrosine

Reagent

Iodine (I_2):	1.0 g	This reagent is stable for several weeks
Potassium iodide (KI):	2.0 g	
Water:	300 ml	
Ammonium hydroxide (28% NH_3; S.G. 0.9):	approximately 2 ml	
Add this gradually until the solution has a pH of about 10		

Decolorizing solution

Potassium metabisulphite:	1.0 g	Keeps for several weeks
Water:	200 ml	

Procedure

De-wax and hydrate paraffin sections. Coating with a film of nitrocellulose will help to keep the sections on the slides. Transfer to the iodination reagent and leave for 24 h. Rinse in water. Immerse in the decolorizing solution until the yellow stain of iodine disappears from the sections. Wash in running tap water for 2 min.

Effects

The Millon reaction for tyrosine is prevented. There is also partial blockade of histochemical reactions for tryptophan.

10.11.14.9. Reduction of cystine to cysteine

Many reagents are available for the conversion of —S—S— to —SH. Two are given below. The first of these, sodium thioglycollate, is the most widely used. The second, dithiothreitol, is just as effective and can be used in a less strongly alkaline solution, but is more expensive.

Thioglycollate reduction

Take sections to water and immerse for 15 min in:

Thioglycollic acid ($HSCH_2COOH$):	5.0 ml	Use at once
Water:	80 ml	
4% (1.0 N) sodium hydroxide (NaOH): Add dropwise until the pH is 9.5		
Then add: Water:	to 100 ml	

Wash carefully in five changes of water.

Dithiothreitol reduction

Take sections to water and immerse for 30 min in:

Dithiothreitol (Cleland's reagent):	1.5 g	Stable for about 3 days in tightly closed container
0.03 M phosphate buffer, pH 8.0:	50 ml	

Wash in five changes of water

Effects

Disulphides are reduced to thiols, for which histochemical tests become positive.

10.11.14.10. Sulphydryl blocking reagents

Either of the following will block sulphydryl groups effectively. Sections are hydrated and then treated as described below. Both reagents should be freshly dissolved.

N-ethylmaleimide

Dissolve 625 mg of this compound in 50 ml of phosphate buffer, pH 7.4. Treat sections for 4 h at 37°C, rinse in 1% acetic acid, wash thoroughly in water.

Iodoacetate

Dissolve 0.9 g of iodoacetic acid in 40 ml of water. Add 1.0 N (4%) NaOH until the pH is 8.0. Make up to 50 ml with water. Treat sections for 18–24 h at 37°C. Wash thoroughly in water.

10.11.14.11. BLOCKING REACTIONS FOR ALDEHYDES AND KETONES

The following three procedures all effectively block aldehyde groups. The blockades are resistant to subsequent treatment with periodic acid. The first method is generally the most satisfactory. Both aldehydes and ketones are reduced by sodium borohydride and blocked by phenylhydrazine. The third method is probably effective only for aldehydes.

Sodium borohydride

Take sections to water and transfer to a freshly prepared solution:

Sodium phosphate, dibasic (Na_2HPO_4):	0.5 g	Use at once
Water:	50 ml	
Dissolve, then add: Sodium borohydride ($NaBH_4$):	25 mg	

Leave slides in this for 10 min, with occasional agitation to release bubbles of hydrogen from surfaces of slides. Wash in four changes of water.

Caution. Sodium borohydride releases hydrogen on contact with acids. Do not use this reagent near any naked flame. Do not acidify its solutions. Discard the alkaline solution by flushing down the drain with several litres of tap water.

Phenylhydrazine

Take sections to water and transfer to the following solution in a screw-capped coplin jar:

Phenylhydrazine: (**Caution:** poisonous: avoid contact with skin)	5 ml	Prepare just before using
Glacial acetic acid:	10 ml	
Water:	to 50 ml	

Leave the slides in this reagent, in the wax oven (about 60°C) for 3 h. Rinse in three changes of 70% alcohol and then in water.

Amine–aldehyde condensation

Take sections to absolute ethanol and then into glacial acetic acid. Transfer to the following solution:

m-aminophenol:	5.5 g
Glacial acetic acid:	50 ml

Keeps for about 2 weeks. May be used three or four times

Leave in this, at room temperature, for 1 h. Rinse the slides in 95% alcohol and take to water.

This procedure can be extended to provide a staining method for aldehydes. After reaction with *m*-aminophenol and washing as described above, the slides are immersed for 2 min at 0–4°C in a freshly prepared solution containing 150 mg fast black K salt (see p. 57) in 50 ml of 0.1 M barbitone–HCl buffer, pH 8.0. The slides are then washed in three changes of 0.1 N hydrochloric acid (each 5 min) to remove excess diazonium salt (and perhaps also to decompose triazenes, see p. 127), rinsed in water, dehydrated, cleared, and mounted in a resinous medium. Sites of aldehydes are stained black to deep purple. Background staining due to azo-coupling with proteins (see p. 128) also occurs, in various shades of pink and brown. The theory and practice of this technique are discussed in great detail by Lillie (1962).

10.12. EXERCISES

Theoretical

1. The cytoplasmic granules of the Paneth cells of the intestine are stained strongly by eosin in the H. & E. and the azure–eosin techniques. How would you set about determining which cationic groups are responsible for the acidophilia of the granules?

2. What would you expect to happen when a section of tissue is treated with a proteolytic enzyme of broad specificity such as trypsin? Could such an enzyme be used to prove that an object seen in a routinely stained section is composed of protein?

3. What would be stained in a tissue containing cartilage (whose matrix is rich in sulphated proteoglycans) subjected to the following treatments:
(a) methylation for 24 h at 60°C; (b) sulphation in ether-sulphuric acid reagent; (c) staining with 0.1% azure A at pH 1.0?

Is it possible, using this or any other procedure, to demonstrate selectively the hydroxyl groups of proteins?

4. How may azo dyes be formed at the sites of (a) aromatic amino acids, (b) amino groups, (c) ketones?

Practical

5. Using paraffin sections of suitably fixed tissues, demonstrate: (a) protein-bound carboxyl groups (any tissue); (b) protein-bound amino groups (any tissue); (c) arginine in nuclei; (d) tyrosine (any tissue); (e) tryptophan in salivary or pituitary gland; (f) cystine in skin or pituitary gland.

Carry out an appropriate blocking reaction to confirm the specificity of each technique.

6. Stain sections of external ear and intestine for protein, using both brilliant indocyanine 6B and the coupled tetrazonium reaction. What structures are only faintly stained? Perform other staining methods (see Chapters 6 and 11) and attempt to determine which organic functional groups are present at sites where very little protein can be demonstrated.

11

Carbohydrate Histochemistry

11.1. Constituent sugars of mucosubstances 145
11.1.1. Structural formulae 145
11.1.2. Some monosaccharide units 146
11.2. Classification of mucosubstances 147
11.2.1. Types of mucosubstance 147
11.2.2. Composition of common mucosubstances 148
11.2.2.1. Polysaccharides 148
11.2.2.2. Proteoglycans 148
11.2.2.3. Glycoproteins 149
11.3. Histochemical methodology 150
11.3.1. Use of cationic dyes 150
11.3.1.1. Alcian blue 150
11.3.1.2. Metachromasia 151
11.3.1.3. Colloidal ferric hydroxide 152
11.3.2. Chemical methods—the periodic acid–Schiff method 152
11.3.2.1. Periodate oxidation 153
11.3.2.2. Schiff's reagent 156
11.3.2.3. Artifacts associated with the PAS method 156
11.3.3. Lectins as histochemical reagents 157
11.3.3.1. Properties of lectins 157
11.3.3.2. Technical considerations 157
11.3.3.3. Interpretation of results 160
11.4. Enzymes as reagents in carbohydrate histochemistry 160
11.4.1. Amylase 160
11.4.2. Hyaluronidase 160
11.4.3. Neuraminidase 160
11.5. Chemical blocking procedures 161
11.5.1. Acetylation of hydroxyl groups 161
11.5.2. Methylation and desulphation 161
11.5.3. Mild acid hydrolysis 162
11.6. Individual methods 162
11.6.1. Alcian blue (pH 1.0 and 2.5) 162
11.6.2. Toluidine blue for metachromasia 162
11.6.3. Colloidal ferric hydroxide 163
11.6.4. The periodic acid–Schiff procedure 163
11.6.4.1. Standard PAS method 163
11.6.4.2. Specialized PAS methods 164
11.6.5. The concanavalin A–peroxidase method 165
11.6.6. Enzymatic extractions 166
11.6.6.1. Amylase 166
11.6.6.2. Hyaluronidase 166
11.6.6.3. Neuraminidase 166
11.6.7. Chemical blocking and unblocking procedures 167
11.6.7.1. Acetylation of hydroxyl groups 167
11.6.7.2. Methylation 167
11.6.7.3. Saponification 167
11.6.7.4. Mild acid hydrolysis 167
11.7. Exercises 167

Although the term "carbohydrate" embraces all the sugars and their derivatives, the only such substances available in sections of fixed tissue are those in which the sugars form parts of macromolecules or of lipids. The latter (the glycolipids) are more properly considered together with other lipids in Chapter 12 and only the former will be discussed at this stage. The compounds concerned are the **polysaccharides**, the **glycoproteins**, and the **proteoglycans**. Collectively, these compounds are known as **mucosubstances**. Histochemical methods for mucosubstances are numerous and only those which are widely used or are of theoretical interest will be discussed here.

11.1. CONSTITUENT SUGARS OF MUCOSUBSTANCES

Mucosubstances are identified histochemically by virtue of the properties of their constituent sugars. The monosaccharide residues most commonly encountered in polysaccharides, glycoproteins, and proteoglycans will now be described. The chemistry is explained at greater length by Brimacombe & Webber (1964), Gottschalk (1972), and Cook & Stoddart (1973). Textbooks of biochemistry also contain accounts of carbohydrate chemistry, and an introductory review intended for histochemists is given by Barrett (1971).

11.1.1. Structural formulae

The structural formulae are shown as rings that lie in a plane perpendicular to the paper. The thickened lines represent bonds in the side of the ring that is nearer to the reader. All these sugars exist as six-membered pyranose rings. The numbering system is shown below for β-D-glucose. In the α-anomers of sugars of the D-series, the hydroxyl group at position C1 is directed downwards. The formulae for the L-enantiomers are obtained by envisaging the images in a mirror held **in the plane of the ring**. In α-L-sugars, therefore, the hydroxyl at position C1 is directed upwards and C6 lies below rather than above C5. Anomers differ only in the configuration at position C1. Epimers differ in the direction of the hydroxyl group at one carbon atom other than C1. Thus, α-D-glucose and β-D-glucose are anomers. Galactose and mannose are epimers of glucose, but not of one another.

In the complex carbohydrates, monosaccharide units are joined together by **glycosidic** linkages. Each unit or residue may be called a glycosyl group. Carbon atom C1 of one sugar is connected via an oxygen atom to one of the carbon atoms (most commonly C3, C4, or C6) of another sugar. In the names of glycosides, the linkage is indicated in an abbreviated form such as α-1→4, which shows which carbon atoms are joined and what the configuration is at C1. It should be noted that the formulae as they are printed here do not accurately represent the shapes of either the sugar units or the larger molecules. Many of the bonds have to be bent or distorted in order to make the structures fit tidily onto the page.

11.1.2. Some monosaccharide units

All but xylose and NANA are hexoses and their carbon atoms are numbered as shown for β-D-glucose.

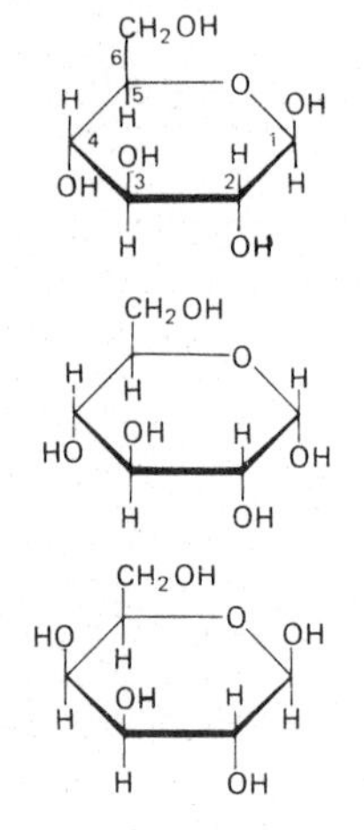

β-D-glucose. β-D-Glc. This is the easiest structure to remember since the hydroxyl groups are placed alternately above and below the ring

α-D-glucose. α-D-Glc. Differs from the β-anomer only in that the hydroxyl at C1 is directed downwards

β-D-galactose. β-D-Gal. The C4 epimer of β-D-glucose. A common component of glycoproteins

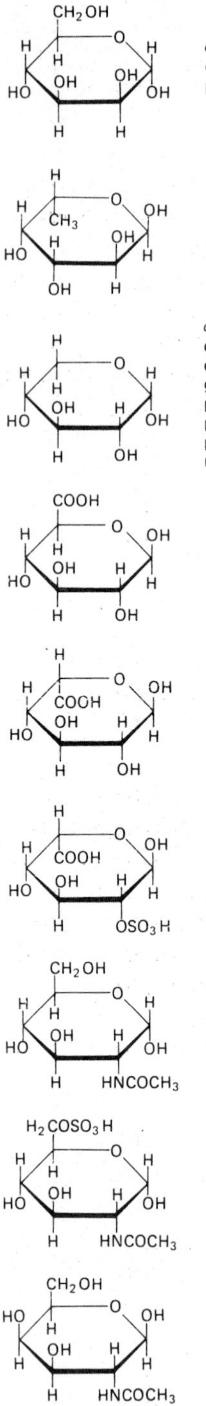

α-D-mannose. α-D-Man. The C2 epimer of α-D-glucose. Abundant in glycoproteins

α-L-fucose. α-L-Fuc. Note the orientation at C1 and C5 and the absence of a hydroxyl group on C6. Occurs in glycoproteins

α-D-xylose. α-D-Xyl. This sugar commonly joins long polysaccharide chains to the side-chain hydroxyl group of serine or threonine of the protein core of glycoprotein and proteoglycan molecules. It is never present in large quantities in mucosubstances

β-D-glucuronic acid. β-D-GlcUA. This is a β-D-glucose in which the primary alcohol group at C6 has been oxidized to carboxyl group, which ionizes as a weak acid: $-COOH \rightleftharpoons -COO^- + H^+$. Occurs in some proteoglycans

α-L-iduronic acid. α-L-IdUA. Closely similar to β-D-glucuronic acid, differing only in the configuration at C5. Note that the D-enantiomer would have a very different structure from that of any of the four glucuronic acids

α-L-iduronic acid-2-sulphate. α-L-IdUA-2-OSO_3H. The hydroxyl group at position C2 of α-L-iduronic acid is esterified by sulphuric acid

α-*N*-acetylglucosamine. α-D-GlcNAc. Also known as 2-acetamido-2-deoxy-α-D-glucose. The group $-NHCOCH_3$ replaces $-OH$ at position C2 of α-D-glucose

β-*N*-acetyl glucosamine-6-sulphate. β-D-GlcNAc-6-OSO_3H. Esterified by sulphuric acid at C6

β-*N*-acetyl galactosamine. β-D-GalNAc. Also known as 2-acetamido-2-deoxy-β-D-galactose. The group $-NHCOCH_3$ replaces $-OH$ at position C2 of β-D-galactose

β-*N*-acetylgalactosamine-4-sulphate. β-D-GalNAc-4-OSO_3H. The hydroxyl group at C4 of the preceding sugar is esterified by sulphuric acid. This sulphate ester group ionizes as a strong acid: $-OSO_3H \longrightarrow -OSO_3^- + H^+$

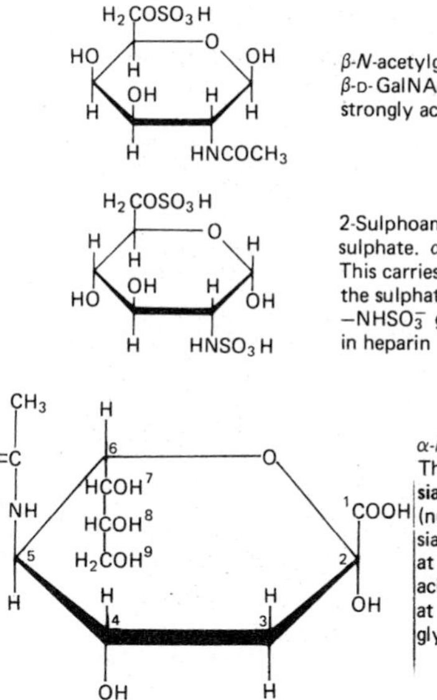

β-*N*-acetylgalactosamine-6-sulphate. β-D-GalNAc-6-OSO_3H. Another strongly acid sulphated sugar

2-Sulphoamino-2-deoxy-α-D-glucose-6-sulphate. α-D-GlcNSO_3H-6-OSO_3H. This carries two strong acid groups, the sulphate ester on C6 and the $-NHSO_3^-$ group on C2. It occurs in heparin

α-*N*-acetylneuraminic acid (NANA). This is the most abundant of the **sialic acids**, which are nonoses (numbering system shown). Other sialic acids differ in the substituents at C5. Like the uronic acids, sialic acids are weak acids. The hydroxyl at C2 is the only one involved in glycosidic linkages

11.2. CLASSIFICATION OF MUCOSUBSTANCES

The macromolecular carbohydrates may be classified on the basis of their chemistry, their behaviour in selected histochemical tests, or their distribution in nature. Pearse (1968) presents a system of classification in which an attempt is made to correlate chemical compositions with histochemically discernible properties, but with the advent of techniques for the demonstration of individual monosaccharides residues (see Section 11.3.3) this scheme is now somewhat outdated. Classifications based solely on staining properties, such as those of Cook (1974) and Culling (1974), are valuable in diagnostic pathology, but they lead to identifications which do not correspond very closely with known chemical entities. Indeed, the mucosubstances known to biochemists cannot all be distinguished by means of histochemical procedures. The differing approaches of the two disciplines have also led to a wealth of confusing terminology. In this account, the recommendations of Reid & Clamp (1978) are followed as far as possible. Ambiguous terms such as "mucin", "mucoid", "mucopolysaccharide", "mucoprotein", "sialomucin", "sulphomucin", and several others that abound in the literature are avoided.†

The following scheme is not so much a classification as a descriptive list. It includes the major mucosubstances of vertebrate animals and a few others. For each mucosubstance the constituent sugars and the forms of glycosidic linkage are given, together with the principal organic functional groups that are present and the histochemically relevant physical properties.

11.2.1. Types of mucosubstance

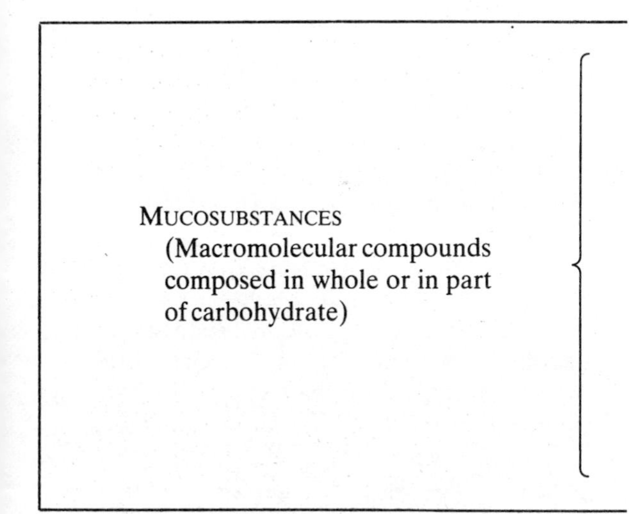

POLYSACCHARIDES
Composed entirely of carbohydrate (polyglycosides)

PROTEOGLYCANS
Long polysaccharide chains, covalently attached to a relatively small protein core

GLYCOPROTEINS
Proteins bearing numerous covalently bound oligosaccharide chains

† The term "mucosubstance", although not recommended by Reid & Clamp (1978), is retained here to embrace all carbohydrate-containing macromolecular compounds.

11.2.2. Composition of common mucosubstances

11.2.2.1. POLYSACCHARIDES

The following polysaccharides are all "homopolysaccharides", composed of single monosaccharide units. The chains may be unbranched, when the same glycosidic linkage occurs throughout the molecule, or branched when some of the units are connected through more than one hydroxyl group.

Glycogen. ---D-Glc(α-1→4)D-Glc---. The chain is branched because there are some ---D-Glc(α-1→6)D-Glc--- linkages in each molecule of glycogen. This polysaccharide is fairly soluble in water, so it is best preserved by alcoholic fixatives.

The only important functional group in glycogen is the hydroxyl group. Adjacent hydroxyls, \C(OH)(H)—C(H)(OH)/ occur in every monosaccharide unit. This is known as the **glycol** formation (also called *vic*-glycol or 1,2-diol) and it is present in many other mucosubstances.

Starch. This plant polysaccharide has the same general chemical structure as glycogen, differing only in the size and conformation of its molecules.

Cellulose. ---D-Glc(β-1→4)-D-Glc---. This, the principal component of plant cell-walls, is absent from animal tissues, with the exception of the exoskeleton in tunicates. It differs from starch and glycogen in being a β- rather than an α-polyglucoside. Cellulose is insoluble in all the reagents commonly used in microtechnique.

Chitin. ---D-GlcNAc(β-1-→4)D-GlcNAc---. This is the principal component of the exoskeleton in insects, crustacea, and various other invertebrates. It also occurs in some fungi and algae. Chitin never occurs in a "pure" state. It constitutes about half the dry weight of exoskeletal material, the balance consisting largely of protein and of insoluble salts of calcium.

11.2.2.2. PROTEOGLYCANS

The carbohydrate components of proteoglycans are "heteropolysaccharides", each long chain being composed of repeating units of two or more different monosaccharide units. The polysaccharide chains of proteoglycans, considered in isolation, are often called "glycosaminoglycuronans" or "glycosaminoglycans". The term "mucopolysaccharide" is often used as a synonym for "proteoglycan" or for the heteropolysaccharide component. The repeating units always include a nitrogen-containing sugar and a sugar-acid. The latter confers affinity for cationic dyes. The repeating units given below for individual proteoglycans are not completely constant: smaller quantities of other monosaccharides are usually present in the molecule, and the distribution of sulphate-ester groups is somewhat variable. For more complete accounts of these compounds, see Brimacombe & Webber (1964) and Jaques (1978). The latter author has proposed a new system of nomenclature, but this is not adopted in the present text.

Proteoglycans are preserved chemically intact by all non-oxidizing fixatives, though there may be extraction of some heteropolysaccharide material, such as the lower polymers of hyaluronic acid. It is advisable, when investigating a previously unstudied tissue, to try more than one fixative mixture. A proteoglycan in the rat's brain, for example, has been shown to be histochemically detectable after fixation in formalin–acetic acid or Bouin, but not after fixation in pure formaldehyde solutions (Lai *et al.*, 1975). Cetylpyridinium chloride (Williams & Jackson, 1956; see also Exercise 6 at the end of Chapter 2) has been proposed as a special fixative for the precipitation of proteoglycans, though it can depress staining by competition with cationic dyes. Cyanuric chloride (Goland *et al.*, 1967), which combines covalently with hydroxyl groups, has also been used as a fixative for mucosubstances, though it would be expected to prevent many histochemical reactions.

Hyaluronic acid. The repeating unit is :---D-GlcUA(β-1→3)D-GlcNAc(β-1→4)---. This proteoglycan exists in a variable state of polymerization. The polymers of low molecular weight are soluble in water and have low viscosity. The high polymers are more viscous and are less easily extracted by water. Glycol and carboxyl groups are present in the β-D-glucuronic acid component of each repeating unit.

Chondroitin-4-sulphate (synonym: chondroitin sulphate A). The repeating unit is: ---D-GlcUA(β-1→3)D-GalNAc-6-OSO_3H(β-1→4)---. Chemi-insoluble in all commonly used histological reagents and has glycol, carboxyl, and sulphate-ester groups.

Chondroitin-6-sulphate (synonym: chondroitin sulphate C). The repeating unit is: --- D-GlcUA(β-1$\rightarrow$3)D-GalNAc-6-OSO_3H(β-1$\rightarrow$4)---. Chemically and physically this is closely similar to chondroitin-4-sulphate, differing only in the position of the sulphate-ester group.

Dermatan sulphate (formerly "chondroitin sulphate B"). The repeating unit is: --- L-IdUA(α-1$\rightarrow$3)D-GalNAc-4-OSO_3(β-1$\rightarrow$4)---. This is another sulphated proteoglycan, differing from chondroitin-4-sulphate only in the configuration at one carbon atom of the repeating disaccharide unit.

Keratan sulphate I. The repeating unit is: --- D-Gal(β-1$\rightarrow$4)D-GlcNAc-6-OSO_3H(β-1$\rightarrow$3)---.

Keratan sulphate II. The repeating unit is: --- D-Gal(β-1$\rightarrow$4)D-GalNAc-6-OSO_3H(β-1$\rightarrow$3)---. These closely related proteoglycans are insoluble in all commonly used histological reagents. Sulphate-ester groups are present but glycol formations are absent, since position C3 of the D-galactose is involved in the glycosidic linkage.

Heparin. The composition of heparin is still not fully known. The protein core of the molecule bears several long heteropolysaccharide chains, the principal repeating unit of which is

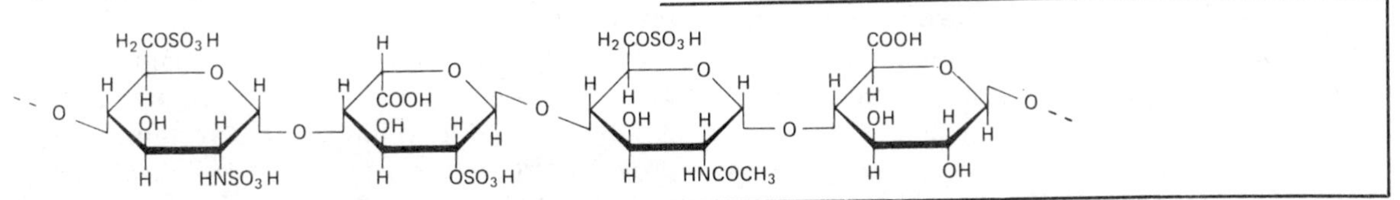

All the glycosidic linkages are α-1$\rightarrow$4. The monosaccharides units are variously sulphated derivatives of L-iduronic and D-glucuronic acids and *N*-acetyl-D-glucosamine. Small amounts of D-galactose and D-xylose are also present (see Jeanloz, 1975; Jansson *et al.*, 1975). All the organic functional groups that occur in mucosubstances are present in heparin, but sulphate groups predominate.

Heparin occurs principally in the cytoplasmic granules of mast cells, but it is possible that lower concentrations are also present in other cells and in connective tissues (Jaques *et al.*, 1977).

Heparitin sulphate. This substance may occur both within and on the surface of animal cells, but it has been studied mainly by chemical and pharmacological methods, since it is a by-product of the manufacture of heparin from animal tissues. At least four different compounds are embraced by the name "heparitin sulphate". They are similar to heparin but contain fewer sulphate and more *N*-acetyl groups. Heparitin sulphate was once thought to be a metabolic precursor of heparin, but this is now thought to be unlikely (Silbert *et al.*, 1975; Backstrom *et al.*, 1975). Heparitin sulphate has also been called heparan sulphate and heparin monosulphuric acid.

11.2.2.3. Glycoproteins

These mucosubstances are more numerous and more varied than the polysaccharides and proteoglycans. The oligosaccharide chains may consist of from 2 to 6 monosaccharide units. The commonest sugars in glycoproteins are β-D-galactose, α-D-mannose, α- and β-*N*-acetylglucosamine, α- and β-*N*-acetylgalactosamine, α-L-fucose and sialic acids. The two last-named always occupy terminal positions (farthest from the protein). Examples of glycoproteins are:

Serum proteins (including immunoglobulins).

Blood-group specific substances. These occur on the surfaces of erythrocytes.

Secretory products. Both exocrine and endocrine glands secrete glycoproteins. Some of these, especially in the alimentary canal, contain sulphated sugars, uronic acids, and sialic acids. Histochemical methods cannot determine whether such substances are proteoglycans, glycoproteins or mixtures.

Constituents of the glycocalyx. Glycoproteins form the "cell coat" on the outside surface of the plasmalemma of every cell. The glycocalyx is an integral part of the cell membrane.

Collagen. This is unusual in that α-D-glucosyl units form a major part of the carbohydrate component. See also Chapter 8.

Amyloid. Deposits of amyloid accumulate in various organs as a consequence of any of a variety of chronic inflammatory diseases. A rare condition known as primary amyloidosis is also encountered occasionally by pathologists. The deposits are of a glycoprotein which probably contains sialic acids, sulphated sugars, and ordinary neutral monosaccharide residues. Amyloid is usually identified by virtue of its affinity for certain anionic dyes of the "direct cotton" type, which are probably bound by hydrogen bonding to the carbohydrate moiety (Puchtler *et al.*, 1964). The histochemistry of amyloid will not be further discussed in this book. An excellent account of the staining methods used for its identification is given by Culling (1974).

Glycoproteins usually have carboxyl and glycol groups, and sulphate-ester groups may also be present. They are preserved by most fixatives, but those containing oxidizing agents (which would attack the glycol formation) should not be used. The carbohydrate moieties of glycolipids are similar to those of glycoproteins. The two groups of substances have similar histochemical properties and can be confused with one another, especially when frozen sections are used. Lipid-extraction procedures (see Chapter 12) are sometimes necessary in order to distinguish between glycoproteins and glycolipids. It is probable that in many sites, especially in mucus that lubricates epithelial surfaces, several different glycoproteins are present. The histochemist can identify and localize various functional groups and some individual monosaccharide residues, but cannot determine whether these occur in the same or different molecules.

11.3. HISTOCHEMICAL METHODOLOGY

The techniques of carbohydrate histochemistry fall into three categories: those in which cationic dyes and related reagents are used to identify the sugar-acids, chemical methods, and methods involving use of lectins. All types of method can also be used in conjunction with chemical blocking procedures for reactive groups and with enzymatic extractions of specific mucosubstances. Several techniques exist other than the ones described in this chapter (see Pearse, 1968, and Cook, 1974, for more information).

11.3.1. Use of cationic dyes

Three types of technique will be discussed: the use of alcian blue, the observation of metachromasia and staining with colloidal ferric hydroxide.

11.3.1.1. Alcian blue

This dye (see Chapter 5, p. 74) binds to carboxyl **and** sulphate-ester groups at pH 2.5, but only to the latter at pH 1.0. With the more strongly acid solution, carboxyl groups are not ionized and therefore cannot electrostatically attract the cations of the dye. It is therefore possible to identify with some degree of certainty mucosubstances which owe their acidity wholly to carboxyl groups, but it is not possible to tell whether a substance stained at both pH levels owes its acidity only to sulphate groups or to both sulphate and carboxyl groups. If, however, a section is subjected before staining to an adequate treatment with hot, acidified methanol (see Chapter 10, p. 141, also this chapter, p. 161) the sulphate-ester groups will be removed and the carboxyl groups will be esterified. Nothing will be stained by alcian blue. The section can next be subjected to a "saponification" procedure (see p. 119), which will cause hydrolysis of the methyl esters and restore the carboxyl groups. These will regain their stainability by alcian blue at pH 2.5. The sulphate esters, however, will have been irreversibly removed and will no longer be detectable. Thus, provided that at least six sections of the same tissue are available it is possible to determine whether affinity for alcian blue at pH 1.0 and 2.5 is due solely to sulphate esters or to both sulphate ester and carboxyl groups coexisting at the same site.

Alcian blue has also been used mixed with an inorganic salt (usually magnesium chloride) at various ionic strengths. The cations of the salt compete with those of the dye for binding sites in the tissue. The dye is used at a high pH (5–6) so that the results will not be complicated by incomplete ionization of carboxyl groups. The highly dissociated acids (—OSO_3H in this case) can bind the dye in the presence of high concentrations of the salt, while the weaker acids (—COOH) cannot. However, the truth of this assertion, the basis of the "critical electrolyte concentration" method, has been questioned (Horobin & Goldstein, 1974).

It is quite feasible to study acid mucosubstances by staining with cationic dyes other than alcian blue. Some effects of pH on the affinities of thiazine dyes for different anions were briefly reviewed in Chapter 6. The main reason for using alcian blue in carbohydrate histochemistry is the fact that this dye does not usually stain nuclei or cytoplasmic deposits of RNA in sections, though it is known to be able to bind to nucleic acids in solution (Scott *et al.*, 1964).

Studies with molecular models (Scott, 1972b) have revealed that the four tetramethylisothiouronium groups attached to the phthalocyanine rings of alcian blue 8GX make the dye molecule too big to fit between the coils of the DNA helix, so that electrostatic attraction between the phosphate groups and the auxochromes is weakened by distance and there can be no close-range interaction (van der Waals forces, etc., see Chapter 5, Section 5.5) between the aromatic rings of the dye and the purine and pyrimidine rings of the DNA. Possibly the access of the large molecules of alcian blue to the phosphate groups of nucleic acids in a fixed tissue is also hindered by the presence of the associated nucleoproteins. Steric hindrance may also explain the failure of this dye to stain RNA. Sometimes, especially in the tissues of very young animals, nuclei are stained by alcian blue pH 2.5. Such staining may be due at least in part to acid mucosubstances present in the chromosomes of dividing cells (Ohnishi *et al.*, 1973). Another advantage of alcian blue over most other cationic dyes is that it is not extracted from stained sections by water, alcohol, weak acids, or solutions of other dyes used for counterstaining. The PAS procedure (see below) is frequently applied to tissues already stained with alcian blue.

Ordinary red cationic dyes, especially safranine (Spicer, 1960), may also be used in conjunction with alcian blue. Differential staining effects have been described in carbohydrate-containing structures such as mast-cell granules (Jasmin & Bois, 1961; Combs *et al.*, 1965), but the tinctorial variations have been shown by Tas (1977) to have no histochemical significance.

For further information concerning the histochemical properties of alcian blue, see Scott *et al.* (1964) and Quintarelli *et al.* (1964a, b).

11.3.1.2. Metachromasia

When a cationic dye imparts its own colour to an object, the staining is said to be *orthochromatic*. In some circumstances the dye ions bind to a substrate in such a way as to alter the wavelength of the absorbed light such that the observed colour of the stained object is different from that of the dye. This phenomenon is known as **metachromasia** and substrates which are stained metachromatically are said to be **chromotropic**. In most cases the metachromatic colour of the dye is of a longer wavelength than the orthochromatic colour. Blue dyes, such as thionine and toluidine blue, stain chromotropic materials in shades of red and purple. The red and purple colours produced by such dyes have been called γ- and β-metachromasia, respectively, but this distinction is probably of little importance. Only γ-metachromasia is of interest in the histochemistry of mucosubstances. The change in colour is due to a shift to shorter wavelengths (a hypsochromic shift) in the absorption spectrum of the dye and to an associated reduction in the intensity of the colour.

The metachromatic effect is produced when the coloured ions of a dye are brought in close proximity to one another. This occurs when anionic radicals of the substrate are close together, as in some proteoglycans. The dye-substrate complex of a metachromatically stained object has the form

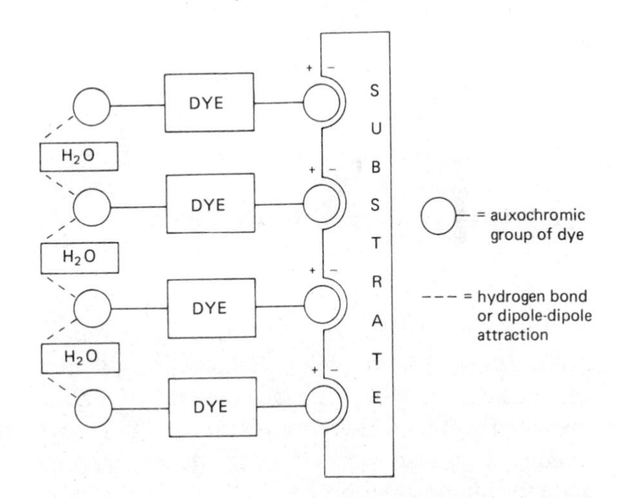

The water molecules interposed between the stacked dye ions are believed to be necessary for modifying the distribution of electrons in the chromophoric system in such a way as to reduce the

wavelength at which light is maximally absorbed (Bergeron & Singer, 1958). The dyes with which metachromatic effects can be obtained are thiazines, oxazines, azines, and xanthenes (see Chapter 5). These have planar molecules and can be formulated with their positively charged auxochromic groups on either side of the systems of fused rings. When the auxochromes are more bulky than the $-\overset{+}{N}(CH_3)_3$ group, metachromatic staining does not occur (Taylor, 1961), probably because there is not room for the interposition of water molecules. The bound water may resist extraction by dehydrating agents: indeed, it is generally agreed that metachromasia must persist in dehydrated, cleared preparations if it is to have any significance in relation to macromolecular carbohydrates (Kramer & Windrum, 1955).

Another cause of metachromasia may be the formation of dimers of dye molecules in the staining solution and the subsequent attachment of these dimers to anionic sites in the tissue (see Schubert & Hamerman, 1956; Wollin & Jaques, 1973 for discussion). This does not seem a very probable mechanism, however, since dye solutions display metachromasia only when they are concentrated, but metachromatic histological staining is easily obtained from very dilute solutions.

Acid mucosubstances are not the only sources of metachromasia found in animal tissues. Nucleic acids are also chromotropic in some circumstances, though their metachromasia usually reverts to orthochromasia after dehydration. Despite the low histochemical specificity of metachromatic staining with cationic dyes, the method is useful for the morphological study of structures which are known to contain proteoglycans, such as mast-cell granules and the intercellular matrices of cartilage and other connective tissues.

11.3.1.3. Colloidal ferric hydroxide

This technique is commonly known as "Hale's dialysed iron method", though many modifications and improvements are now incorporated into the original procedure. The reagent is made by adding a small volume of aqueous ferric chloride to a large volume of boiling water:

$$FeCl_3 + 3H_2O \rightarrow Fe(OH)_3 + 3H^+ + 3Cl^-$$

The ferric hydroxide is not precipitated, but forms a colloidal solution or "sol". The particles in the sol are positively charged, so they have properties similar to those of a cationic dye of high molecular weight. The sol is freed of small ions such as H^+, Cl^+, and any unreacted iron ions by dialysing it against water. For use as a histochemical reagent, the dialysed sol is mixed with aqueous acetic acid to give a pH of 1.8. The colloidally suspended particles in such a solution bind to sulphate-ester groups and to uronic and sialic acids of proteoglycans (see Curran, 1964) and glycoproteins (Mareel *et al.*, 1976).

After the sections have been exposed to the colloidal ferric hydroxide they are washed with water or dilute acetic acid and then treated with a solution of potassium ferrocyanide. Prussian blue is precipitated at the sites of the bound ferric hydroxide. The chemistry of this second reaction has already been reviewed in Chapter 10 (p. 125). The potassium ferrocyanide will, of course, also give a positive Prussian blue reaction with any ferric iron (e.g. ferritin, haemosiderin; see Chapter 13) present in the tissue, but such staining can be allowed for in control sections that have not been treated with the colloidal iron solution.

The results obtained with the colloidal ferric hydroxide method closely resemble those seen with alcian blue at pH 2.5, but the colour of the Prussian blue is deeper and stronger than that of the phthalocyanine dye. Colloidal ferric hydroxide also has the advantage of being a reagent of known composition which can easily be made in the laboratory, while alcian blue is a notoriously variable product whose full chemical structure is a trade secret (see Chapter 5, p. 74). The chief disadvantage of the colloidal ferric hydroxide technique is that it usually gives pale blue staining of nuclei and sometimes also of proteins in addition to the dark blue colour imparted to acid mucosubstances. Although the "background" staining can be largely suppressed by applying contrasting counterstains, it undoubtedly detracts from the value of the technique as a histochemical procedure. It is probably for this reason that alcian blue, despite its shortcomings, is preferred by most histochemists to colloidal ferric hydroxide.

11.3.2. Chemical methods—the periodic acid–Schiff method

The chemical technique most extensively used in

carbohydrate histochemistry is the periodic acid–Schiff (PAS) reaction which, as will be seen below, is positive with structures containing neutral† hexose sugars. Chemical methods are also available for sulphate-ester groups, for sialic acids, and for some amino sugars (see Pearse, 1968; Culling, 1974), though the methods are often not of high specificity. These latter techniques will not be discussed here.

In the PAS method sections are treated with periodic acid, which oxidizes certain hydroxyl groups to aldehydes. The aldehydes are then rendered visible by reaction with Schiff's reagent.

11.3.2.1. Periodate Oxidation

Periodic acid, $HIO_4 . 2H_2O$, is used in histochemistry as a 1% aqueous solution (0.05 M) and is normally allowed to act upon sections of tissue for 5–10 min at room temperature. The sodium salt of the acid may also be used. The effect of this treatment is a selective oxidation by the periodate ion of hydroxyl groups attached to adjacent carbon atoms (i.e. glycols), with fission of the intervening carbon-to-carbon bond and production of two aldehydes:

$$\begin{array}{c} | \\ H{-}C{-}OH \\ | \\ H{-}C{-}OH \\ | \end{array} + IO_4^- \rightarrow \begin{array}{c} | \\ H{-}C{=}O \\ \\ H{-}C{=}O \\ | \end{array} + IO_3^- + H_2O$$

Where there are three neighbouring carbon atoms bearing hydroxyl groups a similar reaction occurs but with the elimination of a molecule of formic acid:

$$\begin{array}{c} | \\ H{-}C{-}OH \\ | \\ H{-}C{-}OH \\ | \\ H{-}C{-}OH \\ | \end{array} + 2IO_4^- \rightarrow \begin{array}{c} | \\ H{-}C{=}O \\ \\ \\ \\ H{-}C{=}O \\ | \end{array} + HCOOH + 2IO_3^- + H_2O$$

Glycol groupings are present in the neutral sugars and in the uronic and sialic acids and some of the *N*-acetylamino sugars.

Periodic acid can also oxidize the α-amino-alcohol formation

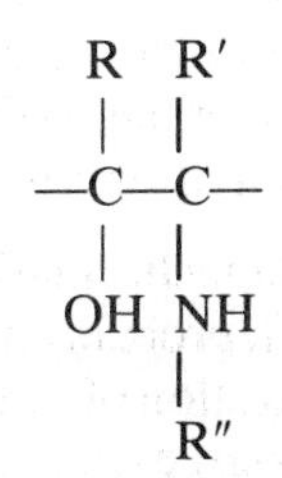

where R, R', and R'' may be hydrogen or alkyl radicals (Nicolet & Shinn, 1939). This configuration occurs in the amino acids serine and threonine (but not when these are incorporated into peptide linkages), in the side-chain of hydroxylysine, and also in sphingosine, an amino alcohol present in certain lipids. The only common sugars that are α-amino alcohols are glucosamine and galactosamine, but these do not occur in significant quantities in tissues. The nitrogen-containing sugars of proteoglycans and glycoproteins, *N*-acetylglucosamine and *N*-acetylgalactosamine, contain the structural arrangement

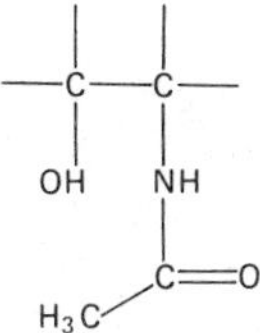

They are α-*N*-acylamino alcohols and, as such, are almost completely resistant to oxidation by periodate (Nicolet & Shinn, 1939; Carter *et al.*, 1947). The *N*-acetyl hexosamines could be oxidized by periodate only if they formed glycosidic linkages at position C6, leaving C3 and C4 with free hydroxyl groups. In mucosubstances these sugars are con-

† The term "neutral sugar", though not strictly accurate in the chemical sense, is used by histochemists for monosaccharide residues which do not have sulphate-ester, carboxylic acid, or nitrogen-containing functional groups. Glucose, galactose, mannose, and fucose are the principal neutral sugars present in mucosubstances.

nected to others through glycosidic linkages involving position C3 or C4, so they are never responsible for stainability by the PAS method.

Lead tetraacetate (Glegg *et al.*, 1952) and sodium bismuthate (Lhotka, 1952) have been used for the same purposes as periodic acid. They act in the same way, but as histochemical reagents they have not been investigated as thoroughly as periodic acid.

If the histochemical situation were as simple as the foregoing account might lead one to believe, all carbohydrate-containing structures would be expected to yield aldehydes with periodic acid. This is not the case, however. It has been shown that proteoglycans (consisting of uronic acids and acetylhexosamines, sulphated or not) are **not** PAS-positive (Hooghwinkel & Smits, 1957). In sections of fixed tissue, periodic acid probably produces aldehydes only from glucosyl, galactosyl, mannosyl, and fucosyl residues (Leblond *et al.*, 1957), which are components of glycogen and of glycoproteins.

Proteoglycans are PAS-negative because the usual treatment with periodic acid fails to oxidize the glycol formations at C2–C3 in glucuronic and iduronic acids. Scott & Harbinson (1969) have shown that uronic acids are not attacked by periodic acid on account of repulsion of the periodate ion by the carboxylate anions and perhaps also by the nearby sulphate-ester anions, when these are present. This electrostatic effect can be overcome by using periodate for a much longer time and at a higher temperature than usual. Although the uronic acid components of mucopolysaccharides yield aldehydes under these more rigorous conditions, the *N*-acetyl hexosamines are still unaffected. Scott & Dorling (1969) have developed a modified PAS technique, based on the principle outlined above, for the selective demonstration of uronic acid-containing mucosubstances (i.e. proteoglycans). A preliminary treatment with periodate is followed by reduction with sodium borohydride:

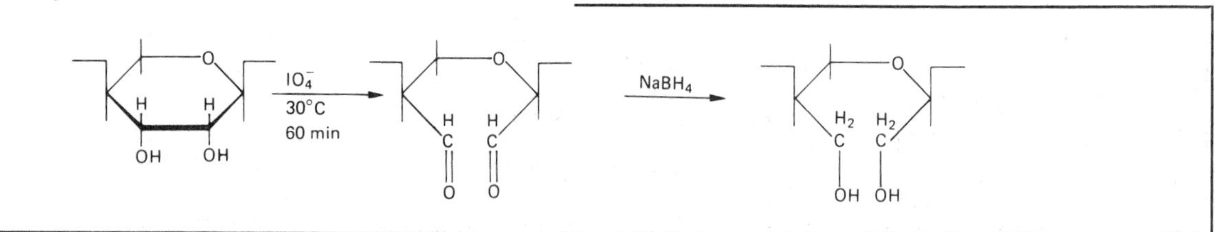

This procedure changes all neutral hexoses into compounds that can no longer yield aldehydes by reaction with periodate. In the next stage of the technique, the sections are again exposed to periodate, but for a time sufficient to cause oxidation of the glycol groups of uronic acids:

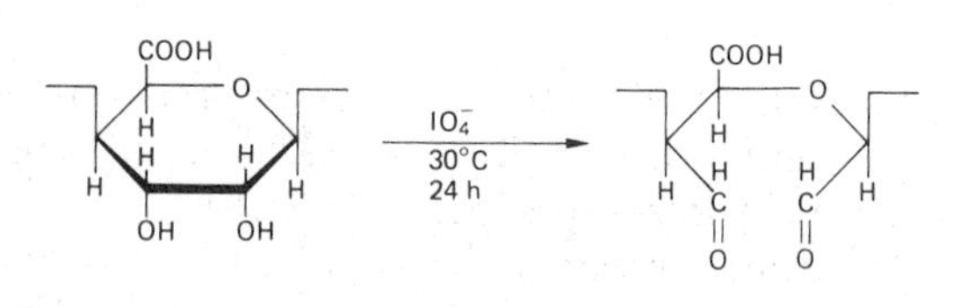

The aldehyde groups produced by the second oxidation are demonstrated with Schiff's reagent. Scott & Dorling's modified PAS method therefore demonstrates mucosubstances whose uronic acid residues have free hydroxyl groups at positions C2 and C3. From Section 11.2.2 of this chapter it can be seen that these are hyaluronic acid, chondroitin-4-sulphate, chondroitin-6-sulphate, dermatan sulphate, and heparin. The polysaccharides, the glycoproteins, and the keratan sulphates will not be stained.

The versatility of periodic acid as a histochemical reagent is further exemplified in its ability, under appropriate conditions, to produce aldehydes selectively from sialic and residues. The only potentially periodate-reactive part of a sialic acid glycosidically linked at position C2 (see formula for NANA, p. 147) is the side-chain (C7, C8, and C9) attached at C6, which bears three adjacent hydroxyl groups. This side-chain reacts with periodate much more rapidly than do the glycols of hexoses (see Hughes, 1976). By using very dilute periodic acid for a short time it is possible to oxidize only the sialic acid residues of glycoproteins:

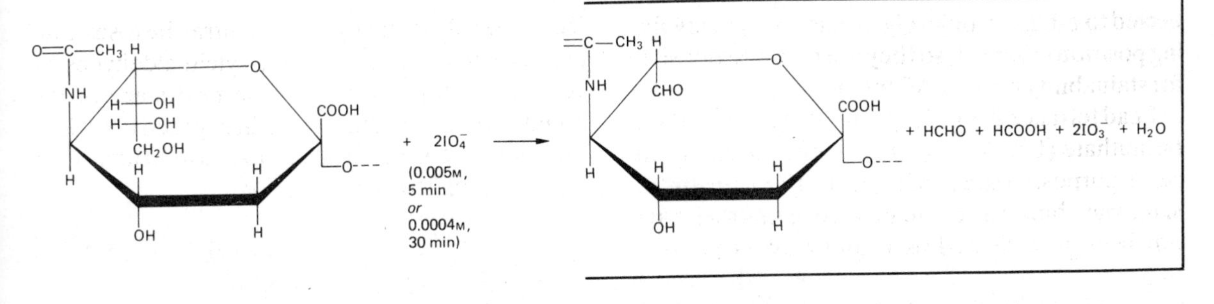

The resultant aldehyde can be demonstrated either with Schiff's reagent (Roberts, 1977) or with a fluorescent hydrazine (Weber *et al.*, 1975).

In some intestinal glycoproteins, *O*-acetyl groups replace the hydroxyls at position C7, C8, or C9 of NANA. These acetylated sialic acids can be recognized histochemically by means of a sequence of treatments:

(1) oxidation by periodic acid (glycol→ aldehyde);

(2) reduction by borohydride (aldehyde→primary alcohol);

(3) saponification with alcoholic potassium hydroxide (acetyl ester→hydroxyl);

(4) second oxidation with periodate (glycol→ aldehyde).

The production of aldehydes is sought, with Schiff's reagent, after stages 1 and 4.

The side-chains of the sialic acids react as follows:

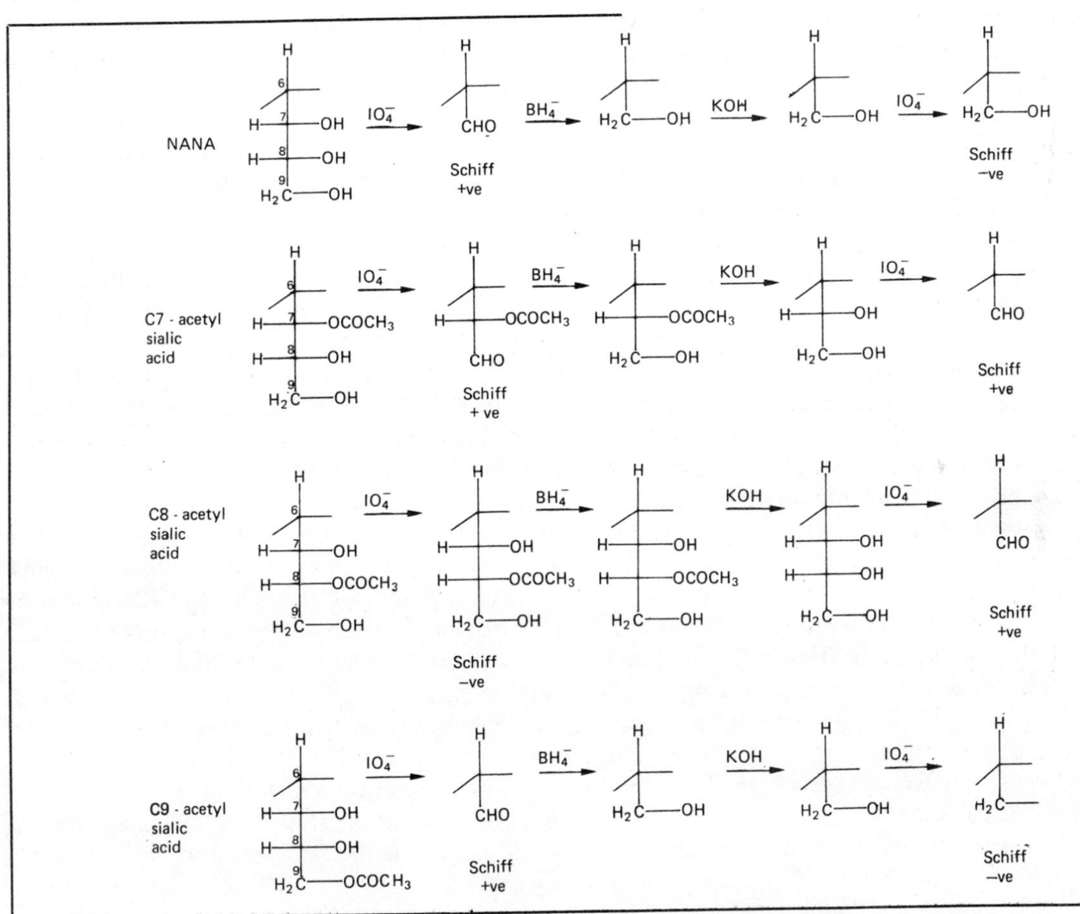

Unsubstituted and C9-acetylated sialic acids are Schiff-positive only after the first periodate oxidation and cannot be distinguished from one another. The C7-acetylated sialic acid is Schiff-positive after both oxidations and the C8-acetylated compound is Schiff-positive only after the second oxidation. If acetyl groups are present on any two or all three of the positions C7, C8, and C9, positive Schiff reactions will be obtained only after the second oxidation with periodate. It must be emphasized that this histochemical analysis is applicable only when it has already been proved that sialic acids are the sole substances responsible for stainability by the PAS method at the sites being investigated. The validity of the technique involving periodate oxidation in conjunction with saponification was demonstrated in a combined biochemical and histochemical investigation by Reid *et al.* (1978), whose excellent paper should be consulted for further information.

Another method, related to the procedure described above, is that of Culling *et al.* (1976). In this, the first oxidation with periodic acid is followed by staining with a "pseudo-Schiff" reagent (see Chapter 10, p. 129) made from thionine. The sections are then saponified, oxidized for a second time with periodic acid, and stained with conventional Schiff's reagent. Sialic acids which bear no *O*-acetyl groups or are acetylated only at C9 are stained blue. Those acetylated at C8 are red and those acetylated at C7 or at more than one of the positions C7, C8, and C9 are purple. In the studies of Culling *et al.* (1976) and Reid *et al.* (1978), the conditions of oxidation by periodate were the same as those commonly used for the oxidation of hexose residues. It would be interesting to know whether very mild oxidation (see p. 154) would increase the selectivity of staining of the various *O*-acetylated sialic acids.

11.3.2.2. Schiff's Reagent

The aldehydes produced by the action of periodic acid are detected with Schiff's reagent, with which they form a stable product, pink to purple in colour. The chromogen is covalently bound to the stained macromolecule and cannot be extracted by water, dilute acids, or alkalis or by alcohol.

The chemistry of Schiff's reagent and its reaction with aldehydes were discussed in Chapter 10. Other aldehyde-detecting reactions may also be used for the demonstration of periodate-oxidized mucosubstances, but Schiff's reagent is adequate for most purposes.

11.3.2.3. Artifacts Associated with the PAS Method

A structure coloured pink or purple by the PAS technique cannot be assumed to contain periodate-reactive sugars unless the following causes of false-positive staining have been excluded.

1. Aldehydes may be initially present in the tissue. These will be stained by Schiff's reagent without prior oxidation by periodic acid and may be derived from the fixative (especially glutaraldehyde) or from atmospheric oxidation of olefinic linkages in unsaturated lipids (the **pseudoplasmal** reaction). Fixatives containing mercuric chloride may produce aldehydes from plasmalogen phospholipids (the **plasmal** reaction). Lipids (see Chapter 12) rarely interfere with the interpretation of PAS staining in paraffin sections.

Blocking procedures for pre-existing aldehyde groups have been described in Chapter 10. Sodium borohydride is the aldehyde-blocking agent of first choice for use in carbohydrate histochemistry.

2. Periodic acid can produce aldehydes from olefinic bonds of lipids as well as from glycols. Consequently, a genuinely positive PAS reaction in a frozen section may not be due to glycoprotein or glycogen, or even to glycolipid, especially if an aqueous mounting medium is used. Adams (1965) describes a modified PAS method in which the reactivity of unsaturated linkages is chemically suppressed before the oxidation of carbohydrates by periodic acid. Sphingosine, an amino alcohol present in several lipids, is also oxidized by periodic acid (see p. 170).

Glycolipids (which are commonly PAS-positive on account of their content of galactose and sialic acids) are distinguished from mucosubstances by virtue of the solubility of the former in a hot mixture of methanol and chloroform. Some glycolipids are largely extracted by water.

3. Hydroxylysine, an amino acid which occurs only in collagen (see Chapter 8) might be expected to be PAS-positive, but Bangle & Alford (1954) have shown that the stainability of collagen by the PAS method is due almost entirely to its carbohydrate content. The hydroxylysine may not be present in sufficient quantity to be histochemically

detectable, or its amino groups may be involved in an amide linkage with some part of the collagen molecule.

4. Some batches of Schiff's reagent prove unsatisfactory and may give either false positive or false negative results. An untried sample of the reagent should be tested on sections in which the correct localizations of PAS-positive structures are already known. An aldehyde-blocking reaction applied after oxidation with periodate will serve as a control for false positive coloration by an unsatisfactory sample of Schiff's reagent.

11.3.3. Lectins as histochemical reagents

11.3.3.1. Properties of lectins

Lectins are proteins extracted from plants (especially from the seeds of many *Leguminosae*) which have the property of causing agglutination of erythrocytes. They are also known as phytohaemagglutinins. It has been known for many years that while some lectins agglutinate all types of red cells, others are selective for particular blood groups in much the same ways as are circulating antibodies. The blood group determinant substances are glycoproteins located on the surfaces of the erythrocytes and it has been shown that lectins bind to specific carbohydrate moieties of these glycoproteins. The haematological properties and uses of lectins have been reviewed briefly by Boyd (1970).

The lectins can bind to appropriate carbohydrates whether these be located on the surfaces of erythrocytes or elsewhere. Måkelå (1957) showed that the specific agglutination of red cells of particular blood groups was due to affinity of the lectins for the terminal monosaccharide residues of the determinant glycoproteins. The blood group O glycoprotein, for example, has terminal α-L-fucosyl residues, and the lectins which agglutinate group O erythrocytes are also those which bind to L-fucose and its α-glycosides. The lectins, which agglutinate erythrocytes of all blood groups, are those which bind to sugars such as mannose and *N*-acetylglucosamine, which are universally distributed on the surfaces of cells. Irrespective of specificity for human blood groups, all the lectins are remarkably selective with respect to the types of sugar to which they will bind. The properties of some of the better-known lectins are reviewed briefly by Cook & Stoddart (1973) and at length by Toms & Western (1971) and by Nicolson (1974).

The binding of a lectin molecule to a carbohydrate does not involve the formation of covalent bonds. It is similar in nature to the attachment of an antigen to its specific antibody (see Chapter 19). The molecules of some lectins incorporate ions of calcium and of a transition metal (usually manganese), and the presence of these metal ions is essential for the carbohydrate-binding activity.

Since lectins are proteins, with molecular weights ranging from 20,000 to 300,000, it is possible to attach dye molecules to some of their free amino groups without interfering with the carbohydrate-binding properties. The label most frequently used has been fluorescein isothiocyanate, a fluorochrome which can be covalently bound to proteins, mainly by combination with the ε-amino groups of lysine (see also Chapter 5, p. 67, and Chapter 19). The sites of attachment of the molecules of the labelled lectin to a section of tissue can then be seen with the fluorescence microscope. Other covalently bound labelling substances that have been used with lectins include various fluorochromes, ferritin, and horseradish peroxidase. Lectins can also be labelled with ^{3}H-acetyl groups, for subsequent detection by autoradiography. A few investigators have identified bound lectins by immunohistochemical methods (see Chapter 19), using anti-lectin antisera.

The names and specificities of a few lectins (and of two carbohydrate-binding proteins of animal origin) are given in Table 11.1.

11.3.3.2. Technical considerations

Histochemical methods involving the use of lectins have much in common with immunohistochemical techniques (see Chapter 19). In the simplest type of procedure, a solution of the fluorescently labelled lectin, usually at a concentration of about 1.0 mg per ml in water or saline, usually buffered to pH 7.0–7.6, is applied to the sections of tissue for 15–60 min. Excess reagent is washed off and the preparation is either mounted in an aqueous medium or dehydrated, cleared, and mounted in a non-fluorescent resinous medium. Fluorescence is observed at sites of binding of the lectin.

Certain precautions are necessary in interpreting

TABLE 11.1. *Some carbohydrate-binding proteins used as histochemical reagents*

Source and name (common abbreviation)[a]	Monosaccharide residues with specific affinity[b]	References[c]	
		Biochemical (for specificity)	Histochemical[d] (techniques and applications)
Canavalia ensiformis (jackbean): concanavalin A (con A)	α-D-Man and α-D-Glc	Goldstein *et al.* (1965)	Nicolson & Singer (1971) Nieland (1973) Stoddart & Kiernan (1973a) Cotman & Taylor (1974) Bittiger & Schnebli (1976)
Dolichos biflorus (horsegram): *Dolichus* lectin (DBA)	α-D-GalNAc	Etzler & Kabat (1970)	Etzler & Branstrator (1974) Essner *et al.* (1978)
Glycine max (soya bean): soybean lectin (SBA)	D-GalNAc	Lis *et al.* (1970)	Stoddart *et al.* (1974) Essner *et al.* (1978) Whyte *et al.* (1978)
Lens culinaris (lentil): *Lens culinaris* lectin (LCA)	α-D-Man and α-D-Glc	Toyoshima *et al.* (1970) Young *et al.* (1971)	Roth *et al.* (1978) Whyte *et al.* (1978)
Lotus tetragonobolus: *Lotus* lectin (LA)	α-D-Fuc	Yariv *et al.* (1967)	Etzler & Branstrator (1974) Whyte *et al.* (1978)
Phaseolus vulgaris (kidney bean): *Phaseolus* lectin (PHA)	D-GalNAc	Sharon & Lis (1972)	Berman & Andrews (1970) Nieland (1973)
Pisum sativum (pea): pea lectin (PA)	D-Glc and D-Man	Paulova *et al.* (1970)	
Ricinus communis (castor bean): *Ricinus* lectin I **or** Ricin-I (RCA_1 **or** RCA_{120})	β-D-Gal	Drysdale *et al.* (1968)	Yamada & Shimizu (1977) Dabelsteen *et al.* (1978) Roth *et al.* (1978)
Triticum vulgaris (wheat): wheat germ lectin (WGA)	D-GlcNAc	Burger & Goldberg (1967) Allen *et al.* (1973)	Gros *et al.* (1977) Essner *et al.* (1978) Roth *et al.* (1978)
Ulex europaeus (gorse): *Ulex* lectin I (UEA_1)	L-Fuc	Matsumoto & Osawa (1969)	Kent (1964) Essner *et al.* (1978)
Limulus polyphemus (horseshoe crab): *Limulus* agglutinin; limulin	Sialic acids	Marchalonis & Edelman (1968) Sharon & Lis (1972)	Mazzuca *et al.* (1977) Yamada & Shimizu (1979)
Bovine and porcine lung, pancreas, salivary glands, etc.: aprotinin; "pancreatic trypsin inhibitor"; Trasylol (trade-name) (A)	Uronic and sialic acids	Stoddart & Kiernan (1973b)	Kiernan & Stoddart (1973)

(a), (b), (c), (d) Footnotes are at bottom of page 159.

the histological appearances of sections stained with fluorescent-labelled lectins:

1. The autofluorescence of the section must be minimal. It may be suppressed by treatment with dilute aqueous osmium tetroxide before staining.
2. The specificity of attachment of the lectin must be determined. This is done by staining control slides in the presence of high concentrations of sugars or glycosides which will occupy binding sites on the lectin molecules before the latter can attach to the tissue. For example, staining with concanavalin A should not occur from a solution containing α-methyl-D-glucoside, but should occur normally in the presence of β-methyl-D-glucoside.
3. Trace amounts of metal ions (especially Ca^{2+} and Mn^{2+}) are necessary for the functioning of many lectins. They may be removed in the course of preparation of fluorescent derivatives and will have to be replaced. Chelating agents prevent the binding of labelled concanavalin A to glucosyl and mannosyl residues.

Most lectins are irreversibly changed, with loss of specific carbohydrate-binding properties, by exposure to strongly alkaline solutions. Since alkalinity is a prerequisite for efficient labelling of proteins with fluorescein isothiocyanate and related compounds, it is often difficult to obtain satisfactory preparations of labelled lectins. The conjugation must be carried out at a neutral pH in the presence of a large excess of the fluorochrome. Suitable conjugation procedures are described by Roth *et al.* (1978). Some fluorescein-labelled lectins are now commercially available.

In the case of concanavalin A, the difficulties of covalent labelling have been circumvented by making use of the fact that the molecules of the lectin are bivalent: each can bind to two glucosyl or mannosyl groups. When one end of a concanavalin A molecule is attached to a component of a sectioned tissue, it is possible to attach a suitable glycoprotein to the other binding site:

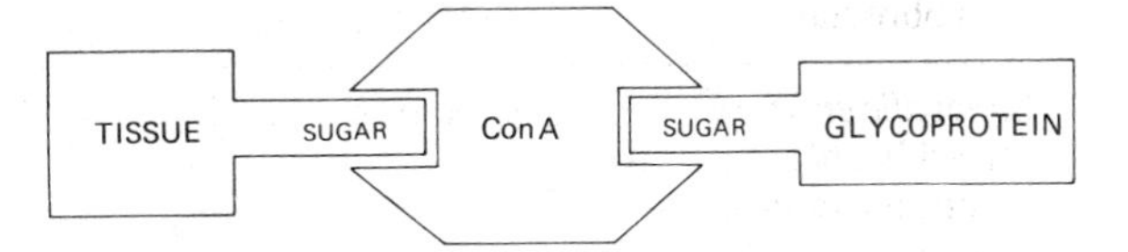

A suitable glycoprotein is horseradish perioxidase, which is rich in mannose and is easily localized by a histochemical method. The binding of concanavalin A has been studied in this way in electron (Bernhard & Avrameas, 1971; Parmley *et al.*, 1973) and light (Kiernan, 1975) microscopy. It is, of course, necessary to control for the specificity of binding of the indicator-enzyme and for any enzymatic activity which was originally present in the tissue. Practical instructions for the concanavalin A–peroxidase method are given in Section 11.6.5 of this chapter. This technique employs reagents which are all readily available from commercial suppliers. Methods involving the use of covalently conjugated lectins, which usually have to be prepared and purified in the laboratory, are still confined to specialized applications in research. For practical details, the interested reader should consult appropriate references from Table 11.1.

Another method for the demonstration of sites of binding of unlabelled lectins to sections has been described by Thoss & Roth (1976). The unattached binding sites of concanavalin A and *Ricinus* lectin I were made visible by treatment with fluorescently labelled serum proteins (which are glycoproteins). Sites of binding of *Lens culinaris* and pea lectins could not, however, be demonstrated. Alroy *et al.* (1978) used a suspension of group O erythrocytes (the determinant glycoprotein of which has oligosaccharide chains that terminate with α-L-fucosyl units) to demonstrate the sites of attachment of *Ulex europaeus* lectin to paraffin sections. The res-

(Footnotes to Table 11.1, p. 158)

[a] The terminal A in most of the abbreviated names of lectins stands for "agglutinin". A subscript number indicates one of a number of lectins extracted from the same source (e.g. UEA_1), or sometimes the M.W. × 10^{-3} (e.g., RCA_1 = RCA_{120}, whose M.W. is 120,000).

[b] See Section 11.1.2 of this chapter for the structures of the sugars corresponding to these abbreviations. For more detailed structural specifications of the affinities of lectins, consult the references under "Biochemical".

[c] These publications are but a tiny selection from a rapidly growing body of literature.

[d] Where no entry appears in this column, the lectin concerned has not been used as a histochemical reagent. Some labelled lectins have been used only in electron microscopy.

olution afforded by this method would, however, be too low for most purposes.

11.3.3.3. INTERPRETATION OF RESULTS

Two considerations are important in assessing the significance of observed binding of a lectin to a component of a tissue.

1. The specificity of the binding must be established by means of suitable controls, as outlined earlier (p. 159). Inhibition with appropriate sugars or glycosides is the most important control procedure. It may be supplemented by chemical blocking methods that destroy or modify the glycosyl residues to which the lectin is expected to bind. Acetylation, for example, prevents the binding of concanavalin A, but not of aprotinin, while methylation prevents staining by fluorescently labelled aprotinin but not by concanavalin A and most other lectins. The reasons for these effects can be determined by consulting Table 11.1. Oxidation by periodic acid prevents the binding of concanavalin A at most sites in tissues, but some structures paradoxically acquire the ability to bind this lectin, even though all glucosyl and mannosyl residues would be expected to be destroyed. The reasons for this effect are not yet understood, but it has been suggested that certain mannosyl units of glycoproteins, previously inaccessible to the large molecules of concanavalin A, are "unmasked" but not oxidized as a result of the treatment with periodic acid (Katsuyama & Spicer, 1978).

2. It may not be assumed that the amount of lectin bound at any site (as judged by intensity of staining) is proportional to the local concentration of the appropriate monosaccharide in the tissue. The binding of any lectin is profoundly affected by neighbouring sugar residues in the oligosaccharide chains of glycoproteins, even though these may not be ones which attach themselves directly to the lectin. The histochemical recognition of individual monosaccharide units forms only a small part of the study of the interactions between lectins and glycoproteins (see Nicolson, 1974).

11.4. ENZYMES AS REAGENTS IN CARBOHYDRATE HISTOCHEMISTRY

Three enzymes are commonly used in the histochemical study of mucosubstances. These are amylase, hyaluronidase, and neuraminidase. The precautions discussed in Chapter 9 (p. 107) apply equally to the use of these enzymes.

11.4.1. Amylase

(A collective name for α-1,4-glucan hydrolases, E.C. 3.2.1.1, 3.2.1.2 and 3.2.1.3.) Amylase catalyses the hydrolysis of the glucosidic linkages of starch and glycogen. The enzyme usually used in histochemistry is β-amylase (E.C. 3.2.1.2), the action of which yields the soluble disaccharide maltose. Glycogen (or starch) can therefore be identified as a substance whose stainability by the PAS method is prevented by digestion of the sections with amylase.

11.4.2. Hyaluronidase

(Hyaluronate lyase, E.C. 4.2.99.1.) Hyaluronidase attacks the glycosidic linkages of some proteoglycans, including hyaluronic acid, causing depolymerization of these substances. Hyaluronidases of different types are available, the one most often used by histochemists being that extracted from ovine testes. The specificities of the enzymes have been reviewed by Pearse (1972) who considers that the testicular enzyme can remove hyaluronic acid, chondroitin-4-sulphate, and chondroitin-6-sulphate from fixed tissues. Streptococcal hyaluronidase is specific for hyaluronic acid.

11.4.3. Neuraminidase

(*N*-acetylneuraminate glycohydrolase, E.C. 3.2.1.18; the enzyme from *Vibrio cholerae* is usually employed.) Neuraminidase removes the terminal NANA residues of glycoproteins. It is therefore possible to determine whether staining is due to the presence of these sugars. However, not all sialic acid groups are removed by the enzyme, so no significance can be attached to a failure to prevent staining by pre-treating the sections with the enzyme. Prior saponification makes some previously neuraminidase-resistant sialic acid residues susceptible to the action of the enzyme, probably by converting O-acetylated sialic acids to NANA. The type of fixative used does not usually affect digestibility by neuraminidase, but cyanuric chloride, which cross-links hydroxyl and amino groups, is exceptional in that it prevents the subsequent action of the enzyme (Sorvari & Laurén, 1973). The identity of the sugar to which NANA is glycosidically bound can affect the susceptibility of the linkage to attach by neuraminidases from different sources. The *Vibrio* enzyme, however, catalysed the hydrolysis of all the types of linkage studied by Drzeniek (1973). Acid hydrolysis (see p. 162) will detach all sialic acid residues from glycoproteins.

Many other enzymes attack specific glycosidic linkages, but so far they have been little used by histochemists. The use of such enzymes in conjunction with labelled lectins might be particularly informative (Whyte *et al.*, 1978).

11.5. CHEMICAL BLOCKING PROCEDURES

The reactive groups of carbohydrates are hydroxyls, carboxyls, and sulphate esters. These can be chemically "blocked" to prevent their subsequent histochemical demonstration. The significance of a negative result produced in this way is dependent upon the selectivity of the blocking procedure. The blocking reactions are discussed more fully in Chapter 10 and the following notes apply only to their applications in the histochemical study of carbohydrates.

11.5.1. Acetylation of hydroxyl groups

Sections are treated with acetic anhydride in dry pyridine. Hydroxyl groups are converted to acetyl esters:

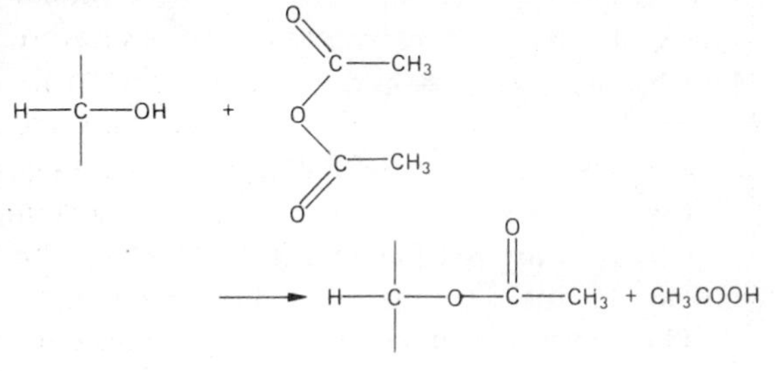

Histochemical reactions due to the hydroxyl groups of carbohydrates are prevented. Acetic anhydride also causes N-acetylation of amines and can react with carboxyl groups (of proteins, but probably not of carbohydrates) to form carbonyl compounds, though the optimum conditions for these reactions are different from those used for acetylating sugars. The acetyl esters of carbohydrates are hydrolysed by treatment with alkali (saponification), which restores the reactivity of the hydroxyl groups.

11.5.2. Methylation and desulphation

When sections are treated with a solution of either HCl or thionyl chloride in methanol (or by any of several alternative procedures), methyl esters are formed with carboxyl groups:

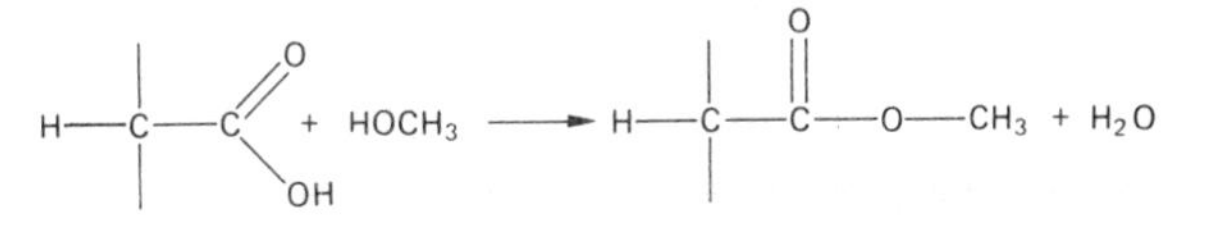

(In the case of the uronic acids it is likely that other reactions also occur—see Sorvari & Stoward, 1970.)

The same reagent, if applied for sufficient time, also **removes** O-sulphate and N-sulphate groups and methylates amino groups. The phosphate anions of nucleic acids are also blocked by esterification. Another effect of the methylating reagent is that it causes acid hydrolysis of most of the glycosidic linkages of sialic acid residues. The loss of basophilia of sialic acid-containing glycoproteins therefore cannot, for the most part, be restored by saponification. Since this last effect is rather unpredictable, methylation is more useful for the study of uronic acids of proteoglycans than for the investigation of glycoproteins.

As with acetylation, the esters can be saponified with restoration of carboxyl groups, but this treatment does not, of course, restore any tinctorial properties which were due to the presence of sulphate esters. Hydroxyl groups are not blocked by methanolic HCl under the conditions in which the reagent is normally used. Free hydroxyls at position C1 of a monosaccharide unit would be converted to methyl glycosides, but in mucosubstances almost all the sugars are already linked at this position. The formation of methyl ethers with the alcoholic hydroxyl groups of carbohydrates requires methyl iodide or dimethyl sulphate with a basic rather than an acid catalyst and does not occur in histochemical methylations.

11.5.3. Mild acid hydrolysis

The α-2-glycosidic linkage by which sialic acids are attached to the subjacent sugars of the oligosaccharide chains of glycoproteins is easily broken by acid-catalysed hydrolysis. Sections are treated with an aqueous acid (pH 2.5 or lower) for 1 or 2 h at 70–80°C. All sialic acid residues are removed but other glycosidic linkages remain intact.

11.6. INDIVIDUAL METHODS

11.6.1. Alcian blue (pH 1.0 and 2.5)

Not all batches of dye are satisfactory. Use a sample designated 8GX, 8GS, or 8GN. A solution of alcian blue which is ineffective at room temperature will sometimes perform adequately at 58–60°C.

Solutions required

A. Alcian blue, pH 1.0

Alcian blue:	1.0 g
0.1 N hydrochloric acid:	100 ml

Stability variable. Sometimes the dye precipitates after 2 or 3 weeks. Other batches are stable for at least 1 year. Filter before using.

B. Alcian blue, pH 2.5

Alcian blue:	1.0 g
3% aqueous acetic acid:	100 ml

Keeps for several months.

Procedure

1. De-wax and hydrate paraffin sections.
2. Stain in solution A **or** solution B for 30 min.
3. Wash in running tap water for 3 min.
4. (Optional.) Apply a pink or red counterstain if desired.
5. Dehydrate in graded alcohols. Alcian blue is not removed by alcohol, but the counterstain may be differentiated.
6. Clear in xylene and mount in a resinous medium.

Result

All acid mucosubstances are stained at pH 2.5. Only sulphated mucosubstances are stained at pH 1.0.

Note

Mucosubstances with carboxyl or sulphate-ester groups may be distinguished from one another if control sections are methylated and saponified. The rationale of this procedure is explained in Section 11.3.1.1 of this chapter. Neuraminidase and mild acid hydrolysis may be used in order to identify glycoproteins that owe their acidity to sialic acids.

11.6.2. Toluidine blue for metachromasia

Solution required

Toluidine blue

Toluidine blue O (C.I. 52040):	0.5 g	Keeps indefinitely
Water:	200 ml	
Glacial acetic acid:	1.0 ml	

Filter before use. (See also *Note 1* below.)

Procedure

1. De-wax and hydrate paraffin sections.
2. Stain in toluidine blue for 1–5 min (see *Note 1*).
3. Wash in water.
4. Dehydrate in 70%, 95%, and two changes of absolute ethanol (see *Note 2*).
5. Clear in xylene and mount in a resinous medium.

Result

Orthochromatic colour (nuclei, cytoplasm of some cells, Nissl substance of neurons)—blue. Metachromatic colour—red.

Notes

1. If staining is excessive, add more acetic acid to the toluidine blue. The pH of the solution should be approximately 4.0.
2. The dye is differentiated by the 70% ethanol. The sections should be **pale** blue before being

cleared. If the staining is initially weak, blot the sections after washing and dehydrate in two changes (each 3–5 min) of *n*-butanol.

11.6.3. Colloidal ferric hydroxide

Solutions required

A. Colloidal ferric hydroxide stock solution

Dissolve 3.0 g of ferric chloride ($FeCl_3 . 6H_2O$) in 6 ml water. Boil 250 ml water and add the concentrated ferric chloride solution. Continue boiling for a further 1 to 2 min until the liquid turns dark red, then remove from heat and allow to cool to room temperature. Transfer to cellulose tubing and dialyse against two changes (each 12–24 h) of 1 litre of water. Filter the dialysed sol twice: first through ordinary Whatman No. 1 paper and then through the finest available filter paper (e.g. Whatman No. 50). Keep it in a glass bottle at room temperature. It is stable for several months. Deterioration is apparent when the colour is weakened and a considerable precipitate has formed.

B. Working colloidal ferric hydroxide solution

Stock solution (A), freshly filtered:	20 ml	Stable for a few hours. Use only once
Water:	36 ml	
Glacial acetic acid:	24 ml	

C. Potassium ferrocyanide solution

Potassium ferrocyanide, $K_4Fe(CN)_6.3H_2O$:	2.0 g
Water:	99 ml
Concentrated hydrochloric acid:	1.0 ml

This should be made and used on the same day. If the hydrochloric acid is contaminated with ferric ions, a little Prussian blue may be precipitated. This is removed by filtration.

Procedure

1. De-wax and hydrate paraffin sections.
2. Immerse in the working colloidal ferric hydroxide solution (B) for 10 min.
3. Wash in running tap water for 5 min, then rinse in two changes of distilled or deionized water.
4. Immerse in the potassium ferrocyanide solution (C) for 10 min.
5. Rinse in two changes of distilled or deionized water, then in running tap water for 3 min.
6. Apply a counterstain, if desired. (See *Note 1* below.)
7. Dehydrate, clear, and mount in a resinous medium.

Result

Proteoglycans and acid glycoproteins are shown in a strong Prussian blue colour. Usually there is also pale blue coloration of nuclei and collagen. Deposits of iron (ferritin, haemosiderin), if present, are also stained dark blue.

Notes

1. The counterstaining solutions and any other reagents applied at this stage must be at least slightly acid, since alkalis dissolve Prussian blue. Alum–brazilin (see p. 83) is a suitable red nuclear counterstain. The PAS method may also be used.
2. *Controls.* (a) Omit stage 2. Only iron deposits will be stained; they may be extracted (see p. 197) if likely to cause confusion, but the extraction may also remove sialic acids. (b) The effects of methylation, saponification, enzymes, etc., may be examined to assist in the identification of stained mucosubstances.

11.6.4. The periodic acid–Schiff procedure

11.6.4.1. STANDARD PAS METHOD

Solutions required

A. Periodic acid solution

Periodic acid ($HIO_4 . 2H_2O$):	2.0 g
Water:	200 ml

Keeps for several weeks and may be used repeatedly, but should be discarded if it goes brown.

B. Schiff's reagent

See under the Feulgen technique (p. 110).

C. Solutions for counterstaining

A suitable sequence: alum–haematoxylin for nuclei, with fast green FCF as a counterstain. For critical evaluation of the PAS reaction it is preferable to use no counterstain.

Procedure

1. Allow the Schiff's reagent to warm to room temperature.
2. De-wax and hydrate paraffin sections. (See *Note 1*.)
3. Oxidize for 10 min in periodic acid (solution A).
4. Wash in running tap water for 3 min.
5. Immerse in Schiff's reagent (solution B) for 20 min.
6. Transfer to copiously running tap water and leave to wash for 10 min (see *Note 2*).
7. Apply counterstains, as desired.
8. Dehydrate in graded alcohols, clear in xylene, and cover, using a resinous mounting medium.

Result

Hexose-containing and sialic-acid-containing mucosubstances pink to bright purplish red. In an adequately stained preparation reticulin, basement membranes, and mucous glands should stand out sharply against the background of faintly coloured or unstained components of the tissue.

Notes

1. Remove mercurial deposits if introduced by the fixative. A control slide should be treated with Schiff's reagent **without** prior oxidation by HIO_4. This will indicate whether any reactive aldehyde groups (from the fixative, for example) were previously present in the tissue. If found, such aldehydes must be chemically blocked **before** the treatment with HIO_4. If pink staining occurs in the absence of aldehydes, the Schiff's reagent has deteriorated.
2. A bisulphite rinse is often recommended between stages 5 and 6, but Demalsy & Callebaut (1967) have shown that this causes some fading of the stain and is not necessary provided that the Schiff's reagent is washed away very rapidly.
3. Glycogen can be identified if the PAS reaction is applied in conjunction with amylase-digested control sections (see p. 166). Neuraminidase and/or mild acid hydrolysis will assist in the identification of sialic acids. The PAS reaction is blocked by acetylation, though this procedure has little analytical value. (See also Section 11.6.4.2.)
4. If frozen sections are used, some glycolipids are stained. A pseudoplasmal reaction is also usually obtained with frozen sections (see Chapter 12), which are therefore unsuitable for the study of glycoproteins unless lipids are extracted before staining.

11.6.4.2. Specialized PAS methods

The theoretical basis of these procedures, which should be used in conjunction with appropriate enzymatic digestions and other controls, is explained and discussed in Section 11.3.2.1 of this chapter.

(a) *Sialic acids*

To generate aldehydes from sialic acids (NANA and possibly others), oxidize in a freshly prepared solution of **either** sodium metaperiodate ($NaIO_4$; 0.0086% aqueous, for 30 min) **or** periodic acid (0.114% aqueous, for 5 min). Transfer the slides directly into 1.3% sodium sulphite (Na_2SO_3) for 5 min to arrest the oxidation. Wash in running tap water for 5 min and then proceed with stages 5–8 of the standard technique.

(b) *O-acetylated sialic acids*

Treat hydrated sections as follows:

1. Oxidize in 1% periodic acid ($HIO_4.2H_2O$), 30 min.
2. Wash in running tap water for 10 min.
3. Stain some of the slides with Schiff's reagent (stages 5–8 of standard technique).
4. Immerse the remaining slides in 0.05% sodium borohydride ($NaBH_4$) in 1% Na_2HPO_4 for 10 min. Use this solution within 30 min of dissolving the $NaBH_4$. **Caution:** see p. 143.
5. Wash in four changes of tap water.
6. Immerse in 0.5% potassium hydroxide (KOH) in 70% ethanol for 30 min.
7. Wash gently in four changes of tap water.
8. Oxidize for a second time with 1% periodic acid for 10 min.
9. Stain with Schiff's reagent (stages 5–8 of standard technique).

If positive reactions due to hexoses have been excluded, staining at stage 3 above is due to NANA or to sialic acids acetylated at position C7 or C9. Staining obtained at stage 9 but not at stage 3 is due to sialic acids acetylated at C8 or at more than one of the positions C7, C8, and C9. A positive result at stages 3 **and** 9 indicates a sialic acid acetylated at C7. The results may be difficult to interpret if mucosubstances containing different sialic acids occur at the same site.

(c) *Uronic acids*

Treat hydrated paraffin sections as follows:

1. Oxidize for 1 h at 30°C in 2% aqueous sodium metaperiodate ($NaIO_4$).
2. Wash in running tap water for 10 min.
3. Treat with freshly dissolved 0.1% sodium borohydride in 1% Na_2HPO_4 for 10 min.
4. Wash in running tap water for 10 min.
5. Oxidize for a second time in 2% $NaIO_4$ for 24 h at 30°C.
6. Wash in running tap water for 10 min.
7. Stain with Schiff's reagent (stages 5–8 of the standard technique).

A positive result is seen at the sites of proteoglycans with free hydroxyl groups at positions C2 and C3 of their uronic acid residues. Hexoses and sialic acids are unstained.

11.6.5. The concanavalin A–peroxidase method

This technique (Kiernan, 1975) is one in which the carbohydrate-binding properties of a lectin are exploited. The three control procedures exclude the possibility of non-specific staining of substances other than those sugar residues to which the lectin is selectively bound.

Solutions required

A. Con A solution

Concanavalin A (Sigma Chemical Co., St. Louis, Mo, U.S.A., Grade IV; lyophilized and free of carbohydrate):	15 mg
0.06 M sodium phosphate buffer, pH 7.2:	30 ml

This solution is used at room temperature. It may be stored frozen and used repeatedly for about 6 months.

B. Con A—α-methyl glucoside solution

This is the same as solution A but containing in addition 200 mg per ml of α-methyl-D-glucopyranoside (or, alternatively, 400 mg per ml of sucrose).

C. HRP solution

Peroxidase (from horseradish; Sigma Chemical Co., St. Louis, Mo, U.S.A., Type II):	2.0 mg
0.06 M sodium phosphate buffer, pH 7.2:	50 ml

Stored frozen; may be used repeatedly, at room temperature. Keeps for about 6 months.

D. Peroxidase substrate mixture

The solid reagent (DAB) should be weighed beforehand and dissolved immediately before stage 6 of the procedure. The 0.5% solution of H_2O_2 should be prepared from a more concentrated stock solution not more than 24 h before using.

3,3′-diaminobenzidine tetrahydrochloride: (DAB) (**caution:** possibly carcinogenic)	50 mg
0.06 M sodium phosphate buffer, pH 7.2:	50 ml

Dissolve. Immediately before use, add:

0.5% aqueous hydrogen peroxide:	0.5 ml

E. Osmium tetroxide solution

Osmium tetroxide (OsO_4):	1.0 g
Water:	200 ml

Clean the ampoule of OsO_4 and break it under water in a bottle. Complete solvation may take a few hours. This solution is stored at 4°C and may be used repeatedly until it contains a black precipitate.

Procedure

1. De-wax and hydrate paraffin sections.
2. Immerse in con A solution (A) for 10 min.
3. Rinse in three changes of water.
4. Immerse in HRP solution (C) for 10 min.
5. Rinse in three changes of water.
6. Incubate for 10 min in the peroxidase substrate mixture (solution D). (See *Note 1*.)
7. Rinse in three changes of water.
8. Immerse in 0.5% OsO_4 (solution E) for 10 min in a closed container.
9. Wash in three changes of water, 3–4 min in each.
10. Dehydrate through graded alcohols, clear, and cover.

Control procedures

(a) At stage 2, immerse in con A-α-methyl glucoside (solution B) instead of solution A. This will greatly reduce or completely prevent the specific binding of the lectin to α-D-glucosyl and α-D-mannosyl residues of mucosubstances. (See *Note 3*.)

(b) Omit stages 2 and 3. A positive result indicates non-specific attachment of the horseradish peroxidase to the sections (see *Note 3*).

(c) Omit stages 2, 3, 4, and 5. Any endogenous peroxidase activity of the tissue will now be demonstrated. (See *Note 2*.)

Result

Sites of mucosubstances containing α-D-mannosyl or α-D-glucosyl residues: dark brown. Controls (a), (b), and (c) should all be completely unstained. (See *Note 3*.)

Notes

1. The procedure may be stopped at stage 7. The purpose of the treatment with OsO_4 is to intensify the brown colour of the insoluble product of the enzymatically catalysed oxidation of DAB by H_2O_2.
2. Red blood cells commonly retain their peroxidatic properties in paraffin sections of fixed tissue. Endogenous peroxidase and peroxidase-like activity may be inhibited (between stages 1 and 2 of the procedure) by immersing the sections for 10 min in 100 ml of absolute ethanol containing 0.2 ml of concentrated hydrochloric acid. The sections should then be washed thoroughly before proceeding with the method.
3. For action to be taken in the event of positive staining in controls (a) and (b), consult the original account of the method (Kiernan, 1975).
4. If staining is weak or absent, add calcium chloride ($CaCl_2$) and manganese chloride ($MnCl_2$) to the con A (solution A), both at concentrations of 10^{-3} M.
5. Yamada and Shimizu (1976) have described a similar technique in which alcian blue (pH 2.5) is used as a counterstain, so that proteoglycans and con-A-binding glycoproteins are shown in contrasting colours.

11.6.6. Enzymatic extractions

The solutions of enzymes may be used in coplin jars or applied as drops to horizontal slides in a closed petri dish with wet filter paper in the bottom to ensure adequate humidity and prevent drying. It is essential to incubate control sections in the solvent alone.

11.6.6.1. Amylase

Used as a solution containing 1.0 mg of enzyme per ml of water. Ideally a pure preparation of β-amylase should be used. An effective alternative is human saliva, though this also contains various proteolytic enzymes and a ribonuclease. An adequate supply of saliva may be obtained by thinking of lemons and drooling into a small beaker. Bubbles should be removed by stroking the surface of the collected liquid with filter paper before applying it to the sections.

Incubate for 30 min at 37°C. The treatment is somewhat destructive to the sections, which may need to be covered with a film of celloidin after applying the enzyme but before staining.

11.6.6.2. Hyaluronidase

Testicular hyaluronidase removes hyaluronic acid, chondroitin-4-sulphate and chondroitin-6-sulphate. Streptococcal hyaluronidase removes only hyaluronic acid.

Both hyaluronidases are used as solutions containing 1.0 mg of enzyme per ml of 0.9% aqueous NaCl.

Incubate overnight at 37°C.

11.6.6.3. Neuraminidase

The enzyme from *Vibrio cholerae* is used.

Solution

0.1 M acetate buffer, pH 5.5:	1.0 ml
Calcium chloride ($CaCl_2$):	1.0 mg (or 0.1 ml of a 1% aqueous solution)
Neuraminidase:	100 units

The sections may be saponified (see below) before exposure to neuraminidase. This will make some otherwise resistant sialic acid residues susceptible to removal by treatment with the enzyme. Incubate overnight at 37°C.

11.6.7. Chemical blocking and unblocking procedures

11.6.7.1. Acetylation of hydroxyl groups

1. Take sections to absolute alcohol.
2. Immerse overnight at room temperature in a mixture of acetic anhydride (2 volumes) and pyridine (3 volumes) in a tightly closed container. Exercise care in handling acetic anhydride, which reacts with water to form acetic acid.
3. Wash in two changes of absolute alcohol and then pass through 70% alcohol to water.

It is often said that the pyridine must be anhydrous, but ordinary reagent-grade pyridine is perfectly satisfactory. This is not surprising since the traces of water in it will react rapidly with acetic anhydride to form acetic acid, though not in quantities sufficient to interfere with the function of the pyridine as a basic catalyst.

11.6.7.2. Methylation

This procedure methylates carboxyl groups and removes sulphate-ester groups.

1. Take sections to absolute methanol.
2. Immerse in methanol containing 1% by volume of concentrated hydrochloric acid in a tightly closed container at 58–60°C for 48 h.
3. Rinse in absolute methanol and pass through 70% alcohol to water.

Methylation overnight (16–18 h) or for 24 h is usually adequate for blocking carboxyl groups and removing sulphate-ester groups. The completeness of the desired consequences of methylation is easily determined by staining with alcian blue at pH 1.0 and 2.5.

11.6.7.3. Saponification

This is, as explained earlier, an unblocking procedure which can be used after acetylation or methylation.

1. Take sections to absolute ethanol.
2. Cover with a film of celloidin (optional), drain, dry, and harden the film in 70% ethanol.
3. Immerse for 20 min in a freshly prepared 0.5% solution of potassium hydroxide (KOH) in 70% ethanol.
4. Rinse in 70% ethanol and take sections to water.

11.6.7.4. Mild acid hydrolysis

This procedure removes all sialic acid residues of glycoproteins, including those that resist digestion by neuraminidase.

1. Take sections to water.
2. Heat some 0.1 N sulphuric acid (add 2.8 ml of concentrated sulphuric acid, S.G. 1.84, 96% H_2SO_4, to about 800 ml of water. Mix thoroughly and add water to obtain 1000 ml) to 80°C. Place the slides in the hot, dilute acid and maintain at 80°C for 1 h.
3. Wash in running tap water for 2 min, rinse in distilled or deionized water, and proceed with the histochemical testing.

11.7. EXERCISES

Theoretical

1. Which mucosubstances are stained by (a) the periodic acid–Schiff reaction, (b) fluorescent-labelled concanavalin A?

2. Hydroxyl groups of sugars can be **sulphated** (i.e. converted to sulphate esters) by treating sections with sulphuric or chlorosulphonic acid. What effect would you expect this procedure to have on the pattern of staining by thionine or toluidine blue?

3. What substances (if any) would you expect to be PAS-positive in a section which had been (a) acetylated, (b) acetylated and then saponified, (c) methylated, (d) methylated and then saponified?

4. What substances (if any) would you expect to stain with alcian blue (pH 1.0 or 2.5) after (a) acetylation, (b) acetylation followed by saponification, (c) methylation, (d) methylation followed by saponification, (d) mild acid hydrolysis?

5. Which proteoglycans and glycoproteins can bind fluorescent-labelled aprotinin? How would the pattern of staining be affected by prior treatment of the sections with (a) amylase, (b) hyaluronidase, (c) neuraminidase?

6. Which of the histochemical techniques discussed here would be suitable for the demonstration of glycogen? Describe exactly how you would proceed to demonstrate glycogen in the cells of a piece of liver freshly removed from an animal.

7. To what extent would chemical blocking procedures be useful in confirming the specificity of binding of a fluorescently labelled lectin believed to be specific for (a) the α-L-fucosyl configuration, (b) *N*-acetylgalactosamine? How else can one ascertain the specificity of a labelled lectin used as a histochemical reagent?

8. With which (if any) of the methods described in this chapter would you expect to obtain a positive reaction with chitin? How do you account for the fact that the exoskeletal tissues of invertebrates are commonly PAS-positive?

9. A glycoprotein secreted by mucous cells of mammalian submandibular glands bears oligosaccharide chains with the structure

$$\text{NANA}\,(\alpha\text{-}2\rightarrow 6)\text{D-Gal}(\beta\text{-}1\rightarrow)\,\text{PROTEIN}$$

How much of the information contained in this formula could be obtained by histochemical analysis?

Practical

10. Stain paraffin sections of liver for glycogen and establish the specificity of the result by the judicious use of amylase.

11. Take some paraffin sections of skin, tongue, salivary gland, and intestine. Treat some of the sections by the procedure described for methylation and acetylation. Stain methylated, acetylated, and control sections with (a) PAS, (b) alcian blue, pH 2.5, (c) alcian blue, pH 1.0. What mucosubstances can be identified by these procedures and in what locations?

12. Stain a section of the external ear to demonstrate metachromasia. Which structures are metachromatically stained and why?

13. Stain sections of (a) tongue, (b) salivary gland, (c) external ear by the concanavalin A–peroxidase procedure. Perform all the necessary controls to establish the specificity of staining.

14. Stain sections of liver and intestine by the colloidal ferric hydroxide method. Identify cells that contain glycoproteins or proteoglycans and cells responsible for false-positive reactions.

15. Using sections of skin or tongue, demonstrate that the cytoplasmic granules of mast cells contain a heavily sulphated proteoglycan with some uronic acid residues, but that sialic acids are absent.

12

Lipids

12.1. Components of lipids 169
12.2. Classification of lipids 171
12.2.1. Free fatty acids 171
12.2.2. Terpenes 171
12.2.3. Steroids 172
12.2.4. Neutral fats 172
12.2.5. Waxes 172
12.2.6. Cholesterol esters 172
12.2.7. Phosphoglycerides 173
12.2.8. Spingomyelins 174
12.2.9. Ceramides 174
12.2.10. Glycosphingosides 174
12.2.11. Lipids conjugated to protein 174
12.3. Histochemical methodology 174
12.4. Extractive and hydrolytic methods 176
12.5. Solvent dyes 176
12.6. Tests for unsaturation 177
12.6.1. Osmium tetroxide 178
12.6.2. Performic acid–Schiff method 178
12.6.3. Palladium chloride 179
12.6.4. Bromination 179
12.7. Glycolipids 179
12.8. Free fatty acids 180
12.9. The plasmal reaction 180
12.10. Cholesterol and its esters 181
12.11. Choline-containing lipids 182
12.12. Individual methods 182
12.12.1. Sudan IV 183
12.12.2. Sudan black B methods 183
12.12.3. Osmium tetroxide method 184
12.12.4. Palladium chloride method for hydrophilic unsaturated lipids 184
12.12.5. Holczinger's method for free fatty acids 185
12.12.6. The plasmal reaction 186
12.12.7. Method for cholesterol and its esters 186
12.12.8. The acid–haematein test 187
12.12.9. Solvent extractions 188
12.12.10. Hydrolysis of esters 188
12.12.11. Masked lipids 188
12.12.12. Bromination 188
12.13. Exercises 188

The term "lipid" is applied to a chemically heterogeneous group of substances which are extracted from tissues by non-polar organic solvents such as chloroform and ether. These substances vary greatly in structural complexity but are built from a limited number of simpler molecules joined together in different ways. These component substances do not occur in large amounts in the free state in living tissue but are present as metabolic precursors of the lipids themselves.

12.1. COMPONENTS OF LIPIDS

In the following brief account the symbols R, R′ and R″ indicate alkyl radicals, mostly of 16–20 carbon atoms. Simplified structural formulae are used for whole lipids: most carbon atoms are represented only as junctions of bonds, and hydrogen atoms attached to carbon are omitted.

(a) *Aliphatic alcohols*

ROH. Present in waxes as their esters. An example is cetyl alcohol, $CH_3(CH_2)_{14}CH_2OH$. Long-chain alkyl groups are also present as glyceryl ethers in the ether phosphatides.

(b) *Fatty acids*

RCOOH. Present as acyl groups in all those lipids which are esters or amides. Most of them have unbranched chains of even numbers of carbon atoms. Olefinic (unsaturated; CH═CH) linkages may or may not be present. In the following list of the commonest fatty acids of animals, the length of chain is shown as C16, C18, etc., and the number of double bonds is indicated by the symbol △.

Saturated:	Myristic acid	C14
	Palmitic acid	C16
	Stearic acid	C18
	Lignoceric acid	C24
Unsaturated:	Palmitoleic acid	C16, △1
	Oleic acid	C18, △1
	Linoleic acid	C18, △2
	Linolenic acid	C18, △3
	Arachidonic acid	C20, △4
	Clupanodonic acid	C22, △5

In normal animals, all lipids (with the exception of some cholesterol esters and sulphatides) contain at least one unsaturated acyl group per molecule.

(c) *Glycerol*

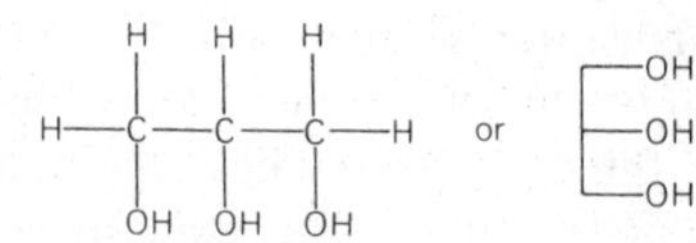

A trihydric alcohol, present as its esters in most lipids.

(d) *Phosphoric acid*

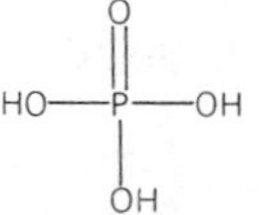

This formulation, though not strictly in accord with modern chemical notation, is the one usually used by biochemists. In phospholipids the H_3PO_4 is esterified through one or two of its hydroxyl groups. The unesterified hydroxyls ionize as acids.

(e) *Choline, ethanolamine, and serine*

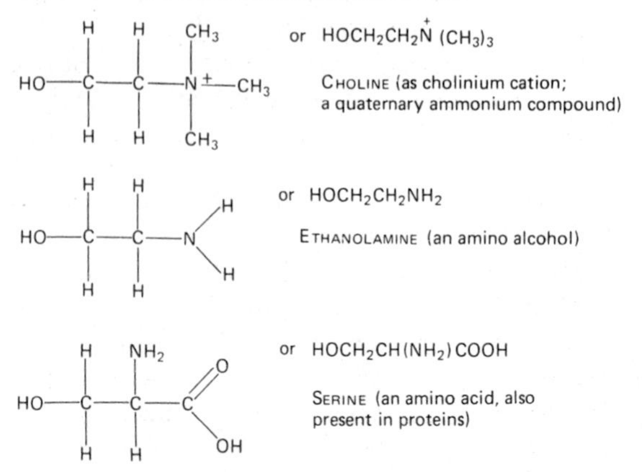

These three bases are esterified through their hydroxyl groups with phosphoric acid in phosphoglycerides and sphingomyelins.

(f) myo-*Inositol*

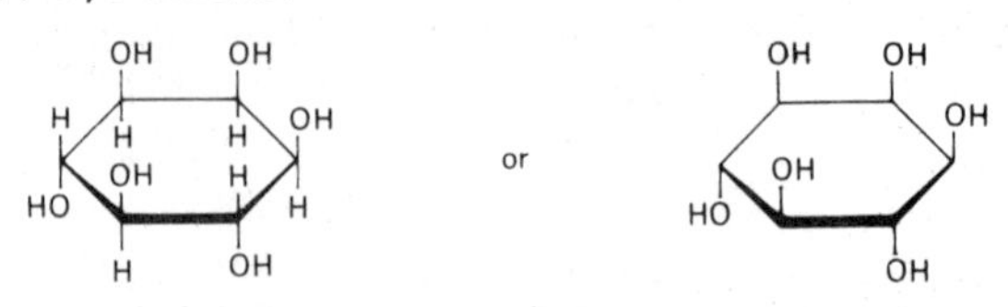

A cyclic alcohol, esterified through phosphoric acid to glycerol in the phosphoinositides, in which some other hydroxyl groups of the inositol are also phosphorylated.

(g) *Sugars*

Galactose, *N*-acetylgalactosamine, *N*-acetylglucosamine, glucose, *N*-acetylneuraminic acid, and other monosaccharides, sometimes carrying sulphate-ester groups, are present in the glycosphingolipids. See Chapter 11 for structures and abbreviated formulae.

(h) *Sphingosine*

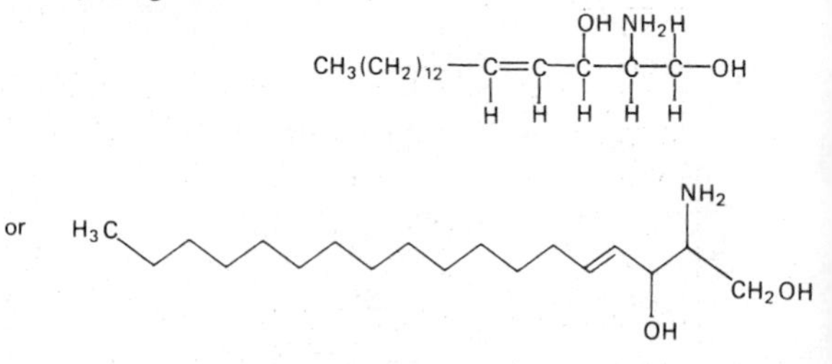

This long-chain (C18) unsaturated amino alcohol is joined in amide linkage to fatty acids and is esterified with phosphoric acid through the hydroxyl group shown at the right-hand side of the formula. In the related compound dihydrosphingosine the bond between C4 and C5 is saturated. In the dehydrosphingosines, more than one unsaturated linkage is present in the hydrocarbon chain.

(i) *Cholesterol*

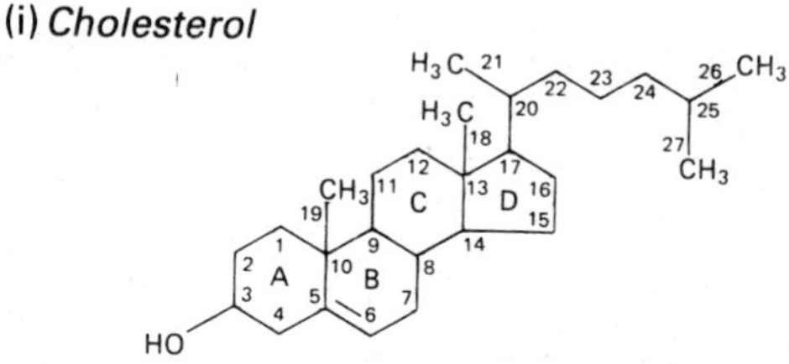

This is a **sterol** (a steroid alcohol). It is insoluble in water or cold ethanol, but freely soluble in ether or acetone, so it is classified as a hydrophobic lipid (see Table 12.1).

The fused ring system (cyclopentanoperhydrophenanthrene), with a wide variety of substituents, is characteristic of all steroids. The numbering system and the letters identifying the rings are applicable to all steroids.

(j) *Isoprene*

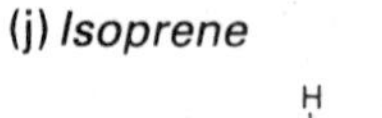

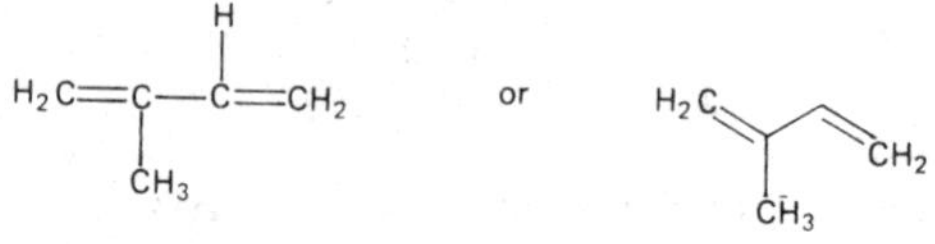

This unsaturated hydrocarbon is the monomer from which the carbon skeletons of most of the terpenes are constructed.

(k) *Proteins*

These are frequently conjugated with phospholipids in proteolipids and lipoproteins.

12.2. CLASSIFICATION OF LIPIDS

The following scheme, which includes only the lipids of vertebrate animals, differs slightly in its arrangement from the classifications used by biochemists. Substances with similar histochemical properties are, as far as possible, grouped together.

The classification is summarized in Table 12.1. Brief descriptions of the main groups of lipids follow.

12.2.1. Free fatty acids

These occur only in traces in normal tissues, but the amounts are increased in some pathological states. Crystal-like deposits of free fatty acids may also form in lipid-rich tissues that have stood for long periods of time in formaldehyde solutions. The fatty acids form water-soluble salts with sodium and potassium and insoluble salts, known as soaps, with calcium and many other metal cations.

The prostaglandins are polyunsaturated fatty acids (C20, △ 2–3) produced by all animal cells, but in minute quantities which are unlikely to be histochemically detectable.

12.2.2. Terpenes

Numerous terpenes occur in plants. The only important one in higher animals is squalene:

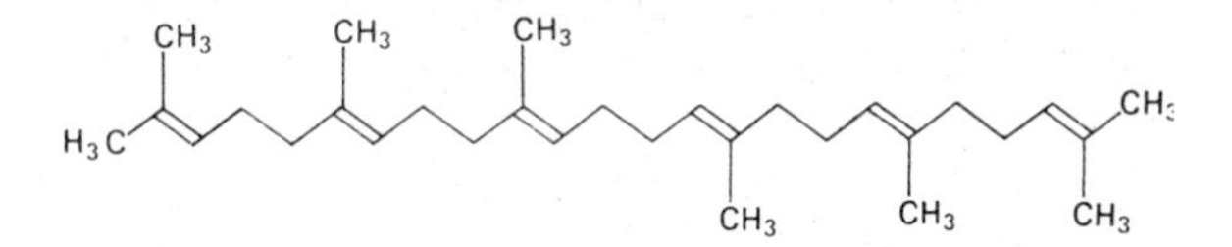

This hydrocarbon occurs in sebum and is also a metabolic precursor of cholesterol.

TABLE 12.1. *A classification of the major groups of lipids*

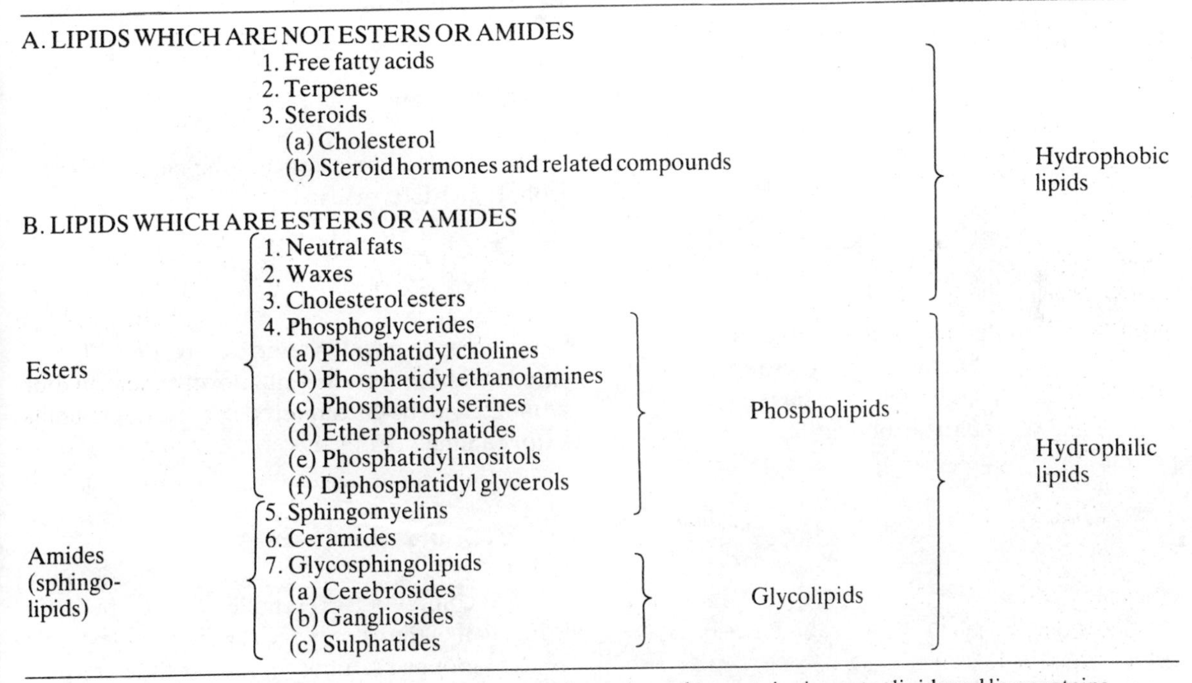

	Lipid	Group	Character
A. LIPIDS WHICH ARE NOT ESTERS OR AMIDES			
	1. Free fatty acids		Hydrophobic lipids (A1 to B3)
	2. Terpenes		
	3. Steroids		
	(a) Cholesterol		
	(b) Steroid hormones and related compounds		
B. LIPIDS WHICH ARE ESTERS OR AMIDES			
Esters	1. Neutral fats		
Esters	2. Waxes		
Esters	3. Cholesterol esters		
Esters	4. Phosphoglycerides	Phospholipids (4 and 5)	Hydrophilic lipids (4 to 7)
Esters	(a) Phosphatidyl cholines		
Esters	(b) Phosphatidyl ethanolamines		
Esters	(c) Phosphatidyl serines		
Esters	(d) Ether phosphatides		
Esters	(e) Phosphatidyl inositols		
Esters	(f) Diphosphatidyl glycerols		
Amides (sphingolipids)	5. Sphingomyelins		
Amides (sphingolipids)	6. Ceramides		
Amides (sphingolipids)	7. Glycosphingolipids	Glycolipids (7)	
Amides (sphingolipids)	(a) Cerebrosides		
Amides (sphingolipids)	(b) Gangliosides		
Amides (sphingolipids)	(c) Sulphatides		

Note. Phospholipids and sphingolipids are often covalently bound to proteins in proteolipids and lipoproteins.

12.2.3. Steroids

Cholesterol (see p. 170) is a waxy solid with a much higher melting point (150°C) than most other lipids. Inspection of its structural formula shows that it is an unsaturated alcohol. It dissolves easily in most organic solvents but is insoluble in water.

The steroid hormones are numerous but it is doubtful whether they can be identified histochemically. They are stored only in minute quantities in the glands that secrete them, though metabolic precursors of similar chemical structure may exist in detectable concentrations in steroid-secreting endocrine cells. Many of the steroid hormones are ketones.

The corticosteroids have two ketone groups per molecule, as well as three alcoholic hydroxyl groups. An example is cortisol (hydrocortisone):

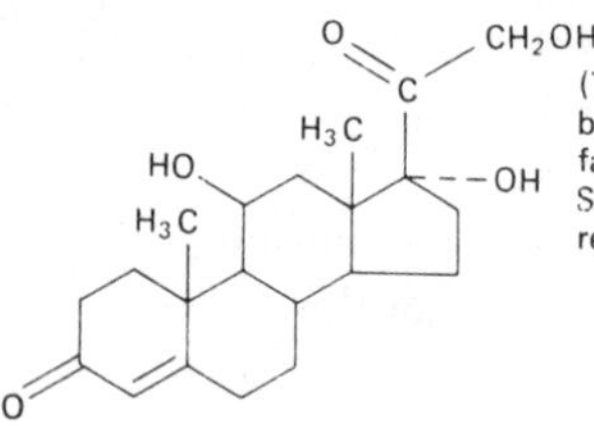

(The dotted line indicates that the bond lies on the side of the molecule facing away from the reader. Substituents directed towards the reader are joined by intact bonds)

Androgens are also ketones but lack the side-chain at position 17. Examples are:

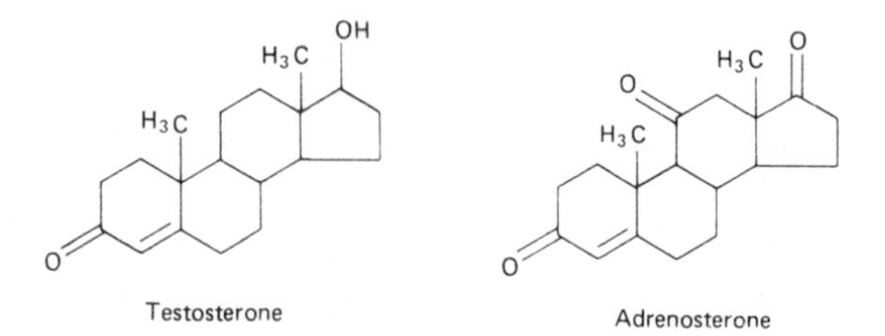

Testosterone Adrenosterone

Oestrogens differ from other steroids in that ring A is aromatic, so that the hydroxyl group attached to it confers a phenolic character on the molecule. Ketone and secondary alcohol groups may also be present.

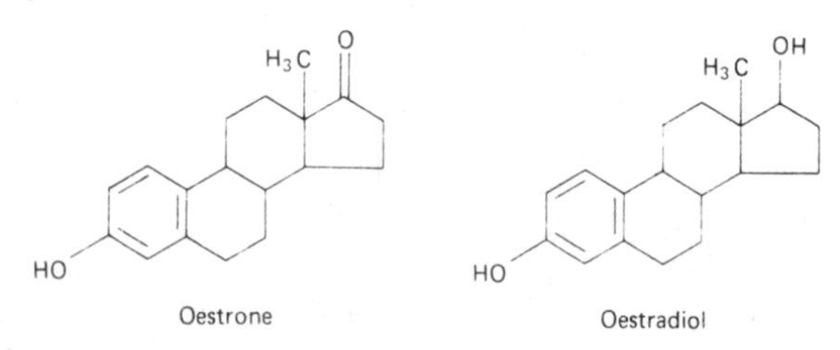

Oestrone Oestradiol

Progesterone, the principal progestogen secreted by the corpus luteum, is also a metabolic precursor of other steroid hormones. It is chemically quite similar to the corticosteroids:

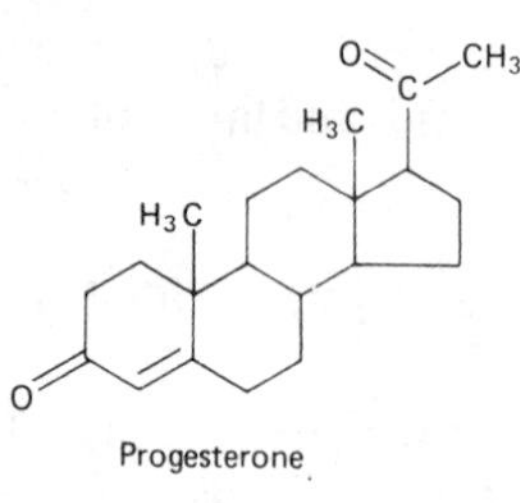

Progesterone

12.2.4. Neutral fats

These may be mono-, di-, or triglycerides:

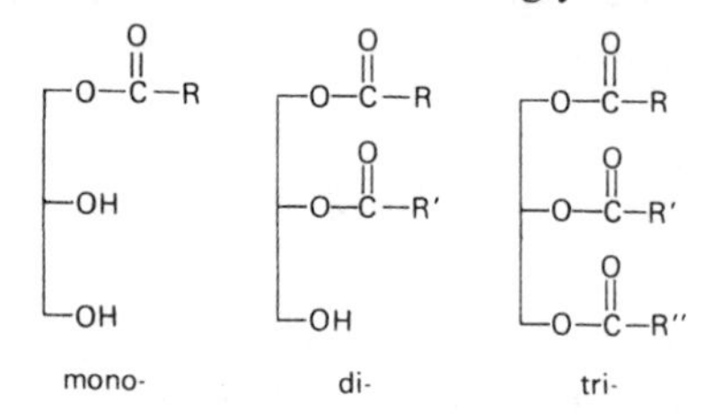

The last-named type is the most abundant. Neutral fats occur principally in adipose connective tissue.

12.2.5. Waxes

Natural waxes are esters of long-chain aliphatic alcohols with fatty acids:

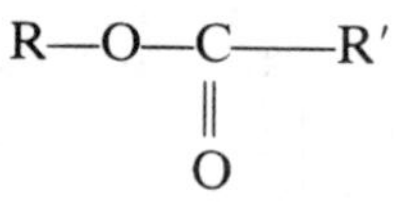

Waxes occur in a wide variety of organisms (spermaceti and beeswax are well-known examples) but are unlikely to be encountered in the tissues of laboratory animals or of man.

12.2.6. Cholesterol esters

These compounds resemble waxes, but the alcohol half of the ester is derived from cholesterol. The acyl groups are unusual in that they are predominantly those of saturated fatty acids.

12.2.7. Phosphoglycerides

These lipids are hydrophilic on account of their content of polar groups such as phosphate ester, hydroxyl, and primary or quaternary amino. They fall into six groups:

(a) *Phosphatidyl cholines*

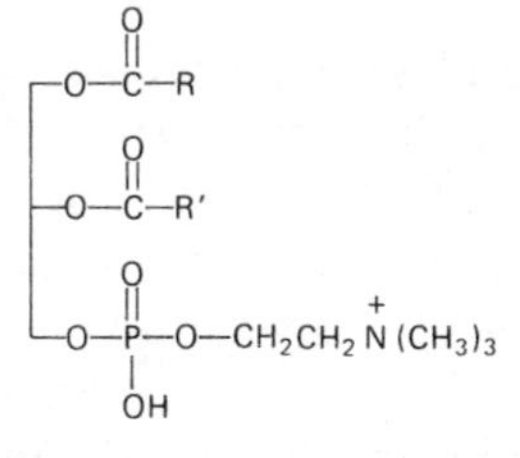

These phospholipids are commonly known as "lecithins". They are soluble in all lipid solvents, including ethanol, with the notable exception of acetone.

(b) *Phosphatidyl ethanolamines*

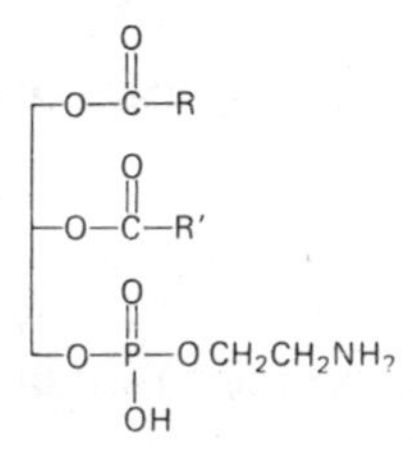

An old synonym is "cephalins". The ethanolamine has sometimes been called "cholamine" or "colamine". In the pure state, phosphatidyl ethanolamines are, like the lecithins, soluble in ethanol but insoluble in acetone. However, the crude mixture of cephalins isolated from tissues, which is a mixture of phosphatidyl ethanolamines, serines, and inositols, is insoluble in ethanol. Extractions with solvents are of little value when attempting to distinguish histochemically between different phosphoglycerides in sections of tissues.

(c) *Phosphatidyl serines*

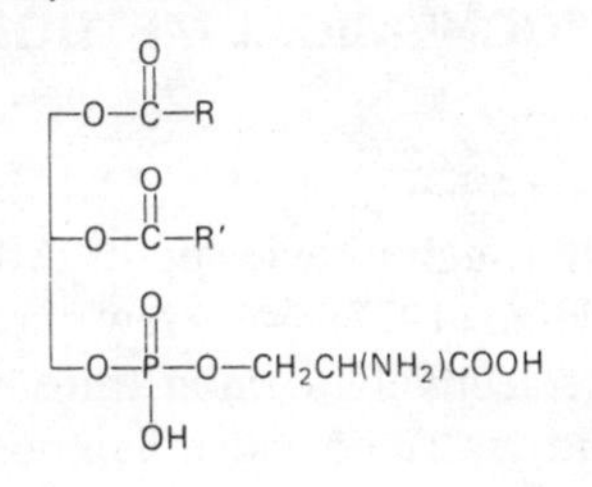

(d) *Ether phosphatides*

In these phospholipids one of the oxygen atoms of glycerol is joined to a long-chain alkyl group to give an ether. The most important ether phosphatides to the histochemist are those in which there is an unsaturated linkage adjacent to the ether oxygen. These are the **plasmalogens**:

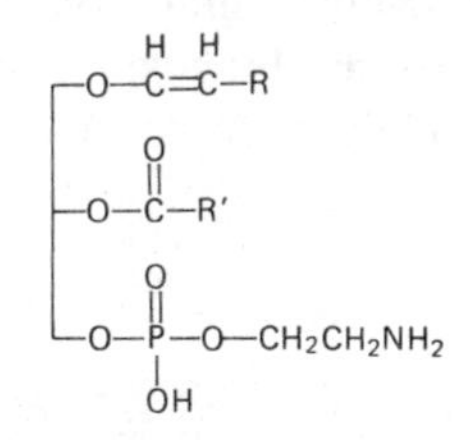

It was once thought that plasmalogens were acetals rather than ethers, hence the old name "acetal lipids". The basic component is most commonly ethanolamine but sometimes choline. Plasmalogens are extracted rapidly from sections of tissue by 80% ethanol but only slowly by 60% ethanol.

(e) *Phosphatidyl inositols*

The simplest lipids of this type, which are also known as "inositol cephalins", have the structure

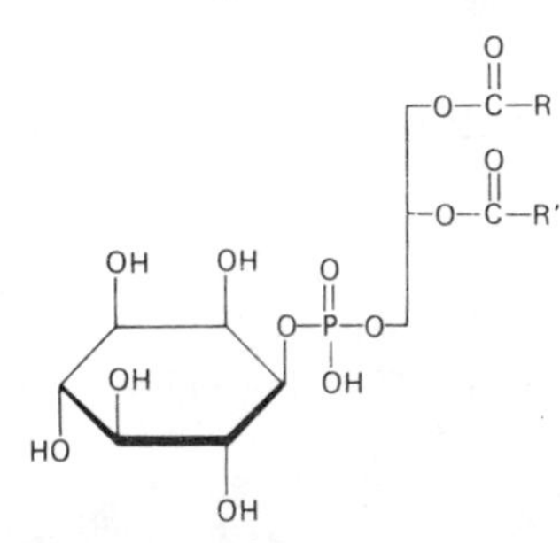

Commonly one or two of the hydroxyl groups shown in the *myo*-inositol ring are esterified by phosphoric acid.

(f) *Diphosphatidyl glycerols*

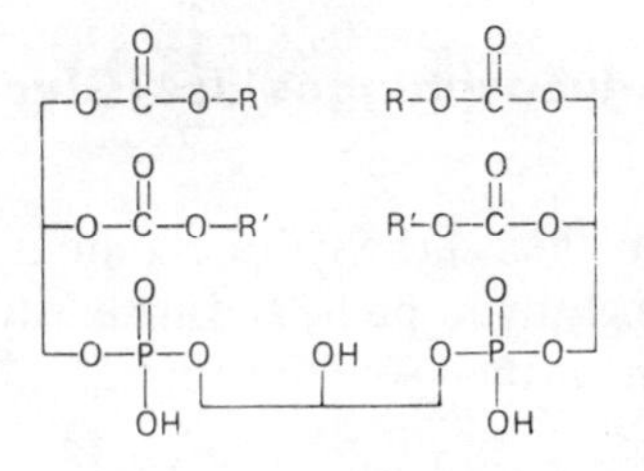

The most important lipid of this type is cardiolipin, a component of the inner mitochondrial membrane. Cardiolipin is unusual among lipids in that it can serve as an antigen.

12.2.8. Sphingomyelins

These choline-containing phospholipids are amides derived from sphingosine, choline and fatty acids, the last named being connected by an amide linkage:

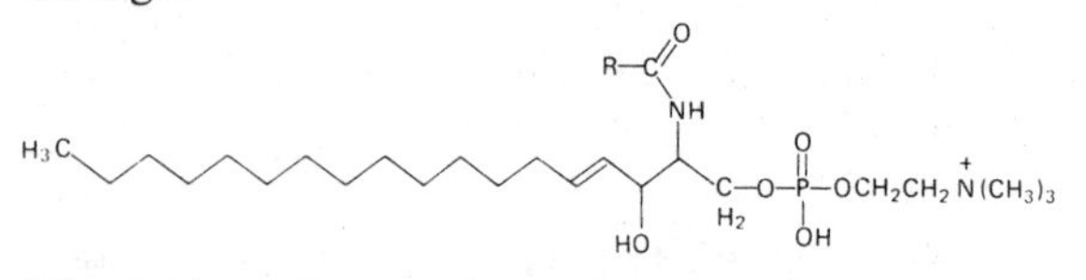

The sphingosine may be replaced by dihydrosphingosine or dehydrosphingosine. The amide linkage of sphingomyelins is much more resistant to alkali-catalysed hydrolysis than are the ester linkages of other phospholipids.

Sphingomyelins, which dissolve in hot ethanol but not in ether or acetone, have more hydrophilic character than do the phosphoglycerides, which are all soluble in ether.

12.2.9. Ceramides

These lipids, which are simple amides of sphingosine containing no phosphorus, are widely distributed but are never present in high concentration in tissues.

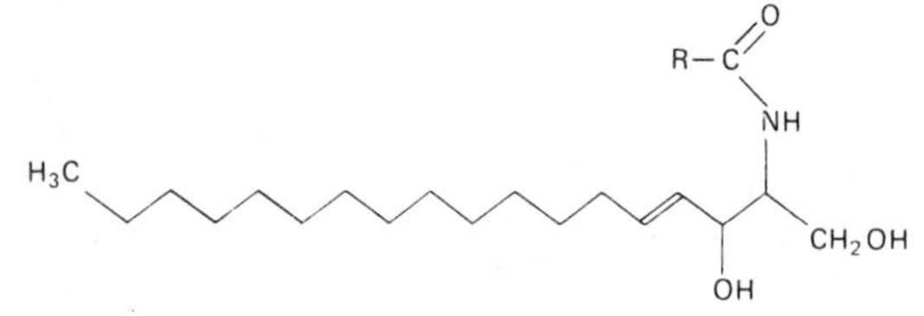

12.2.10. Glycosphingosides (Glycolipids)

Because of their carbohydrate content, these lipids are strongly hydrophilic and are easily lost from tissues owing to their solubility in water. There are three types:

(a) Cerebrosides. These are ceramides in which the terminal hydroxyl group of sphingosine is joined by a glycosidic linkage to a hexose sugar, which is most commonly a β-D-galactosyl residue:

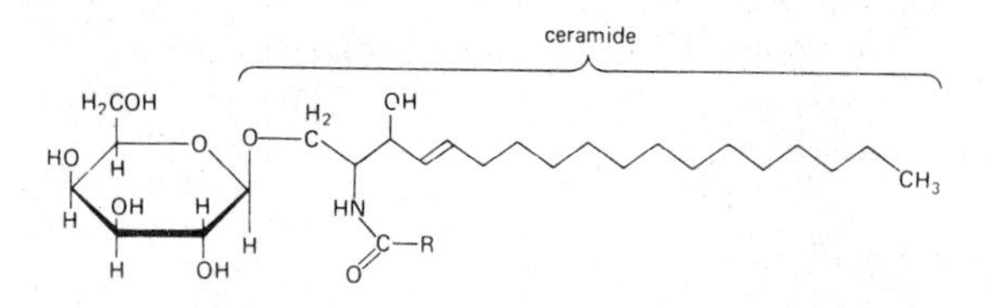

(b) Gangliosides resemble the cerebrosides, but have an oligosaccharide chain in place of the single hexose residue. Sialic acids and *N*-acetyl hexoses are always present. They have structures such as: NANA(α-2 → 4)DGalNAc(β-1 → 4)D-Gal(β-1→ ceramide).

(c) Sulphatides are cerebrosides in which one of the alcoholic hydroxyl groups of the hexose is esterified by sulphuric acid. In the commonest sulphatides of mammalian nervous tissue a β-galactosyl residue is sulphated at position C3.

12.2.11. Lipids conjugated to protein

Proteolipids are compounds in which each protein molecule is combined with several of lipid, so that the complete molecule is soluble in non-polar solvents.

Lipoproteins are large protein molecules with some bound lipid. They are insoluble in non-polar solvents and are also insoluble in most polar solvents, so they are not extracted from tissues by the process of embedding in paraffin wax. The lipid moieties of lipoproteins, which are phospholipids, can be dissolved out of sections only if the bonds to protein are broken. The cleavage can be brought about by adding a strong acid to a suitable solvent. The chemical nature of the bonds between lipid and protein is imperfectly understood.

12.3. HISTOCHEMICAL METHODOLOGY

The many techniques of lipid histochemistry have been thoroughly reviewed by Adams (1965) and Bayliss High (1977). Very few of the available procedures selectively demonstrate any of the classes of lipid described above. It is possible, however, to obtain information about several physical properties and chemical constituents of these substances. Methods are available which will detect with reasonable certainty the presence of:

1. Any lipids.
2. The hydrophobic or hydrophilic nature of a lipid.
3. Unsaturation.
4. Carbohydrates.
5. Free fatty acids.
6. The 1,2-unsaturated ether groups of plasmalogens.
7. Cholesterol and its esters.
8. Choline in phospholipids.
9. Amide rather than ester linkages.

Both the physical and chemical properties have to be taken into consideration when attempting to identify and localize lipids in tissues. The specificities of techniques for the demonstration of lipids have been determined mainly by experiments using pure lipids incorporated into thin paper, which is then treated as if it were a section being stained. Baker (1946) and Adams (1965) have carried out thorough studies of this kind.

Lipids are mostly unaffected by chemical fixation of tissue (Heslinga & Deierkauf, 1961; Deierkauf & Heslinga, 1962), but the presence of calcium ions in the fixative (as in Baker's formal–calcium) increases the preservation of the hydrophilic phospholipids, possibly by forming insoluble calcium–phosphate–lipid complexes.

Bayliss High (1977) has shown that fixation in formal–calcium should not exceed 2–3 days' duration for blocks of tissue or 1 h for cryostat sections of fresh tissue. Longer fixation impairs the staining of phosphoglycerides. Calcium ions derived from the fixative form salts (calcium soaps), insoluble in non-polar solvents, with free fatty acids. Treatment with a strong acid (e.g., 1.0 N HCl for 60 min) will regenerate the fatty acids.

Frozen sections are used for the histochemical examination of lipids, but appreciable quantities of phospholipids persist in sections of paraffin-embedded tissue. They occur mainly in myelin sheaths of nerve fibres and in erythrocytes. These lipids are stained in some of the histological techniques for the demonstration of myelin (see Chapter 18). They can be extracted from paraffin sections by suitable mixtures of solvents.

Phospholipids are rendered completely insoluble in organic solvents by prolonged treatment with potassium dichromate (known as "chromation") or with some metallic cations, though the chemistry of the process is not yet understood. Lillie (1969) has shown that the chromating reagent reacts with double bonds in lipids, though other functional groups such as hydroxyls may also be involved in the binding of chromium, especially if chromation is carried out at 60°C rather than at 3° or 24°C. He has suggested that a cyclic ester is produced:

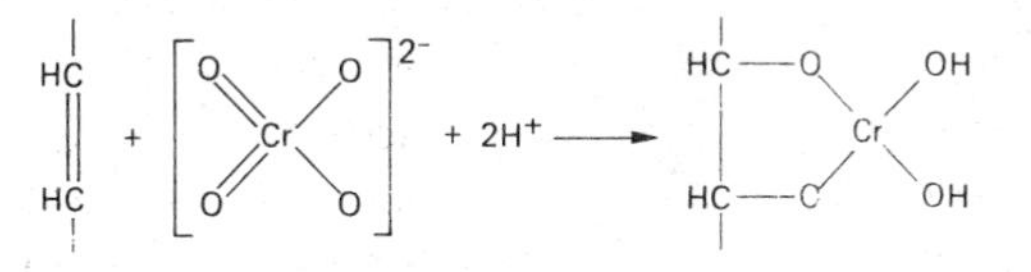

The oxidation number of the chromium in the proposed product is +4. Compounds of Cr(IV) are rare, though some stable organometallic complexes are known. The free hydroxyl groups of such a complex would be expected to participate in the formation of dye–metal complexes (e.g. with haematein) if the coordination number of the chromium atom were 6. Coordination numbers 4 and 6 exist in known Cr(IV) complexes. Lillie's speculations do not take into account the fact that the chromating reagent, between pH 2 and pH 6, exists as an equilibrium mixture of hydrogen chromate and dichromate ions:

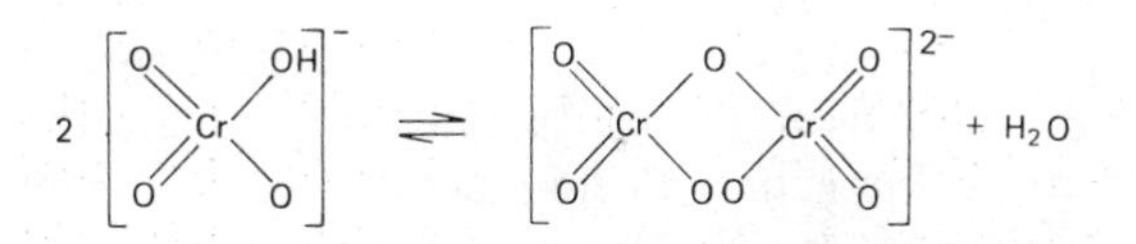

The chromate ion $[CrO_4]^{2-}$ required by Lillie's equation is formed when the pH of the solution is above 6 (see Cotton & Wilkinson, 1972, for chemistry of chromium). A dichromate ion might be able to combine with two unsaturated sites to give a Cr(IV) bis-ester. Such cross-linking of lipid molecules might account for their insolubilization. Further investigation of the mechanism of chromation should involve chemical studies of the reactions of chromates and dichromates with pure lipids and with artificial membranes containing phospholipids of known structure.

12.4. EXTRACTIVE AND HYDROLYTIC METHODS

The hydrophobic lipids, of which the neutral fats of adipose tissue are the most abundant, are extracted by cold acetone. It is important that the acetone be anhydrous. If traces of water are present, partial extraction of hydrophilic lipids occurs (Elleder & Lojda, 1971).

Either pyridine or a mixture of chloroform and methanol will extract all lipids except those which are firmly bound to protein. Proteolipids are soluble in methanol-chloroform but lipoproteins are not. In order to extract the firmly bound lipids of lipoproteins it is necessary to use an acidified solvent, which will hydrolyse the protein–lipid linkages and then dissolve the lipids (Adams & Bayliss, 1962).

Extractive methods may be used in conjunction with any staining method in order to confirm the specificity for the demonstration of lipids.

With some lipoproteins it is necessary to employ acid hydrolysis in an aqueous medium to release the lipid components for staining with solvent dyes (see below). Without such treatment, the hydrophilic protein prevents access of hydrophobic dye molecules to the lipids.

The ester linkages between fatty acids and glycerol in fats and phosphoglycerides can be broken by alkaline hydrolysis (saponification). The fatty acids are liberated as their soluble sodium soaps. Amide linkages (with sphingosine) are not hydrolysed under the same conditions of time, temperature, and concentration of alkali. Consequently the fatty acid moieties of ceramides and sphingomyelins remain insoluble in water after saponification of the other lipids in the tissue. Glycosphingolipids may also be presumed to resist saponification, though they are largely extracted from tissues by aqueous reagents.

12.5. SOLVENT DYES

Coloured, non-polar substances dissolve in lipids and render them visible under the microscope. Since such coloured substances are not necessarily dyes, Baker (1958) prefers to call them **lysochromes.** These coloured substances stain fat because they are more soluble in it than in the solvents from which they are applied. The first solvent dyes to be used for staining fat were Sudan III and Sudan IV:

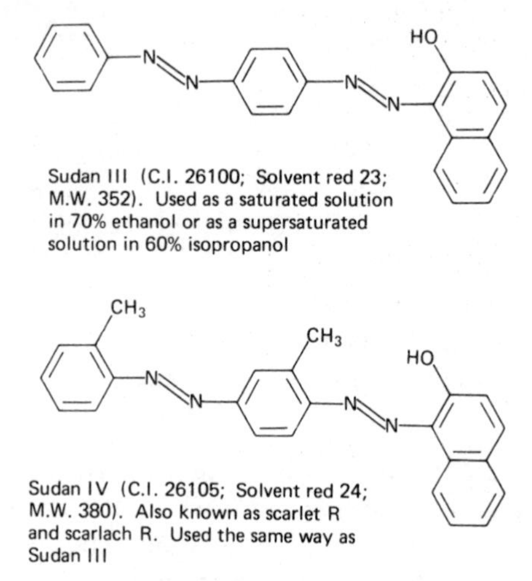

Sudan III (C.I. 26100; Solvent red 23; M.W. 352). Used as a saturated solution in 70% ethanol or as a supersaturated solution in 60% isopropanol

Sudan IV (C.I. 26105; Solvent red 24; M.W. 380). Also known as scarlet R and scarlach R. Used the same way as Sudan III

The latter is more deeply coloured than the former. Both dyes are found by chromatography to be mixtures. Sudan III has one or two major and three or four minor coloured components, while samples of Sudan IV are mixtures of four major components with three minor ones (Marshall, 1977). A major component of some samples of both these dyes is oil red O:

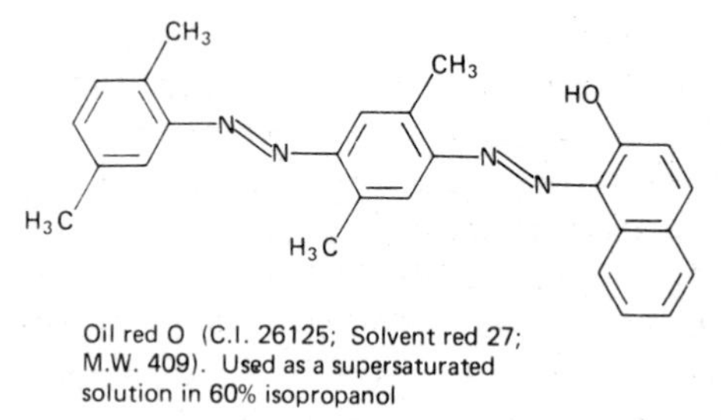

Oil red O (C.I. 26125; Solvent red 27; M.W. 409). Used as a supersaturated solution in 60% isopropanol

Commercial samples of oil red O contain one major component and three coloured contaminants (Marshall, 1977). A very hydrophobic compound, such as Sudan IV, stains only the more hydrophobic lipids such as neutral fats and those esters of cholesterol whose acyl groups are unsaturated.

A more useful reagent for colouring all types of lipid is Sudan black B. This dye is marketed as a mixture of two main components with 8–11 minor coloured contaminants (Marshall, 1977). Several other unwanted dyes are formed in solutions that are more than one month old. Some of these products of deterioration stain proteins and nucleic acids but not lipids (Frederiks, 1977). Even in fresh solutions, contaminating anionic dyes can give rise to false-positive identification of lipids (Malinin, 1977). Chemical analysis of Sudan black B (Pfüller *et al.*, 1977) has revealed that the two major components of the dyestuff have the structures

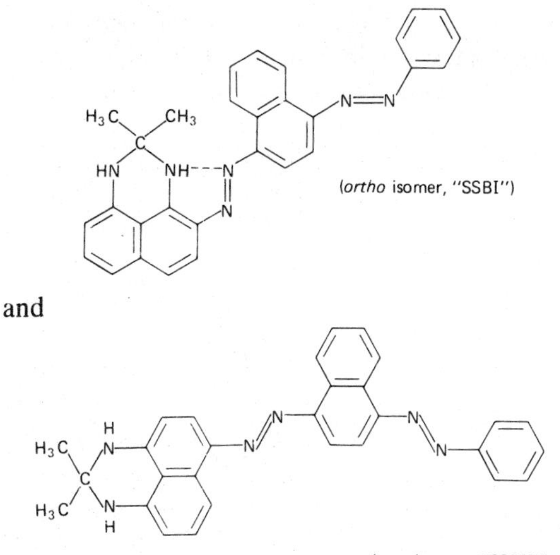

The second formula (SSBII) corresponds to the Sudan black B (C.I. 26150; Solvent black 3; M.W. 457) of *Conn's Biological Stains* (Lillie, 1977). There is no satisfactory assay method for Sudan black B. Batches of the dye are evaluated by the Biological Stains Commission on the basis of their absorption spectra and their performance in methods for staining lipids.

Sudan black B can dissolve in hydrophobic lipids but, by virtue of its two potentially ionizable nitrogen atoms, may also behave like a cationic dye and bind to the hydrophilic phosphate-ester groups of phospholipids, in which it is more soluble than the other Sudan dyes. The *para* isomer (SSBII) is more strongly basic than the *ortho* (SSBI) because in the latter an internal hydrogen bond between one of the ionizable nitrogens and an azo nitrogen tends to inhibit ionization. A *bis-N*-acetylated derivative of Sudan black B can be prepared in the laboratory and some workers prefer this reagent to the commercially supplied dyestuff. Acetylated Sudan black B should not be able to form cations, but Marshall (1977) was unable to find in it any chromatographically distinct components that were not present in the parent dye.

Adams (1965) points out that the staining of lipids is affected by the solvent used for the lysochrome. Thus phospholipids are more likely to be stained from a solution in 70% than from one in absolute alcohol. Phosphoglycerides and free fatty acids are generally supposed to be more prone to extraction by the solvent than are lipids of other types. Some lipoproteins cannot be stained by solvent dyes without prior acid hydrolysis (see p. 176). In such substances the lipid is said to be "masked" by the protein.

Lysochromes can only dissolve in lipids at temperatures above the melting points of the latter. This is a point of some importance, since lipids in which all the fatty acid chains are saturated melt above 60°C. Unsaturated lipids are mostly liquid at room temperature for a reason to be given shortly. Cholesterol melts at 150°C and its saturated esters also have high melting points. However, it is possible to stain these lipids with Sudan black B if the sections have first been treated with bromine water. This reagent reacts with unsaturated linkages in fatty acids (see below) and Bayliss and Adams (1972) have shown that it also reacts with cholesterol to form an oily derivative which is liquid at room temperature. This derivative is not 5,6-dibromo-cholesterol (which has a higher melting point) and its chemical nature is unknown. The bromine–Sudan black method stains virtually all lipids. The only ones not stained would be fully saturated triglycerides or saturated free fatty acids.

12.6. TESTS FOR UNSATURATION

Olefinic linkages occur in isoprene, cholesterol, and sphingosine, and in the very widely distributed unsaturated fatty acids. Consequently, the histochemical demonstration of the carbon–carbon double bond is tantamount to the staining of all lipids. Unsaturated linkages in fatty acids can be of the *cis* or *trans* type. The *cis* configuration is the more abundant in fatty acids of animal tissues.

cis unsaturation

trans unsaturation

It can be seen that a *cis* double bond produces a bend in the chain of carbon atoms. This impairs the packing of the molecules in the solid state and in consequence *cis*-unsaturated lipids have lower melting points than do saturated or *trans*-unsaturated ones. Thus glyceryl tripalmitate melts at 65.1°C while glyceryl trioleate melts at −4°C. Since lysochromes dissolve only in lipids that are in the liquid phase, these dyes impart their colours only to structures containing unsaturated lipids. The ordinary lysochromes, such as Sudan III, Sudan IV, and oil red O stain only those unsaturated lipids that are hydrophobic. Sudan black B (see p. 177) is different and can also enter hydrophilic domains. The methods based on the use of oil-soluble dyes demonstrate lipids by virtue of their physical properties and may reasonably be called "histophysical" techniques (Adams, 1965; Bayliss High, 1977). There are, however, several genuinely histochemical reactions for the demonstration of unsaturation. Some of these will now be discussed.

12.6.1. Osmium tetroxide

The reactions of osmium tetroxide with various substances present in tissues were discussed in some detail in Chapter 2. There it was shown that this compound oxidizes the —CH=CH— bond and is reduced to a black substance, probably osmium dioxide. Since osmium tetroxide is soluble in both polar and non-polar solvents, it serves as a stain for both hydrophilic and hydrophobic lipids. It is possible, however, to distinguish between the two types by mixing the osmium tetroxide with an oxidizing agent that dissolves only in polar liquids. Potassium chlorate is a suitable oxidizing agent for this purpose. When frozen sections are treated with such a mixture, the osmium tetroxide will form cyclic esters (see p. 18) at both hydrophilic and hydrophobic sites. The other product of the reaction, the unstable osmium trioxide, will then disproportionate to give osmium tetroxide (soluble) and dioxide (insoluble and black). However, at hydrophilic sites in the tissue, the newly formed osmium dioxide will immediately be oxidized by chlorate ions:

$$3OsO_2 + 2ClO_3^- \rightarrow 3OsO_4 + 2Cl^-$$

Hydrophilic unsaturated lipids will therefore remain converted to cyclic osmium esters, which may be colourless or brown and are barely visible in the sections. At hydrophobic sites, the precipitation of osmium dioxide will be unimpeded, so that intense black staining will be observed. The cyclic esters of the unsaturated hydrophilic lipids can be stained subsequently by treating the sections with α-naphthylamine, which forms a red or orange complex with the bound osmium. These reactions form the basis of the OTAN (osmium tetroxide-α-naphthylamine) method for differential staining of hydrophobic and hydrophilic lipids. The Marchi method for degenerating myelin (see Chapter 18) is similar. The chemical reactions described above follow Adams *et al.* (1967) and Adams & Bayliss (1968), but have been slightly modified from these accounts in view of the work of Korn (1967).

In practice, osmium tetroxide seems to be a highly sensitive and specific reagent for the demonstration of unsaturation, though the OTAN reaction may sometimes fail to distinguish correctly between hydrophilic and hydrophobic lipids (see Bayliss High, 1977). The reactions of osmium tetroxide with phenolic compounds and with proteins (Nielson & Griffith, 1978, 1979; see also Chapter 2) should not be ignored, however. The specificity can be checked by staining control sections in which unsaturated linkages have been blocked by bromination (see p. 179) and others from which the lipids have been extracted by solvents.

12.6.2. Performic acid–Schiff method

Double bonds are oxidized by performic acid to yield aldehydes (Lillie, 1952):

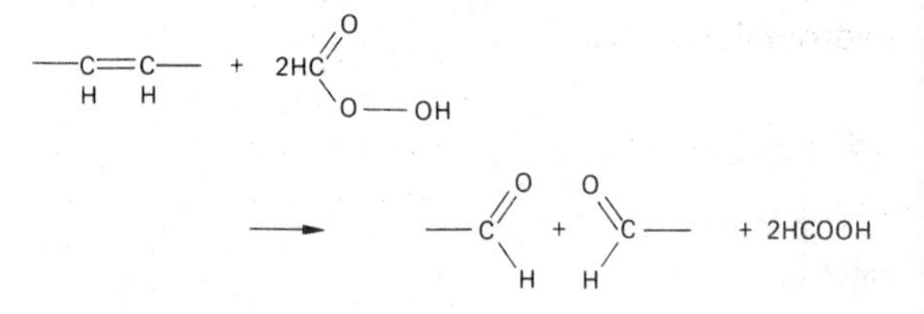

The former sites of unsaturation may then be demonstrated with Schiff's reagent (see Chapter 10, p. 128). In ordinary formalin-fixed frozen sections, appreciable numbers of double bonds are oxidized to aldehydes by atmospheric oxygen, so that many lipid-containing objects are Schiff-positive even without treatment with performic acid. These spontaneously oxidized lipids can confuse the interpretation of the plasmal reaction (see p. 180).

The performic acid–Schiff method is rarely used because the oxidizing agent is damaging to the sections and technically simpler methods are available.

12.6.3. Palladium chloride

Unsaturated hydrophilic lipids can be demonstrated by virtue of their ability to reduce the chloropalladite ion $[PdCl_4]^{2-}$ to metallic palladium. The chloropalladite ion is formed when palladious chloride is dissolved in aqueous hydrochloric acid:

$$PdCl_2 + 2H^+ + 2Cl^- \rightarrow 2H^+ + [PdCl_4]^{2-}$$

(H_2PdCl_4 = chloropalladious acid)

An unstable complex is formed with olefinic compounds and the complex is reduced to the metal, which is visible as a black deposit. Since the reagent is used in aqueous solution, it is reduced only by hydrophilic unsaturated lipids. The chloropalladite ion also behaves as an anionic dye, imparting a yellow background colour to non-lipid substances. This non-specific staining can be largely removed by treating the sections with pyridine, with which are formed complexes of the type $[Pd(pyr)_2Cl_2]$. The chemistry of the technique is discussed in greater detail by Kiernan (1977a).

Palladium chloride has been used in empirical staining procedures for the nervous system (Paladino, 1890) and in electron microscopy to impart electron density to elastin (Morris *et al.*, 1978). The reaction with elastin has not been studied chemically, but it does not result in the formation of black products, so there is no possibility of confusion with hydrophilic lipids in light microscopy.

12.6.4. Bromination

Unsaturated linkages are **blocked** by bromination, either by exposure to bromine vapour

or by treatment with bromine water or an aqueous solution of bromine in potassium bromide ($KBr + Br_2 \rightleftharpoons KBr_3$)

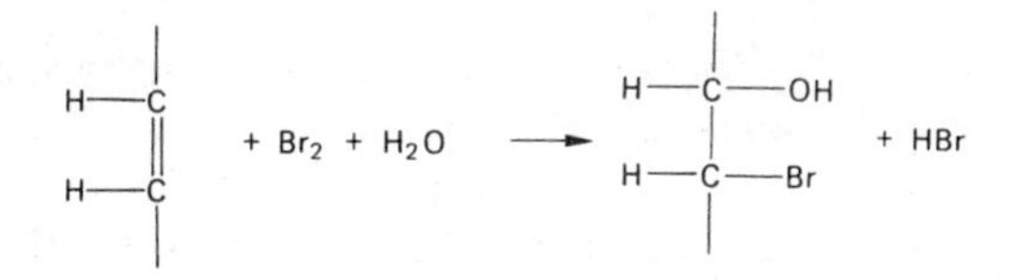

This test, though not entirely specific, may be used to confirm the unsaturated nature of substances stained by the methods discussed above.

The bromo derivatives of unsaturated lipids are decomposed with liberation of bromide ions when treated with a dilute mineral acid. The bromide ions can be precipitated as silver bromide, which can then be reduced to black metallic silver. These reactions constitute the bromine–silver method for the histochemical detection of unsaturation (Norton *et al.*, 1962):

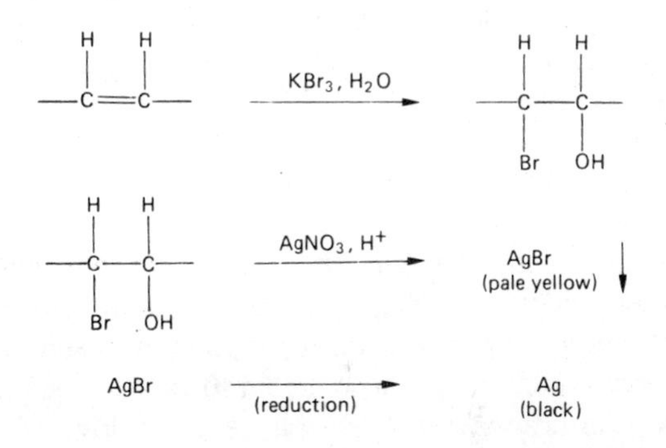

The specificity of this technique is marred, however, by the occasional occurrence of non-specific deposition of silver at sites in the tissue that do not contain lipids.

12.7. GLYCOLIPIDS

Carbohydrates in glycosidic combination with lipids are demonstrated by the techniques of carbohydrate histochemistry. It should be remembered that inositol, though not a sugar, has an arrangement of hydroxyl groups similar to that found in some hexoses. Phosphatidyl inositols will contain the glycol configuration unless three or more of their hydroxyl groups are phosphorylated. Distinction between glycolipids and mucosubstances is made by using extractive procedures for the lipids. It is not possible to draw satisfactory conclusions when the two types of substance are present in the same place.

When the PAS reaction is used with frozen sections, allowance must be made for the possible presence of aldehydes generated by atmospheric oxidation of unsaturated fatty acids. Direct positive staining with Schiff's reagent from this cause is known as the **pseudoplasmal** reaction. The periodate ion also oxidizes a proportion of the unsaturated linkages to pairs of aldehyde groups, resulting in a truly positive PAS reaction that is not due to carbohydrate.

A modified PAS method, devised to circumvent these reactions of olefinic bonds, is described by Bayliss High (1977). Primary amine groups are first oxidatively deaminated (producing aldehydes), unsaturated linkages are next oxidized to aldehydes, with performic acid. All the free aldehyde groups are then blocked with 2,4-dinitrophenylhydrazine. A conventional PAS procedure is then carried out. Since all the aqueous reagents used extract gangliosides, the only lipids stained by this method are cerebrosides (and possibly phosphatidyl inositols).

Lipids containing acid sugars (i.e. gangliosides and sulphatides) are conspicuous only in the tissues of patients with certain lipid-storage diseases. Special methods, not used in ordinary carbohydrate histochemistry, are usually employed for the histopathological diagnosis of these conditions.

12.8. FREE FATTY ACIDS

The soaps formed from fatty acids and heavy metals are insoluble in water. The cationic component of such a soap can be demonstrated by any suitable chromogenic reaction.

In Holczinger's method, sections are treated with a dilute solution of cupric acetate. Loosely bound copper is then removed by a brief treatment with a chelating agent, EDTA. The residual metal, which has been shown by Adams (1965) to be associated only with fatty acids, is made visible by forming an insoluble dark-green compound with dithiooxamide (see Chapter 13).

Elleder & Lojda (1972) were less impressed with the specificity of this technique than was Adams (1965). They found that free fatty acids were sometimes weakly stained when known to be present in considerable quantities in some tissues and that there was false-positive coloration of some phospholipids and of calcified material as well as background staining of proteinaceous and carbohydrate-containing material. They showed that pre-treatment with 1.0 N hydrochloric acid enhanced the staining of free fatty acids, probably by causing hydrolysis of the calcium soaps formed during fixation in formal-calcium. This pre-treatment also dissolved calcium phosphates and carbonate. The differentiation in EDTA could usefully be extended beyond the time suggested in the original method, to minimize the staining of copper bound to "background" structures, but extraction of control sections with cold acetone was necessary in order to ensure that positive results were due to free fatty acids and not to phospholipids. The modifications recommended by Elleder & Lojda (1972) are incorporated into the practical instructions for Holczinger's method given later in this chapter.

12.9. THE PLASMAL REACTION

In frozen sections, plasmalogens yield aldehydes following a brief treatment with a 1% aqueous solution of mercuric chloride. The aldehydes are then demonstrated by means of Schiff's reagent (see Chapter 10, p. 128). The chemistry of the plasmal reaction has been worked out by Terner & Hayes (1961), whose paper should be consulted for the experimental evidence upon which the following account is based.†

†In most publications dealing with plasmalogens, the double bond of the vinyl ether is said to be between the α- and β-carbon atoms. This is an incorrect usage of the Greek letters (see "Conventions and Abbreviations", page ix of this book). These two carbons should be designated as 1 and 2, as in the present account. The letters α and β, used correctly, would refer to carbon atoms 2 and 3.

Mercuric chloride adds to the double bond of the vinyl (1,2-unsaturated) ether linkage of plasmalogens:

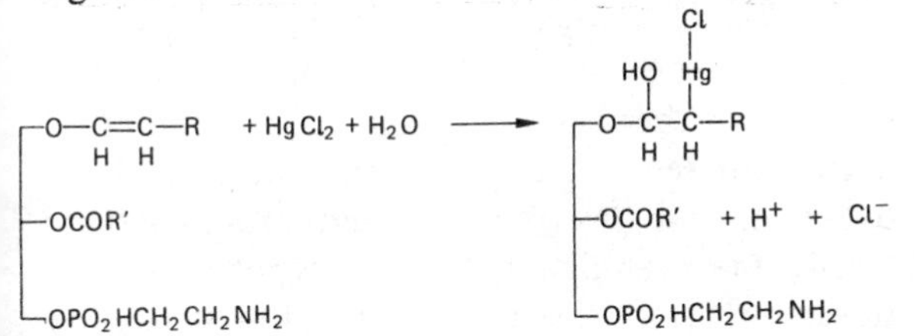

The initial product, a hemiacetal, is unstable and immediately dissociates into an alcohol and an aldehyde:

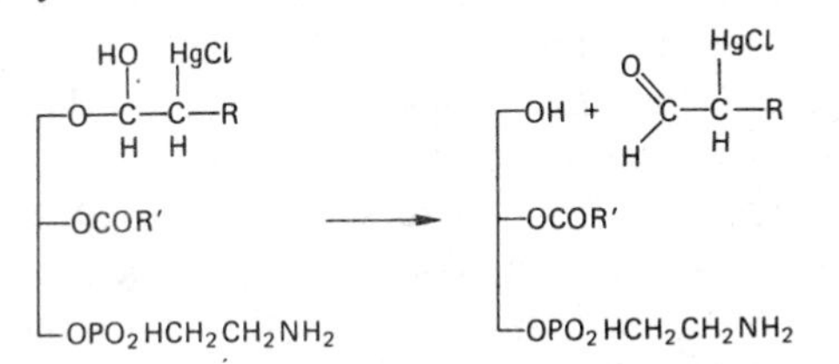

The mercury atom remains attached to carbon 2 of the aldehyde; its presence there can be demonstrated histochemically.

Vinyl ethers are much more easily hydrolysed in the presence of acids than are ordinary ethers, but the pH of the mercuric chloride solution used in the plasmal reaction (3.5) is not low enough to catalyse the hydrolysis. However, 6 N hydrochloric acid is just as effective as 1% mercuric chloride in generating aldehydes from plasmalogens:

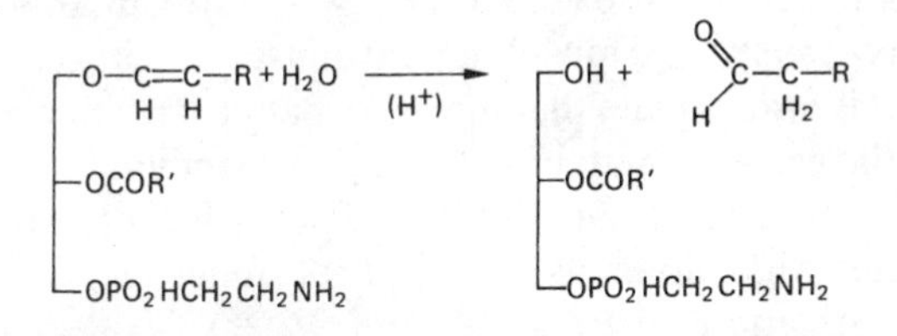

(The obvious product of hydrolysis of this ether would be an enolic "vinyl alcohol", RCH=CHOH. These compounds do not exist as such but as the tautomeric aldehydes.)

In practice, 6 N HCl is not used because it is more injurious to the sections than aqueous mercuric chloride. However, acid-catalysed hydrolysis of plasmalogens can occur in Schiff's reagent (pH 2.5), especially if the sections are immersed in it for more than 20 min (Elleder & Lojda, 1970). Consequently, omission of the treatment with mercuric chloride is not always an adequate control procedure for the plasmal reaction. Staining with Schiff's reagent alone could be due to plasmalogens as well as to a "pseudoplasmal" reaction (see p. 180). If, however, there is no staining with Schiff's reagent alone, a positive reaction after treatment with mercuric chloride certainly indicates the presence of plasmalogens.

If aldehydes are found to be already present in the tissue (positive pseudoplasmal reaction), they should be blocked by reduction with a neutral or alkaline solution of sodium borohydride (see Chapter 10, p. 132) before exposing the sections to mercuric chloride. As a further control, sodium borohydride may be used after the mercuric chloride. If any staining occurs when there are no aldehyde groups in the tissue, the Schiff's reagent is not working properly and must be replaced.

12.10. CHOLESTEROL AND ITS ESTERS

Of the various histochemical tests for steroids which have been devised, the one with the most certain specificity is the perchloric acid-α-naphthoquinone reaction. Adams (1965) has shown that this method gives a positive result with no lipids other than cholesterol, its esters, and a few other closely related sterols. The sections are heated in a solution containing perchloric acid, 1,2-naphthoquinone-4-sulphonic acid, formaldehyde, and ethanol.

Perchloric acid is thought to convert cholesterol to a conjugated diene:

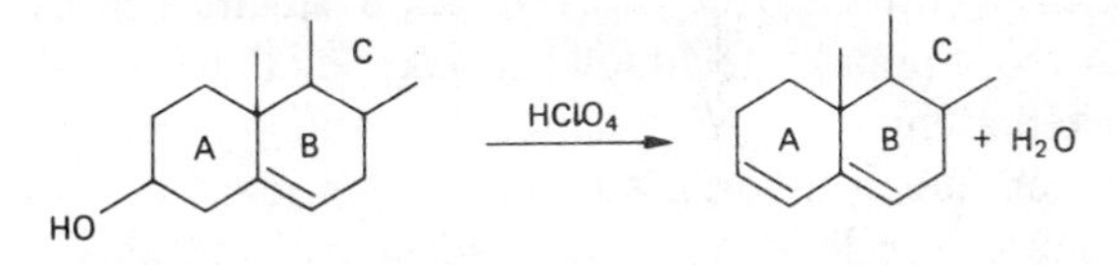

(Only the A and B rings of the cholesterol molecule are shown. The remainder of the structure does not take part in the reaction.)

This reaction is analogous to the well-known preparative technique whereby ethylene is formed by elimination of water from ethanol in the presence of concentrated sulphuric acid. Esters of cholesterol may be hydrolysed in the strongly acid reagent and the resultant cholesterol then converted to the diene. In the next phase of the method, the diene reacts with 1,2-naphthoquinone-4-sulphonic acid (see p. 121) to form a blue compound. The chemistry of this latter reaction and the nature of the end-product are not understood. Neither are the roles of the ethanol and the formaldehyde contained in the reagent.

Free cholesterol may be distinguished from its esters by treating sections with a solution of digitonin before staining. Digitonin is a glycoside of a plant sterol and it forms with cholesterol an adduct which is insoluble in cold acetone. Esters of cholesterol remain soluble in cold acetone, along with other hydrophobic lipids. Digitonin has also been incorporated into fixatives for the purpose of making cholesterol insoluble and also osmiophilic for histochemical studies with the electron microscope. However, Vermeer *et al.* (1978) have found that digitonin added to aqueous fixatives fails to immobilize cholesterol on filter paper and that it accelerates the diffusion of cholesterol esters. The specificity of histochemical tests involving the use of digitonin to insolubilize cholesterol is therefore uncertain.

12.11. CHOLINE-CONTAINING LIPIDS

Various techniques are available for the selective demonstration of phosphatidyl cholines and sphingomyelins. Some of them (e.g. Bottcher & Boelsma-van Houte, 1964; Hadler & Silveira, 1978) are derived from analytical chemical methods for choline and its derivatives. They will not be discussed here since they are not very frequently used. A more popular technique is Baker's (1946) acid–haematein test. Although the chemistry of this technique is poorly understood, model experiments with lipids and other substances on paper have revealed that it is specific for choline-containing phospholipids (Adams, 1965) if the instructions are followed faithfully. Baker's (1946) investigation indicated a lower degree of specificity, but it is likely that his specimens of phosphatidyl ethanolamines ("cephalin") and cerebrosides ("brain galactolipine") were less pure than those available to Adams (1965).

In the first stage of the acid–haematein procedure, sections fixed in either formal–calcium or a mixture similar to Bouin's fluid are exposed to a solution containing potassium dichromate. As explained earlier (p. 175), this reagent probably combines covalently with unsaturated linkages and also with nearby hydroxyl groups. The chromated sections are then stained with an acidic solution of haematein (see Chapter 5), which forms a dark blue dye–metal complex with the bound chromium. Differentiation in an alkaline solution of potassium ferricyanide leaves the strong colour only in structures such as mitochondria and myelin sheaths which contain phosphatidyl cholines and sphingomyelin. This coloured product is insoluble in alcohols and xylene, so the preparations can be mounted permanently in resinous media. Aside from its histochemical value, the acid–haematein method is one of the best techniques for staining mitochondria. Other methods for the demonstration of these organelles by light microscopy of fixed tissues (see Gabe, 1976, for detailed descriptions) are purely empirical. A yellow background coloration is probably due to unmordanted haematein acting as a simple anionic dye.

The mode of action of the differentiator is unknown. Baker (1958) suggested that it might oxidize haematein to more faintly coloured compounds (see Chapter 5, p. 61).

Baker (1946) prescribed an extraction with pyridine as a control procedure to confirm that stained structures were lipids. Nowadays other solvents are preferred and it is also possible to employ a saponification procedure (see p. 174) so that the method becomes specific for sphingomyelins.

It is unfortunate that the histochemical rationale of the acid–haematein test is so poorly understood. The chemistry of the reactions of potassium dichromate with lipids is in need of much further investigation.

12.12. INDIVIDUAL METHODS

Fixation in formal–calcium (preferred) or neutral buffered 4% formaldehyde is recommended for all these methods.

12.12.1. Sudan IV

Preparation of stain

Prepare a saturated solution of Sudan IV in 70% ethanol in a tightly stoppered bottle. Allow to stand for 2 or 3 days before using the supernatant solution. Keeps indefinitely.

Procedure

1. Cut frozen sections and rinse them in 70% ethanol.
2. Stain in Sudan IV solution for 1 min.
3. Transfer to 50% ethanol for a few seconds, until no more clouds of dye leave the section.
4. Wash in two changes of water.
5. (Optional.) Counterstain nuclei progressively with alum–haematoxylin.
6. Wash in water.
7. Mount in an aqueous medium.

Result

Lipids (especially neutral fats)—orange to red. Small intracellular droplets cannot be resolved and hydrophilic lipids are only weakly stained, but the method is quite adequate for adipose tissue and for the detection of isolated fat cells.

12.12.2. Sudan black B methods

This is the bromine–Sudan black B method of Bayliss & Adams (1972) for the demonstration of all types of lipid. If the bromination is omitted, free cholesterol is not stained and phospholipids are less strongly coloured. (See *Note 1* below.)

Solutions required

A. Bromine water

Bromine:	5.0 ml
Water:	200 ml

Bromine is caustic, and its brown, pungent vapour is injurious to the respiratory system. It should be kept in a fume cupboard and must not be pipetted by mouth. The aqueous solution is no more hazardous than most other laboratory chemicals. Bromine water is stable for about 3 months at room temperature in a glass-stoppered bottle and may be used repeatedly. It should be replaced when its colour becomes noticeably weaker.

B. Sodium metabisulphite

0.5% aqueous $Na_2S_2O_5$. Dissolve on the day it is to be used.

C. Sudan black B

Add 600 mg of Sudan black B (C.I. 26150) to 200 ml of 70% ethanol. Place on a magnetic stirrer for 2 h, then pour into a screw-capped bottle. Leave to stand overnight. To use the solution, filter it into a coplin jar and try not to disturb the sediment of undissolved dye in the bottom of the bottle.

The solution of Sudan black B may be kept (and used repeatedly) for 4 weeks.

D. A counterstain

Mayer's carmalum (see p. 83) is suitable nuclear counterstain because it acts progressively and is stable in aqueous mounting media.

Procedure

1. Cut frozen sections, mount them onto slides, and allow them to dry.
2. Immerse slides in bromine water (solution A) for 1 h.
3. Rinse in water.
4. Immerse in sodium metabisulphite (solution B) for about 1 min until the yellow colour of bromine has been removed from the sections. (See *Note 2* below.)
5. Wash in four changes of water.
6. Rinse in 70% ethanol.
7. Stain in Sudan black B (solution C) for 10 min with occasional agitation of the slides.
8. Rinse in 70% ethanol, 5–10 s (see *Note 3* below), with agitation, then wash in water.
9. Apply counterstain if desired.
10. Wash in water and mount in an aqueous medium.

Result

Lipids appear in shades of deep grey, very dark blue, and black.

Notes

1. Stages 2–5 may be omitted if it does not matter that cholesterol will be unstained.

2. In the original account of this method, "Sodium bisulphate" was specified as the reagent for decolorizing the brominated sections: obviously an error. Dilute aqueous solutions of sodium or potassium sulphite, bisulphite, or thiosulphate may be used instead of sodium metabisulphite.
3. This differentiation is the only critical step in the method. Ideally a lipid-extracted control section should be included with those being stained. When the lipid-free section is completely decolorized the end-point of the differentiation has been reached.

12.12.3. Osmium tetroxide method

Solution required

Osmium tetroxide solution

Water:	100 ml
Osmium tetroxide:	1.0 g

To prepare this solution, carefully clean the outside of the sealed glass ampoule in which the OsO_4 is supplied, removing all traces of the label and any gum with which it was attached. Score the glass with a diamond scriber, clean off sebum from your skin with acetone, and allow the ampoule to dry. Drop the scored, cleaned ampoule into a very clean bottle containing 100 ml of the purest available water. Insert the glass stopper (no grease may be used) and shake the bottle until the ampoule breaks. If necessary, the ampoule may be broken by striking it with a clean, degreased glass rod. The OsO_4 often takes several hours to dissolve completely.

In a clean, tightly stoppered bottle, this solution keeps for a few months at 4°C. The solution may be used several times provided that all glassware is very clean. Debris derived from sections may be removed by filtration (in fume cupboard) when the solution is poured back into its bottle. The used filter paper should be soaked overnight in 10% alcohol to reduce OsO_4 to $OsO_2.2H_2O$, before throwing it away.

Osmium tetroxide should always be used in a fume cupboard.

Deterioration is indicated by the presence of a blue–grey colour in the solution, due to colloidal osmium dioxide. Old solutions should be pooled in a tightly screw-capped bottle and kept for recycling (see Kiernan, 1978).

Procedure

1. Wash frozen sections in water. They may be mounted onto slides, preferably without any adhesive.
2. Transfer to the OsO_4 solution and leave in a tightly closed container for 1 h.
3. Wash free-floating sections in five changes of water (at least 10 ml per section), 2 min in each change. Mounted sections may be washed in running tap water if they are still firmly adherent to their slides.
4. Dry free-floating sections onto slides or blot mounted sections with filter paper.
5. Mount in an aqueous medium. Alternatively, dehydrate in dioxane (two changes, each 4 min with occasional agitation), clear in carbon tetrachloride (two changes, each 1 min), and mount in a resinous medium.

Result

Unsaturated lipids—black.

Notes

1. The hydrophobic lipids may be extracted with acetone (see below). This may reveal hydrophilic lipids in the same sites.
2. Thorough washing is necessary to remove excess OsO_4.
3. Alcohols are avoided for dehydration since they would reduce any OsO_4 not removed by aqueous washing. The clearing agent recommended is one in which OsO_4 is extremely soluble. **Caution:** toxic vapours from dioxane and carbon tetrachloride.

12.12.4. Palladium chloride method for hydrophilic unsaturated lipids

In addition to its use as a histochemical test, this method (Kiernan, 1977a) is suitable for the histological demonstration of myelinated nerve fibres in the peripheral nervous system. In the central nervous system, myelin is also coloured, but lipids in the grey matter reduce the contrast.

Solutions required

A. Chloropalladious acid solution

Stock solution

Palladium chloride ($PdCl_2$):	1.0 g
Concentrated hydrochloric acid:	1.0 ml
Water:	5.0 ml

Mix thoroughly to disperse the $PdCl_2$, then add:

Water:	to 50 ml

Keeps for several months

Working solution

Stock solution:	1.0 ml
Water:	9.0 ml

This diluted solution can be used repeatedly until it becomes cloudy.

B. 20% aqueous pyridine

Pyridine:	20 ml	Keeps indefinitely
Water:	80 ml	

Procedure

1. Wash frozen sections in water. (See *Note 1*.)
2. Transfer to working solution of chloropalladious acid for 2 h at 37°C. (See *Note 2* below.)
3. Rinse in two changes of water, 1 min in each.
4. Immerse sections in 20% aqueous pyridine for 1 min.
5. Rinse in water.
6. Dehydrate through graded alcohols, clear in xylene, and mount in a resinous medium.

Result

Unsaturated hydrophilic lipids—dark brown or black. Background pale yellow. (See *Note 3* below.)

Notes

1. Either free-floating sections or sections dried onto slides may be used.
2. Alternatively, leave the sections in chloropalladious acid overnight at room temperature, or for 30 min at 58°C.
3. Bromination or extraction of hydrophilic lipids prevents the reaction but the yellow background coloration is unaffected.

12.12.5. Holczinger's method for free fatty acids

The original procedure has been modified as recommended by Elleder & Lojda (1972).

Use unfixed cryostat sections or frozen sections of tissue fixed in formal–calcium. Acetone-extracted control sections must also be examined.

Solutions required

A. 1.0 N hydrochloric acid

Concentrated hydrochloric acid (S.G. 1.19):	40 ml	Keeps indefinitely
Water:	to 500 ml	

B. Copper acetate

Cupric acetate $(CH_3COO)_2Cu.H_2O$:	5.0 mg	Prepare before using
Water:	100 ml	

C. 0.1% EDTA

Disodium ethylenediamine tetraacetate ($Na_2(EDTA).2H_2O$):	100 mg	Prepare before using
Water:	80 ml	
Adjust to pH 7.1 with drops of 1.0 N (=4%) NaOH, then add:		
Water:	to 100 ml	

D. Dithiooxamide solution

Dithiooxamide:	100 mg	Keeps for several months
Ethanol:	70 ml	
Dissolve and then add:		
Water:	30 ml	

Procedure

1. Affix frozen sections to slides.
2. Immerse in 1.0 N HCl (solution A) for 1 h.
3. Rinse in three changes of water. Drain and allow sections to dry. Carry out acetone-extraction of control sections (see p. 188).
4. Immerse sections in copper acetate (solution B) for 3–4 h.
5. Transfer to two changes of 0.1% EDTA (solution C), 30 s in the first, 60 s in the second.
6. Wash in two changes of water.

7. Immerse in dithiooxamide solution (D) for 30 min. A shorter time (e.g. 10 min) will usually suffice. This stage need not be extended after the sections have stopped darkening.
8. Rinse in 70% ethanol, two changes, 1 min in each.
9. Wash in water and mount in an aqueous medium.

Result

Free fatty acids—dark green to black. Any staining seen in acetone-extracted control sections is **not** due to free fatty acids.

12.12.6. The plasmal reaction

It is important to control for aldehydes already present in the tissue. If found these must be chemically blocked with sodium borohydride (see p. 143). Use either cryostat sections of unfixed tissue or frozen sections of small specimens fixed in formaldehyde for no more than 6 h. The sections should be used within an hour of being cut.

Solutions required

A. 1% mercuric chloride

Mercuric chloride ($HgCl_2$):	1.0 g	Keeps indefinitely
Water:	100 ml	

B. Schiff's reagent

See under the Feulgen reaction (p. 110). Allow the Schiff's reagent to warm to room temperature before use.

C. Bisulphite water

Potassium metabisulphite ($K_2S_2O_5$):	5.0 g	Prepare before using
Water:	1000 ml	
Concentrated hydrochloric acid (S.G. 1.19):	5.0 ml	

Procedure

1. Wash sections (mounted on slides or coverslips) in three changes of water. This step is omitted for sections of unfixed tissue.
2. Immerse in 1% $HgCl_2$ (solution A), 1 min.
3. Transfer slides directly to Schiff's reagent for 5 min.
4. Transfer to bisulphite water (solution C), three changes, 2 min in each.
5. Wash in three changes of water.
6. A progressive aluminium–haematein counterstain (e.g., Baker's haematal-16; see p. 82) may be applied at this stage, if desired. Wash in running tap water after counterstaining.
7. Mount in an aqueous medium.

Result

Plasmalogens pink to purple, provided that a positive reaction in the absence of treatment with $HgCl_2$ is not obtained. Nuclei blue if counterstained as suggested.

12.12.7. Method for cholesterol and its esters

Fixation in formal–calcium is recommended.

Preparation of reagents

A. Perchloric acid–naphthoquinone (PAN) solution

Ethanol:	20 ml
60% aqueous perchloric acid (S.G. 1.54): **(Handle with care)**	10 ml
Formalin (40% HCHO):	1.0 ml
Water:	9.0 ml
1,2-naphthoquinone-4-sulphonic acid:	40 mg

B. 60% aqueous perchloric acid (S.G. 1.54)

Procedure

1. Cut frozen sections and leave in 4% aqueous formaldehyde (from formalin) for 1 week. Before starting the staining procedure, preheat a hotplate to approximately 65°C.
2. Wash sections in water and dry them onto slides.
3. Paint the sections with a thin layer of the PAN (solution A) and place on the hotplate for 10 min. Use the brush to apply more PAN solution at intervals to prevent drying. Colour changes from red to blue.
4. Place a drop of 60% $HClO_4$ (reagent B) on the section and apply a coverslip. Carefully

remove excess perchloric acid from the edges of the coverslip, with filter paper.

Result

Cholesterol, its esters and a few closely related steroids are stained dark blue. (See *Note* below.) Pink background colours are not due to lipids. The colour is not stable in water or in ordinary mounting media.

Note

To demonstrate free cholesterol alone, proceed as follows:

(a) Carry out stages 1 and 2 of the above method.
(b) Place slides in a 0.5% solution of digitonin in 40% ethanol for 3 h.
(c) Immerse slides in acetone for 1 h at room temperature.
(d) Rinse in water.
(e) Proceed with stages 3 and 4 of the above method. Esters of cholesterol are extracted by the acetone, but free cholesterol is rendered insoluble by combination with digitonin.

12.12.8. The acid–haematein test

This is the method of Baker (1946). It is important that all the instructions be followed to the letter. Specimens are fixed in formal–calcium (see p. 22) for 6 h at room temperature. See *Note 1* below for shorter methods.

Solutions required

A. Dichromate–calcium

Potassium dichromate ($K_2Cr_2O_7$):	15 g	Keeps indefinitely
Calcium chloride ($CaCl_2$):	3.0 g	
Water:	300 ml	

B. Acid–haematein

Haematoxylin (C.I. 75290):	50 mg
Water:	48 ml
1% aqueous sodium iodate ($NaIO_3$):	1.0 ml

Heat until it just boils, then allow to cool to room temperature and add:

Glacial acetic acid:	1.0 ml

This reagent is used on the day it is prepared.

C. Borax–ferricyanide differentiator

Potassium ferricyanide, $K_3Fe(CN)_6$:	0.75 g
Borax (sodium tetraborate, $Na_2B_4O_7 . IOH_2O$):	0.75 g
Water:	300 ml

Stable indefinitely at 4°C

Procedure

1. Transfer the specimen directly from formal-calcium into dichromate–calcium (solution A) for 18 h at room temperature.
2. Transfer to a second change of dichromate–calcium for 24 h at about 60°C (wax oven).
3. Wash overnight in running tap water. (At this stage delicate specimens may be embedded in gelatine, which is hardened in formal-calcium.)
4. Cut frozen sections 10 µm thick. Collect them into water, mount onto slides, and allow to dry.
5. Immerse in dichromate–calcium (solution A) for 1 h at 60°C.
6. Wash in five changes of water, 1 min in each.
7. Immerse in acid–haematein (solution B) for 5 h at 37°C.
8. Wash in water until excess dye is removed.
9. Differentiate in borax–ferricyanide (solution C) for 18 h at 38°C.
10. Wash in water (three or four changes).
11. **Either** mount in an aqueous medium **or** dehydrate through graded alcohols, clear in xylene, and mount in a resinous medium.

Result

Certain phospholipids (phosphatidyl cholines and sphingomyelins) are coloured blue, blue-black, or grey. Background yellow. Blue-stained objects may only be assumed to contain phospholipid if they are unstained after appropriate solvent extraction. (See *Note 2* below.) Mitochondria are shown well by the acid–haematein method.

Notes

1. Several simplified versions of this method are available and are quite satisfactory. Thus stages 1–3 of Baker's procedure may be omitted and the sections chromated for 4–24 h at 60°C. When the technique is used solely for staining mitochondria, the differentiation (stage 9) may be shortened to about 8 h. It is

also possible to reduce both the staining (stage 7) and differentiation (stage 9) times to about 2 h.

2. The extraction test used by Baker (1946) involved fixing a second piece of the same tissue in a "weak Bouin" mixture, extracting it with hot pyridine, rehydrating, and then proceeding with the acid–haematein method. Current practice favours extraction of frozen sections with methanol–chloroform or acidified methanol–chloroform (see Section 12.12.9) prior to chromation, which is carried out as described in *Note 1* above.
3. Sections may be subjected to alkaline hydrolysis (see Section 12.2.10) before chromation. Sphingomyelins will then be the only lipids stained by acid–haematein.

12.12.9. Solvent extractions

Times given are for frozen sections. Use solvents in tightly closed containers. The sections are mounted onto slides and allowed to dry before extraction.

Cold acetone. Acetone (4°C, 1 h, but sometimes may need to be left overnight) extracts hydrophobic lipids only. It is important that no water be present in the acetone. Ordinary acetone is dehydrated by adding anhydrous calcium chloride, one-fifth of the total volume of $CaCl_2$ + acetone and allowing to stand for 2 days. Use a bottle of at least 500 ml capacity, since much acetone is lost by absorption into the desiccant. Glassware must be dry and the sections must be allowed to dry in air before and after immersion of the slides in the anhydrous acetone (Elleder & Lojda, 1971).

Hot methanol–chloroform. A mixture of methanol (2 volumes) and chloroform (1 volume), at 58–60°C for 18 h (2 h is sometimes sufficient) removes all lipids except those firmly bound to protein.

Methanol–chloroform–HCl. Methanol, 66 ml; chloroform, 33 ml; concentrated hydrochloric acid, 1.0 ml; 18 h at room temperature (25°C) or 2 h at 58°C. Removes all lipids, including the phospholipid moieties of lipoproteins (Adams & Bayliss, 1962).

12.12.10. Hydrolysis of esters

Esters (but not amides) are hydrolysed by treating free-floating sections for 1 h at 37°C with 2 N (8%) sodium hydroxide (NaOH). The sodium soaps of the liberated fatty acids are dissolved out. After hydrolysis the sections are fragile and should be gently washed in water, followed by 1% aqueous acetic acid, and then returned to water.

12.12.11. Masked lipids

Lipids firmly bound to protein are released by acid hydrolysis prior to staining with a solvent dye. Sections are treated with 25% aqueous acetic acid for 2 min and are then washed in five changes of water. The released lipids are then amenable to extraction by solvents and to staining with lysochromes.

12.12.12. Bromination

Potassium bromide, 6 g; water, 300 ml; bromine, 1.0 ml (**caution:** use fume cupboard; see p. 183). Keeps for a few months. Replace when the colour has faded.

Treat sections with this solution for 5 min at room temperature. Rinse in a bisulphite or thiosulphate solution to remove yellow stain of bromine and wash thoroughly with water.

This treatment prevents reactions due to unsaturated (—CH=CH—) linkages.

12.13. EXERCISES

Theoretical

1. How would you determine whether a structure stained by Sudan black B in 70% ethanol was composed of triglycerides or of phosphoglycerides?

2. Tristearin is a neutral fat in which glycerol is esterified by three molecules of stearic acid. Which of the following methods would and which would not give a positive reaction with tristearin? Give reasons for your answers.

(a) Sudan IV at room temperature.
(b) Sudan IV at 70°C.
(c) Osmium tetroxide.
(d) The plasmal reaction.

3. Answer question 2 for **triolein**, in which glycerol is esterified by three molecules of oleic acid.

4. Myelin, in frozen sections, commonly gives a direct-positive reaction with Schiff's reagent. Why? Describe exactly how you would proceed to demonstrate the presence in myelin of (a) plasmalogens, (b) fatty acid amides (ceramides and sphingomyelins).

5. Aqueous solutions of the dye Nile blue contain two important coloured substances:

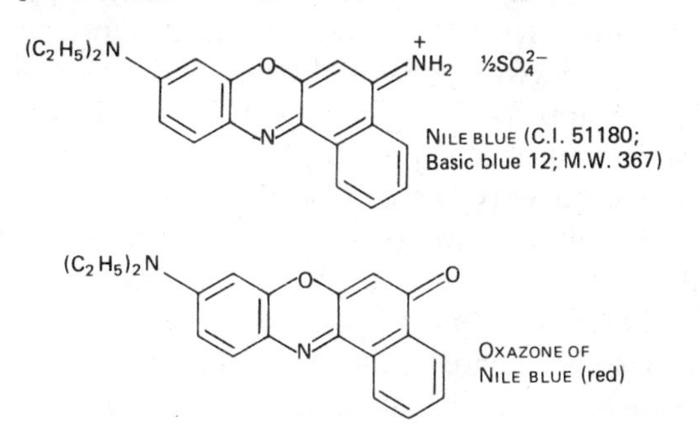

The oxazone, though insoluble in water, dissolves in solutions of Nile blue sulphate. When frozen sections are stained at 60°C, some lipids are stained red and others blue. Which types of lipid would be expected to take up each main component of the mixture? What non-lipid structures would also be stained and in which colour?

It has been shown by Malinin (1980) that different types of intracellular lipid inclusions may be distinguished from one another by red staining with Nile blue at various temperatures from 19 to 65°C. What does this tell you about the mechanism of coloration?

6. What histochemical evidence could be obtained to support the contention that lipid droplets in the adrenal cortex contain cholesterol and steroid hormones rather than neutral fats or phospholipids?

7. In metachromatic leucodystrophy there are accumulations in the brain of **sulphatide** in which the fatty-acid residues are largely saturated. Using histochemical methods, how would you distinguish these pathological deposits from (a) proteoglycans, (b) glycoproteins, (c) cerebroside, (d) ganglioside, (e) neutral fats, (f) any unsaturated lipids?

8. If wax is not adequately removed from paraffin sections, birefringent crystal-like bodies are seen in the tissue, within nuclei. Using histochemical methods, how would you show that these objects are composed of a hydrocarbon rather than of a naturally occurring lipid, lipoprotein, protein, or mineral material? See Nedzel (1951) for further information concerning this interesting artifact.

Practical

9. Cut frozen sections of brain, adrenal gland, tongue, and skin, fixed in neutral, buffered formaldehyde or (preferably) in formal–calcium, for use in this and subsequent exercises. Stain sections with (a) Sudan IV, (b) Sudan black B. Account for different appearances in the same tissue.

10. Stain frozen sections of brain by the periodic acid–Schiff method, using appropriate control procedures, in order to demonstrate (a) glycolipids, (b) glycoproteins.

11. Stain sections with palladium chloride and observe the dark coloration of myelinated nerve fibres. What else is stained? Using extractive and blocking methods, show that the staining is due to the presence of unsaturated lipids. Why are the contents of fat cells not shown by this technique?

12. Carry out the plasmal reaction on sections of brain and tongue. A pseudoplasmal reaction will almost certainly also be obtained and must be prevented in order to determine the true distribution of plasmalogens.

Show that the production of a positive plasmal reaction is dependent upon the presence of unsaturation in the stained lipids.

13. Using frozen sections of rats' adrenal glands (fixed in neutral buffered formaldehyde or in formal–calcium), obtain as much information as possible about the nature of the lipid droplets in the cells of the adrenal cortex.

14. Demonstrate either myelin in the brain or mitochondria in muscle or the kidney, using the acid–haematein method. Identify the type of lipid responsible for the staining by means of appropriate control procedures.

13

Methods for Inorganic Ions

13.1. General considerations	190
13.2. Calcium	190
13.3. Phosphate and carbonate	192
13.4. Iron	192
13.5. Zinc	193
13.6. Copper	193
13.7. Lead	194
13.8. Individual methods	194
13.8.1. Alizarin red S for calcium	194
13.8.2. Detection of calcium with GBHA	195
13.8.3. The von Kossa method	196
13.8.4. Perls' Prussian blue method for iron	196
13.8.5. Detection of zinc with 8-hydroxy-quinoline	197
13.8.6. Dithizone method for zinc	197
13.8.7. Dithiooxamide method for copper	198
13.8.8. Rhodizonate method for lead	199
13.9. Exercises	199

13.1. GENERAL CONSIDERATIONS

Inorganic ions are present in all parts of all tissues, but owing to their solubility they are not generally amenable to detection by histochemical methods. It has been possible to demonstrate some soluble ions, such as potassium and chloride, by making use of freeze-dried material or by including in the fixative a reagent which forms an insoluble salt with the ion concerned. It is, however, difficult to determine the extent to which a soluble ion has diffused during the course of the procedure from its location *in vivo* to the site at which the end-product of the histochemical reaction is seen.

When inorganic substances are present in tissues as insoluble compounds, be these salts such as calcium carbonate or complexes with protein, such as haemosiderin, they remain *in situ* while the specimens are being processed and can therefore be localized accurately by histochemical methods. Some metals, such as calcium, iron, and zinc, are present in animal tissues in sufficient quantity to permit their demonstration in normal material. Others, such as copper, are normally present but can only be detected histochemically when large amounts accumulate as the result of disease or experimental manipulation. Yet other metals, such as aluminium and lead, are not involved in normal mammalian metabolism but may be found in the tissues following intoxication or excessive environmental exposure. Of the inorganic anions, phosphate and carbonate of mineralized tissue are the only ones which are normally present in an insoluble form.

Histochemical techniques for inorganic ions fall into two categories: those in which the visible products are black or coloured inorganic substances, and those in which coloured compounds are formed by chelation with organic ligands. The chemistry has already been introduced in connection with mordant dyes (Chapter 5, Section 5.5.7, p. 44). It is important to realize that the methods described in this chapter will only detect metals present as ions in the tissue. Metals which are already bound by strong coordinate bonds to organic ligands cannot react with the reagents used to demonstrate their ions. For example, the iron in erythrocytes is so firmly combined with the haem group of haemoglobin that it will not give a positive reaction in histochemical tests for iron.

As examples of histochemical techniques for inorganic substances, we will consider methods for calcium, phosphate (and carbonate), iron, zinc, copper, and lead.

13.2. CALCIUM

This metal is demonstrated by virtue of its ability to form coloured chelates with dyes and other organic compounds. Dyes are chosen which do not form complexes of similar colour with other naturally occurring metals such as magnesium and iron. Suitable reagents for calcium are **alizarin red S**, a synthetic anthraquinone dye, and **morin**, a coloured (yellow) polyphenolic ketone obtained from certain tropical plants. Morin is used as a fluorescent stain for calcium and it also forms fluorescent complexes with many other metals not normally present in animal tissue.

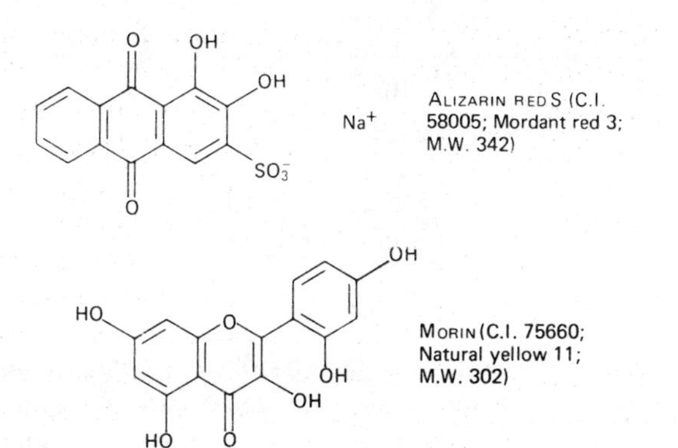

Alizarin red S is the more popular of these reagents. Like other hydroxyanthraquinone dyes, alizarin red S forms chelates in which a quinone oxygen and a phenolic oxygen serve as electron donors to the metal atom, so that a stable six-membered ring is formed. Calcium, with coordination number 4, combines with two molecules of the dye:

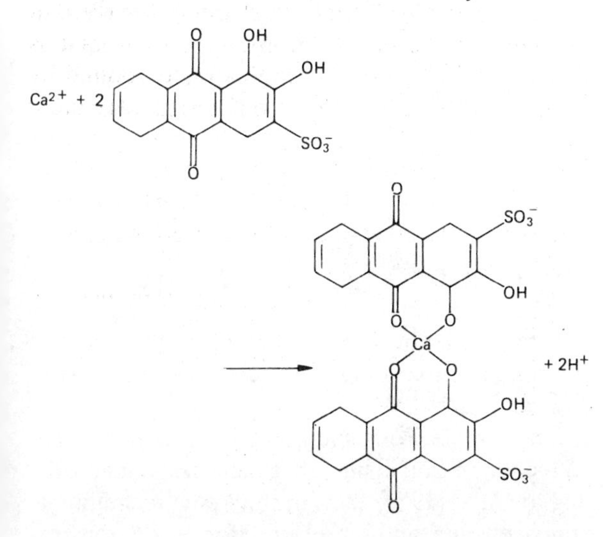

Alizarin red S is also an anionic dye in its own right and, as such, imparts non-specific pink "background" coloration to the tissue.

The calcium ions must be released from the insoluble deposits in which they occur before they can combine with the dye to form an insoluble chelate. Consequently, some diffusion from the original site of deposition of calcium is inevitable. The histological picture is greatly influenced by the pH of the staining solution. An acid solution will extract much calcium and produce a conspicuous coloured deposit, but this will be diffusely localized. An alkaline solution will liberate fewer calcium ions, so that although the accuracy of localization will be greater the sensitivity of the method will be less. Different authors have recommended pH values ranging from 4 to 9 for solutions of alizarin red S.

A more sensitive histochemical method for calcium involves a chelating agent which is not a dye. This is glyoxal-*bis*-(2-hydroxyanil), also known as GBHA:

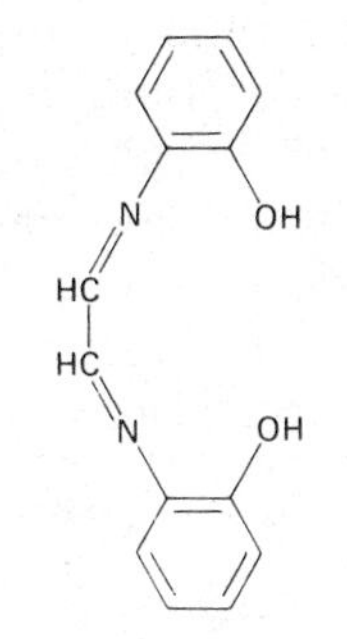

In the presence of a strong base such as sodium hydroxide, the phenolic hydroxyl groups ionize. The anion of GBHA forms insoluble chelates with several metals, including calcium:

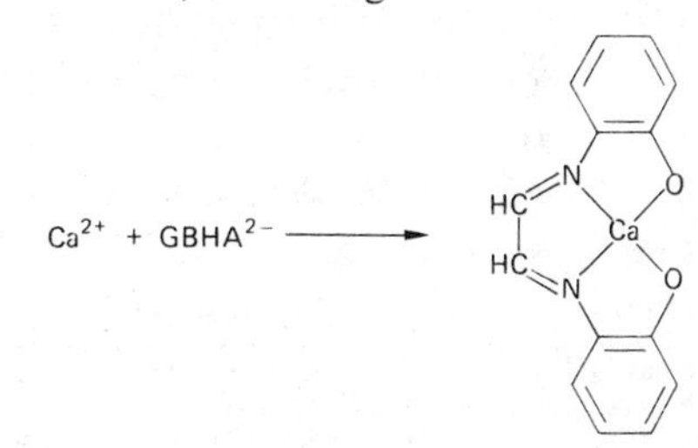

Metal chelates are also formed with strontium, barium, cadmium, cobalt, and nickel. If any of these metals are likely to be present in the tissue their chelates can be decomposed by treating the stained sections with an alkaline solution containing cyanide ions.

The GBHA method is very sensitive and can be used to detect calcium ions other than those of insoluble salts of the metal provided that the tissue has been freeze-dried or freeze-substituted. After fixation in an aqueous reagent, the soluble calcium diffuses and some of it can be detected in nuclei of cells by the GBHA method. This artifact is presumably due to electrostatic attraction of calcium ions by the phosphate groups of nucleic acids. As with alizarin red S, the solution of GBHA used for staining deposits of insoluble calcium salts must contain some water and must not be too strongly alkaline. "Soluble" calcium salts, however, are best stained with a solution of GBHA in absolute ethanol containing some sodium hydroxide.

13.3. PHOSPHATE AND CARBONATE

The von Kossa technique is often designated as a histochemical method for calcium, but it is really a method for phosphate and carbonate, the anions with which the metal is associated in normal and pathological calcified tissues.

The sections are treated with silver nitrate. The calcium cations are replaced by silver:

$$CaCO_3 + 2Ag^+ \rightarrow Ag_2CO_3 + Ca^{2+}$$

$$Ca_3(PO_4)_2 + 6Ag^+ \rightarrow 2Ag_3PO_4 + 3Ca^{2+}$$

(The phosphates are really more complicated compounds than those indicated in this simple equation.)

The reaction is carried out in bright light, which promotes the reduction to metal of silver ions in the crystals of the insoluble silver phosphate and carbonate. Alternatively, the silver nitrate may be applied in darkness or subdued light and the reduction accomplished chemically with a solution of hydroquinone, metol or another photographic developing agent. Finely divided metallic silver is black. Any unreduced ionized silver is finally removed by treatment with sodium thiosulphate, which dissolves otherwise insoluble salts by forming complex anions such as $[Ag(S_2O_3)_3]^{5-}$.

The von Kossa procedure provides sharp and accurate localization of calcified material in tissues, but it is less sensitive than the alizarin red S or the GBHA methods for calcium.

13.4. IRON

The major deposits of iron in mammals are in the red blood cells as haemoglobin, and in the phagocytes of the reticulo-endothelial system as ferritin and haemosiderin. In these compounds, the metal is present in the ferric (oxidation state +3) form, but since it is complexed to protein it is not available as ions to participate in simple chemical reactions.

The iron of ferritin and haemosiderin is readily released as Fe^{3+} ions by treatment with a dilute mineral acid. In the presence of ferrocyanide ions, Prussian blue is immediately precipitated:

$$4Fe^{3+} + 3[Fe(CN)_6]^{4-} \rightarrow Fe_4[Fe(CN)_6]_3 \downarrow$$

Prussian blue is actually a more complicated compound than the simple ferric ferrocyanide implied above (see Chapter 10, p. 125). This is the simplest histochemical test for ion; it is known as the Prussian blue reaction of Perls.

An alternative strategy is to reduce all the ferric ions in the section to ferrous by treatment with a dilute aqueous solution of ammonium sulphide:

$$2Fe^{3+} + 3S^{2-} \rightarrow 2FeS + S$$

The sulphur dissolves in the excess of aqueous ammonium sulphide. Ferrous sulphide is soluble in dilute mineral acids, so when the sections are treated with acidified potassium ferricyanide solution:

$$FeS + 2H^+ \rightarrow Fe^{2+} + H_2S$$

$$3Fe^{2+} + 2[Fe(CN)_6]^{3-} \rightarrow Fe_3[Fe(CN)_6]_2 \downarrow$$

As explained in Chapter 10, p. 125, the precipitate of "Turnbull's blue" is not really ferrous ferricyanide, but is identical to Prussian blue. This method would, of course, detect any ferrous ions that were present in a tissue.

The simple Perls method is adequate for the detection of iron in higher vertebrates, but according to Gabe (1976) the Turnbull blue technique is more sensitive and is preferred for the demonstration of iron in the tissues of cold-blooded vertebrates and invertebrates. The Perls reaction is also suitable for the identification of exogenous ferritin introduced into animals as an experimental tracer protein (Parmley *et al.*, 1978).

It should be noted that the iron of haemoglobin cannot be released as ions by any method that does not totally destroy the section, so it cannot be demonstrated histochemically.

The stainable iron can be **removed** from the sections by treatment with acid in the absence of a precipitant anion. Aqueous solutions of oxalic acid or sodium dithionite are also suitable for this purpose

(Morton, 1978) and act more rapidly than dilute mineral acids.

13.5. ZINC

Zinc is associated with insulin in the β-cells of the pancreatic islets and with various enzymes (notably carbonic anhydrase in exocrine glands) and also occurs in the cytoplasmic granules of leukocytes. Since, in all these situations, the proteins which bind the metal are rather labile, it is desirable to use freeze-dried material, cryostat sections, or air-dried smears. If a liquid fixative must be used, cold absolute ethanol is probably the least objectionable one.

In one of the methods for zinc, the preparation is treated with an alkaline (pH 8.0) solution of a chelating agent, 8-hydroxyquinoline:

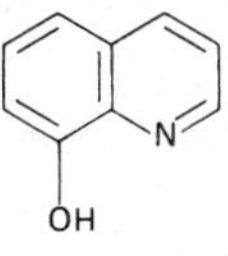

This reagent is also known as 8-quinolinol or oxine. The chelate formed with zinc ions under alkaline conditions is fluorescent. Each atom of zinc (coordination number, 6) binds two molecules of 8-hydroxyquinoline and two of water (see Dwyer & Mellor, 1964):

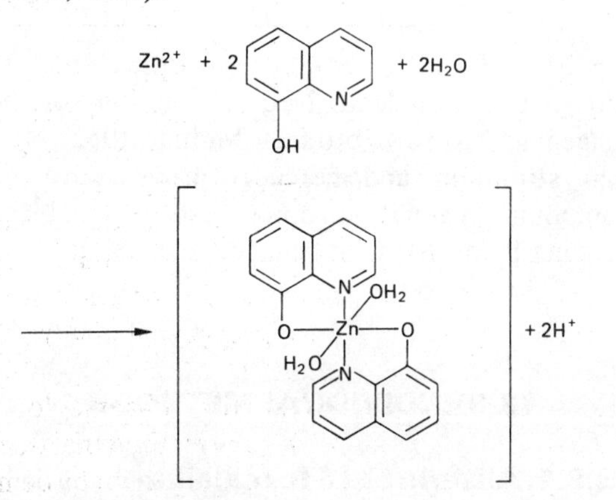

8-Hydroxyquinoline forms chelates with many metals, but only those with calcium, magnesium, and zinc are fluorescent. The calcium and magnesium chelates are unstable at pH 8.0, so the method is specific for zinc (Smith *et al.*, 1969).

Another reagent for the detection of zinc is dithizone (diphenylthiocarbazone; 3-mercapto-1,5-diphenylformazan).

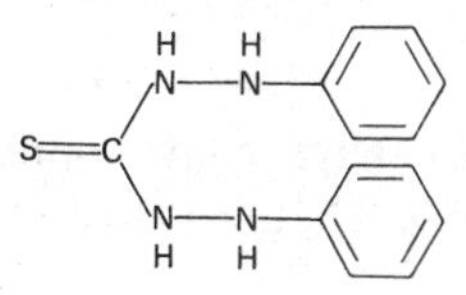

This compound forms a red chelate with zinc. Many other metal ions form differently coloured chelates with dithizone. The reagent has also been used for the histochemical detection of mercury, which forms a reddish-purple complex with the structure

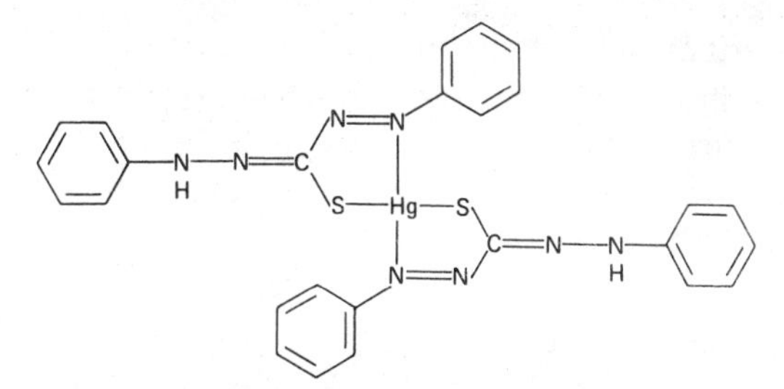

(See Harris & Livingstone, 1964.) The zinc chelate may have a similar structure.

13.6. COPPER

Normal mammalian tissues do not contain enough copper to allow histochemical detection of the metal, but in Wilson's disease (hepato-lenticular degeneration) large quantities of the metal accumulate in phagocytic cells in the brain, liver, and cornea. It is also possible, of course, to demonstrate copper which has been artificially introduced into an animal. In some arthropods and molluscs the haemocyanins, respiratory pigments equivalent to the haemoglobin of vertebrate animals, are copper-containing proteins dissolved in the plasma. Copper ions can be released from combination with protein by exposing sections to the fumes of concentrated hydrochloric acid.

The most satisfactory histochemical method for the detection of copper (as the cupric ion, oxidation state +2) makes use of a chelating agent, dithiooxamide (also called rubeanic acid):

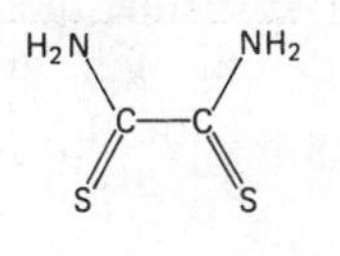

This reagent forms an insoluble dark-green chelate with copper in the presence of ethanol and acetate in alkaline solution. Other metals, such as nickel, cobalt, and silver, also form coloured complexes with dithiooxamide, but under these conditions they are unreactive or give soluble complexes. The copper chelate is believed to be a polymer with the structure

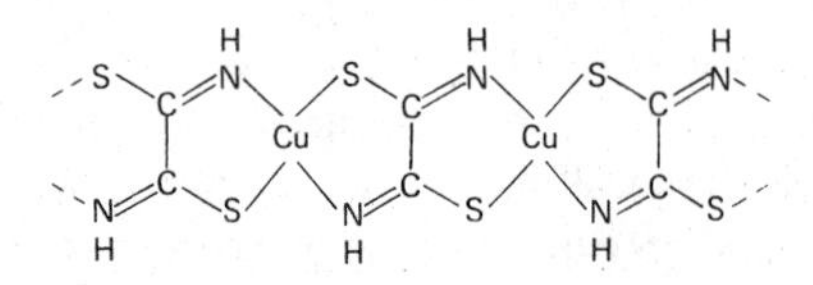

(Harris & Livingstone, 1964).

Dithiooxamide combines only with cupric ions. A reagent which forms a coloured complex with cuprous ions is *p*-dimethylaminobenzylidenerhodanine:

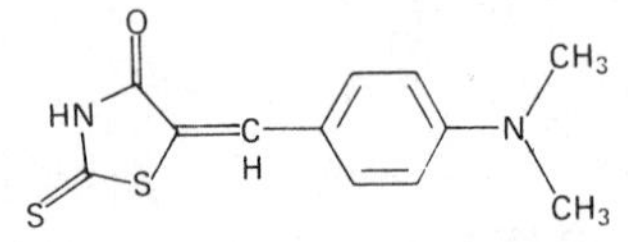

Copper probably replaces the hydrogen attached to the nitrogen atom in the five-membered ring to form a red product. Brown, red, and purple compounds are formed with several other metals. Irons *et al.* (1977) have shown that while dithiooxamide and *p*-dimethylaminobenzylidenerhodanine are equally sensitive reagents for the histochemical detection of copper, only the latter gives an intensity of staining proportional to the concentration of the metal in the tissue.

Little attention has been paid by histochemists to the oxidation states of copper in tissues. The invertebrate haemocyanins contain Cu(I), while caeruloplasmin, the copper-binding globulin of mammalian blood plasma, contains both Cu(I) and Cu(II). The copper-containing deposits in the tissues of patients with Wilson's disease are stainable with dithiooxamide, so they must consist partly or entirely of a compound capable of releasing cupric ions. Further study of the interconversion of Cu(I) and Cu(II) in tissues, which is readily brought about by treatment with oxidizing and reducing agents, might lead to broadening of the applicability of histochemical techniques for the demonstration of copper.

13.7. LEAD

Lead has no known physiological function, but can be demonstrated histochemically in the tissues of animals poisoned by its salts. The element is accumulated in intranuclear inclusion bodies, especially in the kidney (Goyer & Cherian, 1977).

The simplest histochemical method for lead is based on the formation of the insoluble chromate, $PbCrO_4$, which is yellow. In order to ensure precipitation of lead ions, it is probably prudent to include sulphate ions in the fixative. Lead sulphate is insoluble in water (solubility product = 1.6×10^{-8}), but the chromate is even less soluble (solubility product = 2.8×10^{-13}), so the reaction

$$PbSO_4 + CrO_4^{2-} \rightarrow PbCrO_4 + SO_4^{2-}$$

can proceed, though slowly, in the direction indicated.

Though this is a fairly specific test for lead (Ba, Sm, Sr, Tl, and Zn also have insoluble yellow chromates but are unlikely to cause confusion), the product is not strongly coloured. A bright coloration is obtained by using a chelating agent, sodium, or potassium rhodizonate:

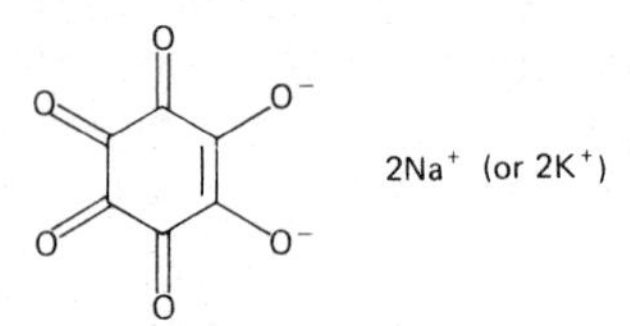

Under mildly acid conditions this forms a pink to red chelate with lead. Neutral solutions of the reagent give a brown product (Molnar, 1952). Barium, strontium, and mercury can also form red compounds with rhodizonate, and a blue–black product is formed with iron (Pearse, 1972).

13.8. INDIVIDUAL METHODS

13.8.1. Alizarin red S for calcium

An alcoholic or a neutral aqueous fixative should be used. Maximum preservation of calcium is probably achieved by fixation in 80% ethanol, though this is a rather poor fixative for histological purposes.

Solutions required

A. Alizarin red S

Alizarin red S (C.I. 58005):	1.0 g	Keeps for 4 weeks
Water:	90 ml	
Dilute NH_4OH (28% ammonia diluted 100 times with water) Add in small aliquots until the pH is 6.4—approximately 10 ml		

B. Differentiating fluid

Ethanol (95%):	500 ml	Mix before using
Concentrated hydrochloric acid:	0.05 ml	

Procedure

1. De-wax and hydrate paraffin sections.
2. Stain in solution A for 2 min.
3. Wash in water for 5–10 s.
4. Differentiate in solution B for 15 s.
5. Complete dehydration in two changes of absolute ethanol, clear in xylene, and mount in a resinous medium.

Result

Calcium—orange to red. Background—dull pink.

13.8.2. Detection of calcium with GBHA

These techniques (Kashiwa & Atkinson, 1963; Kashiwa & House, 1964) are extremely sensitive. It is essential to prepare the tissues as described and to avoid the use of any reagents (including water) that might contain traces of calcium.

Solutions required

A. GBHA stock solution

Glyoxal *bis*-(2-hydroxyanil):	200 mg	Keeps for several weeks at 4°C
Absolute ethanol:	50 ml	

B. 10% NaOH solution

Sodium hydroxide (NaOH):	10 g	Keeps for several weeks
Water:	to 100 ml	

C. GBHA working solution I

GBHA stock solution (A):	2.0 ml	Mix just before using
10% NaOH (solution B):	0.15 ml	
Water:	0.15 ml	

D. GBHA working solution II

GBHA stock solution (A):	2.0 ml	Mix just before using
10% NaOH (solution B):	0.6 ml	

E. Alkaline cyanide solution

Absolute ethanol:	45 ml	Stable for a few weeks at room temperature
Water:	5.0 ml	
Sodium carbonate (Na_2CO_3): to saturation (approx. 1.0 g)		
Potassium cyanide (KCN): to saturation (approx. 1.0 g)		

(**Caution**: potassium cyanide is poisonous. Do not allow it to come into contact with acids. Before throwing away this solution add an excess of sodium hypochlorite solution and wait for 10 min. Then flush down sink with plenty of water.)

F. Fast green FCF counterstain

Fast green FCF (C.I. 42053):	0.24 g
95% ethanol:	300 ml

Keeps indefinitely and may be used repeatedly

Preparation of specimens

Freeze small blocks, no more than 2.0 mm^3, in isopentane cooled in liquid nitrogen. **Either** freeze dry and embed in paraffin wax **or** freeze substitute in acetone at −80°C, clear in xylene, and embed in paraffin wax. Wolters *et al.* (1979) state that superior results are obtained by freeze substitution for 10 days at −80°C in acetone containing 1% oxalic acid (presumably as $H_2C_2O_4.2H_2O$). Cut sections at 7 μ and mount them onto slides without using water or any adhesive. The sections are not de-waxed before staining. (See also *Note 2* below.)

Staining procedure

1. Place the slides bearing the sections on a horizontal staining rack. **Do not remove wax.**
2. **Either**: flood with GBHA working solution I(C) for 3 min, **or** apply one or two drops of GBHA working solution II(D) per section and allow to evaporate to dryness.
3. Rinse in 70% ethanol, then in 95% ethanol.
4. (Optional, see *Note 1* below.) Immerse for 15 min in alkaline cyanide solution (E).
5. Rinse in three changes of 95% ethanol.
6. Apply a counterstain, if desired: rinse in absolute ethanol; de-wax in xylene; rinse in absolute, followed by 95% ethanol. Stain in alcoholic fast green FCF (solution F) for 3 min. Rinse in three changes of 95% ethanol, then proceed to stage 7 below.
7. Dehydrate in four changes of absolute ethanol, clear in xylene, and mount in a resinous medium.

Results

When GBHA working solution I is used, a red colour is seen at sites of soluble calcium salts, but insoluble calcified material is largely unstained. With GBHA working solution II, the red colour is seen predominantly at sites of insoluble calcium salts. If the fast green FCF counterstain is used, the background is bluish green. The red colour fades after 2 or 3 days, so the sections should be photographed to obtain a permanent record.

Notes

1. The alkaline cyanide reagent decolorizes the chelates of metals other than calcium (Sr^{2+}, Ba^{2+}, Cd^{2+}, Co^{2+}, Ni^{2+}). Stage 4 may be omitted if the presence of other metals is unlikely.

2. It is necessary to use freeze-dried or freeze-substituted tissue in order to minimize the diffusion of the calcium ions dissolved in the cytoplasm and extracellular fluid. If the method is applied to sections of conventionally fixed tissue, calcified material is stained but there is also red staining of nuclei, cytoplasmic RNA, and sites of proteoglycans (see Chapter 11). These artifacts are due to electrostatic attraction of diffused calcium ions to macromolecular anions.

3. A green filter in the illuminating system of the microscope enhances the contrast of the red-stained calcium deposits and is recommended for photomicrography.

13.8.3. The von Kossa method

Solutions required

A. 1% aqueous silver nitrate ($AgNO_3$)

B. 5% aqueous sodium thiosulphate ($Na_2S_2O_3.5H_2O$)

These can be used repeatedly until precipitates form in them.

C. A counterstain (0.5% aqueous safranine or neutral red)

Procedure

1. De-wax and hydrate paraffin sections.
2. Immerse in silver nitrate (solution A) in bright sunlight or directly underneath a 100 W electric light bulb for 15 min.
3. Rinse in two changes of water.
4. Immerse in sodium thiosulphate (solution B) for 2 min.
5. Wash in three changes of water.
6. Counterstain nuclei (solution C), 1 min.
7. Rinse briefly in water.
8. Dehydrate (and differentiate counterstain) in 95% and two changes of absolute alcohol.
9. Clear in xylene and mount in a resinous medium.

Result

Sites of insoluble phosphates and carbonates—black. Nuclei—pink or red.

13.8.4. Perls' Prussian blue method for iron

The fixative must not be acidic and must not contain chromium. Hadler *et al.* (1969) have compared several fixatives and found that the largest amounts of histochemically detectable iron were preserved by immersion of tissues for 24 h in 6% aqueous formaldehyde containing 0.27 M calcium chloride, with the pH adjusted to 4.0 by addition of 0.1 N NaOH or 0.1 N HCl. Alcoholic fixatives or buffered aqueous formaldehyde solutions were somewhat less satisfactory.

Solutions required

A. Acid ferrocyanide reagent

Potassium ferrocyanide:	2.0 g	Prepare just before using
Water:	100 ml	
Dissolve and add:		
Concentrated hydrochloric acid:	2.0 ml	

B. Counterstain for nuclei

0.5% aqueous safranine or neutral red.

Procedure

1. De-wax and hydrate paraffin sections.
2. Immerse in acid ferrocyanide reagent (solution A) for 30 min.
3. Wash in four changes of water.
4. Counterstain nuclei (solution B), 1 min.
5. Rinse briefly in water.
6. Dehydrate (and differentiate counterstain) in 95% and two changes of absolute alcohol.
7. Clear in xylene and mount in a resinous medium.

Result

Blue precipitate with Fe^{3+} liberated from ferritin and haemosiderin. Nuclei pink or red. Haemoglobin is not stained. See *Note* below for a control procedure.

Note

Iron can be removed, before staining, by treatment of the hydrated sections with 5% aqueous oxalic acid ($H_2C_2O_4 . 2H_2O$) for 6 h or with freshly prepared 1% sodium dithionite (=sodium hydrosulphite, $Na_2S_2O_4$) in acetate buffer, pH 4.5, for 5 min. Perls' method is also applicable to sections of plastic-embedded tissue, but the times required for extraction of iron are longer: 12 h for oxalic acid or 15 min for sodium dithionite (Morton, 1978). A negative result after this extraction shows that the sections were not coloured artifactually as a consequence of the presence of unwanted traces of iron in the acid ferrocyanide reagent.

13.8.5. Detection of zinc with 8-hydroxyquinoline

This technique is a slight modification of that of Smith *et al.* (1969).

Solutions required

A. 8-hydroxyquinoline, stock solution

8-hydroxyquinoline- (8-quinolinol):	3 g	Stable for several weeks
Absolute ethanol:	100 ml	

B. 0.2 M borate buffer, pH 8.0

Working reagent: Add 0.2 ml of solution A to 50 ml of solution B.

Procedure

1. Immerse air-dried blood-films or unfixed cryostat sections (mounted on coverslips) in the working reagent for 15 min.
2. Drain and rinse in water, two dips.
3. Allow to dry and then mount in non-fluorescent immersion oil.

Result

(Fluorescence microscopy; optimum excitation by blue light, wavelength 385 nm.) Pale green–yellow fluorescence indicates the presence of zinc.

Control

Zinc is extracted by 1% aqueous acetic acid (5 min). This control is necessary in order to check for autofluorescence not due to zinc.

13.8.6. Dithizone method for zinc

This method should be used with paraffin sections of freeze-dried tissue or with cryostat sections of unfixed tissue. Cryostat sections may be fixed for 10 min in absolute ethanol at 4°C for improvement of morphological preservation. If a liquid fixative is unavoidable, fix small pieces for 1 h at 4°C in absolute methanol or ethanol, clear in benzene or xylene, and prepare paraffin sections. Extraction of zinc may be reduced if sections are not flattened on water when mounting onto slides.

Solutions required

A. Dithizone stock solution

Dithizone:	100 mg	Keeps for 1 or 2 months in darkness at 4°C
Absolute acetone (see Chapter 12, p. 188):	100 ml	

B. Complexing solution

Sodium thiosulphate ($Na_2S_2O_3 . 5H_2O$): 55 g
Sodium acetate (anhydrous): 5.4 g
Potassium cyanide (KCN): 1.0 g
(**Caution**: poisonous. See p. 195)
Water: 100 ml

Dissolve the salts in the water. Dissolve a little dithizone in 200 ml of carbon tetrachloride. Shake the aqueous solution in a separatory funnel with successive 50 ml aliquots of the solution of dithizone in CCl_4 until the CCl_4 layer is a clear green colour. This manipulation extracts traces of zinc from the reagents. The carbon tetrachloride fractions are discarded. The vapour of CCl_4 is toxic. Ideally, this solvent should be collected into a metal drum or canister designated for the disposal of "chlorinated solvents", not a drum for ordinary "waste solvents" such as alcohol and acetone. The purified aqueous complexing solution is stable for a few months.

C. 1.0 M acetic acid

Glacial acetic acid: 60 ml
Water: to 1000 ml
} Keeps indefinitely

D. Sodium potassium tartrate solution

$NaKC_4H_4O_6 . 4H_2O$: 20 g
Water: to 100 ml
} Keeps indefinitely

E. Working dithizone solution (See also *Note 1* below)

This is mixed when needed and used immediately.

Solution A: 24 ml
Water: 18 ml
1.0 M acetic acid (solution C): add in 0.1 ml aliquots until the pH is 3.7 (about 2 ml required)
Solution B: 5.8 ml
Solution D: 0.2 ml

F. Chloroform

Required for rinsing.

Procedure

1. Cryostat sections are allowed to dry on coverslips or slides. Paraffin sections are de-waxed in xylene (three changes) and allowed to dry by evaporation.
2. Immerse the slides or coverslips bearing the sections in freshly mixed working dithizone solution (E) for 10 min.
3. Rinse the slides or coverslips in two changes of chloroform (solution F), with agitation, for 30 s.
4. Pour off the chloroform and allow the slides or coverslips to drain but do not let the solvent evaporate completely (to avoid cracking of the sections).
5. Rinse in water, allow most of the water to drain off onto filter paper, and put a drop of an aqueous mounting medium onto each section. Apply slides or coverslips according to the type of preparation.

Result

A red to purple colour is formed where zinc is present in the tissue. See *Notes 1* and *2* below.

Notes

1. The complexing solution is mixed with the dithizone to prevent the formation of coloured chelates with metals other than zinc. 24 ml of solution A may be diluted with 16 ml of water instead of with solutions B, C, and D as described above. The simple dithizone solution will form coloured complexes with Ag, Au, Bi, Cd, Co, Cu, Fe, Hg, In, Mn, Ni, Pb, Pd, Pt, Sn, and Tl, as well as with Zn.
2. Zinc is removed by treating the sections for 5 min with 1% aqueous acetic acid before staining.

13.8.7. Dithiooxamide method for copper

This technique is applicable to frozen or paraffin sections of specimens fixed in phosphate-buffered formaldehyde, to cryostat sections of unfixed tissue, and to de-waxed paraffin sections of freeze-dried material. Fixatives other than formaldehyde are probably acceptable but have not been investigated, but it is probably desirable for the fixative to contain anions such as phosphate, hydroxide, or carbonate which form insoluble cupric salts.

Solution required

Dithiooxamide (rubeanic acid):	0.2 g	Keeps for several months
70% aqueous ethanol:	200 ml	
Sodium acetate (CH_3COONa):	0.4 g	

Procedure

1. Take frozen or paraffin sections to water. (See *Note 1*.)
2. Immerse in the dithiooxamide reagent for 30 min.
3. Rinse in two changes of 70% ethanol. (See *Note 2*.)
4. Dehydrate, clear, and mount in a resinous medium.

Result

Sites of copper deposits—dark green to black.

Notes

1. According to Pearse (1972), protein-bound copper can be released by placing the slides (after de-waxing) face downwards over a beaker of concentrated hydrochloric acid for 15 min followed by washing for 15 min in absolute ethanol.
2. A counterstain may be applied after stage 3.

13.8.8. Rhodizonate method for lead

This method is applicable to frozen or paraffin sections of formaldehyde-fixed specimens. An anion such as sulphate or phosphate should be present in the fixative to ensure precipitation of salts of lead. If the technique is to be applied to bone, sodium sulphate ($Na_2SO_4 . 10H_2O$; 5 to 10% w/v) should be added to the decalcifying fluid.

Solution required

Sodium (or potassium) rhodizonate:	0.2 g	Prepare just before using
Water:	99 ml	
Glacial acetic acid:	1.0 ml	

Procedure

1. Take sections to water.
2. Immerse in rhodizonate solution for 30 min.
3. Wash in water.
4. Mount in a water-soluble medium.

Result

Sites of salts of lead—pink to red, or brown.

Note

Several other metals form coloured rhodizonate complexes, notably Ag, Ba, Bi, Ca, Cd, Hg_2^{2+}, Sn, Sr, Tl and UO_2^{2+}. Confusion with Pb is unlikely in tissues from experimental animals. See Feigl & Anger (1972) and Lillie & Fullmer (1976) for suggested methods for increasing the specificity of the method.

13.9. EXERCISES

Theoretical

1. No satisfactory method has ever been devised for the histochemical localization of sodium ions. Why should this be?

2. Devise a simple technique for the histochemical demonstration of chloride in animal tissues, based on the use of the reactions:

(a) $Ag^+ + Cl^- \rightarrow AgCl \downarrow$

(b) $2AgCl \xrightarrow[\text{(light)}]{} 2Ag + Cl_2$

What factors would limit the accuracy of localization of chloride ions by this method?

3. How would you proceed to show that a brown pigment found in sections of a tissue contains iron?

4. Classify the types of technique for histochemical demonstration of metallic ions according to the types of reagents used. What factors limit the chemical specificity and the accuracy of localization of metals?

Practical

5. Mix 100 ml of 0.5 M NaH_2PO_4 with 400 ml of 0.5 M Na_2HPO_4 (to give pH 7.4). Inject a rat intraperitoneally, once daily, with 1.0 ml per 100 g body weight of this solution for 8 days. This procedure induces calcification of renal basement membranes (Haase, 1975).

Twelve hours after the last injection, kill the rat with an intraperitoneal injection of pentobarbitone. Remove the kidneys and fix slices about 2 mm thick in 80% ethanol overnight. Trim the specimens and prepare paraffin sections 7 μm thick.

Stain some sections with alizarin red S and some by the von Kossa technique. Counterstain some of the von

Kossa sections with the PAS method and alum–haematoxylin. Observe the distribution of the calcified deposits. Stain some more of the sections with GBHA and compare the result with those obtained by the other two methods. Account for the differences.

6. Inject each of four mice (weight approximately 30 g) intraperitoneally with one of the following, dissolved in 0.5 ml of water:

Copper sulphate ($CuSO_4 . 5H_2O$):	2.0 mg (i.e. 0.4 ml of 0.5% solution)
Ferric chloride ($FeCl_3 . 6H_2O$):	5.0 mg (i.e. 0.5 ml of 1% solution)
Lead acetate ($(CH_3COO)_2Pb . 3H_2O$:	2.5 mg (i.e. 0.5 ml of 0.5% solution)
Zinc sulphate ($ZnSO_4 . 7H_2O$):	2.5 mg (i.e. 0.5 ml of 0.5% solution)

After 24 h, kill these mice and one normal animal. Remove the livers and kidneys and fix or freeze pieces of tissue appropriately for the histochemical detection of the injected metal cations. Carry out histochemical methods for Cu^{2+}, Fe^{3+}, Pb^{2+}, and Zn^{2+} on sections of liver and kidney from all animals.

This experiment can be made into a diagnostic exercise by allocating code numbers to the different specimens.

7. Using the histochemical methods described, demonstrate the physiological presence of (a) iron in paraffin sections of liver, (b) zinc in leukocytes in a freshly prepared blood-film.

8. Stain a paraffin section of decalcified rat's sternum for calcium and for phosphate–carbonate and check the completeness of decalcification.

14

Enzyme Histochemistry: General Considerations

14.1. Some properties of enzymes 201
14.2. Names of enzymes 201
14.3. Scope and limitations of enzyme histochemistry 202
14.4. Methods not based on enzymatic activity 202
14.5. Methods based on enzymatic activity 203
    14.5.1. Substrate-film methods 203
    14.5.2. Dissolved substrate methods 203
14.6. Types of enzymes 204
14.7. Technical considerations 204
    14.7.1. Preparation of tissue 204
    14.7.2. Conditions of reaction 205
14.8. Exercises 205

This chapter introduces the large subject of enzyme histochemistry. A short account of the general properties of enzymes is followed by a discussion of the physical and chemical principles underlying the methods used for their localization in tissues. These techniques are exemplified by more detailed treatments of the histochemistry of some hydrolytic enzymes (Chapter 15) and some oxidative enzymes (Chapter 16).

14.1. SOME PROPERTIES OF ENZYMES

An enzyme is a protein which functions as a catalyst. The catalysis is brought about as a result of combination of a molecule of the enzyme with a molecule of one of the reactants, known as the **substrate**. The substrate, which is nearly always an organic compound or ion, is thereby made more chemically active than it would otherwise be towards another reactant, which may be an inorganic ion or molecule or another organic compound. The products of the reaction are released from the enzyme molecule, which will then be free to bind another molecule of the substrate. It is an important feature of enzymatic catalysis that reactions are enabled to occur at ambient or body temperature, usually in a medium whose pH is near to neutrality. In most instances the same chemical reactions could, in the absence of enzymes, occur only at unphysiologically high temperatures, in strongly acid or alkaline solutions, or in non-aqueous solvents.

Individual enzymes are highly specific for their substrates and for the types of reaction they catalyse. However, the substrate specificity is not always absolute and this is fortunate for the biochemist or histochemist. By providing an artificial substrate, similar but not identical to the natural one, it is often possible to study the enzymatic catalysis of a reaction that yields products more easily detectable than those generated from the physiological substrate.

For more information about enzymes and their physiological functions the reader should consult a textbook of biochemistry or enzymology. The works of West & Todd (1956), Dixon & Webb (1964) and White *et al.* (1973) are especially recommended. The histochemistry of enzymes is dealt with in depth by Burstone (1962) and Pearse (1968, 1972).

14.2. NAMES OF ENZYMES

Knowledge of enzymes and their properties has, like all other scientific information, been acquired gradually. The recognition of more and more enzymes has resulted in a bewildering profusion of names for these substances. A systematic scheme of nomenclature is available (International Union of Biochemistry, 1961) in which each enzyme is named according to the reaction it catalyses. The system also embraces a classification of enzymes and allows the use of approved trivial names when the full systematic names are cumbersome. Each enzyme is given a number (the Enzyme Commission, E.C. number) which establishes its place in the classification. For example, E.C. 3.1.1.3 is one of the major group (Group 3) of hydrolases. It acts on ester bonds (3.1) of carboxylic acids (3.1.1) and is the third member (3.1.1.3) of the list of carboxylic ester hydrolases. Its systematic name is glycerol ester hydrolase and its trivial name is lipase. It catalyses the reaction:

A triglyceride + $H_2O \rightarrow$ a diglyceride + a fatty acid

In this and the following two chapters, the approved trivial names of enzymes will be used. The E.C. numbers and systematic names will be given at the first mention in the text. Some non-approved names will also have to be used and discussed. This is an unfortunate consequence of the fact that the histochemical study of enzymes is still a much less precise science than is biochemical enzymology.

14.3. SCOPE AND LIMITATIONS OF ENZYME HISTOCHEMISTRY

The accurate identification of the cellular and subcellular locations of many enzymes has been one of the most conspicuous achievements of modern histochemistry. When using a method for the detection of a particular enzyme, a histochemist seeks answers to two questions.

1. *Is the activity detected in the tissue the same as the activity of the purified enzyme identified by biochemical methods?*

This is essentially a re-statement of the requirement for chemical specificity associated with every histochemical technique. It is important to remember that the biochemist purifies enzymes from homogenized tissue and usually studies their properties in solution. In a section of tissue, many enzymes are present in great variety, and all of them may be available to act upon the histochemical reagents which have been chosen for the purpose of demonstrating just one of their number.

The sequestration of an enzyme in an organelle can prevent or suppress the activity of that enzyme in a section. When this is the case, the failure of a histochemical reaction may be a more informative indication of enzymatic activity *in vivo* than the biochemical detection of plentiful activity in an extract of the same tissue, in which all the cells and their organelles have been thoroughly destroyed by homogenization. In practice, however, histochemical techniques are usually devised in such a way as to detect as much enzymatic action as possible.

2. *Where in the tissue is the enzyme located* in vivo?

This is usually the more difficult question to answer. The problems associated with accuracy of localization are different for the three principal types of method which are used for the histochemical demonstration of enzymes. These will now be discussed.

14.4. METHODS NOT BASED ON ENZYMATIC ACTIVITY

Enzymes are proteins and are antigenic. It is therefore possible to prepare antibodies to purified enzymes and to use these antibodies in immunofluorescent and other immunohistochemical techniques. For accurate localization it is necessary that the antigenic (though not necessarily the enzymatic) properties of the enzyme be unaffected by fixation or other pre-treatment of the tissue and that the enzyme does not diffuse away from its normal position before forming an insoluble complex with its antibody. Increasing numbers of enzymes are being localized in this way, but since the method is immunological rather than strictly histochemical, it will not be considered here. Immunohistochemical techniques are discussed in Chapter 19.

Another technique which may be used more extensively in the future is that of affinity labelling with specific inhibitors of enzymes. The first method of this type to be described involved the binding of radioactively labelled DFP (see p. 209) to sections. The bound inhibitor was then detected by autoradiography. Unfortunately this inhibitor attaches to and inhibits many enzymes that have serine residues at or near their active sites, so the specificity of the technique is low. More recently sodium, potassium-adenosine triphosphatase (Na, K-ATPase; a variant of E.C. 3.6.1.4, ATP phosphohydrolase) has been detected by virtue of its binding of ouabain, a highly specific inhibitor of the enzyme. The ouabain was covalently coupled to a peptide with peroxidase activity. The peptide could then be identified by a histochemical method for peroxidase (see Chapter 16). This gave an osmiophilic product, visible with the electron microscope (Mazurkiewicz *et al.*, 1978).

Carbonic anhydrase (E.C. 4.2.2.1) has been demonstrated by means of a fluorescent inhibitor, dimethylaminonaphthalene-5-sulphonamide, by Pochhammer *et al.* (1979). This inhibitor was given orally to animals, which were killed 7–12 h later. Cryostat sections of unfixed tissue were examined and it was possible, with appropriate filters, to distinguish the fluorescence of the enzyme-bound from that of the unbound inhibitor.

14.5. METHODS BASED ON ENZYMATIC ACTIVITY

14.5.1. Substrate-film methods

In this type of technique, the enzyme acts upon a substrate carried in a film of some suitable material which is closely applied to the section of tissue. After incubation for an appropriate time, the change in the substrate is detected by any expedient means. The sites of change in the film can then be seen to correspond to sites in the underlying (or overlying) section. A method of this type, for proteolytic enzymes, is described in Chapter 15.

While substrate-film techniques are potentially very versatile, they have two important limitations. Firstly, **the enzyme must diffuse** from its original locus to the film. Since diffusion takes place in all directions, not just straight up or down, the change in the film will occur at some distance from the place in which the enzyme was located in the section. The inaccuracy from this cause will be minimized if the contact between section and film is uniformly very close and if the film is as thin as is compatible with seeing the changes in it. Secondly, the change produced in the substrate must be immediate and irreversible: otherwise lateral diffusion of products of enzymatic action within the film would render the method useless. Probably the highest resolution that can be expected with a substrate-film technique is the identification of the cells in which an enzyme is contained.

14.5.2. Dissolved substrate methods

In these methods all reagents are used in solution and the end-products of enzymatic action are deposited within the sectioned tissue. The great majority of histochemical methods for enzymes fall into this category. In all such techniques, the enzyme in the section acts upon a substrate (often not the natural substrate of the enzyme) which is provided in an incubation medium. Depending upon the enzyme, the substrate may be hydrolysed, or it may be oxidized or reduced by another substance provided in the incubation medium or already present in the section. Other chemical reactions are also possible. The enzymatic reaction results in the formation of **products** (of hydrolysis, oxidation, etc.), one of which must be immobilized at its site of production and eventually made visible under the microscope. This immobilization of the product is accomplished by means of a **trapping agent**, usually a compound which reacts rapidly with the product to form an insoluble, coloured **final deposit**. Sometimes the precipitate produced by the trapping agent is colourless and must be made visible by a third chemical reaction. Every histochemical method in which a dissolved substrate is used entails, therefore, at least two chemical processes: the enzyme-catalysed reaction and the trapping of a product. Techniques of this type are subject to five possible sources of artifact.

(i) Diffusion of the enzyme may occur, either during preparation of the tissue or during incubation with the substrate and trapping agent. The diffusion of many soluble enzymes can be greatly reduced by incorporating an inert synthetic polymer, such as polyvinyl alcohol, in the incubation medium. Fixation of the tissue will also immobilize enzymes, but most fixatives also cause sufficient denaturation to destroy the catalytic properties. If it is possible the tissue should always be fixed: many histochemical methods will work even when only a small fraction of the original enzymatic activity survives.

(ii) The product of the enzymatic reaction may diffuse away from its site of production before it is precipitated by the trapping agent. The diffused product may then attach itself to nearby structures which are different from the sites in which the enzyme is located. For example, if the product is a positively charged ion, it may be bound to anionic sites in the nucleus of the cell in whose cytoplasm it was formed. In order to minimize diffusion of the primary product of the enzyme-catalysed reaction, it is important that: (a) the trapping reaction occur very rapidly, (b) the final deposit have a very low solubility in the incubation medium. The latter condition depends, in some cases, on the solubility product of the final deposit. For example, if phosphate ions, released by a phosphate ester hydrolase, are to be trapped by precipitation of calcium phosphate, the incubation medium must contain as high a concentration of calcium ions as possible. In this way the product: $[Ca^{2+}]^3 [PO_4^{3-}]^2$ is likely to exceed the very low solubility product of $Ca_3(PO_4)_2$ in the presence of only minute traces of phosphate (see also Exercise 5 at the end of this chapter).

(iii) The final deposit is necessarily precipitated somewhere in the section. If the deposit is insoluble in water but somewhat soluble in lipids, there may be erroneous localization of the enzyme in the latter. Since protein is ubiquitously present in cells, it is preferable for the final deposit to be bound to protein once it has been precipitated. The term "substantivity" is used to denote the affinities of different types of final deposit for lipid or protein. High substantivity for protein is clearly a desirable property.

(iv) If a third reaction is necessary to make the trapped product visible, diffusion can occur at this stage. This is rarely a cause of artifact in light microscopy, but may assume greater significance in enzyme histochemistry at the higher levels of resolution of the electron microscope.

(v) At any stage of the procedure, coloured deposits may be formed which do not result from activity of the enzyme being studied. False-positive reactions of this type may be due to (a) other enzymes acting upon constituents of the incubation medium, (b) spontaneous occurrence of the reaction catalysed by the enzyme, or (c) unwanted non-enzymatic reactions of other kinds. As controls against this type of artifact it is necessary to try out the method (a) in the absence of the substrate, and (b) on sections in which specific enzymatic activity is prevented (e.g. by heating or, preferably, by a specific inhibitor of the enzyme).

14.6. TYPES OF ENZYMES

Biochemists classify enzymes according to the types of chemical reaction they catalyse. The systematic classification of enzymes is, however, too thorough for the histochemist, whose repertoire of techniques is much more limited than that of the enzymologist. Histochemically demonstrable enzymes fall into the following rather broad categories:

(a) *Hydrolytic enzymes*

These catalyse the reactions of their substrates with water. The substrate molecule is split into two parts, one of which can be trapped.

1. Phosphatases. Catalyse the hydrolysis of esters and amides of phosphoric acid.
2. Sulphatases. Catalyse the hydrolysis of sulphate esters.
3. Carboxylic esterases. Catalyse the hydrolysis of ester linkages between carboxylic acids and any of a variety of alcohols or phenols.
4. Glycosidases, which catalyse the hydrolysis of glycosidic linkages.
5. Proteolytic enzymes, which are of several types and catalyse the splitting of peptide bonds.

(b) *Oxidoreductases*

These enzymes, which catalyse oxidation–reduction reactions, are discussed in Chapter 16.

(c) *Transferases*

A transferase is an enzyme which removes part of a molecule to form a new compound. Since a number of chemical reactions (degradation, synthesis, changes in coenzymes) are associated with the activities of transferases, there is a varied assortment of histochemical methods for enzymes of this type.

(d) *Other enzymes*

Several enzymes which do not fall into one of the above major categories can be demonstrated histochemically.

14.7. TECHNICAL CONSIDERATIONS

14.7.1. Preparation of tissue

The preservation of an enzyme in a tissue may be achieved in one of two ways.

(a) The tissue may be rapidly frozen and sectioned with a cryostat. The unfixed sections can then be incubated to detect enzymatic activity. This procedure is strongly recommended by Chayen *et al.* (1973) for all histochemical methods for enzymes. However, most histochemists prefer to fix the tissue if possible (see Pearse, 1968, 1972).

(b) A chemical fixative may be used which does not inhibit or denature the enzyme. While not many enzymes resist fixation, this procedure is convenient for those which do. Thus, most carboxylic ester hydrolases can be demonstrated in frozen sections of formaldehyde-fixed tissue and some phosphatases survive fixation in acetone followed by paraffin embedding. Cryostat sections of unfixed tissue may be fixed for a few minutes (usually in cold formaldehyde or acetone) before incubation and several enzymes will survive this treatment. Fixation will limit diffusion of enzymes. While it is easier to prepare frozen sections of fixed tissue than cryostat sections of unfixed material, the latter method is capable of providing thinner sections and more satisfactory morphological preservation, if carried out skilfully.

14.7.2. Conditions of reaction

Most incubations are carried out at 37°C, though room temperature or even 4°C is preferred in some methods. The enzymes of homoiothermic animals generally function optimally at 37°C; those of poikilotherms and of plants at 25°C. However, there is less diffusion of the enzymes and of the products of reaction at lower temperatures. Sections mounted on slides or coverslips are most conveniently incubated in coplin jars. When only a small volume of medium is available (e.g., if an expensive ingredient is used), drops may be placed over individual sections. The slides are then enclosed in a humid container, such as a petri dish containing wet gauze, in order to prevent evaporation. Free-floating frozen sections are incubated in small petri dishes, glass cavity-blocks or haemagglutination trays.

Clean glassware is essential — even more so when dealing with enzymes than in other branches of histochemistry, since contaminating substances may inhibit the enzymes or react with constituents of the incubation media.

Some enzyme inhibitors (e.g., potassium cyanide, diethyl-*p*-nitrophenyl phosphate and many others) are dangerously toxic and must be handled with care. It should also be noted that some reagents (e.g., benzidine and related compounds) are carcinogenic and that the possible hazards associated with the use of many substances are unknown.

14.8. EXERCISES

Theoretical

1. Explain the principles underlying the techniques for histochemical localization of enzymes.

2. What factors limit the accuracy of localization of an enzyme by a technique which is known to be chemically perfectly specific?

3. The final coloured deposit formed in a histochemical method for an enzyme is sometimes seen to be in the form of well-defined intracytoplasmic granules. Explain why this result, though aesthetically satisfying, does not necessarily prove that the enzyme of the living cell is identically distributed.

4. In histochemical methods for dehydrogenases, hydrogen and electrons are transferred from the physiological substrate to a tetrazolium salt, which is thereby reduced to an insoluble pigment. Identify, in general terms, the substrate, the products of enzymatic action, the trapping agent, and the final product.

5. A section is incubated in 0.1 M calcium chloride together with a substrate which releases phosphate ions under the influence of an enzyme. How many phosphate ions would have to be released in order to precipitate calcium phosphate, $Ca_3(PO_4)_2$: (a) in an organelle with a volume of 1.0 μ^3; (b) in a spherical cell 8.0 μ in diameter? The solubility product $[Ca^{2+}]^3\ [PO_4^{3-}]^2$ is 2.0×10^{-29}. Avogadro's number, $N_A = 6.022 \times 10^{23}$ molecules per mole. Assume that the enzymatically liberated phosphate ions do not diffuse outside the organelles or cells in which they are formed and ignore any effects of pH and of formation of other phosphates of calcium.

[*Answers*: (a) 85 ions; (b) 22,780 ions.]

15

Hydrolytic Enzymes

15.1. Phosphatases 206
15.1.1. Acid phosphatase 206
15.1.2. Alkaline phosphatase 207
15.2. Carboxylic esterases 208
15.2.1. Classification 208
15.2.2. Substrates and inhibitors 209
15.2.3. Indigogenic method for carboxylic esterases 210
15.2.4. Choline esterases 211
15.3. Proteinases 212
15.4. Individual methods 213
15.4.1. Acid phosphatase 213
15.4.2. Alkaline phosphatase 213
15.4.3. Inhibition of carboxylic esterases 214
15.4.3.1. Preparation of inhibitor solutions 214
15.4.3.2. Methods of application 215
15.4.4. Indigogenic method for carboxylic esterases 215
15.4.5. Acetylcholinesterase and cholinesterase 216
15.4.6. Substrate-film method for proteinases 217
15.5. Exercises 218

Methods for five different enzymes will be discussed in this chapter. The techniques have been selected with a view to illustrating a wide variety of histochemical principles. It should be remembered that many other methods are available for these and related hydrolytic enzymes.

15.1. PHOSPHATASES

(Phosphoric monoester hydrolases; E.C. 3.1.3.)

These enzymes catalyse the hydrolysis of esters of phosphoric acid. Acid and alkaline phosphatases are distinguished by their widely separated pH optima. The former are typical lysosomal constituents, while the latter group includes enzymes which hydrolyse specific substrates (e.g. adenosine triphosphate, thiamine pyrophosphate) as well as ones which can attack a variety of substrates.

15.1.1. Acid phosphatase

(Orthophosphoric monoester phosphohydrolase-acid pH optimum; E.C. 3.1.3.2.)

This name is applied to a family of enzymes contained in lysosomes which catalyse the hydrolysis of phosphate esters optimally at around pH 5.0.

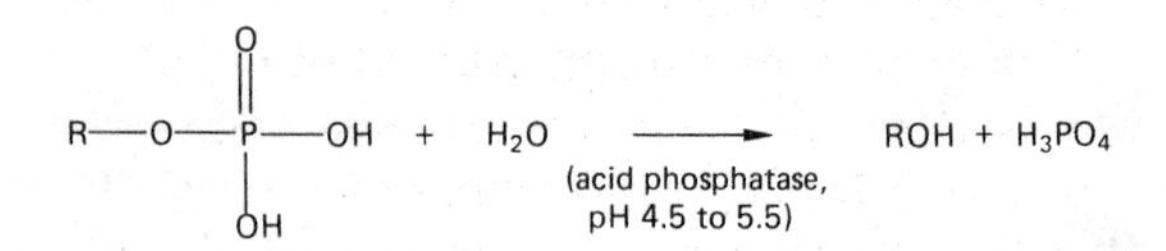

R may be an alkyl or aryl radical. Only one hydroxyl group of phosphoric acid may be esterified in a substrate for this enzyme.

In the lysosomes acid phosphatase is associated with several other enzymes, all with acid pH optima, which catalyse the hydrolysis of a wide variety of esters, amides, and proteins. These enzymes occur in high concentrations in cells of the renal tubules and the prostate gland and in phagocytes and cells undergoing degeneration. Smaller numbers of lysosomes are found in most other types of cells.

In Gomori's lead phosphate method, which has been improved upon by many later investigators, the phosphate ions released by hydrolysis of the substrate (sodium β-glycerophosphate) are trapped by lead ions, with which they combine to form insoluble lead phosphate. The latter is white, but the colour can be changed to black by treatment with hydrogen sulphide or a solution of sodium or ammonium sulphide.

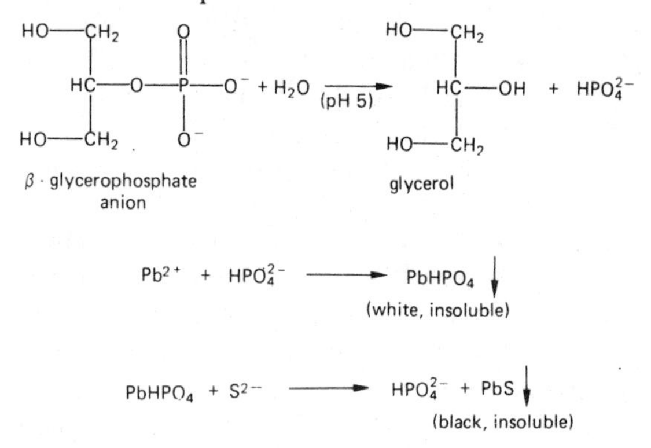

Although lysosomal enzymes are enclosed in membranous organelles, the substrate can penetrate the membranes at the optimum pH of acid phosphatase.

The concentration of Pb^{2+} ions is critical and the incubation medium has to be mixed well in advance in order to allow any spontaneous precipitation of insoluble lead salts to occur before the sections are introduced. The effects of variations in concentration of Pb^{2+} and substrate, the pH, and the overall ionic strength of the incubation medium have all been examined (see Pearse, 1968; Chayen *et al.*, 1973, for references to original literature). The mixture used is formulated so as to minimize the binding of lead ions to nuclei and other structures, including the lysosomes that contain the enzyme.

Acid phosphatase is inhibited by fluoride ions.

The enzyme will survive brief fixation in cold (4°C) acetone or aqueous formaldehyde, and it is even possible to demonstrate it in paraffin sections if a wax with a low melting point is used. The technique is described on p. 213.

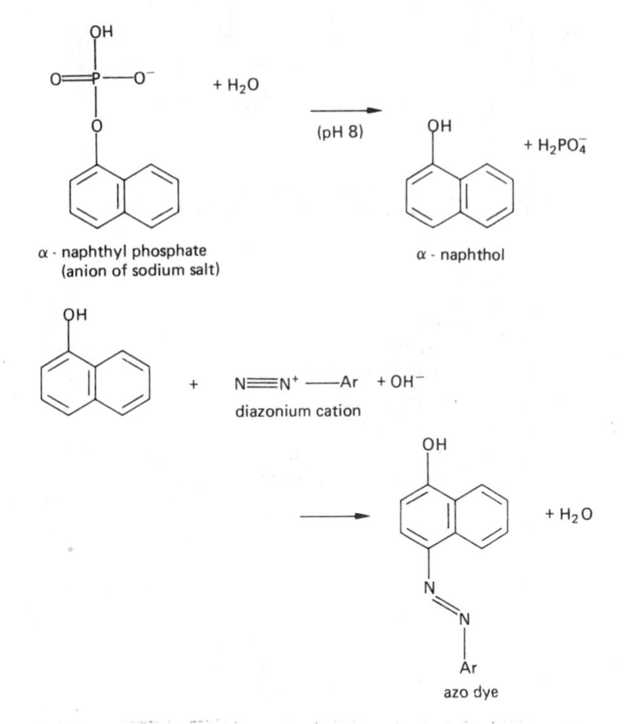

15.1.2. Alkaline phosphatase

(Orthophosphoric monoester phosphohydrolase-alkaline pH optimum; E.C. 3.1.3.1.)

The reaction catalysed by this group of enzymes (which excludes, in the present context, those phosphomonoesterases that have specific substrates) is the same as for acid phosphatase except that the pH optimum lies around 8.0. Alkaline phosphatase occurs in many types of cell, especially in regions specialized for endocytosis and pinocytosis.

Although it is possible to demonstrate alkaline phosphatases by a method similar to the one described above, which produces an eventual precipitate of a metal sulphide, we shall consider instead another technique in which the organic product of hydrolysis is trapped.

The substrate is the monobasic sodium salt of α-naphthyl phosphate, the monoester of α-naphthol, and phosphoric acid. The α-naphthol freed by hydrolysis is a phenolic compound and can therefore couple with a diazonium salt, which is included in the incubation medium. Coupling occurs rapidly at an alkaline pH and an insoluble, coloured azoic dye is produced.

It will be noticed that the azo compound is also phenolic and is therefore an acid dye, albeit one which is insoluble in water and not appreciably ionized at pH 8. This property may confer some substantivity for protein. However, the azo dyes formed from naphthols are usually soluble in non-polar organic liquids. A phosphate ester of naphthol-AS (see p. 58) can also be used as a substrate. Naphthol-AS forms more intensely coloured azoic dyes than does α-naphthol, but its rate of coupling with diazonium salts is slower (Burstone, 1962). The diazonium salt used as a trapping agent should be as stable as possible and must not inhibit the enzyme. The dye formed as the end-product of the histochemical reaction should precipitate as exceedingly small particles, with substantivity for protein. The perfect diazonium salt has not yet been found, but fast blue RR salt (see p. 57) is one of the best available. The relative merits and faults of many stabilized diazonium salts used in enzyme histochemistry are discussed by Pearse (1968) and Lillie (1977).

There are no inhibitors of high specificity for alkaline phosphatase, but enzymatic activity is prevented by prior treatment of the sections with a solution of iodine in potassium iodide. Some alkaline phosphatases are inhibited when cysteine is added to the incubation medium. Like acid phosphatase, alkaline phosphatase can be detected after fixation in cold acetone followed by careful paraffin embedding. Greatly superior preservation is obtained, however, in frozen sections of tissues fixed at 4°C in neutralized aqueous formaldehyde. The technique is described on p. 213.

15.2. CARBOXYLIC ESTERASES

(Carboxylic ester hydrolases; E.C. 3.1.1.)

The carboxylic esterases catalyse the general reaction

$$R{-}O{-}\underset{\underset{O}{\|}}{C}{-}R' + H_2O \longrightarrow ROH + HOOC{-}R'$$

The alkyl or aryl radical R may be derived from one of many possible alcohols (including glycerol), or phenols, or from a hydroxylated base such as choline. The acyl group

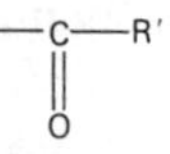

may be derived from a simple carboxylic acid such as acetic or from one of the long-chain fatty acids.

15.2.1. Classification

Many names have been used for the histochemically demonstrable enzymes in this group. The synonymy is explained in Table 15.1. The substrates of the different enzymes and the products of their hydrolysis are listed in Table 15.2.

TABLE 15.1. *Nomenclature of some carboxylic esterases*

Recommended trivial name	E.C. number and systematic name	Synonyms which should no longer be used	
Carboxylesterase	3.1.1.1 Carboxylic ester hydrolase	Ali-esterase B-esterase Organophosphate-sensitive esterase	Non-specific esterases
Arylesterase	3.1.1.2 Aryl ester hydrolase	Arom-esterase A-esterase Organophosphate-resistant esterase E600-esterase DFP-ase	Non-specific esterases
Acetylesterase	3.1.1.6 Acetic ester acetyl-hydrolase	C-esterase Organophosphate-resistant, sulphydryl inhibitor-resistant esterase	Non-specific esterases
Acetylcholinesterase (AChE)	3.1.1.7 Acetylcholine acetyl-hydrolase	Specific cholinesterase True cholinesterase Cholinesterase (term still used by physiologists and pharmacologists)	Choline esterases
Cholinesterase (ChE)	3.1.1.8 Acylcholine acyl-hydrolase	Pseudocholinesterase Butyrylcholinesterase Non-specific cholinesterase	Choline esterases
Lipase	3.1.1.3 Glycerol ester hydrolase		
Phospholipase B	3.1.1.5 Lysolecithin acyl-hydrolase		

TABLE 15.2. *Substrates and actions of some carboxylic esterases*

Enzyme	Substrate whose hydrolysis is catalysed[a]	Products of reaction
Carboxylesterase	A carboxylic ester (may be of an aliphatic or aromatic alcohol or a phenol)	An alcohol or a phenol and a carboxylic acid
Arylesterase	An ester formed from acetic acid and a phenol	A phenol and acetic acid
Acetylesterase	An ester of acetic acid	An alcohol or a phenol and acetic acid
Acetylcholinesterase	Acetylcholine[b]	Choline and acetic acid
Cholinesterase	An acylcholine[b]	Choline and an acid
Lipase	A triglyceride[c]	A diglyceride and a fatty acid
Phospholipase B	A lysolecithin	Glycerolphosphocholine and a fatty acid

[a] These are the substrates which serve to define the biochemical specificities of the enzymes in conjunction with effects of specific inhibitors. The enzymes also act upon other substrates, including those used in histochemistry.

[b] Histochemical substrates include phenolic esters and thioesters of choline.

[c] Histochemical substrates include esters of polyhydric alcohols other than glycerol and phenolic esters of higher fatty acids.

15.2.2. Substrates and inhibitors

In this text only the first five enzymes in Table 15.2 will be considered. These enzymes are detected histochemically by providing them with synthetic substrates and trapping the products of hydrolysis. Unfortunately, no artificial substrate is available that is acted upon exclusively by any one of the individual enzymes. All five catalyse the hydrolysis of the substrates for carboxylesterase, arylesterase, and acetylesterase. The hydrolysis of some other substrates, which are not attacked by these three enzymes, is catalysed by both acetylcholinesterase and cholinesterase. Unlike the two phosphatases discussed earlier in this chapter, the carboxylic esterases cannot be distinguished from one another by taking advantage of their pH optima: they all function efficiently over the range of pH 5.0 to 8.0. It is therefore necessary to make use of toxic substances which specifically inhibit some of the enzymes.

The two choline esterases are inhibited by eserine (also known as physostigmine). This alkaloid competes with esters of choline for substrate-binding sites on the enzyme molecules. Because it is a competitive inhibitor, eserine must be included in the incubation medium together with the substrate. In the presence of eserine, a substrate for all five histochemically detectable carboxylic esterases will be hydrolysed only as a result of the activities of carboxylesterase, arylesterase, and acetylesterase.

The first of these three is inhibited by low concentrations of organophosphorus compounds. These potent inhibitors, which also inactivate both the choline esterases and some proteinases, act by blocking serine residues at or near to the active sites on the enzyme molecules, thus preventing access of the substrate. Typical organophosphorus inhibitors are diisopropylfluorophosphate (DFP) and diethyl-*p*-nitrophenyl phosphate (E600).

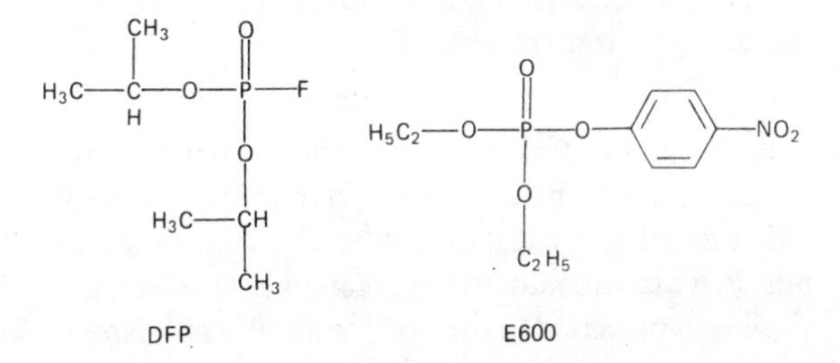

The hydroxyl group of serine displaces the substituent shown on the right-hand side of the phosphorus atom in each of the above formulae. Organophosphorus compounds are dangerously toxic, mainly because they also inhibit acetylcholinesterase. Inhibition is irreversible, so when sections have been pre-incubated in DFP or E600 it is not necessary to add these inhibitors to the incubation medium.

Only arylesterase and acetylesterase are still active after exposure to an organophosphorus compound. The activity of arylesterase is dependent upon the integrity of a free sulphydryl group at the substrate-binding site of the enzyme molecule. It can therefore be inhibited by blocking this sulphydryl group with an organic mercurial compound. The inhibitor usually chosen is the *p*-chloromercuribenzoate ion (PCMB).

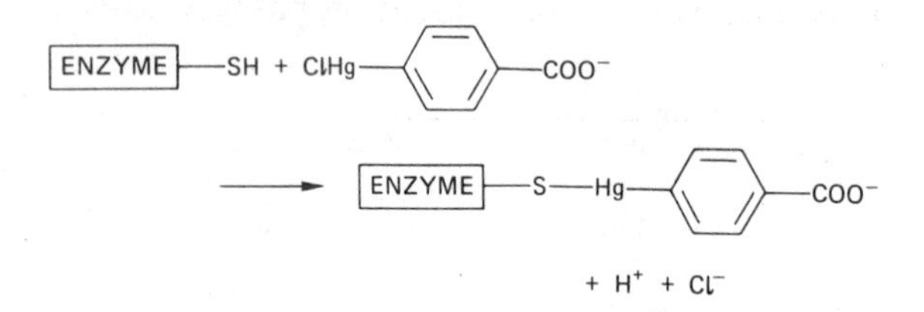

If the sections are first incubated in a dilute solution of E600, then in a solution of PCMB, and then in a substrate-containing medium with added PCMB, acetylesterase will be the only enzyme to give a positive histochemical reaction.

Selective inhibitors of acetylcholinesterase and cholinesterase will be mentioned later in connection with the histochemical methods for these enzymes. For a thorough review of the carboxylic esterases and their inhibitors, the reader is referred to Pearse (1972).

15.2.3. Indigogenic method for carboxylic esterases

Histochemical methods for these enzymes are available in which acetyl esters of naphthols and of indoxyls are used as substrates. The former methods, in which the substrates are naphthyl esters, are similar in principle to the one for alkaline phosphatase discussed earlier, so only the latter type of technique will be described.

The substrate may be one of a selection of halogenated derivatives of indoxyl acetate. 5-bromoindoxyl acetate is a typical example. The unsubstituted indoxyl ester is not used, since the end-product of its hydrolysis forms unduly large crystals.

Hydrolysis yields the halogenated indoxyl:

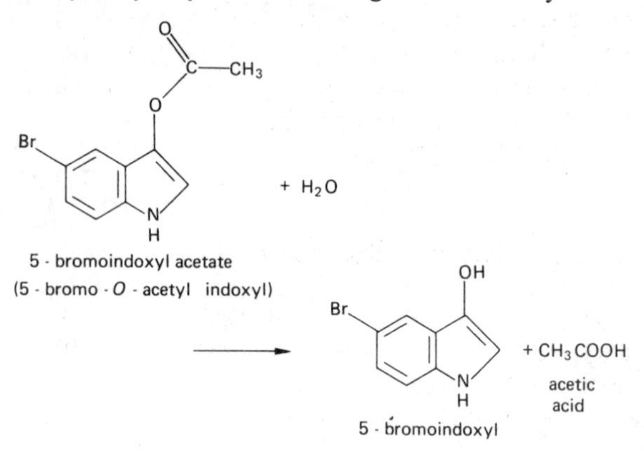

The indoxyl exhibits keto-enol tautomerism:

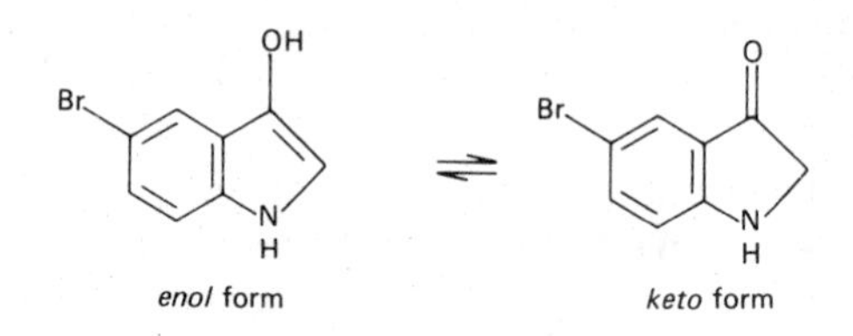

It is colourless and soluble in water, but the *keto* form is rapidly oxidized by a balanced mixture of ferrocyanide and ferricyanide ions present in the incubation medium. The continuous removal of the *keto* tautomer causes the reversible reaction shown above to be driven from left to right, virtually to completion. The product of oxidation of two molecules of the halogenated indoxyl is 5,5′-dibromoindigo:

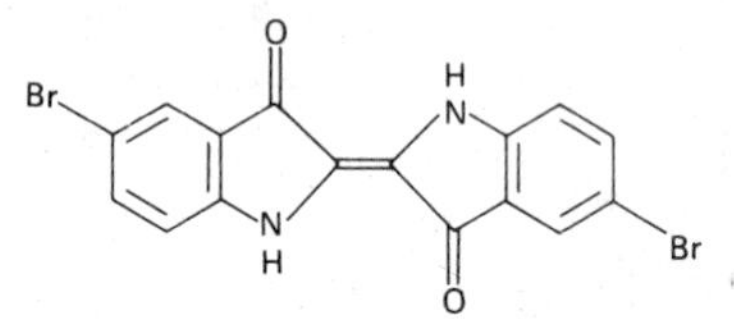

This compound is insoluble in water and also in histological dehydrating and clearing agents. It contains the indigoid chromophore (see Chapter 5) and is coloured blue.

Since all five of the carboxylic esterases under consideration are able to catalyse the hydrolysis of halogenated indoxyl acetates, the individual enzymes must be identified by the judicious use of inhibitors, as outlined in the preceding section (15.2.2) of this chapter. No inhibitors are available that spare only the choline esterases. Consequently, the indigogenic method is used mainly for carboxylesterase, arylesterase, and acetylesterase. It is also useful, however, for the demonstration of acetylcholinesterase at motor end-plates in skeletal striated muscle (see Chapter 18), where other carboxylic esterases are absent. The technique is described on p. 215.

15.2.4. Choline esterases

Mammalian tissues contain two enzymes that catalyse the hydrolysis of esters of choline. Both are inhibited by the alkaloid eserine.

Acetylcholinesterase (AChE; see Table 15.1 for synonyms) is present in erythrocytes and in some neurons. This is the enzyme that terminates the action of acetylcholine at cholinergic synapses and neuromuscular junctions, though it must have other functions as well.

Cholinesterase (ChE; also frequently called "pseudocholinesterase", see Table 15.1) can also catalyse the hydrolysis of acetylcholine, though more slowly than AChE. The preferred substrates are esters of choline with acyl groups containing more carbon atoms than acetyl. ChE occurs in serum, in neuroglia, and in some neurons. It is also present in the endothelial cells of cerebral capillaries in some species, notably the rat.

Histochemical methods for AChE and ChE are valuable in the histological study of the nervous system, since they selectively demonstrate certain groups of neuronal somata and axons. The uses of the methods in neuroanatomy are reviewed by Kiernan and Berry (1975).

The substrate in the most widely employed technique (introduced by Koelle & Friedenwald, 1949) is acetylthiocholine (AThCh), the thio-ester analogous to acetylcholine.

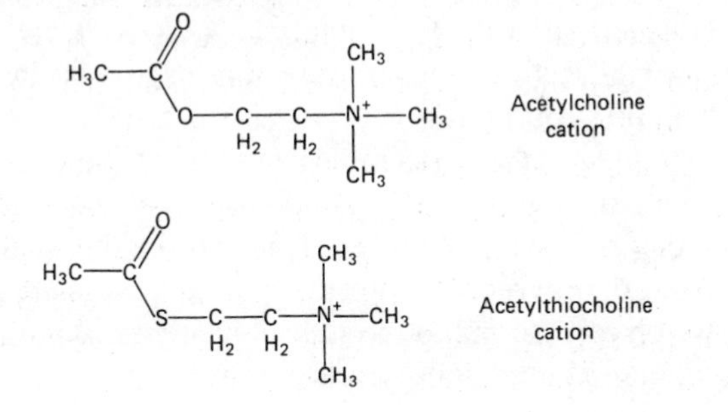

Enzymatic cleavage of AThCh yields acetic acid and the thiocholine cation:

$$CH_3COS(CH_2)_2\overset{+}{N}(CH_3)_3 + H_2O$$
AThCh cation

$$\longrightarrow CH_3COOH + HS—(CH_2)_2—\overset{+}{N}(CH_3)_3$$
Thiocholine cation

Thiocholine has a free sulphydryl group. The incubation medium also contains copper (complexed with glycine, so that the AChE is not inhibited by a high concentration of Cu^{2+}) and sulphate ions. Thiocholine combines with copper and sulphate to form an insoluble crystalline product, copper–thiocholine sulphate, which is probably

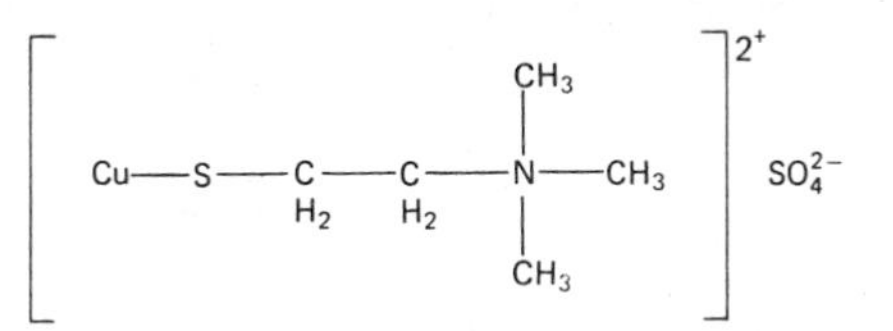

This product is usually made visible in the microscope by treatment with a soluble sulphide (H_2S, Na_2S or $(NH_4)_2S$), which results in the formation of brown cuprous sulphide. While the crystals of copper–thiocholine sulphate are visible in light microscopy, the deposits of cuprous sulphide are amorphous (Malmgren & Sylven, 1955) even under the electron microscope. This re-distribution of the final product of the reaction must be remembered when the supposed localizations of AChE and ChE are studied at the ultrastructural level.

In another widely used technique (Karnovsky & Roots, 1964) the incubation medium contains a thiocholine ester, copper (as a citrate complex), and ferricyanide ions. The latter are reduced to ferrocyanide by the sulphydryl group of the thiocholine released by enzymatic hydrolysis of the substrate. Brown copper ferrocyanide (Hatchett's brown) is formed and is immediately precipitated at the site of the enzymatic activity. This technique is known as a "direct-colouring" method to distinguish it from the procedures in which the product of the reaction has to be converted to cupric sulphide. Since the deposits of $Cu_2Fe(CN)_6$ are rather weakly coloured, it is often advantageous to intensify them by treatment with diaminobenzidine (DAB) and osmium tetroxide (Hanker *et al.*, 1973). The DAB is oxidized by copper ferrocyanide to a dark-brown polymer (see Chapter 16). The polymer reacts with osmium tetroxide to form an even darker product. The chemistry of this latter reaction is not properly understood. There are many other variants of the thiocholine techniques (see Bell, 1966; Koelle & Gromadzki, 1966; Davis & Koelle, 1967; Eranko *et al.*, 1967; Broderson *et al.*, 1974; Tsuji, 1974; Kiernan & Berry, 1975, for details and for further information on the chemistry of the methods).

Since AThCh is hydrolysed, though slowly, under the influence of catalysis by ChE, it is necessary to include a selective inhibitor of this enzyme in the incubation medium if AChE is to be demonstrated in isolation. A more suitable substrate for the deliberate demonstration of ChE is butyrylthiocholine (BuThCh). This is, however, slowly attacked by AChE, so it is necessary to inhibit the latter enzyme. The most generally useful selective inhibitor of AChE is a quaternary ammonium compound, B.W. 284C51 (1,5-*bis*-(4-allyldimethylammoniumphenyl)pentan-3-one dibromide). Ethopropazine hydrochloride is a convenient inhibitor of ChE, but this enzyme is also inhibited by concentrations of DFP, which are too low to inhibit both the choline esterases. It should be remembered that there is considerable variation amongst different species of animals in the susceptibilities of choline esterases and other carboxylic esterases to different inhibitors. Except in well understood species (such as the rat), it is necessary to investigate the effects of several inhibitors before reaching a decision as to the identity of an enzyme. Interspecific variation is discussed by Kiernan (1964) and Pearse (1972). These references and a paper by Pepler and Pearse (1957) may be consulted for lists of substances which inhibit the different carboxylic esterases.

A direct-colouring method for choline esterases is described on p. 216.

15.3. PROTEINASES

The proteolytic enzymes catalyse the hydrolysis of peptide bonds between amino acids in polypeptides and proteins:

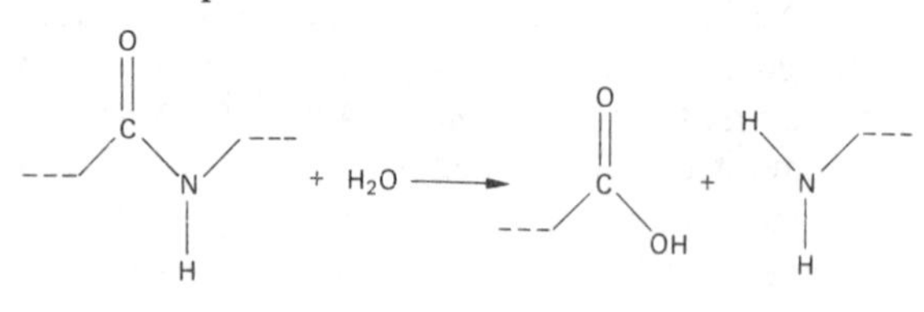

The different enzymes attack peptide linkages adjacent to particular amino acids or at one or other end (the C- or N-terminal amino acids) of a polypeptide chain.

Some of the enzymes catalyse the hydrolysis of synthetic substrates containing the peptide configuration, such as naphthyl amides. Histochemical methods are available in which the naphthyl amines liberated from substrates of this type are trapped by diazonium salts in a similar way to that in which α-naphthol is trapped in the method for alkaline phosphatase described earlier in this chapter. Some proteinases are able to act upon esters and this property has also been utilized histochemically. Unfortunately, considerable uncertainty exists as to the correspondence between the histochemically demonstrated enzymes and those identified by biochemists. The enzymes considered below belong to the broad category of peptide peptidohydrolases (E.C. 3.4.4), which catalyse the hydrolysis of proteins to yield small peptides.

We shall now consider a method which is of low specificity but introduces a type of technique quite different from those in which artificial substrates and trapping agents are used. It has the advantage that the substrate, being a protein, is perhaps not unduly different from the natural one. This is the substrate-film method for proteinases. In the first method of this kind to be described (Adams & Tuqan, 1961) the sections are placed on the emulsion of an over-exposed, developed, fixed, and washed photographic plate. The section and the film are both slightly moist (but not wet) and are equilibrated with a suitable buffer. This preparation is incubated for about an hour in a humid atmosphere and then dried and mounted for examination.

The emulsion on the film contains tiny particles of metallic silver suspended in gelatine (which is a protein of high molecular weight made by boiling collagen in water). Proteolytic activity derived from the section changes the subjacent gelatine into soluble peptides of lower molecular weight which diffuse into the surrounding regions of the emulsion, carrying the suspended silver particles with them. After a suitable time of incubation, the blackened photographic emulsion will have holes in it which correspond to the sites of proteinases in the tissues. The section itself is obscured by the black emulsion, so the positions of the digested holes have to be identified by comparison of the substrate-film preparation with an adjacent, stained section of the same specimen. Colour films have also been used as substrates, but most are unsuitable because their emulsions are made of gelatine that has been excessively cross-linked in the manufacturing process (Hasegawa & Hasegawa, 1977).

A proteinaceous substrate of more certain composition than a photographic emulsion may be prepared by covalently linking a dye to a layer of gelatine on a microscope slide (Cunningham, 1967). A dyed gelatine film is sufficiently transparent to allow examination of the section by phase-contrast microscopy, thereby permitting more accurate localization of the holes produced by the action of proteolytic enzymes. An alternative approach is to use an undyed gelatine film and, after incubation, to stain both the film and the section by a general method for protein (Fried *et al.*, 1976). The colour is then strongest where both section and film are present, so that digested areas are easily seen only at the edges of the tissue.

15.4. INDIVIDUAL METHODS

15.4.1. Acid phosphatase

Cryostat sections of unfixed tissue should be used for maximum sensitivity of the method. Frozen sections of tissues fixed for 12–18 h in neutral buffered formaldehyde at 4°C are also satisfactory. Frozen sections must be washed in three or four changes of water to remove phosphate ions (from the buffered fixative), which would precipitate the Pb^{2+} in the incubation medium.

Solutions required

A. Incubation medium (Waters & Butcher, 1980)

Dissolve 132 mg of lead nitrate, $Pb(NO_3)_2$, in 25 ml of 0.2 M acetate–acetic acid buffer, pH 4.7.

Dissolve 315 mg sodium β-glycerophosphate in 75 ml water.

Combine the two solutions and warm to 37°C before use.

This medium is more stable than were earlier formulations and it does not form a precipitate on standing. It may be used immediately or stored at room temperature for several days.

B. Sulphide solution

Add about 0.5 ml of yellow ammonium sulphide to about 100 ml of water immediately before use. Do not allow fumes from the ammonium sulphide bottle to come into contact with the incubation medium (solution A). Ammonium sulphide is malodorous and toxic and should be used in a fume cupboard. Discard the used solution by washing it down the sink with plenty of running tap water.

Procedures

1. Incubate sections in solution A for 30 min at 37°C.
2. Wash in four changes of water, 1 min in each.
3. Immerse in dilute ammonium sulphide (solution B) for about 30 s.
4. Wash in three changes of water.
5. Mount in a water-miscible medium. (See *Note 1*.)

Result

Black deposits of lead sulphide indicate sites of acid phosphatase activity. (See also *Note 2* below.)

Notes

1. An aqueous mountant is preferred because small amounts of PbS may be dissolved during dehydration and clearing. Saturation of the alcohols and xylene with PbS is sometimes recommended when it is desired to use a resinous mounting medium.
2. Control sections should be incubated (a) without substrate, (b) in the full incubation medium containing, in addition, 0.01 M sodium fluoride, which inhibits the enzyme.

15.4.2. Alkaline phosphatase

This method may be used with cryostat sections (unfixed or fixed for 3–10 min in acetone at 4°C) or with frozen sections of tissue fixed for 12–24 h at 4°C in neutral, buffered formaldehyde. Formaldehyde-fixed blocks may be stored for a few weeks at 4°C in gum sucrose (p. 215). It is also possible to detect the enzyme in paraffin sections of small blocks fixed overnight at 4°C in acetone. It is necessary to use a wax which melts at 42°C (for technical details see Bancroft, 1967). The following description applies to free-floating frozen sections, which must be handled carefully to minimize physical damage in the alkaline incubation medium.

Solutions required

A. Incubation medium

0.05 M TRIS buffer, pH 10.0 (see Chapter 20):	10 ml
α-naphthyl acid phosphate (sodium salt):	10 mg
Magnesium chloride ($MgCl_2.6H_2O$):	10 mg
Fast blue RR salt (C.I. 37155):	10 mg

Mix in the order stated. Filter and use immediately.

B. 1% aqueous acetic acid

Procedure

1. Carry sections through two changes of water to remove fixative, then incubate them in solution A for 20 min at room temperature.
2. Transfer to water (in a large dish) for 1 min.
3. Transfer to 1% acetic acid (solution B) for 1 min.
4. Rinse in water.
5. Mount sections onto slides and cover, using a water-miscible mounting medium.

Result

Sites of alkaline phosphatase activity purple to black.

Notes

1. Incubate sections in solution A **without** the substrate as a control for non-specific staining by the diazonium salt.
2. To control for spontaneous hydrolysis of the substrate, treat some sections with an iodine solution for 3 min followed by a rinse in $Na_2S_2O_3$ (as for removal of mercurial deposits, see p. 36) and four rinses in water, before incubating in the substrate-containing solution A. The treatment with iodine inhibits the enzyme.
3. The magnesium salt may enhance enzymatic activity, but it is often omitted. The rinse in dilute acetic acid neutralizes the alkaline buffer and renders the sections less fragile.

15.4.3. Inhibition of carboxylic esterases

15.4.3.1. Preparation of inhibitor solutions

A. **Eserine.** Used as eserine (= physostigmine) sulphate, M.W. 649, usually as a 10^{-5} M solution. Dissolve 6.5 mg of eserine sulphate in 10 ml of water. Add 0.1 ml of this to 9.9 ml of buffer (for preincubation) or incubation medium (for simultaneous incubation and inhibition). 10^{-5} M eserine inhibits AChE and ChE.

B. **E600.** Diethyl-*p*-nitrophenyl phosphate is sold as an oily liquid in an ampoule. It is exceedingly poisonous and care must be taken to avoid ingestion, contact with the skin, or inhalation of vapour. Ampoules contain 1.0 ml (M.W. 275; S.G. 1.27). Drop an ampoule into 462 ml of propylene glycol in a stoppered reagent bottle under a fume hood. Break the ampoule with a glass rod and insert the stopper. This gives a 10^{-2} M stock solution of E600, which keeps indefinitely at 4°C. The bottle should be stood in a container of absorbent material such as kieselgühr or vermiculite.

For use, dilute with buffer to obtain the desired concentration (usually 10^{-7} to 10^{-3} M). Dispose of old solution as described for DFP.

10^{-5} M E600 inhibits AChE, ChE, and carboxylesterase. Arylesterase and acetylesterase are inhibited only by very high concentrations (e.g. 10^{-2} M).

C. **DFP.** Diisopropylfluorophosphate is a liquid, supplied in a glass ampoule. On account of its volatility, it is even more dangerous than E600. Ampoules contain 1.0 g of DFP (M.W. 208). In a fume hood, drop an ampoule into 480 ml of propylene glycol in a glass-stoppered reagent bottle. Break the ampoule with a glass rod and insert the stopper. This stock solution is 10^{-2} M DFP. Store at 4°C with the bottle standing in a container of absorbant material. Keeps indefinitely. Do not pipette by mouth. For use dilute with buffer to obtain the desired concentration (usually 10^{-7} to 10^{-5} M). Solutions containing 10^{-2} M DFP should be discarded by pouring into an excess of 4% aqueous NaOH and leaving in the fume cupboard for several days before washing down the sink with plenty of water. More dilute solutions, in quantities less than 100 ml, may be discarded without special precautions.

AChE and ChE are inhibited by 10^{-5} to 10^{-4} M DFP. ChE is inhibited by 10^{-7} to 10^{-6} M DFP.

D. **Ethopropazine.** Available as its hydrochloride (M.W. 349), or as its methosulphate (M.W. 359), which is used clinically as a drug for relieving symptoms of Parkinson's disease. Ethopropazine is usually used as a 10^{-4} M solution. Dissolve 34.9 mg (hydrochloride) or 35.9 mg (methosulphate) in 10 ml of water. Add 0.1 ml of this to 9.9 ml of buffer (for pre-incubation) and to 9.9 ml of incubation medium (for simultaneous incubation and inhibition). ChE is inhibited selectively.

E. **B.W. 284C51.** (1,5-*bis*-(4-allyldimethylammoniumphenyl)-pentan-3-one dibromide) (M.W. 560). Dissolve 140 mg in 25 ml of water to give a stock solution which is 10^{-2} M. Keeps for a few months at 4°C. Dilute with buffer and with incubation medium to the desired concentration (10^{-5} to 10^{-4} M). B.W. 284C51 inhibits AChE but not ChE.

F. ***p*-Chloromercuribenzoate** (PCMB). Available as *p*-chloromercuribenzoic acid (M.W. 357) or its sodium salt (M.W. 379). Dissolve 36 mg of the acid in the smallest possible volume (1.0–2.0 ml) of 2 N (8%) NaOH. Add 5 ml of buffer, then adjust to the desired pH by careful addition of 1.0 N HCl. Make up to 10 ml with buffer to obtain a 10^{-2} M solution of the inhibitor. The sodium salt (38 mg) can be dissolved directly in 10 ml of buffer. Dilute with buffer and with incubation medium to 10^{-4} to 10^{-6} M PCMB. Acetylesterase is not inhibited.

15.4.3.2. Methods of Application

The irreversible inhibitors (i.e. the organophosphorus compounds E600 and DFP) are used by **pre-incubation** in a solution with the same pH as the substrate-containing mixture. The sections are placed in a buffered solution containing the desired concentration of the inhibitor for 30 min at 37°C. The sections are rinsed in four changes of buffer before being transferred to the incubation medium. The rinsing is a particularly important part of the procedure when inhibited and uninhibited sections are to be incubated side by side in the same batch of medium.

Other inhibitors work by competition with the substrate for the active sites of the enzymes and must be incorporated in the incubation medium. Usually it is desirable also to pre-incubate the sections with a solution of the inhibitor, so that there will be no chance of enzymatic hydrolysis of the histochemical substrate during the earliest stages of incubation. Pre-incubation in a competitive inhibitor should **not** be followed by washing.

In any attempt to identify carboxylic esterases it is necessary to use several inhibitors over a range of concentrations extending at least two orders of magnitude either side of that generally thought to be optimal for the inhibition of any particular enzyme. Inhibitors may not be necessary when a histochemical procedure is used only as a staining method for some structure which happens to contain one of the enzymes. Inhibitors other than the ones described here are listed by Pearse (1972, pp. 796 and 798).

15.4.4. Indigogenic method for carboxylic esterases

This method (Holt & Withers, 1952) works well with frozen sections of tissues fixed in neutral, buffered formaldehyde for 12–24 h at 4°C. The fixed blocks may be stored for several months at 4°C if they are transferred directly from the fixative to **gum sucrose** (gum acacia (arabic), 2 g; sucrose, 60 g; water, 200 ml). Several hydrolytic enzymes can be immobilized by soaking tissues in gum-sucrose. The large molecules of the gum, which is a polysaccharide, may hinder the diffusion of the enzyme molecules.

Stock solutions

A. TRIS-HCl buffer, pH 7.2 (see Chapter 20)	Keeps for several months at 4°C, unless infected
B. 0.1 M calcium chloride ($CaCl_2$; 1.11%)	
C. 0.05 M potassium ferricyanide ($K_3Fe(CN)_6$; 1.65%)	
D. 0.05 M potassium ferrocyanide ($K_4Fe(CN)_6.3H_2O$; 2.11%)	Keeps for about 4 weeks at 4°C

It is slowly oxidized by air to the ferricyanide, with consequent change of colour from almost colourless to yellow.

Incubation medium

Dissolve 5 mg of *O*-acetyl-5-bromoindoxyl (= 5-bromoindoxyl acetate) in about 0.2 ml of ethanol in a small beaker, then add:

Solution A (buffer):	8.0 ml	Mix just before using
B ($CaCl_2$):	4.0 ml	
C ($K_3Fe(CN)_6$):	4.0 ml	
D ($K_4Fe(CN)_6$):	4.0 ml	

(See also *Note 1*.)

Procedure

1. Cut frozen sections and transfer them without washing to the incubation medium. Leave at room temperature for about 30 min. The time is not critical; some tissues will show an adequate reaction in 5 min while others may require 2 h.
2. Rinse section in two changes of water.
3. (Optional.) Counterstain nuclei with neutral red or safranine. van Gieson's stain (p. 99) is also suitable.
4. Rinse in water, mount onto slides, allow to dry, dehydrate, clear and mount in a resinous medium.

Result

Sites of carboxylic esterase activity—blue. The deposit is finely granular. See *Note 2* for comments on specificity and controls.

Notes

1. Indoxyl acetate is a poor substrate, giving a conspicuously crystalline, diffused end product, but it has been used. *O*-acetyl-4-chloro-5-bromoindoxyl is said to be superior to the substrate suggested here, in that an even more finely granular blue precipitate is formed.
2. This method detects all carboxylic esterases, but AChE and ChE are inhibited by eserine. E600 and PCMB should be used to identify the other enzymes, but in critical histochemical studies it is necessary to determine the effects of other inhibitors as well, especially with species other than the rat.

15.4.5. Acetylcholinesterase and cholinesterase

This method is a slightly modified version of that of Karnovsky & Roots (1964) with the addition of an intensification of the contrast of the product of the reaction based on the procedure introduced by Hanker *et al.* (1973).

Frozen sections are cut from specimens fixed at 4°C for 12–24 h in neutral, buffered formaldehyde. Glutaraldehyde (1.0–3.0%, pH 7.2–7.6) is also a suitable fixative and formal–sucrose–ammonia (formalin (40% HCHO), 100 ml; sucrose, 150 g; strong ammonia (NH_4OH, S.G. 0.880), 10 ml; water to 100 ml) is a mixture devised specially for the preservation of these enzymes (Pearson, 1963).

Reagents required

A. (i) Acetylthiocholine iodide (substrate for AChE)
(ii) Butyrylthiocholine iodide (substrate for ChE)
(iii) Inhibitors, according to the requirements of the investigation.

B. Stock solution for incubating medium

0.1 M acetate buffer, pH 6.0 (see Chapter 20):	65 ml
Sodium citrate (trisodium salt; dihydrate):	147 mg
Cupric sulphate (anhydrous):	48 mg
Water:	to 100 ml
Then add:	
Potassium ferricyanide ($K_3Fe(CN)_6$):	17 mg

This pale green solution keeps for about 1 week at room temperature. It must not be used if it contains a brown precipitate.

C. Incubation medium (working solution)

Dissolve 5.0 mg (approximately) of **either** acetylthiocholine iodide **or** butyrylthiocholine iodide in a drop of water and add 10 ml of solution B. This medium is stable for several hours and may be used repeatedly if not cloudy, but should be discarded at the end of the day.

Inhibitors should be added to this medium as needed.

D. TRIS buffer, pH 7.2 (see Chapter 20)

E. DAB solution

3,3′-diaminobenzidine tetrahydrochloride:	10 mg
Water:	10 ml

Prepare just before using. **Caution**: DAB may be carcinogenic. For disposal, see p. 243.

F. Dilute OsO_4 solution

1% aqueous osmium tetroxide:	1.0 ml
Water:	to 100 ml

Dilute the stock 1% solution just before use. Use a fume hood. The diluted (0.01% solution) keeps for 2 or 3 weeks at 4°C.

Procedure

1. Collect frozen sections into water in which they may remain for up to 1 h. Pre-incubate any control sections in appropriate inhibitors. Wash irreversibly inhibited sections in four changes of water. Do not wash after pre-incubation with competitive inhibitors.
2. Transfer sections to the incubation medium (solution C) for 10–30 min at room temperature. When regions of enzymatic activity go reddish brown, incubation is adequate. Check under a microscope for isolated sites of activity such as motor end-plates.
3. Transfer to TRIS buffer (pH 7.2; solution D) for 5 min. This time is not critical but should not exceed 30 min. (See *Note 1*.)
4. Transfer to water, 1 min (or longer as convenient). (See *Note 3*.)
5. Transfer to DAB solution (E) for 10 min.
6. Wash in two changes of water, each at least 1 min.
7. Mount sections onto slides and allow to dry.
8. Immerse slides in dilute OsO_4 (solution F) for 5 min. (See *Note 2*.)
9. Wash in five changes of water, each at least 3 min.
10. Dehydrate through graded alcohols, clear in xylene, and cover using a resinous mounting medium.

Result

Sites of enzymatic activity dark brown.

Notes

1. The rinse in TRIS buffer has been shown by Hanker *et al.* (1973) to remove occasional non-specific deposits of $Cu_2Fe(CN)_6$ which are liable to form if the incubation is unduly prolonged. Used as described; TRIS buffer does not dissolve the precipitate formed by enzymatic action.
2. Treatment with OsO_4 would be expected to cause blackening of myelin, but this does not occur with a very dilute solution applied for a short time, as in this method. It is important to wash out excess OsO_4, since this would be reduced to black OsO_2 by the alcohol used for dehydration.
3. Stages 5–9 may be omitted when only sites of strong activity, such as AChE at motor end-plates or ChE in cerebral capillaries and some central neurons, are of interest.
4. Although the substrates (especially butyrylthiocholine) show partial specificity, the use of inhibitors is imperative for certain identification of the two enzymes. A control with eserine should also be provided, though non-enzymatic precipitation of the end-product is very unusual with this technique.

15.4.6. Substrate-film method for proteinases

This technique is based on that of Cunningham (1967), differing mainly in the use of a reactive dye rather than a coupled tetrazonium reaction for colouring the protein film. The method is applicable to cryostat sections of tissues fixed for 4 h in neutral buffered formaldehyde or formal–calcium. The blocks may be stored in gum-sucrose (p. 215) at 4°C for at least 24 h.

Preparation of substrate films

Dissolve 2.0 g of gelatine in warm water to make 50 ml. Use this solution at 37–45°C. Pre-warm some glass slides to about 45°C on a hotplate. Have available also a cold plate, such as a metal tray that has been kept at 4°C, and a desiccator containing anhydrous calcium sulphate ("Drierite"), silica-gel, or concentrated sulphuric acid.

Spread approximately 0.1 ml of the warm gelatine solution over the surface of each warm slide, then immediately place the slide on the cold plate so that the gelatine will set. When a convenient number of gelatine-coated slides have been made, place them in the desiccator for 6–24 h (time is not critical) at room temperature. Prepare the following staining solution no more than 4 h before using it:

Procion billiant red M2B (C.I. 18158; Reactive red 1):	150 mg
Water:	25 ml
Dissolve the dye and then add:	
0.05 M borax ($Na_2B_4O_7 . 10H_2O$, 1.91% w/v):	25 ml

This solution may be used several times but should be discarded at the end of the day. It is used in a coplin jar.

Remove slides from the desiccator and place them in the staining solution for 10 min. Wash in gently running tap water for 30 min, then in distilled or deionized water for 5 min. Drain the slides and return them to the desiccator for storage.

Procedure

Collect cryostat sections directly onto the slides bearing the stained films and allow the sections to thaw and adhere. Immerse the slide for 15 s in a buffer solution appropriate to the enzymatic activity sought (e.g. pH 5–6 for lysosomal proteinases; pH 7.5–8.5 for pancreatic proteinases). Drain the slide and blot it gently with filter paper moistened with the same buffer. Place the slides in a warm, humid container such as a wide-necked screw-cap jar with wet cotton wool in the bottom in an oven at 37°C. Leave for 15 min to 6 h, removing slides at intervals for inspection.

The incubation is terminated by immersing the slides in a fixative:

95% ethanol:	85 ml	Mix before using and use only once
Formalin (37–40% HCHO):	10 ml	
Glacial acetic acid:	5 ml	

After fixation for 30–60 min the slides are rinsed in 70% ethanol, washed in water, counterstained, dehydrated, cleared, and mounted in a resinous medium. The counterstain should not be too strong. Baker's Haematal-16 (Chapter 6, p. 82) applied for 5–10 min is suitable.

Result

Clear holes or spaces in the pink gelatine film at sites of proteolytic activity. Nuclei blue if counterstained as suggested above. The sizes of the holes increase and the accuracy of localization decreases as the time of incubation is lengthened.

15.5. EXERCISES

Theoretical

1. Explain in general terms the methods by which the cellular localizations of hydrolytic enzymes are determined. What factors limit (a) the chemical, and (b) the topographical accuracy of such techniques?

2. What are the differences between reversible and irreversible inhibitors of carboxylic esterases? How are the differences reflected in the histochemical utilization of inhibitors?

3. In sections of nervous tissue it was found that the cells of a certain nucleus had the following histochemical properties:

(a) Strong reaction with 5-bromoindoxyl acetate as substrate in the presence of 10^{-5} M diethyl-*p*-nitrophenyl phosphate (E600),

(b) Strong reaction with butyrylthiocholine as substrate; inhibited by 10^{-5} M eserine, 10^{-5} M E600, 10^{-4} M ethopropazine, or 10^{-7} M diisopropyl-fluorophosphate (DFP),

(c) Weak reaction with acetylthiocholine as substrate, inhibited by the same agents which inhibited the hydrolysis of butyrylthiocholine.

Which enzyme (or enzymes) is (are) likely to be responsible for these histochemical findings? What further investigations are indicated?

4. Thiamine pyrophosphatase (TPP-ase), an enzyme found in the Golgi apparatus, releases phosphate ions from the substrate, thiamine pyrophosphate (TPP). The phosphate can be trapped by calcium ions. TPP is also hydrolysed to some extent by alkaline phosphatases, but the latter are inhibited by cysteine. TPPase can function over a wide pH range. Calcium salts are colourless, but the Ca^{2+} ion can be replaced by cobalt, which has an almost colourless phosphate but a black sulphide. Co^{2+} cannot, however, be included in the substrate mixture, since it would inhibit the enzyme.

On the basis of the above facts, devise an incubation medium for TPPase and explain how the sites of enzymatic activity could be made visible under the microscope.

5. It is possible to determine the histological localization of the enzyme **deoxyribonuclease** by placing cryostat sections on a film made from a solution of DNA in aqueous gelatine and incubating for 1 h. Because of the ways in which the section and film are mounted, it is then possible to separate them and obtain two slides, one bearing the section and the other bearing the substrate-film.

How could the sites of DNase activity be demonstrated? What control procedures would be necessary in order to establish the specificity of the method?

Practical

6. This exercise will require two people, working together. Specimens are collected for use in Exercises 7–10. Anaesthetize a rat with pentobarbitone (a lethal dose) and open the thorax. Proceed to perfuse the animal with neutral, buffered 2.5–4% formaldehyde, but before starting the perfusion collect some blood from the right atrium and prepare films on slides as described in Chapter 7. After the rat has been perfused with the fixative, remove the following organs and continue their fixation by immersion for a further 12–24 h at 4°C:

(a) The kidneys.
(b) The small intestine (a few pieces about 1.0 cm long).
(c) The brain. After a few hours of fixation, take two specimens: (i) a piece of the cerebrum 2–3 mm thick cut in the coronal plane and including, on its ventral surface, the optic chiasma, (ii) a piece of the medulla for sectioning in the transverse plane, just rostral to the obex, where the fourth ventricle closes caudally to form the central canal.
(d) A piece of liver.
(e) A loop of duodenum enclosing its mesentery, which contains pancreatic tissue. Trim after 6 h of fixation to obtain a block which, when sectioned, will provide transverse sections of the duodenum with parts of the attached pancreas. Transfer the trimmed tissue to gum-sucrose (p. 215) overnight before sectioning on a cryostat for Exercise 10 below.

The blood-films should be used as soon as they are dry. The fixed material can be used on the following day.

7. Stain two blood-films for acid phosphatase. Lightly counterstain one of the preparations with safranine to show the nuclei of the leukocytes. Stain a third film with a conventional blood stain. Which cells contain acid phosphatase? (*Note*. The mature granulocytes of the rat have annular nuclei rather than the segmented forms found in human blood.)

8. Cut frozen sections of kidney, intestine, and liver. Stain for alkaline phosphatase. Where is the enzyme located?

9. Cut frozen sections of kidney and cerebrum. Preincubate some sections in (a) 10^{-5} M eserine sulphate, (b) 10^{-5} M E600, (c) E600 followed by 10^{-5} M PCMB. For the eserine- and PCMB-treated sections, the inhibitor must also be added to the substrate mixture. Incubate sections according to the indigogenic method for esterases.

By comparing the appearances of inhibited and uninhibited sections, identify regions or cells containing (i) choline esterases, (ii) arylesterase, (iii) carboxylesterase, (iv) acetylesterase. In the sections of brain pay particular attention to pericytes (next to small blood-vessels), the neuropil of the caudate nucleus, and the large neurosecretory cells of the supraoptic nucleus (found on either side of the optic chiasma).

Using the thiocholine method, with appropriate substrates and inhibitors, demonstrate the distributions of the two cholinesterases in the two regions of the brain removed in Exercise 6. Is it possible to show that some neuronal somata in the medulla and hypothalamus contain both AChE and ChE? Which enzyme occurs in axons and which in the endothelial cells of cerebral capillaries?

10. Cut cryostat sections of the duodenum and pancreas (see Exercise 6 (e)) about 10 μ thick and collect them onto slides with dyed gelatine films. Demonstrate sites containing proteolytic enzymes. Proteolytic enzymes may also be sought in salivary gland and kidney. Use a range of pH values from 5.5 to 8.5 and incubate for 15 min to 6 h to find the optimum conditions.

16

Oxidoreductases

16.1. Oxidation and reduction 220
16.2. Oxidation–reduction potentials 221
16.3. Biological oxidations 222
16.4. Histochemistry of dehydrogenases 224
16.4.1. Tetrazolium salts 225
16.4.2. Diaphorases 227
16.4.3. Technical considerations 228
16.4.3.1. Tissue preparation 228
16.4.3.2. Composition of incubation medium 228
16.4.3.3. Conditions of reaction 230
16.4.3.4. Controls 230
16.5. Histochemistry of oxidases 231
16.5.1. Cytochrome oxidase 231
16.5.2. Catechol oxidase 232
16.6. Histochemistry of peroxidases 233
16.6.1. Actions and occurrence 233
16.6.2. Histochemical localization 234
16.6.3. Specificity and accuracy of localization 235
16.7. Individual methods 236
16.7.1. Succinate dehydrogenase 236
16.7.2. Coenzyme-linked dehydrogenases 237
16.7.3. Diaphorases (tetrazolium reductases) 240
16.7.4. Cytochrome oxidase 241
16.7.5. Catechol oxidase 241
16.7.6. Peroxidase 242
16.7.6.1. DAB–hydrogen peroxide method 242
16.7.6.2. DAB–*p*-cresol–hydrogen peroxide method 243
16.8. Exercises 243

The following account is a simplified one and several controversial issues are not taken into consideration. Pearse (1972) discusses the subject in great detail and a shorter treatment is given by Chayen *et al.* (1973). The metabolic functions of the oxidoreductases are described in textbooks of biochemistry.

16.1. OXIDATION AND REDUCTION

An atom or molecule is said to be oxidized when it loses one or more electrons and to be reduced when it gains one or more electrons. A simple example of a reaction of this type is a change in the oxidation state of a metal ion:

$$Fe^{2+} \underset{\text{reduction}}{\overset{\text{oxidation}}{\rightleftharpoons}} Fe^{3+} + \varepsilon^-$$

The position of the equilibrium is determined by the presence of other substances which can accept or donate electrons (**oxidizing** and **reducing agents** respectively). For example, the ferric ion is reduced by hydroquinone in acid conditions:

$$2Fe^{3+} + \text{hydroquinone (HO–C}_6\text{H}_4\text{–OH)} \rightleftharpoons 2Fe^{2+} + p\text{-quinone (O=C}_6\text{H}_4\text{=O)} + 2H^+$$

Here, ferric ion is the oxidizing agent and is itself reduced to the ferrous state. It is equally valid to consider that hydroquinone is a reducing agent which acts upon the ferric ions and is itself oxidized to *p*-quinone.

The reaction between ferric ions and hydroquinone is the algebraic sum of two half-reactions:

$$Fe^{3+} + \varepsilon^- \rightleftharpoons Fe^{2+} \qquad (16.1)$$

$$H_2Q \rightleftharpoons Q + 2H^+ + 2\varepsilon^- \qquad (16.2)$$

($Q = p$-quinone; H_2Q = hydroquinone)

Both sides of equation (16.1) are multiplied by 2 before adding, in order that the number of electrons participating in each half-reaction will be the same. Thus:

$$2Fe^{3+} + 2\varepsilon^- \rightleftharpoons 2Fe^{2+}$$
$$H_2Q \rightleftharpoons Q + 2H^+ + 2\varepsilon^-$$

$$2Fe^{3+} + H_2Q \rightleftharpoons 2Fe^{2+} + Q + 2H^+$$

It will be noticed that the electrons have been cancelled out and do not appear in the equation for the complete reaction. It will also be noticed that the reaction is reversible. The oxidation of hydroquinone in neutral and alkaline solutions is more complicated and leads to the formation of polymeric products.

The net effect of the oxidation of hydroquinone has been the loss of two atoms of hydrogen. The gain or loss of hydrogen is a feature of most oxidation–reduction reactions of biological importance.

According to an earlier definition, a substance was said to be oxidized when it gained oxygen, lost hydrogen, or increased its positive charge and to be reduced when oxygen was lost, hydrogen gained, or negative charge increased. The modern definition, in terms of the transfer of electrons, simplifies the multiple requirements of the older one.

The number of electrons gained or lost by an organic molecule is not always as obvious as it is in the case of an inorganic ion. It may be determined (see Hendrickson *et al.*, 1970 for more information) from the change in the oxidation number of the carbon atom at which oxidation or reduction takes place. The oxidation number of a carbon atom is found by adding the following values for each of its four bonds:

−1 for each —H
0 for each —C
+1 for each bond to an element other than H or C

Thus, for the carbon atom here shown at the top of the *p*-quinone molecule there are two bonds to other carbon atoms ($2 \times 0 = 0$) and two to an oxygen atom ($2 \times +1 = +2$), so the oxidation number is +2. At the equivalent position in hydroquinone there are three bonds to carbon ($3 \times 0 = 0$) and one to oxygen ($1 \times +1 = +1$), giving the oxidation number +1. This carbon atom therefore loses one electron in the process of being reduced. The carbon on the opposite side of the ring behaves identically, so the reduction of the whole molecule of *p*-quinone involves the acquisition of two electrons. Conversely, the oxidation of hydroquinone is accomplished by its losing two electrons. Application of the rule given above will show that each of the other four carbon atoms has oxidation number −1 in both *p*-quinone and hydroquinone.

16.2. OXIDATION–REDUCTION POTENTIALS

It is possible to make an electrical cell in which one electrode is gaseous hydrogen and the other is an inert metal such as platinum, immersed in an electrolyte which is a solution of a substance capable of being oxidized or reduced. Completion of the external circuit by a wire will result in the passage of electrons from one electrode to the other as a consequence of the gain or loss of electrons by the electrolyte at the inert electrode. The potential difference between the electrodes can be measured, in volts, and its magnitude, under standardized conditions of temperature and concentration, is the **oxidation–reduction potential** (E_0) of the electrolyte. It is a measure of the ease with which the electrolyte is oxidized or reduced or, conversely, of the strength of the electrolyte as an oxidizing or reducing agent.

Hydrogen is arbitrarily assigned $E_0 = 0$. The oxidation–reduction potentials of other substances are, by the most widely used convention, applied to half-reactions in which electrons are gained. Thus the half-reaction in which a strong oxidizing agent (which takes up electrons avidly) is reduced will have a high positive value of E_0. The half-reaction in which a weaker oxidizing agent than the hydrogen ion is reduced will have a negative E_0. These half-reactions are reversible. If they are written the other way round, the negative or positive sign of E_0 is changed.

Consider, for example, the following half-reactions, arranged in descending order of their oxidation–reduction potentials:

$$MnO_4^- + 8H^+ + 5\varepsilon^- \rightleftharpoons Mn^{2+} + 4H_2O \quad (E_0 = +1.51\,V)$$
$$Fe^{3+} + \varepsilon^- \rightleftharpoons Fe^{2+} \quad (E_0 = +0.77\,V)$$
$$2CO_2 + 2H^+ + 2\varepsilon^- \rightleftharpoons (COOH)_2 \quad (E_0 = -0.49\,V)$$
$$Na^+ + \varepsilon^- \rightleftharpoons Na \quad (E_0 = -2.71\,V)$$

Of the substances shown, permanganate ion is the strongest oxidizing agent, while sodium metal is the strongest reducing agent. A substance on the left-hand side of one of the above half-reactions can be expected to oxidize a substance on the right-hand side only if the latter has a more negative E_0 than the former. Permanganate ions in acid solution therefore react with oxalic acid to give manganous ions and carbon dioxide, but sodium ions will not react with ferrous ions. The equation for the overall reaction is obtained by reversing the half-reaction that contains the reducing agent and adding it to the half-reaction containing the oxidizing agent. It may be necessary to multiply both sides of one or both of the equations by appropriate integers in order to obtain equal numbers of electrons on the two sides of the final equation. For the reaction between permanganate ion and oxalic acid, one reaction must be multiplied by 2 and the other by 5:

$$2MnO_4^- + 16H^+ + 10\varepsilon^- \rightleftharpoons 2Mn^{2+} + 8H_2O$$
$$5(COOH)_2 \rightleftharpoons 10CO_2 + 10H^+ + 10\varepsilon^-$$
$$2MnO_4^- + 6H^+ + 5(COOH)_2 \rightleftharpoons 2Mn^{2+} + 10CO_2 + 8H_2O$$

Although this reaction is theoretically reversible, it proceeds from left to right, virtually to completion, because there is a large difference between the oxidation–reduction potentials of the two component half-reactions. (The continuous removal of carbon dioxide from the system, as gas or by combination with water to form carbonic acid, also helps to drive the reaction from left to right, in accordance with Le Chatelier's principle and the law of mass action.)

Tables of oxidation–reduction potentials are valuable for showing which oxidations are likely to occur and which are not, but they must be used with caution. Other chemical properties of the reactants are not taken into account and may complicate the overall reaction. The mixing of ferric ions with oxalic acid, for example, will result in the formation not only of ferrous oxalate (a sparingly soluble salt) but also of soluble complexes in which one, two, or three oxalate ions are coordinately bound to iron ions in both oxidation states +2 and +3. These complications could only be predicted by taking into account the chemistry of complex formation as well as the oxidation–reduction potentials.

Of greater importance in biochemical oxidation–reduction reactions is the fact that the value of the potential for any system varies with temperature, with the pH of the medium in which the reactants are dissolved, and with the proportions of the oxidizing and reducing agents present. A constant more useful than E_0 is E_0', the oxidation–reduction potential of the half-reduced system at specified temperature and pH. Table 16.1 (p. 223) gives values of E_0' for some biochemically and histochemically important half-reactions at pH 7 and 25°C.

16.3. BIOLOGICAL OXIDATIONS

The life of every cell depends upon the coordinated oxidation and reduction of many organic compounds. These chemical reactions, collectively known as **cellular respiration**, involve a great number of substrates and enzymes and a much smaller number of **electron carriers**. One electron carrier can function as coenzyme or prosthetic group to many different substrate-specific respiratory enzymes. A **coenzyme** is a soluble substance (itself enzymatically inert and not a protein) which can diffuse in the cytoplasm and attach itself, reversibly, to the proteinaceous **apoenzyme**. A **prosthetic group** is a non-protein organic compound which is covalently bound to the protein molecule constituting the enzyme.

When attached to the apoenzyme, the coenzyme accepts electrons from the substrate and protons from either the substrate or the surrounding medium. Thus the coenzyme is reduced and the substrate is oxidized. Effectively, the coenzyme removes one or two atoms of hydrogen from the substrate, so it is equally valid to call the electron carrier a **hydrogen acceptor**.

The reduced form of the coenzyme can act as an electron donor for the enzymatic reduction of some other substrate. The original coenzyme is thus regenerated. Enzymes which catalyse the re-oxidation of reduced coenzymes are often called **diaphorases**. They are more correctly named as the enzymes that catalyse the reduction of their specific substrates. The diaphorase activity (catalysis of the oxidation of reduced coenzyme) is incidental to the main function of such an enzyme. The histochemist, however, is often primarily interested in localizing the sites of re-oxidation of reduced coenzymes. Since more than one enzyme may catalyse this reaction for a single reduced coenzyme, the conveniently vague term "diaphorase" will continue to be used when the enzymes concerned cannot be accurately identified and named.

Before discussing some of the enzymes that catalyse oxidation and reduction and the histochemical methods for their identification, it is necessary to review the system which transfers electrons and protons from oxidized metabolites to the ultimate oxidizing agent, which is atmospheric oxygen. The transport takes place in several stages and involves the repetitive oxidation and reduction of various prosthetic groups, coenzymes, and cytochromes. The last-named substances are proteins with iron-containing haem groups. The oxidation number of the iron atom is +3 or +2, when the molecule is in the oxidized or reduced state, respectively. Cytochromes are not, strictly speaking, enzymes, but one of them, cytochrome a_3, is commonly known as **cytochrome oxidase**. It transfers electrons and protons to molecular oxygen. Since it is not itself consumed in this reaction, cytochrome a_3 catalyses the reduction of molecular oxygen to water. The reactions of the electron-transport chain occur in an order dictated largely by the oxidation–reduction potentials of the various half-reactions that make up the system. These, together with the names of the more important coenzymes, prosthetic groups and cytochromes, are set out in Table 16.1. The data in the table are taken mainly from Loach (1976).

TABLE 16.1. *Oxidation–reduction potentials in the electron-transport system (abbreviations explained at foot of table)*

Half-reaction			E_0'(V) (at pH 7.0; 25°C)
$H_2O_2 + 2H^+ + 2\varepsilon^-$	$\rightleftharpoons$	$2H_2O$	+1.35
$O_2 + 4H^+ + 4\varepsilon^-$	$\rightleftharpoons$	$2H_2O$	+0.82
cyt. $a_3^{3+} + \varepsilon^-$	$\rightleftharpoons$	cyt. a_3^{2+}	+0.29
cyt. $a^{3+} + \varepsilon^-$	$\rightleftharpoons$	cyt. a^{2+}	+0.29
cyt. $c^{3+} + \varepsilon^-$	$\rightleftharpoons$	cyt. c^{2+}	+0.25
cyt. $c_1^{3+} + \varepsilon^-$	$\rightleftharpoons$	cyt. c_1^{2+}	+0.22
$UQ + 2H^+ + 2\varepsilon^-$	$\rightleftharpoons$	UQH_2	+0.10
cyt. $b^{3+} + \varepsilon^-$	$\rightleftharpoons$	cyt. b^{2+}	+0.08
$FMN + 2H^+ + 2\varepsilon^-$	$\rightleftharpoons$	$FMNH_2$	−0.21
$FAD + 2H^+ + 2\varepsilon^-$	$\rightleftharpoons$	$FADH_2$	−0.22
$NAD^+ + 2H^+ + 2\varepsilon^-$	$\rightleftharpoons$	$NADH + H^+$	−0.32
$NADP^+ + 2H^+ + 2\varepsilon^-$	$\rightleftharpoons$	$NADPH + H^+$	−0.32

Coenzymes: UQ = ubiquinone (coenzyme Q); UQH_2 = reduced form. NAD^+ = nicotinamide adenine dinucleotide (formerly known as coenzyme I or DPN); NADH = reduced form. $NADP^+$ = nicotinamide adenine dinucleotide phosphate (formerly known as coenzyme II or TPN); NADPH = reduced form.

Prosthetic groups: FMN = flavin mononucleotide; $FMNH_2$ = reduced form. FAD = flavin adenine dinucleotide, $FADH_2$ = reduced form. The value of E_0' for a prosthetic group may differ from that given above for different apoenzymes. For example, the FAD of succinic dehydrogenase has $E_0' = -0.03$.

Cytochromes: Designated "cyt." followed by a letter (a, a_3, b, c, c_1) with superscript indicating oxidation state of the iron atom. Thus cyt. c^{3+} and cyt. c^{2+} are the oxidized and reduced forms of cytochrome c.

Not all the substances shown in Table 16.1 are involved in the oxidation of all metabolites. Electrons most often pass from NADH or NADPH to one of the flavoproteins and thence via ubiquinone and the cytochromes to molecular oxygen.

All the components of the electron-transport system are present in the mitochondria of eukaryotic cells and some occur also in the general cytoplasmic matrix, associated with soluble enzymes. The oxidation of any metabolite involves the enzymatically catalysed transfer of protons and electrons from a **substrate** to an **acceptor:**

$$\text{substrate} + \text{acceptor} \underset{\text{(enzyme)}}{\rightleftharpoons} \text{oxidized substrate} + \text{reduced acceptor}$$

Commonly the acceptor is a coenzyme such as NAD^+ or $NADP^+$ or a prosthetic group such as FMN or FAD. The enzyme catalysing the reaction is then known as a **dehydrogenase**. Such an enzyme (consisting of apoenzyme + oxidized form of the coenzyme) combines specifically with its substrate and renders it highly reactive towards the coenzyme. When the reaction has taken place, the oxidized substrate and the reduced coenzyme part company with the apoenzyme and are then free to enter into other chemical reactions. Specificity for the substrate resides in the apoenzyme, though the latter cannot bind to the substrate unless it has first combined with the coenzyme.

A simple example of a reaction catalysed by a dehydrogenase is the oxidation of the lactate ion. The enzyme concerned is known as lactate dehydrogenase.

The lactate ion first combines with an enzyme molecule:

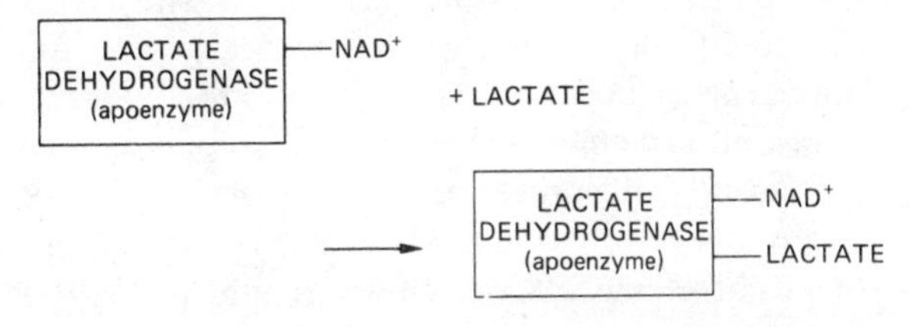

Then, on the surface of the enzyme molecule:

$$\text{LACTATE} + \text{NAD}^+ \rightarrow \text{PYRUVATE} + \text{NADH} + \text{H}^+$$

and, finally,

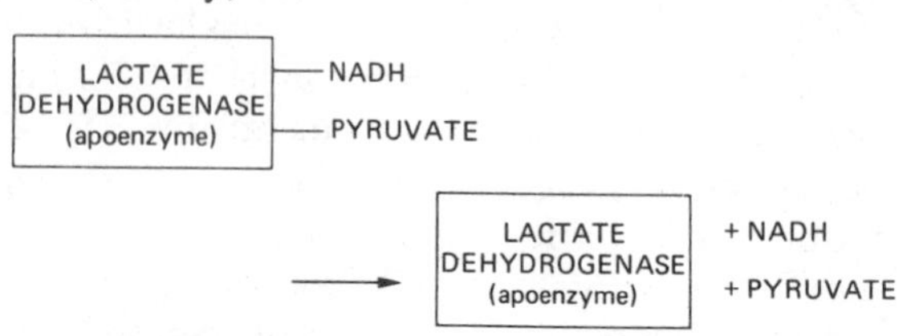

The NADH will subsequently be re-oxidized to NAD^+, when it serves as coenzyme in the enzymatically catalysed reduction of some other substrate. It will be noticed that the acceptor in the oxidation of lactate was NAD^+. The systematic name for lactate dehydrogenase, which identifies both the substrate and the acceptor, is L-lactate: NAD oxidoreductase (E.C. 1.1.1.27). The same enzyme will also catalyse the reduction of pyruvate ions. The direction in which the reversible reaction proceeds is determined by the relative concentrations of the reactants: lactate, pyruvate, NAD^+, NADH, and H^+. Oxidation of lactate occurs when this ion is present in excess and when NADH and H^+ are continuously removed, either by other metabolic activities or by deliberate manipulation of the conditions of the reaction.

The electrons removed from a substrate such as lactate, when it has been oxidized, are incorporated into a reduced coenzyme. When this is re-oxidized, the electrons will be transferred to another acceptor, which has a higher oxidation–reduction potential than the coenzyme. Usually the flow of electrons passes via flavoprotein enzymes, ubiquinone, and the cytochrome system to molecular oxygen. For the histochemist, the importance of all this lies in the fact that the flow of electrons may be interrupted by the introduction of an artificial electron-acceptor with an oxidation–reduction potential intermediate between those of any two of the members of the electron-transport chain. The tetrazolium salts, to be discussed below, are suitable for this purpose since, on reduction, they are converted to insoluble pigments. Thus, whenever a substrate is oxidized in the presence of a tetrazolium salt, the released electrons will not be transported through the usual sequence of cytochromes, etc., but will be trapped in the formation of a stable, coloured substance.

Not all oxidoreductases make use of the coenzymes NAD^+ and $NADP^+$; some dehydrogenases are flavoproteins. The most familiar of these is succinate dehydrogenase, whose prosthetic group is FAD. The physiological acceptors associated with the flavoprotein dehydrogenases are not certainly known, though ubiquinone is a likely candidate. Other oxidoreductases use molecular oxygen as an acceptor, thus bypassing all the intermediate components of the electron-transport system. These enzymes are known as **oxidases**. Tetrazolium salts cannot often be used to detect the activity of oxidases unless they can be made to act as substitutes for oxygen. Other methods are therefore usually needed for these enzymes. The only other oxidoreductases considered in this chapter are the **peroxidases**. These catalyse the oxidation of many substances by hydrogen peroxide and are discussed in Section 16.6.

16.4. HISTOCHEMISTRY OF DEHYDROGENASES

The dehydrogenases catalyse the general reaction

$$\text{Ⓢ}H_2 + \text{Ⓐ} \rightleftharpoons \text{Ⓢ} + \text{Ⓐ}H_2$$

(Ⓢ and ⓈH_2 represent oxidized and reduced forms of the substrate. Ⓐ and ⒶH_2 represent oxidized and reduced forms of the acceptor, which is a substance other than O_2 or H_2O_2.)

When the reaction proceeds from left to right, the net effect is the removal of hydrogen (usually two atoms of it) from the substrate. The acceptor is the coenzyme (NAD^+ or $NADP^+$) in the case of coenzyme-linked dehydrogenases. These enzymes are detected histochemically by substituting an artificial electron acceptor for the naturally occurring substances constituting the electron transport chain. The artificial substance chosen is one which becomes insoluble and coloured in its reduced state. Consequently, a visible precipitate forms at sites where hydrogen is given up by an electron carrier whose oxidation–reduction potential is negative with respect to that of the artificial hydrogen acceptor.

In the histochemical methods for dehydrogenases, the substrates are the physiological ones, and no attempts are made to trap the products of their oxidation. Instead, a special kind of indicator (the artificial hydrogen acceptor or electron carrier) is used to detect the place in which a biological oxidation is taking place. The substrate is provided in large amounts in the incubation medium. An adequate quantity of the acceptor must also be present, either as coenzyme added to the medium or as intermediate electron carriers already present in the tissue.

16.4.1. Tetrazolium salts

The artificial hydrogen acceptors of greatest value to the histochemist are the tetrazolium salts. These are heterocyclic compounds (derivatives of tetrazole, CH_2N_4) which are changed by reduction into insoluble, coloured **formazans**.

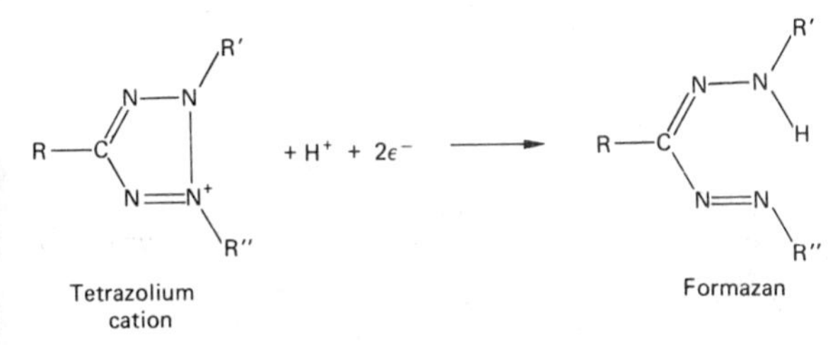

This reaction is irreversible because the formazan is insoluble. Several tetrazolium salts have been used in histochemical methods for dehydrogenases. The ideal one would be stable, not chemically altered by exposure to light, and would be reduced very rapidly to yield a formazan with exceedingly small crystals that were insoluble in lipids and had some substantivity for protein. These properties are most closely approached by some of the ditetrazolium salts, which have the general structure

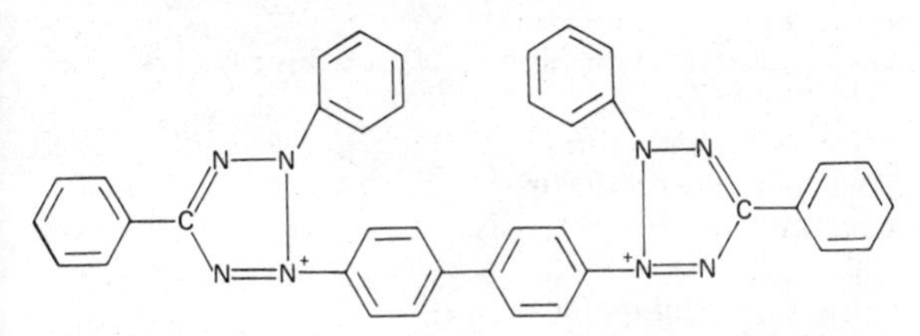

with various substituents on the benzenoid rings. Monotetrazolium salts, which have only one tetrazole ring in the molecule, are generally less suitable, though some have been used as histochemical reagents. When a ditetrazolium salt is reduced, the product may be either a monoformazan in which only one of the tetrazole rings has been opened or a diformazan in which both tetrazole rings have been opened. Monoformazans are usually red, while diformazans are blue, purple, or black. Both the coloured products may be formed in histochemical reactions, though the diformazan is the one desired. Other colours (usually reds) may also result from the presence of monotetrazolium salts as contaminants in samples of ditetrazolium salts.

A list of tetrazolium salts, with some of their properties, is given in Table 16.2.

The most generally useful tetrazolium salt for use in light microscopy is **nitro blue tetrazolium** (nitro-BT), which has the advantage of forming a formazan which is not visibly crystalline, is insoluble in lipids, and is substantive for protein.

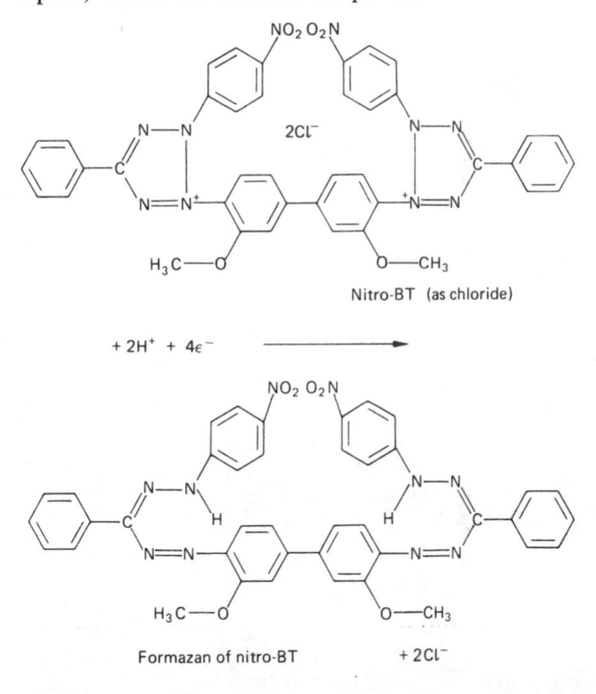

It will be seen from the above formula that nitro-BT is a ditetrazolium salt in which the radical R″ of the general formula is joined to two substituted tetrazole rings. The oxidation–reduction potential of nitro-BT is −0.05 V, which lies between that of FAD and that of ubiquinone. This tetrazolium salt can therefore be expected to accept electrons from

TABLE 16.2. *Properties of some tetrazolium salts and their formazans*
For full chemical names and other information, see Burstone (1962), Pearse (1972), Lillie & Fullmer (1976), and Lillie (1977; Chapter 9, by G. G. Glenner). The oxidation–reduction potentials E'_0 are taken from Pearse (1972). These values of E'_0 may not be accurate and cannot be compared in a meaningful way with the potentials of systems in which both the oxidizing and reducing agents are soluble, for reasons given by Jámbor (1954) and Clark (1972). They are useful, however, for comparing one tetrazolium salt with another.

Trivial name, abbreviation, and M.W.	E'_0 (V) (pH 7.2, 22°C)	Properties of formazan	Carriers from which electrons are accepted in histochemical usage
MONOTETRAZOLIUM SALTS			
Triphenyltetrazolium (chloride). TTC (M.W. 335)	−0.49	Large red crystals. High lipid solubility. Reduction is slow	cyt. a; cyt. a_3
Tetrazolium violet (chloride). TV (M.W. 384)		Large dark blue crystals (pink contaminant). High lipid solubility. Reduction is slow	
Methylthiazolyldiphenyltetrazolium (bromide). MTT (M.W. 414)	−0.11	Co^{2+}-chelate has small, black crystals. Lipid-soluble but also binds to protein. Reduction is rapid	UQ; cyt. b; cyt. c_1
Iodonitrotetrazolium (chloride). INT (M.W. 505)	−0.09	Large dark red crystals (orange contaminant). High lipid solubility. Reduction is rapid	UQ: cyt. b; cyt. c_1
2-(2-benzthiazolyl-5-styryl-2-(4-phthalhydrazidyl) tetrazolium (chloride). BSPT (M.W. 502)		Purple, amorphous. Osmiophilic. Used in EM histochemistry. Reduction is rapid	
DITETRAZOLIUM SALTS			
Neotetrazolium (dichloride). NT (M.W. 668)	−0.17	Dark purple, small crystals, lipid-soluble (red monoformazan or contaminant). Reduction is rapid	UQ; cyt. b; cyt. c
Blue tetrazolium (dichloride). BT (M.W. 728)	−0.16	Small deep blue crystals lipid-soluble (red monoformazan). Reduction is slow	
Nitro blue tetrazolium (dichloride). Nitro-BT (M.W. 818)	−0.05	Dark blue, amorphous, slight lipid-solubility. Binds to protein. Resists organic solvents (red monoformazan and contaminant: lipid- and alcohol-soluble). Reduction is rapid	Flavoproteins; UQ; cyt. b
Tetranitro blue tetrazolium (dichloride). TNBT (M.W. 908)		Brown, amorphous, insoluble in lipids and organic solvents. Binds to proteins. (Pink monoformazan or contaminant.) Reduction is rapid	(Probably closely similar to nitro-BT)
Distyryl nitro blue tetrazolium (dichloride). DS-NBT (M.W. 870)		Amorphous, osmiophilic. Used in EM histochemistry. Reduction is rapid	

NADH, NADPH, or $FADH_2$, but not from dihydroubiquinone or from any of the cytochromes. However, studies in which metabolic inhibitors of various components of the electron-transport chain have been used indicate that within mitochondria tetrazolium salts collect electrons from UQH_2 and even from reduced cytochromes (see Burstone, 1962).

Reduction of tetrazolium salts by systems with higher oxidation–reduction potentials would occur if there were large differences between the concentrations of the products of reaction and those of the reactants. The values of E_0' in Tables 16.1 and 16.2 pertain when [products] = [reactants]. The actual potential for the reduction, at a given temperature and pH, is

$$E = E_0' + \frac{RT}{nF}\log_e \frac{[\text{oxidizing agent}]}{[\text{reducing agent}]}$$

where R is the gas constant, T is the absolute temperature, n is the number of electrons gained by the reduced molecule or ion (usually two), and F is the faraday. $RT/F = 0.026$ at 25°C.

Thus for a tetrazolium salt E will be higher than E_0' when the concentration of this reagent in the medium exceeds that of the formazan. The concentration of formazan is, of course, always very low on account of its very low solubility. It has been pointed out by Clark (1972) that artificial electron acceptors used in biochemical studies of dehydrogenases commonly react at rates completely out of line with their oxidation–reduction potentials. The values of E_0' apply to systems in true thermodynamic equilibrium, which is not likely to be the state of an experimental system, with an excess of oxidizing agent present, or indeed that of the substances in a living cell.

The tetrazolium salt in a histochemical incubation medium is in competition with the naturally occurring electron carriers of the cell. In order to divert electrons from the oxidized substrate to the tetrazolium salt, it is sometimes necessary to inhibit the flow of electrons to oxygen. This may be accomplished either by incubating under strictly anaerobic conditions or, more easily, by adding cyanide ions to the medium. The cyanide inhibits cytochrome oxidase (cyt. a_3). Azide ions act similarly.

16.4.2. Diaphorases

When a tetrazolium salt is reduced by NADH or NADPH, the reaction is catalysed by an enzyme, either NADPH-diaphorase or NADPH-diaphorase.

These enzymes catalyse the reaction

$$\begin{matrix}\text{Reduced}\\ \text{coenzyme}\\ \text{(NADH or}\\ \text{NADPH)}\end{matrix} + \begin{matrix}\text{tetrazolium}\\ \text{salt}\end{matrix} \rightarrow \begin{matrix}\text{Oxidised}\\ \text{coenzyme}\\ (\text{NAD}^+\text{ or}\\ \text{NADP}^+)\end{matrix} + \text{formazan}$$

in which the reduced coenzyme is the substrate and the tetrazolium salt is the acceptor. When a tetrazolium salt is mixed with NADH or NADPH in the absence of a diaphorase apoenzyme, the reaction is very slow. Thus, the coloured product of a histochemical method for a coenzyme-linked dehydrogenase is formed by the catalytic action of another enzyme, the diaphorase. Consequently, a coenzyme-linked dehydrogenase will be accurately localized only if it occurs in the same place as the diaphorase. Fortunately, the diaphorases are present in all cells, in mitochondria, and sometimes also in the cytoplasmic matrix. They are rather "tough" enzymes, unlikely to be inhibited by short fixation in formaldehyde or by other preparative manipulations.

The natural acceptors associated with the two histochemically recognized diaphorases are not known with certainty, and the activities may be shared by several enzymes. The enzymes are believed to be flavoproteins. NADH-diaphorase may well be lipoamide dehydrogenase (NADH: lipoamide oxidoreductase; E.C. 1.6.4.3), which contains FAD as its prosthetic group and catalyses the reaction

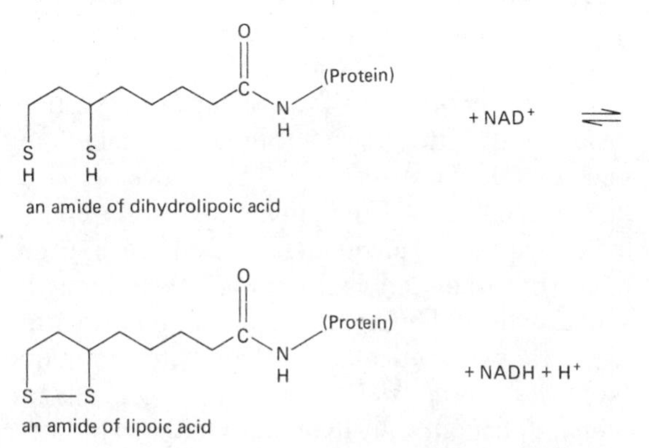

Tetrazolium salts are able to serve as acceptors in the place of lipoic acid when the reaction proceeds from right to left. The properties of NADP-diaphorase are shared by "Warburg's old yellow enzyme" (NADPH: (acceptor) oxidoreductase: E.C. 1.6.99.1), which contains FMN. The physiological acceptor is unknown. Until more is known of the identities of the enzymes that catalyse the oxidation of reduced coenzymes by tetrazolium salts, it is probably best to continue to call them NADH- and NADPH-diaphorases. Names such as "NADH-tetrazolium reductase" are also used and are acceptable.

The deliberate histochemical localization of the diaphorases is a very simple matter. Sections are incubated in a suitably buffered medium containing a tetrazolium salt and the appropriate **reduced form** of the coenzyme. The general methodological principles applicable to dehydrogenase histochemistry (see below) should also be observed.

In the case of the flavoprotein dehydrogenases, which have prosthetic groups rather than coenzymes, diaphorases cannot be responsible for the production of the formazan deposits. The mechanisms of electron-transfer from enzyme-bound $FADH_2$ or $FMNH_2$ to the tetrazolium salt are poorly understood, but may well involve ubiquinone and cytochromes, as discussed earlier (p. 227). Some tetrazolium salts are able to accept electrons directly from reduced flavin nucleotides.

16.4.3. Technical considerations

16.4.3.1. Tissue preparation

Dehydrogenases are generally much more easily inactivated than the hydrolytic enzymes discussed in Chapter 15. It is not possible to fix the tissues thoroughly enough to allow the cutting of sections on an ordinary freezing microtome and embedding in wax is out of the question. Small blocks of tissue may be fixed for 5–10 min in neutral, buffered formaldehyde at 4°C and then sectioned in a cryostat, or fresh frozen sections from the cryostat may be similarly fixed. Unfixed cryostat sections are often used, though meticulous attention to technique (see Chayen *et al.*, 1973) is necessary if the cells and their mitochondria are to remain recognizable after incubation. The two diaphorases and lactate dehydrogenase are notable in that they will survive fixation for several hours in neutral formaldehyde solutions. For the reasons given in Chapter 14, minimal fixation should be employed if possible. Fixation of the sections is more easily controlled than that of blocks. An alternative to formaldehyde is acetone (5–10 min at 4°C), which also extracts some of the cytoplasmic lipids, to which certain of the formazans may be artifactually bound. Acetone also extracts ubiquinone, which may be a necessary intermediate for the reduction of tetrazolium salts in methods for the flavoprotein enzymes (see above). It has been shown in the case of succinate dehydrogenase that it is necessary to apply UQ to sections that have been treated with acetone in order to be able to detect the enzyme at all sites of activity (Contestabile & Andersen, 1978).

16.4.3.2. Composition of incubation medium

The incubation medium for histochemical localization of a dehydrogenase includes the following:

(i) **Buffer.** The pH of the medium should be 7.0–7.2, even if this is not optimum for the enzyme. At pH values more alkaline than this, non-enzymatic reduction of NAD^+ or $NADP^+$ occurs. The NADH or NADPH so produced serves as substrate for its appropriate diaphorase, with consequent meaningless deposition of formazan within the section. This artifact, known as "nothing dehydrogenase", is probably due to reduction of the coenzyme by sulphydryl groups of proteins containing cysteine. Nothing dehydrogenase activity is maximal at pH 9.

Since incubation media for dehydrogenases often contain cations of divalent metals, TRIS buffer is usually used. Phosphate buffer is suitable when no metal ions that form insoluble phosphates are present.

(ii) **Substrate.** Commonly the substrate is an organic anion and is used as its sodium salt, at a concentration of 0.1 M. Addition of the substrate usually changes the pH of the buffer, which must therefore be adjusted to the correct value by adding a few drops of N NaOH or N HCl.

(iii) **Coenzymes.** The amount of coenzyme contained in a section is usually very small, so for coenzyme-linked dehydrogenases it is necessary to provide an excess of NAD^+ or $NADP^+$ in the incubation medium at a concentration of approximately 0.003 M. For demonstration of diaphorases, the reduced forms of the coenzymes (NADH, NADPH) are used.

(iv) **Cofactors.** Many dehydrogenases have requirements for traces of divalent metal cations. It is usual to include magnesium chloride (0.005 M) in the medium. This does no harm, but is not necessary for all the enzymes. Magnesium ions probably also help to prevent rupture of mitochondria during incubation (see also (vii) below).

(v) **Tetrazolium salt.** The concentration of the tetrazolium salt is not very critical and may range from 10^{-4} to 10^{-3}M.

(vi) **Electron transport inhibitors.** To suppress aerobic cellular respiration, sodium or potassium cyanide or sodium azide (0.005–0.01 M) is incorporated in the incubation medium. An alternative, but inconvenient technique is to incubate in the complete absence of oxygen. For many enzymes, these precautions are unnecessary when a rapidly reducible tetrazolium salt such as nitro-BT is used.

(vii) **Protective agents.** The inclusion of a chemically unreactive synthetic polymer in the medium prevents osmotic damage to mitochondria during incubation and limits the diffusion of soluble enzymes. A protective agent is not always needed if the tissue has been partially fixed, but is desirable when sections of unfixed tissues are used. The polymers employed for this purpose are polyvinylpyrollidone (PVP) (7.5% w/v) and polyvinyl alcohol (PVA) (20% w/v). The molecular weight of PVP used for this purpose is not critical. PVA should have a molecular weight of 30,000. Addition of these polymers often acidifies the medium and the pH must be adjusted accordingly.

(viii) **Intermediate electron acceptors.** It is a common practice to add phenazine methosulphate (PMS) (10^{-5}–10^{-3}M) to incubation media for dehydrogenases. This easily reduced substance transfers electrons directly from reduced coenzymes or other acceptors to tetrazolium salts. The addition of PMS accelerates the reaction and gives more intense staining, but sometimes also causes non-specific deposition of formazan in the sections. This is due to spontaneous, non-enzymatic reduction of the

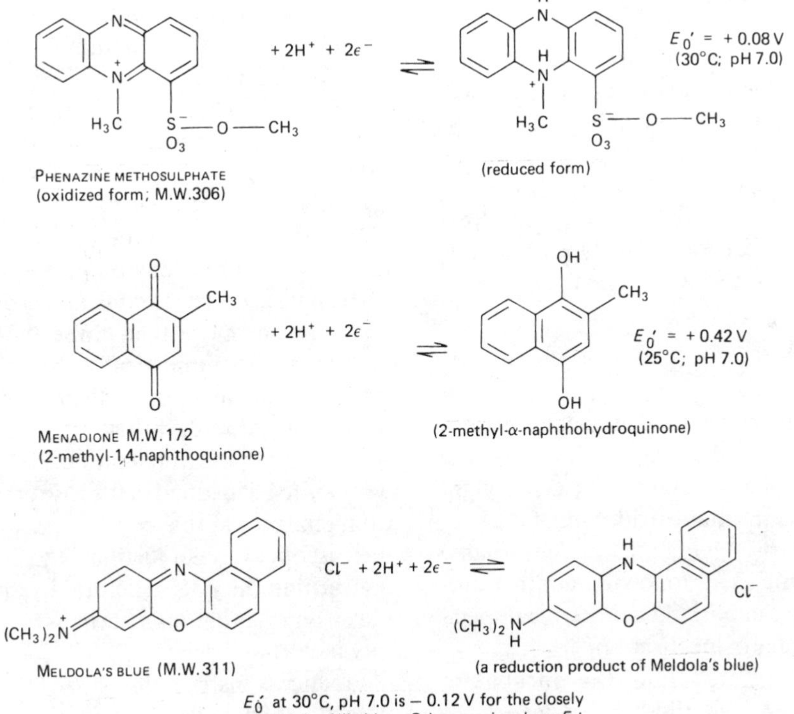

Fig. 16.1. Intermediate electron acceptors.

tetrazolium salt by the reduced form of PMS. Menadione has also been used for the same purpose, though less often. Recently an oxazine dye, Meldola's blue (C.I. 51175; Basic blue 6) has been proposed as an intermediate electron acceptor in dehydrogenase histochemistry (Kugler & Wrobel, 1978). It is used at a concentration of 10^{-4}M. The effects of Meldola's blue are the same as those of PMS, but the dye, unlike PMS, is not rapidly decomposed by light and causes only slight spontaneous reduction of tetrazolium salts.

Several other easily reduced dyes have been tried as intermediate electron acceptors, but with little success. Pearse (1972) recommends that intermediate electron acceptors be avoided except in the methods for a few dehydrogenases that cannot otherwise be demonstrated.

16.4.3.3. Conditions of reaction

Sections, carried on slides or coverslips, are incubated at 37°C for about 20 min. The reaction is terminated by transferring the sections to neutral buffered formalin, which also completes the fixation and stabilizes the tissue for any further manipulations. Counterstains may be applied if desired in colours that contrast with that of the formazan.

When the tetrazolium salt is nitro-BT or TNBT, the preparations may, with advantage, be dehydrated, cleared, and mounted in a resinous medium. The formazans from other tetrazolium salts are extracted by alcohol, so water-miscible mounting media are necessary. The type of mounting medium will, of course, influence the choice of a counterstain.

16.4.3.4. Controls

When a histochemical method for a dehydrogenase is performed, it is necessary to show that the production of the coloured end-product is brought about as a result of enzymatic oxidation of the substrate and that the product is present in the same place as the enzyme. The following control procedures will help to establish the biochemical specificity and the accuracy of localization:

(i) Omit the substrate from the incubation medium. No formazan should be produced.

(ii) Inhibitors are available for some dehydrogenases. They are usually competitive and are used by short pre-incubation followed by addition of the inhibitor to the substrate-containing incubation medium. Fixation of the section in formaldehyde for a few hours inactivates most dehydrogenases but usually spares the diaphorases, but this is a test of low specificity.

(iii) In the case of a coenzyme-linked dehydrogenase, carry out the technique for the appropriate diaphorase. This enzyme should be present at the same sites as the dehydrogenase, and will usually be seen in other places too. If there is deposition of formazan from the dehydrogenase medium at sites where there is no diaphorase, the formazan must have diffused away from its place of production. Lipid-soluble formazans are often falsely localized in cytoplasmic lipid droplets. For this reason, the tetrazolium salts whose formazans are substantive for protein are to be preferred.

(iv) Before accepting a negative result, try the method with PVA or PVP added to the medium, with cyanide or azide added (if not done the first time) and with PMS or Meldola's blue added. A dehydrogenase present at a site not also occupied by an appropriate diaphorase will be detectable only in the presence of an intermediate electron acceptor. Try also with unfixed as well as with briefly fixed material. In the case of flavoprotein dehydrogenases (which have prosthetic groups and do not use coenzymes), replenishment of the section's content of ubiquinone may enable a positive reaction to be obtained.

(v) It must be remembered that in histochemical demonstrations of dehydrogenases, the production of the final product is a consequence of at least three and often of more than three different chemical reactions. It is optimistically assumed that the intermediate reactants, especially the reduced forms of coenzymes, do not diffuse appreciably during the progress of the incubation. This assumption appears to be justified in the case of some mitochondrial enzymes at the level of resolution of the light microscope. It is not justifiable to draw conclusions concerning the fine structural localization of soluble enzymes other than perhaps to identify the cells in which they occur.

Practical instructions for the histochemical detection of dehydrogenases and diaphorases, with notes on some of the individual enzymes, are given in Sections 16.7.1–16.7.3 of this chapter.

16.5. HISTOCHEMISTRY OF OXIDASES

Oxidases catalyse the general reaction

$$2\begin{pmatrix}\text{reduced}\\\text{substrate}\end{pmatrix} + O_2 \underset{\text{(oxidase)}}{\rightleftharpoons} 2H_2O + 2\ (\text{oxidized substrate})$$

(It is assumed for convenience that the oxidation of each molecule of this generalized substrate entails the abstraction from it of two hydrogen atoms.)

The equation can be derived from two half-reactions (see Section 16.1):

$$O_2 + 4H^+ + 4\varepsilon^- \rightleftharpoons 2H_2O \qquad (E_0' = +0.82)$$

$$\begin{pmatrix}\text{oxidized}\\\text{substrate}\end{pmatrix} + 2\varepsilon^- \rightleftharpoons \begin{pmatrix}\text{reduced}\\\text{substrate}\end{pmatrix} \qquad (E_0' < +0.82)$$

If the value of E_0' for the second half-reaction is lower (or only slightly higher) than that for the reduction of a tetrazolium salt, it is possible to use a histochemical method similar to those used for the detection of dehydrogenases. The tetrazolium salt will act as a substitute for oxygen and will be reduced to its formazan. However, the substrates for many oxidases have oxidation–reduction potentials appreciably higher than those of the tetrazolium salts, so it is necessary to use different histochemical techniques. Considerable ingenuity has gone into the devising of methods for the localization of oxidases, and the variety of techniques can be expected to increase in the future. Two representative examples of histochemical methods for oxidases are described in this chapter.

16.5.1. Cytochrome oxidase

(Cytochrome c: O_2 oxidoreductase; E.C. 1.9.3.1.)

The terminal members of the electron-transport chain are cytochromes a and a_3, from which electrons are transferred to oxygen. The electrons are derived from cyt. c^{2+}, the reduced form of cytochrome c. Cytochrome oxidase catalyses the reaction

$$4\ \text{cyt. } c^{2+} + O_2 + 4H^+ \rightarrow 4\ \text{cyt. } c^{3+} + 2H_2O$$

The identity of cytochrome oxidase with cytochromes a and a_3 or with both these substances has not yet been established. It is known, however, that cytochrome oxidase contains iron atoms, tightly bound in haem-like prosthetic groups, and copper atoms. The enzyme is inhibited by cyanide and azide ions and by several other toxic substances, including hydrogen sulphide and carbon monoxide. Cytochrome oxidase occurs in all cells of aerobic organisms and is present in mitochondria.

The earliest histochemical reaction for cytochrome oxidase was the "NADI" (naphthol–diamine) technique, in which the formation of an indoaniline ("indophenol") dye from α-naphthol and *N*-dimethyl-*p*-phenylenediamine is catalysed in the presence of atmospheric oxygen and cytochrome c. The last-named substance is naturally present in the tissue. Two oxidation–reduction reactions are involved:

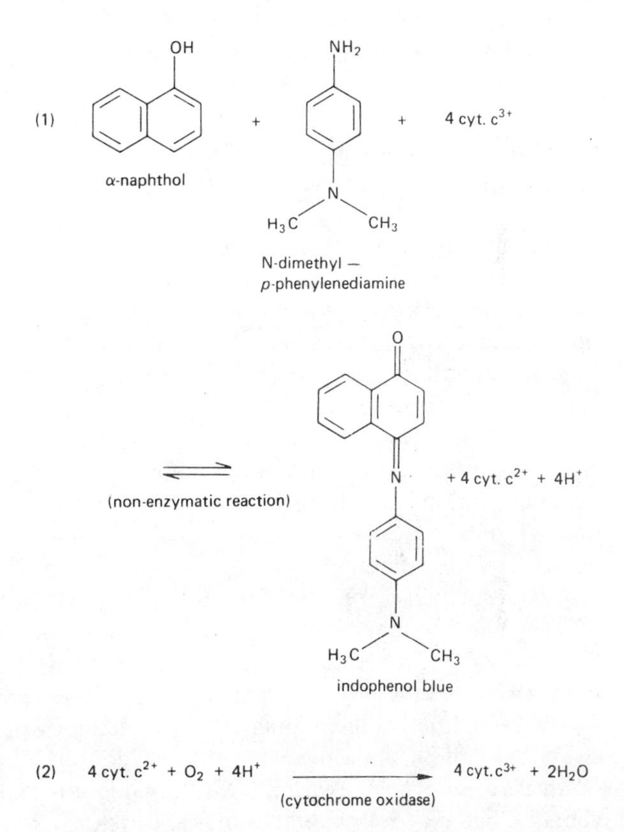

Reaction (2) serves to remove cyt. c^{2+} and H^+ from the products of reaction (1), thereby promoting formation of the indophenol blue and assuring a continued supply of oxidized cytochrome c. It is the diamine, not the naphthol, which is oxidized by cytochrome c. The unstable product of oxidation of the diamine oxidizes and couples with the naphthol to form the dye. The discovery of cytochrome oxidase was intimately linked with the elucidation of the mechanism of the NADI reaction (Elliott & Greig, 1938; Keilin & Hartree, 1938). Inhibition of the NADI reaction by inhibitors of cytochrome oxidase confirms its specificity. The oxidized form of cytochrome c (cyt. c^{3+}) does not, by itself, cause the oxidation and coupling of α-naphthol and *N*-dimethyl-*p*-phenylenediamine to occur as rapidly as is observed in the histochemical NADI reaction. It is probable, therefore, that the substrate, cyt. c^{3+}, is made more active (by being bound to cytochrome oxidase) in the tissue than it would be in solution.

The NADI method in its original form is unsatisfactory as a histochemical technique for various reasons. A positive reaction is seen in some sites, especially leukocyte granules, when cytochrome oxidase is not active, as in fixed tissues. The product of the reaction, indophenol blue, fades quite rapidly on exposure to light. It is highly soluble in lipids and has no substantivity for protein. The production of indophenol blue in myeloid leukocytes (known as the M-NADI reaction, in contrast to the cytochrome oxidase-dependent or G-NADI reaction) is now known to be catalysed by a peroxidase. Improved methods for cytochrome oxidase have been evolved from the original NADI technique (see Burstone, 1962, for a detailed account). They are based on the production of indoaniline, indamine, and indophenol dyes, of uncertain composition, from a variety of naphthols, amines, quinones, and quinolines.

The most satisfactory amine for reduction by cyt. c^{3+} is *N*-phenyl-*p*-phenylenediamine (= *p*-aminodiphenylamine). Several substances have been found that will serve as coupling agents to substitute for the α-naphthol of the original NADI technique (Burstone, 1962). The only one to yield a coloured product resistant to dehydration with organic solvents is 8-hydroxy-1,4-naphthoquinone (Burstone, 1961). The structures of the dyes constituting the end-products of the newer histochemical methods for cytochrome oxidase have not yet been elucidated. The colours can be stabilized by treatment with salts of various metals (Co^{2+}, Ni^{2+}, Fe^{3+}, Cd^{2+}, UO_2^{2+}, Pb^{2+}) which are presumed to form chelates with the dyes. Technical directions are given in Section 16.7.4 of this chapter.

16.5.2. Catechol oxidase

(*o*-diphenol: O_2 oxidoreductase; E.C. 1.10.3.1.)

Several names have been applied to the copper-containing enzymes that catalyse the oxidation of *o*-diphenols by oxygen to yield *o*-quinones. These include the approved trivial name, catechol oxidase, as well as tyrosinase, phenol oxidase, polyphenol oxidase, and DOPA-oxidase. The multiplicity of names is a consequence of the fact that the same enzyme can act upon many phenolic substrates, including monophenols. Related enzymes catalyse the oxidation of *p*-diphenols and aromatic amines. The importance of catechol oxidase in histochemistry derives from the function of the enzyme in the synthesis of melanin. This pigment is produced in melanocytes and some other types of cells by a series of reactions, some enzymatically catalysed and others spontaneous. The initial metabolite in the sequence is the amino acid tyrosine. This is first slowly oxidized to dihydroxyphenylalanine (DOPA):

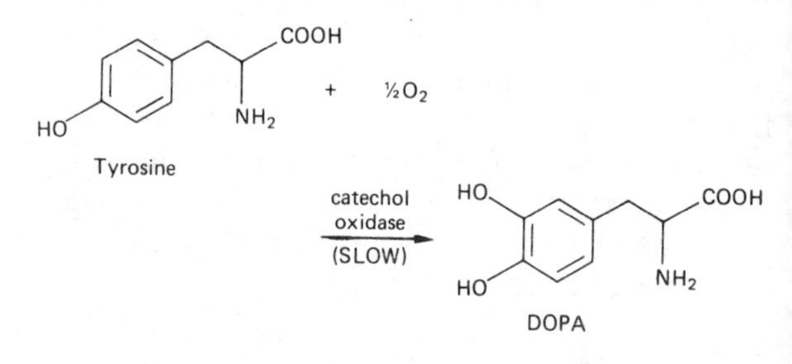

Under the catalytic influence of catechol oxidase, the DOPA is now rapidly oxidized to DOPA quinone.

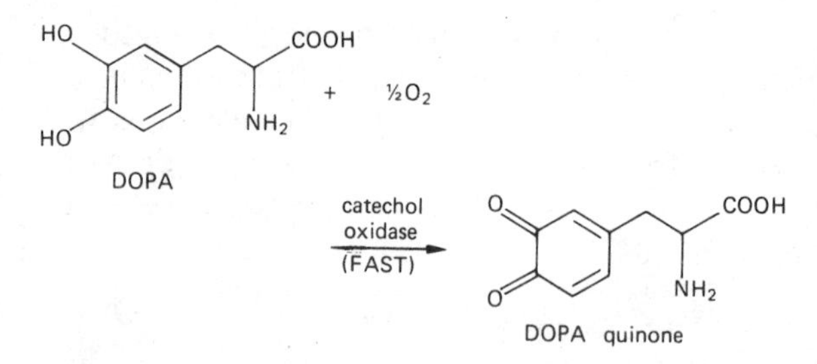

The remaining reactions are believed to occur spontaneously:

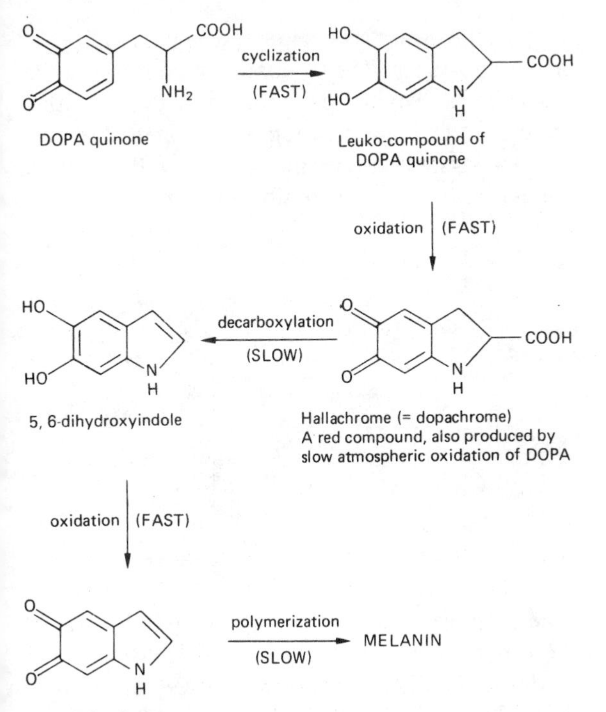

The composition of melanin is not known with certainty, but it is probably a polymer of the form

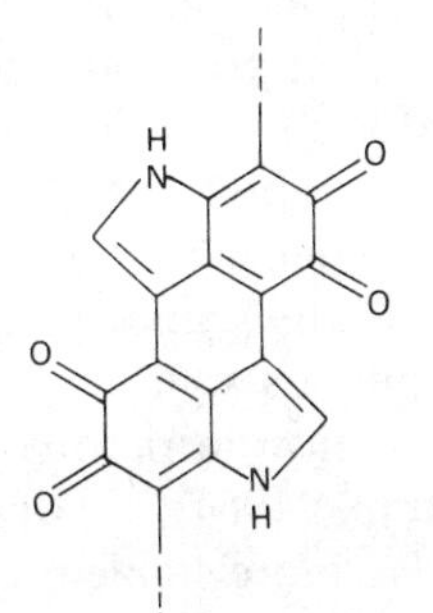

derived from indole-5,6-quinone and bound to the proteinaceous matrix of the granules in which it occurs. It is a stable, black, insoluble substance.

Catechol oxidase is demonstrated histochemically by virtue of its catalysis of the rapid oxidation of DOPA by oxygen, with the ultimate formation of melanin. Several potentially diffusible intermediates are produced in this series of reactions, so it is possible that the final deposits of melanin are not formed in exactly the same sites as those at which the substrate was oxidized. However, the formation of a finely granular pigment in melanocytes, where melanin is normally synthesized, suggests that diffusion does not occur over great distances.

The chemical specificity of the reaction is certainly not complete. Positive staining of erythrocytes and of leukocyte granules is probably due to peroxidase activity, with tissue-derived hydrogen peroxide as the substrate and DOPA as the electron-donor (see Section 16.6). Inhibitors are of little value since both catechol oxidase and the peroxidases are inhibited by cyanide, azide, and sulphide ions, though lower concentrations (10^{-4}–10^{-3}M) are effective with the former enzyme. These inhibitors also block the activity of cytochrome oxidase, but this enzyme is unlikely to be involved in the histochemical oxidation of DOPA when formaldehyde-fixed tissue is used. Lillie & Fullmer (1976) state that sodium dithionite ($Na_2S_2O_4$, 5×10^{-3}M) and cysteine (10^{-3}M) enhance the reaction. They might do this by reducing hallachrome to 5,6-dihydroxyindole, thus speeding up the slowest non-enzymatic reaction in the series leading from DOPA to melanin.

The technique for the histochemical demonstration of catechol oxidase is described in Section 16.7.5. The method is unusual in that the substrate used is a physiological one and that the product is an otherwise naturally occurring pigment.

16.6. HISTOCHEMISTRY OF PEROXIDASES

The peroxidases catalyse the oxidations of various substances, including reduced coenzymes, fatty acids, amino acids, reduced cytochromes, and many other substances by hydrogen peroxide.

16.6.1. Actions and occurrence

The name "peroxidase" (in the singular; donor: H_2O_2 oxidoreductase; E.C. 1.11.1.6) embraces several enzymes of plant and animal origin. They are all iron-containing haemoproteins and they catalyse the reaction

$$\text{donor} + H_2O_2 \rightarrow \begin{matrix}\text{oxidized}\\ \text{donor}\end{matrix} + 2H_2O$$

in which the net effect is the removal of two atoms of hydrogen from each molecule of the donor. Many organic compounds, including amines, phenols, and the leuco-compounds of dyes, can serve as donors. The substrate is hydrogen peroxide which, when it is bound to the enzyme, is made to oxidize other substances much more rapidly than it would if it were acting alone.

In mammals, peroxidase activity is present in the granules of myeloid leukocytes and in some neurons and secretory cells. A positive histochemical reaction is also given by the haemoglobin of erythrocytes, though this is not considered to be due to truly enzymatic catalysis. The animal enzyme is inhibited by cyanide ions at 10^{-2}M, a concentration higher than that which will inhibit cytochrome oxidase. It is also inhibited by pre-treatment of unfixed cryostat sections with methanol (Streefkerk & van der Ploeg, 1974) but not by fixation of tissue in 70% ethanol or 4% formaldehyde (Burstone, 1962). The peroxidase-like property of erythrocytes, which persists in fixed, paraffin-embedded material, is abolished together with that of leukocytes by pre-treatment of sections with 0.024 N HCl in ethanol (Weir *et al.*, 1974).

The importance of histochemical methods for peroxidase has increased in recent years because the enzyme extracted from the root of the horseradish (*Armoracia rusticana*) is extensively used as an intravital tracer protein in studies of vascular permeability and in neuroanatomy for both light and electron microscopy. Horseradish peroxidase (HRP) is also used as a reagent in carbohydrate histochemistry (Chapter 11) and in immunohistochemistry (Chapter 19). Inhibition of the endogenous peroxidase of animal tissues is often necessary before applying methods in which HRP is used as a reagent. The blocking procedures mentioned above cannot, however, be used on tissue containing exogenous HRP, since this enzyme is also inhibited by them.

16.6.2. Histochemical localization

Histochemical methods for peroxidase are based on the catalysed reactions of hydrogen peroxide with substances that yield insoluble coloured products upon oxidation. In one of the oldest techniques, the donor is benzidine, which is oxidized to a blue substance. The incubation medium contains benzidine and hydrogen peroxide. In this and other methods, the concentration of hydrogen peroxide, the substrate, should not exceed 0.03 M, since higher concentrations inhibit the enzyme.

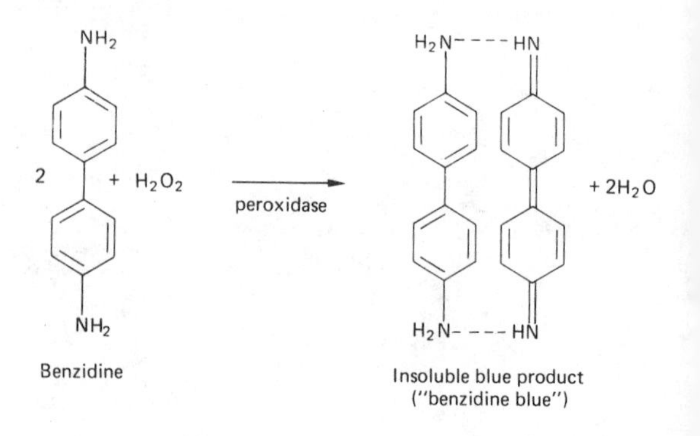

The chemical nature of the product is not certainly known, but it is generally believed to have the quinhydrone-like structure shown above. The simple benzidine technique is not often used because the blue product often forms as unduly large crystals and its colour soon fades to a less conspicuous brown. Under some conditions of reaction (pH >7, temperature >4°C) a brown product is formed in the first instance. It is probably a polymer derived from condensation of benzidine with its unstable quinone–imine. Various methods are available for the stabilization of benzidine blue, the best-known being treatment of the stained preparations with a concentrated aqueous solution of sodium nitroprusside (sodium nitroferricyanide, $Na_2Fe(CN)_5NO.2H_2O$) (Straus, 1964). However, alternative donors which give more stable products are to be preferred to benzidine.

The donor most widely applicable to the histochemical localization of peroxidases is 3,3′-diaminobenzidine tetrahydrochloride (DAB), introduced by Graham & Karnovsky (1966):

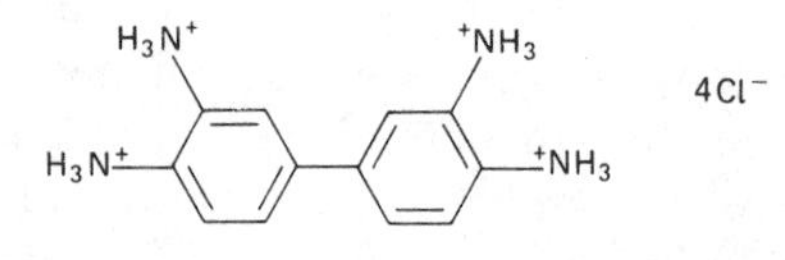

The spontaneous oxidation of this amine by hydrogen peroxide is quite slow, but in the presence of peroxidase an insoluble, amorphous, brown substance is rapidly precipitated. The initial products of oxidation are presumed to be quinone–imines:

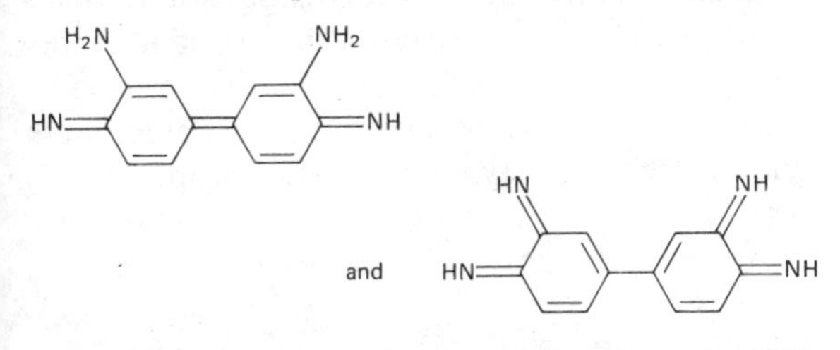

These unstable compounds immediately react with DAB to give polymers, which contain the quinonoid and indamine chromophores. The polymerization, since it involves the elimination of hydrogen atoms attached to aromatic rings, is also an oxidation reaction and may also be brought about by hydrogen peroxide and catalysed by peroxidase. The polymers are thought to contain such structural formations as

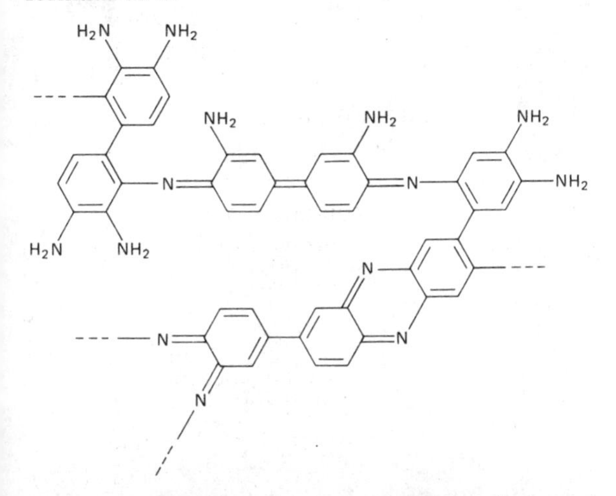

The colour of this product may be further darkened by subsequent treatment with osmium tetroxide, which also makes it electron dense.

Other methods for the localization of peroxidase are based on the oxidation and coupling of amines with phenols, quinones, and other substances. The methods for cytochrome oxidase (Section 16.5.1) will demonstrate peroxidase if hydrogen peroxide is added to the incubation medium (Burstone, 1962). Related methods use combinations of *p*-phenylenediamine with α-naphthol (Hanker *et al.*, 1977) and of DAB with *p*-cresol (Streit & Reubi, 1977). The insoluble products of these methods are probably similar to the "oxidation colours" used in the dyeing of fur and hair (see Chapter 5, p. 50).

The use of HRP as an intravital tracer in neuroanatomical studies has resulted in the development of several techniques for the demonstration of this plant enzyme in animal tissues. The activity of HRP is optimally preserved by vascular perfusion of a cold glutaraldehyde–formaldehyde mixture followed by perfusion of and immersion of the excised tissues in a buffered solution of sucrose (Rosene & Mesulam, 1978). The method using DAB has been widely employed, but much greater sensitivity is attained if tetramethylbenzidine (TMB) is the donor. A blue product, stabilizable by sodium nitroprusside, is formed (Mesulam, 1978). The incubation medium for the TMB method has to be quite strongly acidic (pH 3.3). It disrupts antigen–antibody complexes and cannot therefore be used in immunohistochemical techniques (see Chapter 19). The TMB technique is the most sensitive one yet developed for the demonstration of exogenous HRP in fixed tissues (Mesulam & Rosene, 1979). Preparations made by this method are, however, sometimes marred by the deposition of large blue crystals, both at the sites of enzymatic activity and elsewhere (Reiner & Gamlin, 1980).

16.6.3. Specificity and accuracy of localization

The activities of various enzymes may be expected to cause false-positive reactions in sections stained by histochemical methods for peroxidase. Those most likely to cause confusion are cytochrome oxidase and catalase.

Donors such as DAB can be oxidized by cytochrome c in the presence of cytochrome oxidase and oxygen. However, cytochrome oxidase is inactivated by fixatives such as formaldehyde, which are usually employed in the preparation of tissues for the demonstration of peroxidase. With unfixed tissue intended for the localization of cytochrome oxidase, false-positive results can be due to peroxidase, as in the M-NADI reaction. Enough hydrogen peroxide is generated within the tissue to act as substrate for the enzyme. The endogenous substrate can be destroyed, thus eliminating artifacts of the M-NADI type, by adding purified catalase to the histochemical incubation medium.

Catalase (H_2O_2: H_2O_2 oxidoreductase; E.C. 1.11.1.6) catalyses the reaction in which hydrogen peroxide functions as both an oxidizing and a reducing agent:

$$2H_2O_2 \rightarrow 2H_2O + O_2$$

The enzyme occurs in nearly all cells, in organelles known as "microbodies" or "peroxisomes". Its physiological function is probably to prevent the accumulation of hydrogen peroxide, a potentially toxic metabolite. The decomposition of hydrogen peroxide occurs exceedingly rapidly in the presence of catalase. This enzyme can, however, also function as a peroxidase and catalyse the oxidation of chromogenic donors by hydrogen peroxide. Catalase is specifically inhibited by 3-amino-1,2,4-triazole, but attempts to distinguish between this enzyme and peroxidase are rarely made.

Peroxidase and catalase may also be distinguished by varying the concentration of their substrate. Silveira & Hadler (1978) have shown that catalase cannot be detected (by benzidine) when the concentration of hydrogen peroxide in the incubation medium is less than about 3×10^{-3} M, but remains active when $[H_2O_2]$ is as high as 4.0 M. Peroxidase, on the other hand, is fully active in the presence of 1.5×10^{-3} M H_2O_2, but is inhibited by concentrations greater than about 0.05 M. Silveira & Hadler found, however, that 4.0 M H_2O_2 was necessary for complete inhibition of all peroxidases. The peroxidase-like activity of haemoglobin resembles catalase in that it is not detectable when $[H_2O_2]$ is very low.

16.7. INDIVIDUAL METHODS

Instructions follow for the histochemical localization of some typical oxidoreductases.

16.7.1. Succinate dehydrogenase

(Succinate: (acceptor) oxidoreductase; E.C. 1.3.99.1.)

Fresh, unfixed tissue should be rapidly frozen and then sectioned at 4–10 μm in a cryostat. The sections, carried on slides or coverslips, may be fixed for 10 min at 0–4°C in neutral, buffered formaldehyde, or in acetone. The formaldehyde fixative is washed off by rinsing in three changes of 0.6 M phosphate buffer, pH 7.0. Acetone is allowed to evaporate. See also *Note 1* below.

Since succinate dehydrogenase is a flavoprotein whose electron-carrying prosthetic group, FAD, is part of the enzyme molecule, an exogenous coenzyme does not have to be added to the substrate mixture.

Solutions required

A. Incubation medium

0.06 M phosphate buffer, pH 7.0:	50 ml
Nitro blue tetrazolium (nitro-BT):	20 mg
Disodium succinate (hexahydrate):	0.68 g

Prepare just before using. Warming and stirring are sometimes needed in order to dissolve the nitro-BT. The final solution should be filtered if it contains any undissolved material.

B. Neutral, buffered 4% formaldehyde

Procedure

1. Incubate sections, prepared as described above, in the medium (solution A) for 10–30 min at room temperature or at 37°C. Cellular regions of the sections should become blue or purple to the unaided eye. Check under a microscope for intracellular deposition of formazan.
2. Transfer to fixative (solution B) for 10 min. This will stop the reaction and provide morphological fixation of the tissue.
3. Wash in water, apply a counterstain if desired (e.g. a pink nuclear stain), dehydrate through graded alcohols, clear in xylene, and cover, using a resinous mounting medium.

Result

Sites of enzymatic activity (mitochondria) dark blue to purple. Any red formazans (see p. 225) that form are extracted during dehydration.

Notes

1. Fixation in acetone extracts ubiquinone from the sections. This electron-acceptor may be restored by depositing a thin layer of a 0.1% solution of ubiquinone$_{10}$ (coenzyme Q_{10}) in a

mixture of equal volumes of ether and acetone on the coverslip or slide (Wattenberg & Leong, 1960) or onto the fixed section (Contestabile & Andersen, 1978) and allowing the solvent to evaporate away. Alternatively, an intermediate electron carrier may be added to the incubation medium (see *Note 3*).

2. *Controls.* (a) Omit the substrate. A positive reaction in the absence of succinate ions cannot be due to succinate dehydrogenase. (b) Pre-incubate the sections for 5 min in 0.05 M sodium malonate (0.37 g of the anhydrous disodium salt in 50 ml of buffer) and add sodium malonate at the same concentration to the complete incubation medium. Malonate is a competitive inhibitor of succinate dehydrogenase.
3. The incubation time can be shortened by adding 2.0 mg of phenazine methosulphate (PMS) to 50 ml of the incubation medium. The incubation must be carried out in darkness if PMS is used. Meldola's blue (see p. 230), 1.5 mg per 50 ml of medium, may be preferable to PMS on account of its greater stability. Do not incubate for more than 10 min if any intermediate electron-acceptor is used, or there may be non-specific deposition of formazan. The addition of electron-transport inhibitors (CN^- or N_3^-) to media for succinate dehydrogenase is not necessary when a rapidly reducible tetrazolium salt such as nitro-BT is used.

16.7.2. Coenzyme-linked dehydrogenases

The following procedure, applicable to the histochemical demonstration of several dehydrogenases, is based on the techniques described by Pearse (1972) and Chayen *et al.* (1973), with slight modifications. For individual enzymes the incubation media are made up by adding the tetrazolium salt, the substrate, and the appropriate coenzyme to a previously prepared stock solution containing the stable ingredients. The substrates and coenzymes required by the various enzymes are set out in Table 16.3.

Stock solution

0.2 M TRIS-HCl buffer, pH 7.2:	65 ml
Magnesium chloride ($MgCl_2.6H_2O$):	200 mg
Sodium azide (NaN_3):	15 mg
Water:	85 ml
Add **either:** Polyvinyl alcohol (M.W. 30,000):	40 g
or: Polyvinylpyrollidone (M.W. about 20,000, but not critical):	15 g

(Let the PVA or PVP float on the surface with a magnetic stirrer bar revolving slowly in the bottom of the beaker or flask in which the solution is being prepared. If the powder sinks it will form lumps which take longer to dissolve.)

TABLE 16.3. *Some coenzyme-linked dehydrogenases*

Trivial name	E.C. number	Systematic name (indicating substrate and coenzyme)	Product(s) of oxidation of substrate
Alcohol dehydrogenase	1.1.1.1	Alcohol: NAD oxidoreductase	An aldehyde or ketone
Glycerolphosphate dehydrogenase	1.1.1.8	L-glycerol-3-phosphate: NAD oxidoreductase	Dihydroxyacetone phosphate
UDPG dehydrogenase	1.1.1.22	UDP-glucose: NAD oxidoreductase	UDP-glucuronate
Lactate dehydrogenase	1.1.1.27	L-lactate: NAD oxidoreductase	Pyruvate
Glucose-6-phosphate dehydrogenase	1.1.1.49	D-glucose-6-phosphate: NADP oxidoreductase	D-glucono-8-lactone 6-phosphate
Glutamate dehydrogenase	1.4.1.2	L-glutamate: NAD oxidoreductase (deaminating)	2-oxoglutarate + NH_3
Glutamate dehydrogenase	1.4.1.3	L-glutamate: NADP oxidoreductase (deaminating)	2-oxoglutarate + NH_3

Adjust to pH 7.0–7.2 if necessary by adding 1.0 N sodium hydroxide (4% NaOH).

Add water to bring the volume up to 200 ml.

This solution is stable for several weeks at 4°C. The sodium azide, included as an electron-transport inhibitor (see pp. 227, 229), also serves to check bacterial and fungal growth. Potassium or sodium cyanide (26 or 20 mg) may be substituted for sodium azide, but the cyanides are less stable, and solutions containing them should be used on the day they are made. All three of these substances are poisonous and must be handled carefully, but the small amounts contained in 200 ml of this solution can safely be discarded by flushing down the sink with plenty of water.

Incubation media

Some of the ingredients, especially the coenzymes, are expensive, so it is the usual practice to prepare only small amounts of incubation media. For each section of tissue, 0.1–0.2 ml of medium will be needed. The following instructions are for the preparation of 5 ml volumes of the media.

Since the numbers of atoms of the cation and of molecules of water of crystallization may vary with some of the substrates, always check the M.W. shown by the supplier and ensure that the correct number of moles of substrate is taken.

Alcohol dehydrogenase

Stock solution:	5.0 ml
Absolute ethanol: (5×10^{-4} mole)	0.03 ml
Nitro-blue tetrazolium	1.0 mg
Nicotinamide adenine dinucleotide:	1.0 mg

(The small volume of ethanol is more easily measured out by diluting 10 ml of ethanol to 33 ml with water and adding 0.1 ml of the diluted alcohol to the stock solution.)

Glycerolphosphate dehydrogenase

Stock solution:	5 ml
Glycerol-3-phosphate, disodium salt: (5×10^{-4} mole)	158 mg
Nitro-blue tetrazolium:	1.0 mg
Nicotinamide adenine dinucleotide:	1.0 mg

Check that the pH is 7.0–7.2.
Adjust with drops of 1.0 N HCl if necessary.

UDPG dehydrogenase

Stock solution:	5 ml
Uridine-5-diphosphate glucose trisodium salt: (approx. 1.5×10^{-6} mole)	1.0 mg
Nitro-blue tetrazolium:	1.0 mg
Nicotinamide adenine dinucleotide:	1.0 mg

Lactate dehydrogenase

Stock solution:	5 ml
Sodium DL-lactate ($NaC_3H_5O_3$): (5×10^{-4} mole)	56 mg
Nitro-blue tetrazolium:	1.0 mg
Nicotinamide adenine dinucleotide:	1.0 mg

Glucose-6-phosphate dehydrogenase

Stock solution:	5 ml
Glucose-6-phosphate, disodium salt. $3H_2O$: (5×10^{-4} mole)	179 mg
Nitro-blue tetrazolium:	1.0 mg
Nicotinamide adenine dinucleotide phosphate, sodium salt:	1.0 mg

Check that the pH is 7.0–7.2.
Adjust with drops of 1.0 N HCl if necessary.

Glutamate dehydrogenases

Stock solution:	5 ml
Sodium-L-glutamate: (5×10^{-4} mole)	85 mg
Nitro-blue tetrazolium:	7.0 mg
Either nicotinamide adenine dinucleotide **or** nicotinamide adenine dinucleotide phosphate, sodium salt:	1.0 mg

Check that the pH is 7.0–7.2.
Adjust with drops of 1.0 N HCl if necessary.
(Note that a higher than usual concentration of the tetrazolium salt is needed for the demonstration of these enzymes.)

Procedure

Fresh tissue is rapidly frozen and sectioned on a cryostat (4–10 μm), the sections being collected onto coverslips or slides. The sections may be fixed for 5–10 min in pre-chilled (4°C) acetone or neutral, phosphate-buffered formaldehyde. Unfixed sections may also be used: the reactions will occur more rapidly but the integrity of the tissue will suffer.

Rinses are carried out in small coplin jars or beakers. The incubation takes place in a closed petri dish with a piece of moist filter paper in the bottom to ensure a humid atmosphere and prevent evaporation of the medium. Read the *Notes* below before carrying out this method.

1. (Fixed sections.) Allow acetone to evaporate or rinse the formaldehyde-fixed section in water (10–15 s, with agitation). Drain.
2. Place slides or coverslips, section uppermost, on the damp filter paper in the bottom of the petri dish. Cover each section with a generous drop of freshly prepared incubation medium. Put the lid on the dish and carefully place it in an oven at 37°C. (Often room temperature is satisfactory.)
3. Inspect the sections at 10-min intervals for the formation of blue, intracellular deposits. The time of incubation should not exceed 1 h.
4. When staining is judged to be optimum, pick up the slides or coverslips with forceps, drain them, and place them in neutral buffered formaldehyde for 10 min at room temperature. This will arrest the histochemical reaction and provide further morphological fixation.
5. Rinse in water.
6. Apply a suitable counterstain if desired (e.g. a pink or green nuclear stain: see Chapter 6).
7. Rinse in water, dehydrate, clear, and mount in a resinous medium.

Result

Sites of enzymatic activity—purple to dark blue.

Notes

1. Lactate dehydrogenase is noteworthy in that it is more resistant to fixation in formaldehyde than the other enzymes. Small blocks may be fixed at 4°C in 2.5–4.0% neutral, buffered formaldehyde and the histochemical method performed on ordinary frozen sections.

2. Always remember that the formazan deposit is formed as a result of the activity of a diaphorase, not of the dehydrogenase whose substrate was included in the incubation medium. See also *Note 3*.

3. In order to by-pass the diaphorase, an intermediate electron-acceptor may be added to the incubation medium (see p. 229). Immediately before applying the medium to the sections, dissolve in it 0.15 mg of phenazine methosulphate (PMS) per 5.0 ml and incubate in darkness. The time of incubation should not exceed 10 min if PMS is used, or spontaneous reduction of the tetrazolium salt in solution may cause non-specific precipitation of formazan on the sections. Meldola's blue (0.15 mg per 5.0 ml of medium) may be preferable to PMS, for reasons given on p. 230.

4. *Controls*. It is necessary to control for non-enzymatic deposition of formazan and for production of formazan as a result of the activity of enzymes other than the dehydrogenase in which one is interested. The following control procedures are recommended:

(a) Incubate in a solution containing all the ingredients of the incubation medium except the substrate. No staining should occur. If colour does develop, it may be due to "nothing dehydrogenase" (see p. 228 and check that the medium is not too alkaline).

(b) Omit the coenzyme from the incubation medium. If staining is seen in the absence of the coenzyme, the oxidation of the substrate is being catalysed by a dehydrogenase with a prosthetic group. For example, in addition to the glycerolphosphate dehydrogenase shown in Table 16.3 there is also a mitochondrial flavoprotein (E.C. 1.1.2.1) which catalyses the reaction:

L-glycerol-3-phosphate + (acceptor) $\rightleftharpoons$ dihydroxyacetone phosphate + (reduced acceptor)

No coenzyme is involved, but the reduction of the acceptor (once believed to be cytochrome c, but now thought to be ubiquinone) will trigger the transport of electrons to other intermediates of the respiratory chain and to a tetrazolium salt. The mitochondrial flavoprotein glycerolphosphate dehydrogenase can be demonstrated histochemically by using a medium without a coenzyme, though the inclusion of an intermediate electron-acceptor is desirable.

(c) Inhibitors of high specificity are not available for most dehydrogenases. Some of the enzymes (e.g. alcohol, glycerolphosphate, and UDPG dehydrogenases) have sulphydryl groups at their active sites and are inhibited by SH-blocking agents such as PCMB and *N*-ethylmaleimide (10^{-4} to 10^{-3} M). Metal ions (Mg^{2+}, Mn^{2+}, Zn^{2+}) are cofactors for many dehydrogenases, so that chelating agents such as EDTA and 8-hydroxyquinoline (10^{-3} to 10^{-2} M) are inhibitory. A few of the enzymes (e.g. soluble glycerolphosphate dehydrogenase) display increased activity in the presence of chelators.

(d) If an unexpected negative result is obtained, try again with up to ten times the concentration of the coenzyme. Enzymes that catalyse the hydrolysis of NAD^+ and $NADP^+$ are present in some tissues. These coenzymes can also deteriorate on storage in the laboratory.

5. The histochemical detection of an enzyme requires the penetration of cellular and mitochondrial membranes by all the reagents. Usually freezing and thawing will damage the membranes sufficiently to make them permeable. The glutamate dehydrogenases show increased activity in mitochondria that have been traumatized by rough handling of the tissue (Chayen *et al.*, 1973).

6. Although the intensity of the colour of the final reaction product provides an approximate indication of the activity of the enzyme, the concentration of the latter does not vary in direct proportion with the amount of formazan deposited. (See Exercise 7 at the end of this chapter.)

16.7.3. Diaphorases (tetrazolium reductases)

The presence of these systems of enzymes in tissue is essential for the production of coloured end-products in methods for coenzyme-linked dehydrogenases, except when intermediate electron-acceptors such as PMS are used. In any examination of dehydrogenases whose activities involve NAD^+ or $NADP^+$, the distribution of the appropriate diaphorases should also be ascertained. NADPH-diaphorase is located predominantly in mitochondria, while NADPH-diaphorase is mainly found elsewhere in the cytoplasm.

Incubating medium

Stock solution (see p. 237):	2 ml
Nitro-blue tetrazolium:	0.5 mg
Either nicotinamide adenine dinucleotide, reduced form (disodium salt):	4 mg
Or nicotinamide adenine dinucleotide phosphate, reduced form (tetrasodium salt):	4 mg

For NADH-diaphorase, the "stock solution" may be replaced by TRIS buffer, pH 7.0–7.2. The sodium azide included in the stock solution is also unnecessary, though it does no harm. The incubation medium should be made up immediately before using.

Procedure

1. Cryostat sections are prepared as for histochemical methods for dehydrogenases.
2. Incubate in the above medium for 10–30 min at room temperature or at 37°, as described in stages 2 and 3 of the general method for coenzyme-linked dehydrogenases (p. 239).
3. Drain off incubating medium and transfer the slides or coverslips bearing the sections to neutral buffered formaldehyde (see pp. 22, 230) for 10–15 min.
4. Rinse in water, dehydrate through graded alcohols, clear in xylene, and mount in a resinous medium.

Result

Purple to blue–black deposits indicate sites of formazan deposition due to diaphorase activity.

16.7.4. Cytochrome oxidase

This is one of the methods described by Burstone (1961). The chromogenic reagents are *p*-aminodiphenylamine and 8-hydroxy-1,4-naphthoquinone.

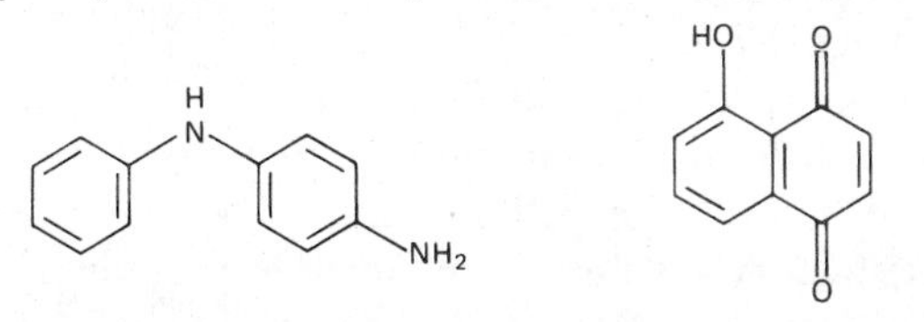

Cytochrome c may be added to the incubation medium (see below) or there may be enough of this substance in the section.

Solutions required

A. Incubating medium (prepare before using)

p-aminodiphenylamine (= *N*-phenyl-*p*-phenylenediamine): 10 to 15 mg
(Must be the amine base, not the hydrochloride)
8-hydroxy-1,4-naphthoquinone: 10 to 15 mg
Absolute ethanol: 0.5 ml
Dissolve the reagents in the ethanol, then add:
Water: 35 ml
TRIS buffer, pH 7.4: 15 ml

Shake the solution, which is cloudy, and filter it.

(**Optional:** Add cytochrome c, 10–20 mg after filtering the solution. See *Note* 1 below.)

B. Chelating and fixing solution

Cobaltous acetate, $(CH_3COO)_2Co.4H_2O$:	5 g	Make on day of use
Formalin (40% HCHO; **not** neutralized):	5 ml	
Water:	to 50 ml	

Procedure

1. Cryostat sections of unfixed tissue are mounted onto slides or coverslips.
2. Incubate the mounted sections in solution A for 15 min to 3 h at room temperature (30 min are often sufficient), until the colour is conspicuous to the unaided eye.
3. Transfer to the chelating and fixing solution B for 1 h.
4. Wash in running tap water for 5–10 min, dehydrate, clear, and mount in a resinous medium. (Water-miscible mounting media may also be used.)

Result

Sites of cytochrome oxidase activity (i.e. mitochondria) are brown.

Notes

1. Exogenous cytochrome c is required only for the demonstration of the enzyme in cells that do not contain enough of their own cytochrome c. Strong staining is obtained in muscle fibres, neurons, and some secretory cells without adding any cytochrome c to the medium.
2. *Controls.* (a) Cytochrome oxidase is inhibited by cyanide or azide ions, which may be added to the incubation medium at a concentration of 10^{-3}M. A positive result in the presence of the inhibitor cannot be due to cytochrome oxidase, but other oxidases are also inhibited. (b) Artifacts due to peroxidase (see p. 235) are prevented by adding catalase (2 μg per ml) to the incubation medium. This will decompose any hydrogen peroxide generated in the tissue.

16.7.5. Catechol oxidase

This enzyme is also known as phenolase, tyrosinase, phenol oxidase, and DOPA oxidase. The method works with cryostat sections of unfixed tissue, with frozen sections of tissue fixed for 6–24 h at 4°C in neutral, buffered 4% formaldehyde and with paraffin sections of freeze-dried material. Small specimens can be stained whole by a slight modification of the technique (see Pearse, 1972, and other texts) and then fixed and embedded in wax.

Incubation medium

0.06 M phosphate buffer, pH 7.4: 100 ml
DL-β-dihydroxyphenylalanine (DOPA): 100 mg

Pre-warm the buffer to 37°C. Add the DOPA and place on a magnetic stirrer or shake vigorously for 10–15 min. Any of the solute that has not dissolved after this time should be removed by filtration.

Procedure

1. Incubate sections in the medium for 60 min at 37°C. After the first 45 min, prepare a fresh batch of incubation medium.
2. Replace the incubation medium with the new batch and incubate for another 60 min at 37°C.
3. Wash in three changes of water.
4. Apply a counterstain (e.g. neutral red or safranine) if desired, and rinse in water.
5. Dehydrate, clear, and mount in a resinous medium.

Result

A dark brown deposit of melanin forms at sites of enzymatic activity (see *Notes* below).

Notes

1. It is important to incubate control sections in buffer without added DOPA in order to detect pigments already present in the tissue. See also Exercise 13 at the end of this chapter.
2. Catechol oxidase is inhibited by cyanide ions, though specificity of the inhibition is low. Sections may be pre-incubated in 10^{-3}M KCN (M.W. 65) or NaCN (M.W. 49) for 5 min. The same concentration should also be included in the incubating medium. (**Caution.** Sodium and potassium cyanides are poisonous. They must be handled carefully and not allowed to come into contact with acids. Their solutions must not be pipetted by mouth. Quantities smaller than 100 mg may safely be discarded by washing down the sink with copious running tap water.)

16.7.6. Peroxidase

This enzyme survives brief fixation in formaldehyde or glutaraldehyde. (See also Section 16.5 and *Note 3* below.) Frozen (or cryostat) sections should be used. The methods described here are suitable for the demonstration of endogenous peroxidase of animal tissues and for horseradish peroxidase used as a reagent in immunohistochemical and other techniques. The first is derived from the technique of Graham & Karnovsky (1966). The second is that of Streit & Reubi (1977).

16.7.6.1. DAB–HYDROGEN PEROXIDE METHOD

Preparation of incubation medium

This solution is prepared just before use. Dissolve 25 mg of 3,3′-diaminobenzidine tetrahydrochloride (DAB) in 50 ml of phosphate or TRIS buffer at pH 7.6. Different batches of DAB vary in their colour, solubility, and suitability for use in this method. The incubation medium may need to be filtered before adding the hydrogen peroxide. If unsatisfactory results are obtained, try obtaining the DAB from a different supplier. (**Caution:** DAB is probably carcinogenic. See *Note 4* below.)

Dilute a stock solution of hydrogen peroxide (e.g. "100 volumes" = 30% w/v) to give a 1% w/v solution of H_2O_2 in water. (**Caution:** avoid contact of the strong H_2O_2 solution with skin or clothing.)

Add 0.5 ml of the 1% H_2O_2 to the 50 ml of DAB solution to give a working incubation medium.

Procedure

1. Incubate sections in the working incubation medium for 15 min at room temperature.
2. Wash in three changes of water, each 1 min (see also *Note 1*).
3. (Optional.) Apply a counterstain if desired.
4. Dehydrate, clear, and mount in a resinous medium.

Result

Sites of peroxidase activity—brown. The colour sometimes fades after a few months.

Notes

1. The colour of the product may be intensified by a brief treatment with dilute osmium tetroxide, as described in the technique for choline esterases (p. 217).
2. Control sections should be incubated with DAB in the absence of H_2O_2. In unfixed sections a positive reaction in the absence of H_2O_2 can be due to cytochrome oxidase.
3. For detection of exogenous horseradish peroxidase, perfuse the animal for 30 min with 0.1 M phosphate buffer, pH 7.4 containing 1% paraformaldehyde and 1.25% glutaraldehyde. Wash out the excess fixative by perfusing cold (4°C) 10% sucrose in 0.1 M phosphate buffer, pH 7.4 for a further 30 min. Store the tissue before sectioning in the same buffered sucrose solution for up to 7 days at 4°C.

4. Before throwing away the incubation medium, add 10 ml of 5% sodium hypochlorite (household bleach is suitable) and leave for 30–60 min. This destroys the DAB.
5. Catalase may be inhibited by adding 3-amino-1,2,4-triazole (10^{-2}M) to the incubation medium. Reduction of the concentration of H_2O_2 in the medium to 0.005% (w/v) should also prevent the formation of coloured products due to activity of catalase.

16.7.6.2. DAB–*p*-CRESOL–HYDROGEN PEROXIDE METHOD

This technique was devised for the detection of exogenous HRP in the nervous systems of experimental animals. It is more sensitive than the preceding method. The method can also be applied to sections subjected to the peroxidase–antiperoxidase (PAP) immunohistochemical technique.

Sections are prepared as for the preceding method.

Incubation medium

A. Buffer solution
(pH 5.0–6.0 recommended by original authors)

Ammonium acetate (CH_3COONH_4):	3.85 g	Stable indefinitely at 4°C unless infected
Citric acid (anhydrous):	1.10 g	
Water:	to 375 ml	

(These quantities give pH 5.0–5.1)
This mixture does not have very high buffering capacity beyond the range pH 4.0–5.5, but is adequate for this method. Other buffers could probably be used but have not been tried.

B. Working incubation medium (mix just before using)

Buffer (solution A):	50 ml
p-cresol (melted by heating to about 60°C):	0.07 ml

Shake until dissolved, then add:

DAB:	10 mg
0.02% hydrogen peroxide (take 1.0 ml of 30% ("100 volumes") H_2O_2 and make up to 1500 ml with water):	50 ml

Filter if any undissolved material persists.

Procedure
1. Incubate the sections for 10–60 min at room temperature.
2. Wash in three changes of water, each approximately 1 min.
3. (Optional.) Apply a counterstain if desired.
4. Dehydrate, clear, and mount in a resinous medium.

Result
Sites of enzymatic activity brown. The notes appended to the account of the simpler DAB–hydrogen peroxide method are also relevant to this technique.

16.8. EXERCISES

Theoretical

1. In the following chemical reactions, what is oxidized and what is reduced?

(a) The reaction at the cathode in the electrolysis of molten sodium hydroxide:

$$Na^+ + \varepsilon^- \rightarrow Na$$

(b) The copper-plating of an iron nail immersed in a solution of cupric sulphate:

$$Cu^{2+} + Fe \rightarrow Fe^{2+} + Cu$$

(c) The conversion of an unsaturated lipid to a saturated one:

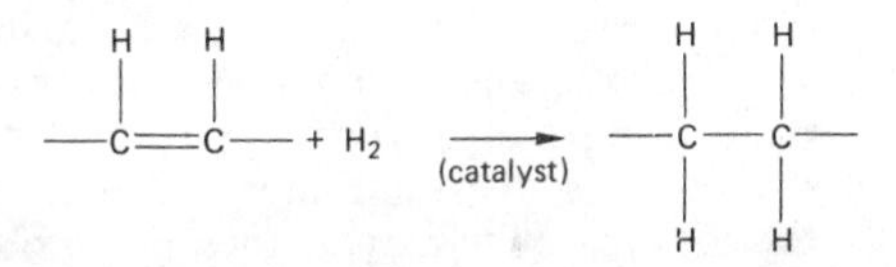

2. Examine the formula for the formazan of nitro blue tetrazolium. Why is this compound coloured? Is it a dye?

3. The physiological action of **catalase** is to accelerate the reaction

$$2H_2O_2 \xrightarrow[\text{(catalase)}]{} 2H_2O + O_2$$

What are the oxidizing and reducing agents in this reaction?

4. If, in the histochemical method for succinate dehydrogenase, one used a tetrazolium salt with an oxidation–reduction potential of +1.0 V, the specificity and accuracy of localization might be lower than when nitro blue tetrazolium was used. Why?

5. Compare the rationales of histochemical methods for (a) NAD^+-linked dehydrogenases; (b) flavoprotein dehydrogenases. Justify the inclusion of all constituents of the incubation media and indicate how the specificity for the substrate can be determined in each case.

6. In a histochemical test for enzymatic oxidation of a substrate XH_2 to X by an enzyme requiring NAD^+, the following results were obtained in a tissue containing two cell types, A and B.

(a) With substrate, NAD^+ and nitro-BT: reaction in mitochondria of all cells of both types.
(b) With NADH and nitro-BT: strong reaction in mitochondria of both cell types.
(c) With substrate and nitro-BT but no coenzyme: moderate reaction in mitochondria of type A cells; nothing in type B.
(d) With nitro-BT but no substrate or coenzyme: no reaction in mitochondria of type A or B cells.

What are the probable localizations of (i) XH_2: NAD^+ oxidoreductase, (ii) NADH-diaphorase? What other enzyme has been detected, and in which cell type?

7. In histochemical methods employing tetrazolium salts, the quantity of a dehydrogenase in a cell is not directly proportional to the intensity of the colour of the deposited formazan, even when the conditions of the reaction are carefully controlled. Discuss this statement.

8. It is possible to use 3,3′-diaminobenzidine (DAB) in histochemical methods for cytochrome oxidase. Devise a suitable technique and explain how it works. What control procedures are necessary to establish the chemical specificity of your technique?

Practical

9. Kill a rat by overdosage with ether vapour or by decapitation under ether anaesthesia. Remove the following organs and process them as indicated. Specimens are frozen onto cryostat chucks either by placing the chuck bearing the specimen into liquid nitrogen (or an acetone–solid carbon dioxide slurry) or by spraying with a proprietary aerosol freezing compound. The chucks with the attached frozen tissues are then put into small, tightly closed plastic bags and stored in the cabinet of the cryostat until they are to be sectioned. Prepare:

(a) Some air-dried blood-films.
(b) A piece of kidney (freeze onto a cryostat chuck). Fix another piece of kidney in neutral, buffered formaldehyde at 4°C, overnight.
(c) A piece of cardiac muscle from the ventricles of the heart (freeze onto a cryostat chuck).
(d) A tibialis anterior muscle. Pinch the lower (narrower) end of the specimen firmly with blunt forceps, wait 10 min, and then freeze onto a cryostat chuck, oriented for cutting of longitudinal sections.
(e) A submandibular salivary gland (freeze onto a cryostat chuck).
(f) A piece of liver (freeze onto a cryostat chuck).
(g) The thyroid gland (fix in neutral, buffered formaldehyde at 4°C, overnight).

Cut cryostat sections of the unfixed tissues at 5 μm; frozen sections of fixed material at 20–30 μm. The sections and films will be used in Exercises 10, 11, 12, and 14.

10. Carry out the method for **succinate dehydrogenase** on cryostat sections of kidney and muscle and on air-dried blood-films. Use sodium malonate as an inhibitor to control for specificity.

11. Carry out methods for a selection of coenzyme-linked dehydrogenases on cryostat sections of kidney, liver, cardiac muscle, and salivary gland. The partly damaged tibialis anterior muscle is suitable for demonstration of the effect of injury on the apparent activity of glutamate dehydrogenases.

12. Demonstrate **cytochrome oxidase** in unfixed cryostat sections of cardiac muscle, salivary gland, and kidney, and in blood-films. Control for non-enzymatic production of colour and for interference by peroxidase. Identify populations of cells that are rich and poor in their intrinsic contents of cytochrome c.

13. Carry out the method for catechol oxidase on formaldehyde-fixed frozen sections of skin with appropriate controls. The skin must not be from an albino animal. Account for the different distributions of the enzyme and of the melanin already present in the skin.

14. Demonstrate the distribution of **peroxidase** in formaldehyde-fixed frozen sections of kidney, thyroid, and skin and in blood-films. Control for non-specific oxidation of the DAB.

17

Methods for Amines

17.1. Nature and occurrence of biogenic amines 245
17.2. Histochemical methods 246
17.2.1. Amines in secretory granules of argentaffin, chromaffin, and mast cells 246
17.2.1.1. Serotonin 246
17.2.1.2. The chromaffin reaction 247
17.2.1.3. Catecholamines in glutaraldehyde-fixed tissue 247
17.2.1.4. Histamine 247
17.2.2. Sensitive methods for amines in neurons 248
17.2.2.1. Formaldehyde-induced fluorescence 248
17.2.2.2. Other fluorescence methods 249
17.3. Individual methods 250
17.3.1. Azo-coupling method for argentaffin cell granules 250
17.3.2. The chromaffin reaction 250
17.3.3. Glutaraldehyde–osmium method for noradrenaline 251
17.3.4. Method for histamine in mast cells 251
17.3.5. Formaldehyde-induced fluorescence 252
17.3.6. Glyoxylic acid method 253
17.3.7. A formaldehyde–glutaraldehyde fluorescence method 253
17.4. Exercises 254

There is a shortage of histochemical techniques for the demonstration of soluble organic compounds of low (<1000) molecular weight. This is due partly to the fact that such substances diffuse rapidly and partly to a lack of suitable chemical reactions. For one group of small molecules, however, there are several satisfactory methods; this is the series of substances known as biogenic amines.

17.1. NATURE AND OCCURRENCE OF BIOGENIC AMINES

Of the amines occurring in mammalian tissues, it is possible to localize dopamine, noradrenaline, adrenaline, serotonin, and histamine in sections of tissue. The structures of these substances are shown below.

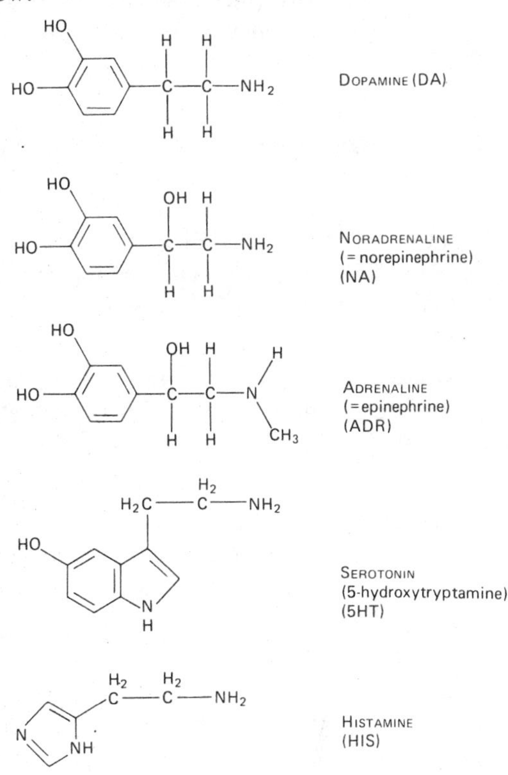

It can be seen that dopamine (DA) and noradrenaline (NA) are primary monoamines derived from phenylethylamine ($C_6H_5CH_2CH_2NH_2$), while adrenaline (ADR) is a secondary amine formed by *N*-methylation of NA. Serotonin (5HT) is another primary monoamine, being an indolylethylamine derivative, while histamine (HIS), an imidazolylethylamine, is usually classified as a diamine on account of the basicity of the imidazole ring. All of these compounds have fully aromatic rings attached to the ethylamine moieties and all except HIS are phenols as well as amines.

All of the above-mentioned amines are readily soluble in water and would therefore be expected to diffuse rapidly from their cellular sites of storage under the ordinary physical conditions of a histochemical technique. Diffusion can be minimized by using free-dried material or by the application of a chemical fixative which reacts rapidly with the amine to produce an insoluble compound. Both of these approaches are used. The amines are found in two types of site. These are:

(a) In large quantities, in secretory granules of endocrine (or similar) cells. Here, binding to a protein or carbohydrate matrix greatly reduces diffusion of the amines.

1. Chromaffin cells of the adrenal medulla and related paraganglia (NA and ADR in separate cell-types).
2. Argentaffin (= enterochromaffin) cells of the intestinal mucosa (5HT).
3. Mast cells (HIS in all mammalian species; also 5HT in rodents and DA in the lungs of ruminants).

(b) In small quantities as transmitter substances in aminergic neurons. The amines in these sites are easily extracted by water and other solvents.

1. Most postganglionic sympathetic neurons, their axons and terminal aborizations (NA).
2. Various systems of neurons in the central nervous system, most conspicuously in the terminal parts of axons (DA, NA, and 5HT).

In situation (a), where the diffusion of amines is hindered by the associated structural macromolecules, the sensitivity of a histochemical method does not need to be as high as for the demonstration of amines in situation (b). Furthermore, the amines of the endocrine cells are fairly easily preserved by chemical fixation, while those of neurons are much more labile. This difference is probably also due to stronger binding of the amines to the proteinaceous matrices of the granules of secretory cells than to the synaptic vesicles of neurons.

17.2. HISTOCHEMICAL METHODS

The histochemical techniques for the biogenic amines (for a review, see Hahn von Dorsche *et al.*, 1975) depend on the detection either of the phenolic groups or of ethylamine with an aromatic substituent on the β-carbon atom.

17.2.1. Amines in secretory granules of argentaffin, chromaffin, and mast cells

17.2.1.1. Serotonin

The serotonin-containing cells of the alimentary tract are named "argentaffin" on account of their ability to reduce ammoniacal or similarly complexed solutions of silver nitrate to the metal. Other presumably endocrine cells found in the gastrointestinal epithelium reduce complex silver ions more weakly and can only be seen if treatment with the silver reagent is followed by the application of a developer which increases the sizes and densities of the initially deposited metal particles. The latter types of cell are termed "argyrophil" and the methods for demonstrating them are reminiscent of the methods used for staining axons in the nervous system (see Chapter 18). Unfortunately, the argentaffin and argyrophil reactions, though useful to histopathologists, have very low histochemical specificities.

The most convenient histochemical techniques for argentaffin cells are based on the detection of the phenolic function of serotonin. Phenols couple with diazonium salts in alkaline solution to form coloured azo compounds. Azo-coupling with phenols occurs preferentially *para* to the hydroxyl groups. Since this position is already occupied in 5HT, one of the *ortho* sites will be used:

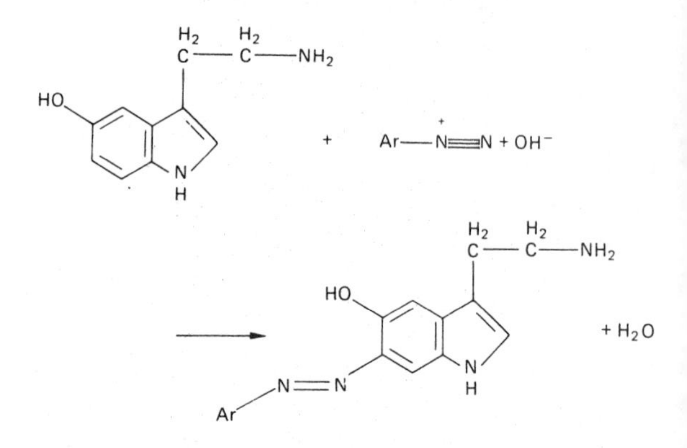

Similar azo-coupling reactions are also to be expected with the aromatic amino acids of all proteins and with any other phenolic compounds present in the tissues. Consequently, generalized "background" staining occurs if the time of exposure to the diazonium salt is unduly prolonged. The secretory granules of argentaffin cells (and of rodent mast cells) are recognized by virtue of their more intense and more rapidly developing coloration.

The argentaffin cells also give a positive chromaffin reaction, though this is not generally used for their identification.

17.2.1.2. THE CHROMAFFIN REACTION

Cytoplasmic granules in the cells of the adrenal medulla assume brown colours after fixation in a solution containing potassium dichromate. Although such cells are said to be "chromaffin", the coloured product is derived not from the fixative but from the adrenaline and noradrenaline present in the cells. The amines are oxidized by dichromate to coloured quinones. The reaction involves oxidation of the catechol (*o*-diphenol) moiety to a quinone and oxidative coupling of the amino group to one of the carbon atoms of the ring

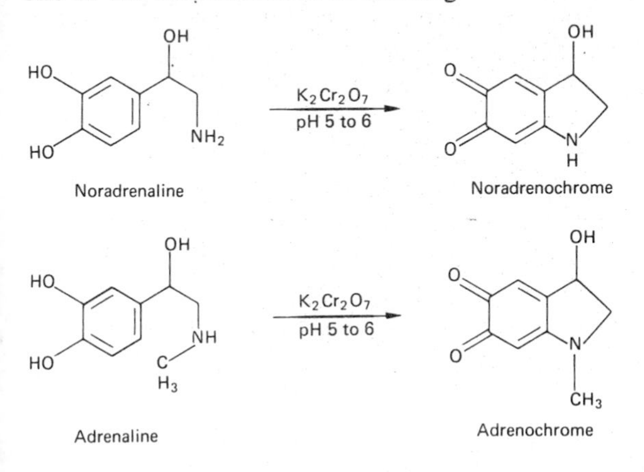

A similar reaction occurs with DA. It is likely that further oxidations and polymerization take place in the formation of the final brown products, which are insoluble in water and in organic solvents. The chromium atoms in the reagent are reduced from oxidation state +6 to +3. The resulting Cr^{3+} ions are bound at the sites of the amine (Lever *et al.*, 1977), and cause increased electron density of the chromaffin granules.

The iodate ion can oxidize catecholamines in the same way as dichromate though it acts much more rapidly upon NA than upon adrenaline. Consequently it is possible, using sodium iodate, to stain noradrenaline-containing cells selectively. Thin slices of tissue are immersed in aqueous sodium iodate prior to fixation in formaldehyde and sectioning on a freezing microtome.

17.2.1.3. CATECHOLAMINES IN GLUTARALDEHYDE-FIXED TISSUE

Glutaraldehyde reacts with amino groups, probably to form imines:

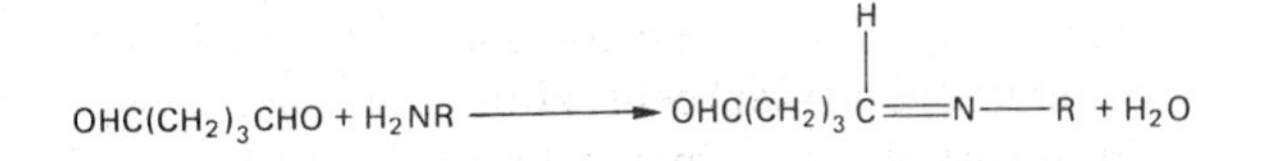

Ordinarily an imine such as that shown above will be unstable when, as in the biogenic amines, R is not an aromatic ring.† However, in a tissue the other end of the glutaraldehyde molecule is likely to combine similarly with a protein-bound amino group. Extensive cross-linking of the proteinaceous matrix of the tissue probably increases the resistance of the linkages to hydrolysis. Primary catecholamines (DA and NA) are bound by their amino groups and are therefore immobilized, but their catechol groups remain free to react with other reagents. Adrenaline, which is a secondary amine, reacts with glutaraldehyde more slowly than NA or DA and is not usually retained in the tissue.

The catechols are strong reducing agents and may be detected by several simple histochemical methods. A simple procedure is to post-fix the glutaraldehyde-fixed specimens in osmium tetroxide (Coupland *et al.*, 1964), which is reduced to a black material (see also Chapter 2, p. 18). Alternatively, the catechol-containing site may be made visible by virtue of its ability to reduce silver diammine ions to the metal (Tramezzani *et al.*, 1964; see also Chapter 10, p. 131). Methods of this type impart optical blackness and electron density to granules containing NA or DA. If adrenaline is to be demonstrated as well, it is necessary to fix the specimens in a solution containing both glutaraldehyde and potassium dichromate at empirically determined optimum concentrations and pH (Coupland *et al.*, 1976; Tranzer & Richards, 1976).

17.2.1.4. HISTAMINE

Various histochemical methods for histamine have been described. The most satisfactory is based on the formation of a fluorescent adduct with *o*-phthaldialdehyde (OPT). The chemistry of the reaction has not yet been elucidated, but Shore *et al.* (1959) have proposed that the reagent condenses with both the primary amine group and the >NH group of the imidazole ring:

† In aromatic amines, NH_2 is attached **directly** to an aromatic ring, not to an aliphatic side-chain.

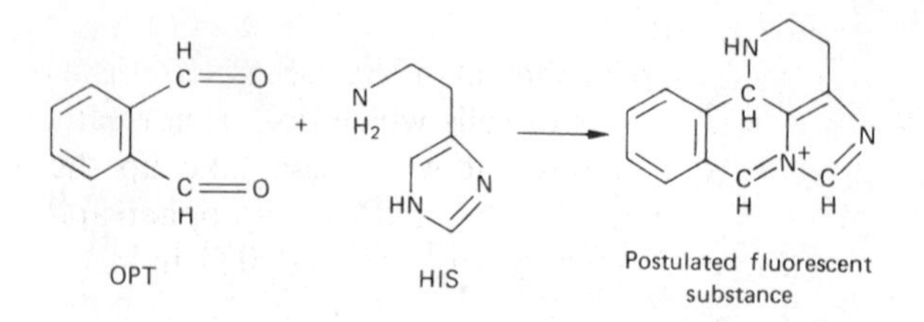

Since the above equation does not balance, some other ions or compounds must participate in it if the product has the structure suggested. Further molecular rearrangement and/or oxidation may occur in the formation of the fluorophore.

With high concentrations of histamine (i.e. in mammalian mast cells), the emitted fluorescence is yellow. With low concentrations (as in certain enterochromaffin-like cells of the gastric mucosa), the colour is blue. Unfortunately, OPT forms fluorescent compounds with other substances, notably with glucagon and secretin (Hakanson *et al.*, 1972). This imperfect specificity of the reagent seriously limits its value for the localization of histamine in sites of low concentration, especially in cells of suspected endocrine function in the gastrointestinal tract.

Where histamine is stored in high concentration (i.e. in mast cells), OPT can be used in solution, on paraffin sections of Carnoy-fixed tissue, provided that the tissue is not at any time exposed to aqueous reagents. For low concentrations of the amine it is necessary to use freeze-dried material and to apply the reagent in gaseous form.

17.2.2. Sensitive methods for amines in neurons

17.2.2.1. Formaldehyde-induced fluorescence

The reactions discussed above for serotonin and noradrenaline are not sufficiently sensitive to demonstrate aminergic neurons. For this purpose it is necessary to make use of the formation of highly fluorescent compounds by reaction of the amines with formaldehyde (or any of a few other reagents; see below). Techniques of this type are exceedingly sensitive and can detect as little as 5×10^{-4} pg of NA or DA.

Formaldehyde reacts with NA, DA, and 5HT by condensation and cyclization to form non-fluorescent compounds. These then undergo either oxidation by atmospheric oxygen or reaction with more formaldehyde, to form brightly fluorescent products. While the first stage of the reaction can occur with formaldehyde in either the liquid or the gaseous phase, the second stage requires an almost dry proteinaceous matrix at 80–100°C. The overall reactions are:

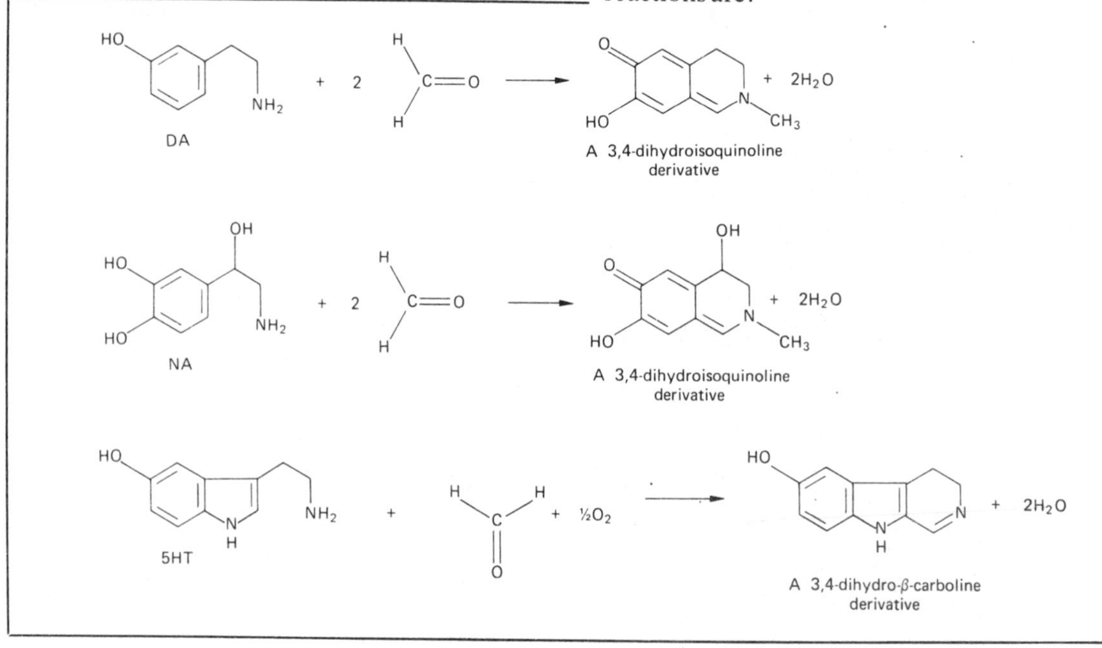

The reactions are more complicated than is indicated by the above equations and the fluorescent products shown are not the only ones formed. Adrenaline yields only a weakly fluorescent product. The chemistry of formaldehyde-induced fluorescence is discussed in detail by Corrodi & Jonsson (1967) and, more simply, by Björklund *et al.* (1975).

It is usual to employ freeze-dried tissue in studies of the biogenic amines of the nervous system, though methods have been devised for cryostat sections and for spreads of membranous tissues such as the iris and mesentery. Glyoxylic acid or, alternatively, a formaldehyde–glutaraldehyde mixture (see below) is a more suitable reagent when facilities for freeze-drying are not available. In the simplest form of the method, freeze-dried blocks of tissue are treated with formaldehyde gas, derived from solid paraformaldehyde, under controlled conditions of humidity at about 80°C. The treated blocks are then embedded directly in paraffin wax. The sections must not be exposed to water or alcohols, which remove the fluorophores. There are several modified techniques (e.g. Laties *et al.*, 1967; Watson & Ellison, 1976) in which a buffered solution of formaldehyde is perfused through the vascular system of the animal prior to either freeze-dry or the cutting of cryostat sections. Heating of the sections or of the freeze-dried blocks, with or without paraformaldehyde vapour, effects the second stage of the reaction in which the fluorophores are formed.

The different amines can be distinguished from one another by microspectrofluorimetric methods (see Björklund *et al.*, 1975).

Formaldehyde-induced fluorescence is valuable not only for the identification and morphological study of cells containing biogenic amines, but also for investigating the pharmacological properties of such cells. For example, reserpine depletes cells of their monoamines, and drugs which inhibit monoamine oxidase produce visible increases in the intensity of fluorescence of aminergic neurons. A pharmacological method for distinguishing between DA and NA has been devised by Hess (1978). Animals are treated first with reserpine and then with DOPA, a metabolic precursor of the catecholamines. After a suitable interval of time, only the DA-containing cells exhibit fluorescence in formaldehyde-treated tissues. A much longer time is needed for replenishment of NA. Pearse (1968) has identified a series of cells which do not normally contain monamines but which can take up the amino acids DOPA and 5-hydroxytryptophan and decarboxylate them to give DA and 5HT in histochemically demonstrable quantities. These cells, most of which have known or suspected endocrine functions, have been called APUD (amine precursor uptake and decarboxylation) cells. The synaptic varicosities of aminergic neurons normally reabsorb their transmitter substances (see Iverson, 1967) and this property is occasionally exploited in order to enhance their formaldehyde-induced fluorescence.

The histochemical demonstration of small quantities of the monamine neurotransmitters has been of great value in the study of aminergic connections in the brain, spinal cord, and peripheral autonomic nervous system. The neuroanatomical applications of formaldehyde-induced fluorescence have been reviewed by Dahlstrom (1971) and Kiernan & Berry (1975).

17.2.2.2. Other Fluorescence Methods

In addition to formaldehyde, several other carbonyl compounds are able to form fluorescent condensation products with biogenic amines. The most valuable is glyoxylic acid:

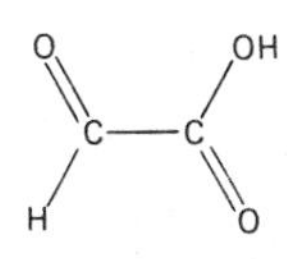

This substance was introduced as an alternative to formaldehyde by Axelsson *et al.* (1973). It reacts in a similar manner, the fluorescent products being closely similar though obtained in higher yield (see Björklund *et al.*, 1975).

A major advantage of methods making use of glyoxylic acid is that it is not necessary to prepare the tissue by freeze-drying. Usually the animal is perfused with a buffered aqueous solution of glyoxylic acid. Sections are cut with a vibrating microtome or a cryostat, dried onto slides and heated, either alone or in glyoxylic acid or formaldehyde vapour. It is also possible to use paraffin sections of freeze-dried specimens. Different variations of the method (e.g. Lindvall & Björklund, 1974; Furness & Costa, 1975; Bloom & Battenberg, 1976; Loren *et al.*, 1976; Watson & Barchas, 1977) have been developed to give optimum results with different tissues. The fluorescence of amine-containing nerve fibres is brighter after reaction with glyoxylic acid than after reaction with formaldehyde.

In another method (Furness *et al.*, 1977, 1978), tissues are treated with an aqueous solution of formaldehyde and glutaraldehyde by immersion or vascular perfusion. Fluorophores are formed with DA, NA, and 5HT, but the chemistry of the technique has not been studied. The fluorescence resists extraction by cold water, alcohols, several other organic solvents, and melted paraffin wax. Unfortunately, the non-specific "background" fluorescence associated with this technique is brighter than that seen with formaldehyde- or glyoxylic acid-induced fluorescence.

17.3. INDIVIDUAL METHODS

17.3.1. Azo-coupling method for argentaffin cell granules

Use paraffin sections of formaldehyde-fixed material.

Solutions required

A. Diazonium salt solution

Fast red B salt (stabilized diazonium salt derived from 5-nitroanisidine) (see Chapter 5):	100 mg
TRIS buffer, pH 9.0:	100 ml

Prepare just before using

B. A counterstain

(An alum–haematoxylin, used progressively, is suitable.)

Procedure

1. De-wax and hydrate section (see *Note* below).
2. Immerse in the diazonium salt solution for 30 s.
3. Carefully rinse in five changes of water, each 20–30 s, without agitation.
4. Counterstain nuclei. If alum–haematoxylin is used for this purpose, do not wash in running water but "blue" the sections in water containing a trace of $Ca(OH)_2$.
5. Dehydrate, clear, and cover, using a resinous mounting medium.

Result

Argentaffin cell granules—orange–red; background (protein)—yellow; nuclei (if counterstained with alum–haematoxylin)—blue.

Note

It may be necessary to cover the sections with a film of celloidin in order to prevent their detachment from the slides after treatment with the alkaline solution A. The film should be removed before clearing. The instructions for stages 3 and 4 allow for more gentle handling of the preparations than is usually necessary.

17.3.2. The chromaffin reaction

This technique is applied to small pieces or thin slices of fresh tissue (e.g. a rat's adrenal gland cut in half).

Solution required

Potassium dichromate ($K_2Cr_2O_7$):	5.0 g
Potassium chromate (K_2CrO_4):	0.45 g
Water to:	100 ml

Keeps indefinitely

Procedure

1. Fix small pieces of tissue in the above solution (see *Note 1* below) for 24 h.
2. Wash in running tap water overnight. (See also *Note 2* below.)
3. Dehydrate, clear, embed in paraffin wax, and cut sections at 4 or 7 μm.
4. De-wax and clear (three changes of xylene) and mount in a resinous medium.

Results

The cytoplasm of cells containing adrenaline (ADR) and noradrenaline (NA) is coloured brown. The product formed from ADR is darker than that from NA.

Notes

1. A chromate–dichromate mixture is used in order to obtain a solution of pH 5.5–5.7. Alternatively, one may fix the tissue in a fixative mixture based on potassium dichromate provided that it does not contain acids or other heavy

metal ions. However, these mixtures are more acid than the optimum pH for the chromaffin reaction.

2. Stage 2 may be followed by a secondary fixation for 12–24 h in 4% formaldehyde if desired. This is necessary if frozen rather than paraffin sections are to be cut.
3. A counterstain may be applied to the sections if desired. Alum–haematoxylin is suitable for nuclei. Many cationic dyes stain the chromaffin cells, thereby increasing the intensity of their colour in addition to demonstrating nuclei.

17.3.3. Glutaraldehyde–osmium method for noradrenaline

This method (Coupland *et al.*, 1964) is for the study of the adrenal medulla.

Solutions required

A. Fixative

25% aqueous glutaraldehyde:	20 ml
0.1 M phosphate buffer, pH 7.3:	80 ml

Prepare on day of use

B. Buffered osmium tetroxide

2% aqueous OsO_4 (stock solution):	5.0 ml
0.1 M phosphate buffer, pH 7.3:	5.0 ml

Stable for about 6 h after addition of the phosphate buffer. A simple 1% aqueous solution of osmium tetroxide, which is stable for several weeks, may be substituted.

Procedure

1. Fix freshly removed pieces of tissue no more than 3 mm thick (e.g. halved adrenal gland of a rat) in solution A for 4 h.
2. Rinse specimen in water. Cut frozen sections 10–40 µm thick and collect them into water. Mount onto slides, blot, and allow to dry for 5–10 min. Position the sections near the ends of the slides so that only a small volume of solution B will be needed.
3. Immerse the sections in solution B in a closed coplin jar for 30 min.
4. Wash slides in running tap water for 30 min.
5. Dehydrate, clear, and mount in a resinous medium.

Result

Noradrenaline-containing cells—black to grey. Osmiophilic lipids (see Chapter 12) are also blackened.

Control

Substitute neutral, buffered formaldehyde for solution A in stage 1 of the method. The osmium tetroxide will now give black products only with lipids.

17.3.4. Method for histamine in mast cells

This technique, for Carnoy-fixed tissue, detects histamine in mast cells but is not sensitive enough for the demonstration of the amine in the enterochromaffin-like cells of the gastric mucosa (Enerback, 1969).

Reagents required

A. *Fixative.* Carnoy's fluid
B. Absolute ethanol
C. Ethylbenzene
D. Ethylbenzene: 20 ml
o-phthaldialdehyde (OPT): 200 mg
Prepare just before using
E. Tetrahydrofurfuryl alcohol

Reaction vessel

Place three sheets of filter paper in the bottom of a large glass petri dish. Wet the paper with water and put the lid on. This provides a humid atmosphere into which the slides can be introduced.

Procedure

1. Fix small pieces of tissue in Carnoy's fluid for 2 h.
2. Complete the dehydration in three changes of absolute ethanol, each 30 min.
3. Clear in ethylbenzene, 1 h.
4. Infiltrate with paraffin wax, two changes, for total of 30 min, in a vacuum-embedding apparatus.
5. Block out. Cut sections at 4–7 µm and mount them onto dry slides, previously warmed on a hotplate. Heat to 60°C to melt wax and flatten the sections onto the slides. (*Note:* Neither water nor albumin is used in mounting the sections.)

6. Put the slides (sections uppermost) into the reaction vessel and pour on sufficient of the OPT solution (D) to cover the sections. Replace the lid of the vessel and wait for 2 min.
7. Remove the slides, drain off the OPT solution, and cover the sections with tetrahydrofurfuryl alcohol (E). Apply coverslips and examine by fluorescence microscopy. Permanent preparations can be made by clearing in tetrahydrofurfuryl alcohol and mounting in DPX.

Result

Sites of storage of histamine in cytoplasm of mast cells—yellow fluorescence.

17.3.5. Formaldehyde-induced fluorescence

The methods described below are suitable, on account of their simplicity, for instructional purposes. For use in research, more sensitive techniques are available, such as that of Loren *et al.* (1980) in which a pre-treatment of the tissue with aluminium sulphate results in greatly enhanced fluorescence.

Preparation of tissues

Alternative methods are prescribed for (a) thin whole mounts (mesentery, iris, etc.), and (b) solid specimens, no more than 2 mm thick.

(a) Stretch the thin tissue (removed immediately after killing the animal) onto glass slides. Place the slides in a glass rack in a desiccator containing a small beaker, one-third filled with phosphorus pentoxide. Partially evacuate the desiccator and wait for 1 hr. Remove the slides and treat with formaldehyde gas as described below.

(**Caution:** P_2O_5 is very corrosive. It must not be allowed to come into contact with liquid water with which it reacts violently to form phosphoric acid. To discard used P_2O_5, place the vessel containing it in a safe place, open to the air for 12–24 h. The white powder will then have deliquesced to a syrupy mass, which may be washed down the sink with copious running water.)

(b) The piece of tissue is rapidly frozen (by immersion in isopentane cooled by liquid nitrogen) and transferred to a freeze-drying apparatus. When freeze-drying is complete, the specimen is treated with gaseous formaldehyde as described below.

Treatment with formaldehyde vapour

Formaldehyde is generated by the action of heat on solid paraformaldehyde. The moisture content of the latter is a critical factor: the paraformaldehyde should be equilibrated with air of 50% relative humidity. This can be achieved by storing the powder for 7 days or longer in a desiccator which contains a mixture of concentrated sulphuric acid (specific gravity 1.84), 32 ml and water, 68 ml. (**Caution:** add the acid, slowly with stirring, to the water.)

About 6 g of the water-equilibrated paraformaldehyde is placed at the bottom of a wide-necked jar fitted with a tightly fitting screw cap. The slides or freeze-dried specimens are put into the jar, which is then closed and maintained at 80°C in an oven for 1 h.

Subsequent processing

The jar containing formaldehyde vapour should be transferred to a fume cupboard before opening.

(a) Slides bearing spreads of tissue are covered, using liquid paraffin (mineral oil, heavy, U.S.P.) as mounting medium. Alternatively, rinse in xylene and mount in a non-fluorescent resinous mounting medium such as DPX.

(b) Freeze-dried specimens are vacuum-embedded in paraffin wax (time of infiltration, 15–30 min). Section are cut and mounted onto slides without using water. The slides can be heated to 60°C to melt the wax and flatten the sections. The sections are then covered with liquid paraffin and coverslips applied. With further warming on the hotplate, the wax is dissolved by this mounting medium. Alternatively, the wax may be removed by xylene or petroleum ether and the cloverslip applied with a non-fluorescent resinous mounting medium such as DPX. Clearing and mounting must be carried out very carefully to avoid loss of the sections.

Result

Green fluorescence from noradrenaline and dopamine; yellow fluorescence from serotonin. The optimum wavelength for excitation is 410 nm (violet–blue).

If the sections are cleared and mounted in a resinous medium, there is slight suppression of the specific fluorescence, but the preparations are optically superior to those mounted in liquid paraffin.

17.3.6. Glyoxylic acid method

This technique (after Watson & Barchas, 1977) is applicable to cryostat sections of unfixed specimens of central nervous tissue. Each piece of fresh tissue is placed on a cryostat chuck, which is then stood in a slush of dry ice (solid CO_2) and acetone or in liquid nitrogen until the tissue has frozen.

Materials required

A. 0.1 M phosphate buffer, pH 7.0 (approximately) at 0–4°C: 45 ml
Glyoxylic acid ($HOOC.CHO.H_2O$): 1.0 g
Magnesium chloride ($MgCl_2.6H_2O$): 0.25 g
1.0 N (4%) sodium hydroxide (NaOH): add drops until the pH is 4.9–5.0
Water (at 0–4°C): to make 50 ml

Prepare immediately before use. Cool to 0°C by standing the coplin jar in iced water

B. A hotplate, maintained at 45°C.

C. A dry screw-capped coplin jar containing approximately 2 g of solid glyoxylic acid ($HOOC.CHO.H_2O$), heated to 100°C on a boiling water bath for 30 min prior to use. The lid should be only half-tightened during heating, to allow for expansion of air.

Procedure

1. Cut cryostat sections at any convenient thickness up to 20 µm. The temperature of the cryostat cabinet should be −17°C. Collect each section onto a warm (20–25°C) slide, where it will thaw and dry at once. Proceed **immediately** to stage 2, handling only 1–3 slides at a time.
2. Place slides in the glyoxylic acid solution (A) and leave in it at 0°C for 12 min (optimum for nerve terminals) **or** 4 min (optimum for cell bodies) **or** 45 s (optimum for some axons).
3. Remove slides from solution A, blot with filter paper, and place on the hotplate at 45°C for about 5 min.
4. Place the slides in the coplin jar containing solid glyoxylic acid at 100°C for approximately 3 min.
5. Remove the slides and apply coverslips, using liquid paraffin (mineral oil, heavy, U.S.P.) as a mounting medium.

Result

Sites of NA and DA emit green fluorescence (excitation by violet–blue light).

Note

Glyoxylic acid is deliquescent and may deteriorate with repeated exposures to air. As soon as a bottle is received from the supplier, divide it into aliquots of 1.0 g and store these in separate vials, **in a desiccator**, below 0°C. Do not allow solid glyoxylic acid to come into contact with the skin: it is irritant and corrosive.

17.3.7. A formaldehyde–glutaraldehyde fluorescence method

This is a technique based on that of Furness *et al.* (1977). It differs from other aldehyde-condensation methods in that no heating of dry sections is needed. The fixative (solution A, below) is a satisfactory one for electron microscopy. See also *Note 1* below.

Solutions required

A. "FAGLU" reagent

Paraformaldehyde: 12 g
0.1 M phosphate buffer, pH 7.0: 290 ml

(Dissolve, with heating if necessary. Cool to room temperature before adding next ingredient)

25% aqueous glutaraldehyde: 6.0 ml
Water: to 300 ml

This solution should be made on the day it is to be used.

Both the paraformaldehyde and the glutaraldehyde should be grades marketed as being suitable for electron microscopy.

B. Acidified DMP

2,2-dimethoxypropane (DMP): 25 ml
Concentrated hydrochloric acid: one drop (about 0.05 ml)

This is prepared as required, but is stable for several hours or possibly for much longer. See Chapter 4 for discussion of DMP as a dehydrating agent.

C. Clearing agent

Chloroform: 50 ml
Benzene: 50 ml

Procedure

1. Remove small pieces of fresh tissue, no more than 2 mm thick, and place them in the FAGLU reagent (solution A) for 3 h. (See also *Note 2* below.)
2. Blot off excess FAGLU solution with filter paper and immerse specimen in acidified DMP (solution B) for 20 min at 20–30°C, with occasional shaking.
3. Transfer to clearing agent (solution C). Shake at intervals of about 3 min until the specimen has sunk to the bottom of the container, then leave for a further 10 min.
4. Infiltrate with paraffin wax, three changes, each 10 min, in a vacuum-embedding apparatus. Block out. Store blocks at 4°C until they are to be cut.
5. Cut sections at a convenient thickness: e.g. 15–25 μm for peripheral tissues; thinner for CNS. Mount the sections onto dry slides **without using water**. Place the slides on a hotplate at about 60°C until the wax melts and the sections flatten, then proceed immediately to stage 6.
6. Remove slides from hotplate and put them into a staining rack. Immerse gently into a tank of previously unused xylene. Agitate very gently for 15–20 s and leave for 5 min.
7. Carefully remove slides individually from the xylene and apply coverslips, using a non-fluorescent mounting medium such as DPX. Place slides on a hotplate at about 60°C for 5 min, then transfer them to a tray. If they are not to be examined immediately, keep them at 4°C for no more than 1 week.

Result

Sites of NA and DA green. Sites of 5HT yellow. Other components of the tissue also fluoresce but less intensely. Optimum excitation is by blue–violet light: the non-specific background fluorescence then appears as green when an orange–yellow barrier is used. If excitation is by broad-band ultraviolet, with a colourless or very pale yellow barrier filter, positively reacting elements appear yellow or greenish yellow against a blue background.

Notes

1. This technique is valuable on account of its technical simplicity, but the non-specific background fluorescence is more intense than with the two preceding methods. Furness *et al.* (1977) did not provide detailed practical instructions for the preparation of paraffin sections. The procedure described here is effective, but alternative schedules for dehydration and clearing may also be possible. It is important to avoid flotation of the paraffin sections on warm water, which results in loss of the specific fluorescence.
2. For the CNS, Furness *et al.* (1978) recommend perfusion of the animal with a buffered solution containing 4% formaldehyde and 1% glutaraldehyde, followed by sectioning with a vibrating microtome or a cryostat.

17.4. EXERCISES

Theoretical

1. What are the principal difficulties associated with the histochemical localization of organic compounds of low molecular weight?

2. Ascorbic acid (M.W. 176) is freely soluble in water and is not precipitated by any fixative. It is possibly the only normal component of mammalian tissues which rapidly reduces silver ions to the metal in darkness at pH 2.5:

$$C_6H_8O_6 + 2Ag^+ \xrightarrow{(60^\circ C,\ 1\ h)} C_6H_6O_6 + 2H^+ + 2Ag\downarrow$$

(ascorbic acid) (5% $AgNO_3$ in 3% acetic acid) (Dehydroascorbic acid)

Silver chloride and complexes of silver with proteins are slowly reduced to the metal in bright light. Sodium thiosulphate (5% aqueous $Na_2S_2O_3.5H_2O$) dissolves AgCl and removes silver from combinations with protein.

Using the above information, devise a histochemical technique for ascorbic acid.

(The specificity of the acid $AgNO_3$ method has been disputed. See Pearse, 1968, for discussion and references.)

3. Why is it necessary to use formaldehyde- or glyoxylic acid-induced fluorescence to demonstrate noradrenaline (NA) in sympathetic neurons when technically simpler methods suffice for the NA-containing cells of the adrenal medulla?

4. What substances are responsible for the "background" staining in sections prepared by the azo-coupling method for argentaffin cells?

5. Amines react with Reinecke salt, NH_4^+ $[Cr(NH_3)_2(SCN)_4]^-.H_2O$, the anion of which forms insoluble salts with the ionized amino radical. Reinecke salt has been used as a fixative for amines in both light and electron microscopy.

Would you expect amines precipitated in this way to exhibit:

(a) Formaldehyde-induced fluorescence?
(b) Positive azo-coupling reactions?
(c) Increased electron-density at their sites of storage?

Give reasons for your answers.

Practical

6. Carry out the azo-coupling reaction for argentaffin cells on paraffin sections of formaldehyde-fixed small intestine, tongue, and skin (from the rat). Identify cells which contain serotonin.

7. Kill a rat and remove the following organs. (The same rat may be used for Exercise 12 if it is a male.)

(a) Some mesentery. Dry onto slides for subsequent demonstration of formaldehyde-induced fluorescence. Proceed to Exercise 8 below.
(b) Both adrenals. Cut one in half and fix for the chromaffin reaction. Fix the other by immersion in buffered glutaraldehyde. Proceed to Exercises 9 and 10 below.
(c) A transversely cut slab, about 3 mm thick, from the tongue and/or a small piece of skin. Fix in Carnoy for 2 h. Clear and embed and section as described for Enerback's OPT method. Remember to mount the sections without using water. Proceed to Exercise 11 below.

8. Process the dried mesenteric spreads to demonstrate formaldehyde-induced fluorescence. Identify the sites of noradrenaline and serotonin in the preparations. If one of the spreads (not for fluorescence microscopy) is lightly stained with toluidine blue, this will assist in recognizing the cells that contain serotonin.

9. Observe the chromaffin reaction in paraffin sections of adrenal gland fixed in the chromate–dichromate mixture. It is advisable to embed both halves of the adrenal gland side by side in the same block, since the small medulla may be in only one of the pieces. In making these preparations you may experience one of the adverse effects of fixation in potassium dichromate.

10. Cut frozen sections of the glutaraldehyde-fixed adrenal gland. Demonstrate the noradrenaline-containing cells in its medulla. What do you see in the adrenal cortex?

11. Carry out the OPT reaction for histamine on the paraffin sections of tongue and/or skin. Compare the distribution of mast cells by this method with that seen in a section stained for acid mucosubstances (see Chapter 12).

12. Kill a male rate by decapitation under ether anaesthesia. Remove the following organs:

(a) The brain. Cut out a piece no more than 2 mm thick, in the coronal plane, including the optic chiasma and adjacent hypothalamus. Rapidly freeze the specimen by placing it onto a flat piece of metal foil and immersing it in isopentane cooled by liquid nitrogen. Freeze-dry the specimen and proceed to Exercise 13 below.
(b) The fundus of the stomach and two pieces, each about 3 mm long, of ductus deferens. Fix these in a formaldehyde-glutaraldehyde (FAGLU) mixture as described in Section 17.3.7 of this chapter. Proceed to Exercise 14 below.

13. Apply the formaldehyde-induced fluorescence technique to the freeze-dried piece of brain tissue. Observe the distribution of monoaminergic axons in the hypothalamus. Other sections may be stained by appropriate methods (see Chapter 18) to display neuronal somata.

14. Process the specimens fixed in FAGLU for the demonstration of monoaminergic axons in the stomach and in transverse and longitudinal sections of vas deferens. What structures other than axons fluoresce? Account for your observations.

18

Neurohistological Techniques

18.1. Classification of techniques 256
18.2. Nissl stains 258
18.3. Myelin 258
  18.3.1. Normal myelin 258
    18.3.1.1. Mordant-dye methods 258
    18.3.1.2. Luxol fast blue 258
    18.3.1.3. Osmium tetroxide 259
  18.3.2. Degenerating myelin 259
18.4. Silver methods for normal axons 260
  18.4.1. Methodology and chemistry 260
  18.4.2. Gold toning 261
  18.4.3. Choice of silver methods 262
18.5. The Golgi methods 262
18.6. Vital methylene blue 263
18.7. Neuroglia 264
18.8. Miscellaneous methods 264
  18.8.1. Demonstration of motor end-plates 264
  18.8.2. Neurosecretory material 264
  18.8.3. Iodide–osmium methods 265
18.9. Individual methods 266
  18.9.1. Chromoxane cyanine R method for myelin 266
  18.9.2. Luxol fast blue for myelin 266
  18.9.3. A Marchi method for degenerating myelin 267
  18.9.4. Holmes's silver method for axons 268
  18.9.5. Mitchell's rapid silver method 269
  18.9.6. Winkelmann and Schmit method for peripheral nerve endings 270
  18.9.7. Gros–Schultze method for axons 271
  18.9.8. A physical developer method for axons 272
  18.9.9. Golgi–Cox method 273
  18.9.10. Vital methylene blue 274
  18.9.11. Cajal's gold–sublimate method for astrocytes 275
  18.9.12. Bromoindigo–silver method for motor end-plates 276
  18.9.13. Aldehyde–fuchsine for neurosecretory material 277
  18.9.14. Iodide–osmium method 277
18.10. Exercises 278

It has long been recognized that the structure of the nervous system is not adequately revealed by the staining methods generally employed for other types of tissue. Special techniques are required and an enormous number of these exists, though many are merely slight variants of other methods. The procedures used in neurohistology are designed to display selectively certain elements characteristic of nervous tissues. These include normal components such as myelin sheaths and neuroglial cells, pathological features such as gliosis and viral inclusion bodies, and experimentally induced abnormalities such as axonal degeneration and substances artificially introduced as tracers.

The following account covers some of the techniques which are used in the elucidation of the normal structure of the nervous system.

18.1. CLASSIFICATION OF TECHNIQUES

This classification embraces most of the methods used in nèuroanatomy and neuropathology. For techniques discussed later in this chapter, the reader is referred to the appropriate numbered sections. For other methods, references are given to other books or papers.

A. *Nissl stains*

For neuronal somata. Nuclei of neuroglial cells are also stained (see Section 18.2).

B. *Myelin stains*

(a) Normal myelin (see Section 18.3.1)
  (i) Mordant-dye methods.
  (ii) Luxol fast blue.
  (iii) Osmium tetroxide.
(b) Degeneration myelin
  Marchi's method (see Section 18.3.2).

C. *Methods for axons*

(These are methods for both myelinated and unmyelinated axons, including nerve-endings.)

(a) Normal axons
  (i) Thiocholine methods for choline esterases (see Chapter 15).
  (ii) Amine histochemistry (see Chapter 17).
  (iii) Vital methylene blue (see Section 18.6).
  (iv) Silver methods (see Section 18.4).

(b) Degenerating axons

(These methods are used in experimental neuroanatomy. Usually they cannot be applied to human post-mortem material, since the time allowed for degeneration is critical.)

(i) Glees method (Glees, 1946; Marsland *et al.*, 1954).
(ii) Nauta and Fink–Heimer procedures (Nauta & Gygax, 1951, 1954; Nauta, 1957; Fink & Heimer, 1967).

D. *Methods for whole neurons*

(Used mainly for the study of dendritic architecture. These methods demonstrate only a small fraction of the neuronal population, so that stained cells stand out clearly.)

(a) The Golgi methods (see Section 18.5).
(b) Vital methylene blue (see Section 18.6).

E. *Tracer methods*

(Used in neuroanatomical and neurophysiological research.)

(a) Radioactive amino acids

These are taken up by dendrites, incorporated into proteins and transported to axonal terminal fields, where they are detected by autoradiography (Cowan *et al.*, 1972; Rogers, 1975).

(b) Proteins, notably horseradish peroxidase (HRP)

These are taken up principally by axonal terminals and transported retrogradely to neuronal somata. HRP is detected histochemically (see Chapter 16; also LaVail *et al.*, 1973; Mesulam, 1978).

(c) Intracellular tracers, notably procion yellow (a fluorescent reactive dye) and cobalt chloride

These are introduced into neurons by micropipette to fill the cytoplasmic processes (Kater & Nicholson, 1973; Fuller & Prior, 1975; Lynch *et al.*, 1973; Zieglänsberger & Reiter, 1974).

F. *Methods for neuroglia*

(a) Normal neuroglial cells

(i) Cajal's gold–sublimate method for astrocytes (see Sections 18.7; 18.9.11).
(ii) Silver carbonate and similar methods, for oligodendrocytes and microglia (Penfield & Cone, 1950). Mechanism of action unknown.
(iii) Golgi methods (see Section 18.5). These demonstrate occasional neuroglial cells as well as neurons.

(b) Abnormal neuroglia

There are several techniques, involving the use of metal salts or dyes, suitable for pathological neuroglia, but not useful for the normal cells (see Clark 1973; Ralis *et al.*, 1973).

G. *Miscellaneous methods*

(a) Motor end-plates

(i) Gold chloride methods. The traditional method of Ranvier and its more recent variants, applicable to teased preparations (see Gray, 1954; Swash & Fox, 1972). Mechanism of action unknown.
(ii) Combined acetylcholinesterase and silver techniques (see Section 18.8.1). These are applicable to frozen sections.

(b) Boutons terminaux (presynaptic axon terminals)

The pre-chromation procedures, some of which demonstrate mitochondria in presynaptic boutons (Armstrong *et al.*, 1956; Rasmussen, 1957; Armstrong & Stephens, 1960; Abadia-Fenoll, 1968; Braak & Jacob, 1973; Desclin, 1973). These have been used in the quantitative histological study of synapses, but are now largely supplanted by electron microscopy.

(c) Neurosecretory material

(i) Cystine histochemistry (see Chapter 10).
(ii) Chrome alum–haematoxylin (Bargmann, 1949).
(iii) Permanganate–aldehyde fuchsine (see Section 18.8.2).

(d) Iodide–osmium methods for peripheral nerve endings (see Section 18.8.3)

Some techniques used in neurohistology have been discussed in earlier chapters while others will not be considered further. The interested student is referred to the references given with the above classification. The remaining methods, most of which do not have sound histochemical rationales, will now be briefly described and practical instructions for some of them will be given. For critical discussions of neurohistological methods and of their uses and limitations, see Nauta & Ebbesson (1970); Kiernan & Berry (1975); Santini (1975).

18.2. NISSL STAINS

The histochemical properties of the nucleic acids have already been discussed in Chapter 9. The nuclei of cells in the nervous system, together with the cytoplasmic Nissl bodies (RNA) of neurons, are easily stained with basic dyes. The procedure described in Chapter 11 for demonstrating metachromasia with toluidine blue gives excellent results when used as a Nissl stain. Many other cationic dyes, including cresyl violet acetate, neutral red, and thionine, are equally suitable for the purpose.

18.3. MYELIN

The myelin sheaths of axons are composed of concentric membranous lamellae derived from the ensheathing cells (Schwann cells in the peripheral nervous system; oligodendrocytes in the central nervous system). The membranes contain lipoproteins and proteolipids, and are stained histologically by virtue of the presence of these substances. The lipids of normal myelin are relatively hydrophilic compared with those of degenerating myelin and in both states unsaturated fatty acid residues are sufficiently abundant to permit their histochemical identification. The phospholipids, being for the most part protein-bound, are only partially extracted in the course of embedding in wax. While many of the methods used in lipid histochemistry will demonstrate myelin, there are some procedures which have been developed specially for the purpose. The latter techniques are valuable for the CNS and a few of them will now be discussed.

18.3.1. Normal myelin

18.3.1.1. Mordant-dye methods

A long-established staining method is the Weigert–Pal procedure in which formaldehyde-fixed blocks of nervous tissue are treated with a mixture of potassium dichromate and chromium fluoride. This renders all the phospholipid components insoluble in dehydrating and clearing agents and probably also acts as a mordant (see Chapter 12). Paraffin or celloidin sections are prepared, stained with a ripened haematoxylin solution, and then differentiated until only the myelin sheaths retain the blue–black chromium lake of haematein. There are many variations of the technique, and in some of them an iron mordant is used. The principal inconveniences of the methods using haematoxylin are the critical differentiation and the fact that some of the procedures take several days. The intensity of staining, however, is higher than can be obtained with the simpler luxol fast blue method.

A more recent mordant-dye method is that of Page (1965) in which iron alum and chromoxane cyanine R are used. This method is applicable to paraffin sections, and prior insolubilization of phospholipids by chromation is not necessary. Presumably the fixed lipoproteins of myelin are retained in the sections in sufficient quantity to allow adequate staining by the dye–metal complex. The differentiation (by a solution of the ferric salt which is also used as the mordant) is technically simpler in Page's method than in the techniques employing haematoxylin.

18.3.1.2. Luxol fast blue

As explained in Chapter 5 (p. 75), the luxol dyes are arylguanidinium salts of anionic chromogens. They are insoluble in water and are used as solutions in alcohol or other moderately polar organic liquids. Phospholipids are stained by these dyes, though probably not very specifically (Salthouse, 1963; Lycette *et al.*, 1970). Luxol dyes probably also enter hydrophobic domains of protein molecules (Clasen *et al.*, 1973). It has been shown that the arylguanidinium gegen-ion is liberated into the solvent during the process of attachment of the dye to its substrate, leaving the coloured anion behind. The staining is then differentiated and made specific for myelin by treatment with dilute aqueous lithium carbonate, followed by 70% ethanol. The mechanism of differentiation is in need of investigation. A red or violet cationic dye subsequently applied as a counterstain binds not only to the nuclei and Nissl substance but also to the luxol fast blue anions present in the myelin sheaths, thereby increasing the intensity of coloration of the latter (Clasen *et al.*, 1973).

Luxol fast blue MBS, G, and ARN, and methasol fast blue 2G are some of the dyes that can be used to stain myelin in the manner described above. The first such technique to be described was that of Klüver and Barrera (1953), which is still widely used. A somewhat simpler version is described below. Luxol fast blue is easy to use but slow in its action. It does not give such a strong colour to the myelin, especially in the PNS, as do the mordant-dye methods. The latter are therefore preferred for the demonstration of myelinated axons of narrow calibre.

18.3.1.3. Osmium tetroxide

The chemical reactions of OsO_4 with unsaturated lipids have already been described in connection with fixation (Chapter 2) and lipid histochemistry (Chapter 12). See also under "Degenerating myelin" (Section 18.3.2) below.

Osmium tetroxide is of little value for the staining of myelinated fibres in the CNS because it also causes blackening of all the other membranous components of this region. It is most valuable in morphometric studies of peripheral nerves. Small nerves may be teased in OsO_4 to display individual fibres. The nodes of Ranvier are clearly shown and internodal lengths can be measured. If nerves are fixed in OsO_4 or post-fixed in it after primary fixation by glutaraldehyde, the diameters and thicknesses of myelin sheaths can be measured in transverse sections.

18.3.2. Degenerating myelin

When an axon is severed, the part distal to the neuronal cell-body degenerates into a string of tiny fragments. The myelin sheath also undergoes fragmentation, and this process is associated with a change in the component lipids. Phospholipids, which are hydrophilic, predominate in the normal myelin sheath, while esters of cholesterol are the principal lipids of the degenerate fragments. In peripheral nerves, the remains of the axon and its sheath are rapidly phagocytosed by the surrounding Schwann cells, but in the CNS the fragments remain *in situ* as extracellular objects for some weeks (Demêmes *et al.*, 1974) before being ingested by neuroglial cells.

The Marchi method demonstrates the degenerated remains of myelin sheaths. There are several variations of the technique, but in all of them the tissue is treated simultaneously with OsO_4 (soluble in polar and non-polar substances) and an oxidizing agent (which must be one soluble only in polar substances). The oxidizing agent is most commonly potassium dichromate (Marchi, 1892) or potassium chlorate (Swank & Davenport, 1935). In degenerating myelin, OsO_4 is bound by the unsaturated fatty-acid residues of the hydrophobic cholesterol esters and then reduced to an insoluble black compound, probably $OsO_2.2H_2O$ (see Chapter 2, p. 18, for chemistry). In normal myelin, which is hydrophilic, the oxidizing agent prevents the reduction of bound osmium.

Important differences exist between the early and late stages of degeneration of myelin. Strich (1968) and Fraser (1972) have shown that in the **early phase** (up to about 10 days after injury for PNS or about 100 days for CNS) the products are destroyed by freezing and thawing or by storage for more than 3 weeks in a formaldehyde-containing fixative. They can therefore be demonstrated only by treating blocks of tissue with OsO_4–oxidant mixtures. The products of the **late phase** of degeneration, however, persist much longer after axonal transection (up to 30 days in the PNS and at least 2 years in the CNS), and are not destroyed by freezing and thawing or removed by prolonged storage of tissues in solutions of formaldehyde. The degenerating myelin of the late phase can therefore be stained either in the block or in frozen sections. If the latter are to be used, the method of choice is the osmium tetroxide–α-naphthylamine procedure of Adams (see Chapter 12, p. 178). For staining in the block, the method of Swank & Davenport (1935) is recommended and is described below.

The chief advantage of the Marchi method for neuroanatomical tracing is that the length of the time of survival after a destructive lesion is not very critical, so it is possible to apply the technique to human post-mortem material or to tissues taken from animals used in experiments which have not been designed solely for the purpose of identifying degenerating pathways. The disadvantages are the impossibility of demonstrating degenerating unmyelinated axons (including all pre-synaptic branches) and a tendency for sporadic blackening of normal myelin sheaths.

18.4. SILVER METHODS FOR NORMAL AXONS

The multitude of available methods in this category is sufficient testimony to the fact that there is no entirely satisfactory histological technique for staining all types of axon. Histological techniques of this type have been developed empirically. The argyrophilic component of the axon is contained in the axoplasm and is proteinaceous (Kiernan, 1970). Nothing else is certainly known of its nature, though there is some evidence to support the contention that in normal nerve fibres silver may be precipitated in relation to neurofilaments, which are organelles characteristic of the axon. The argyrophilia of degenerating axons, however, is probably not due to neurofilaments (see Kiernan & Berry, 1975, for discussion and references).

18.4.1. Methodology and chemistry

The silver methods all have two features in common. These are:

1. Treatment of the tissue (blocks or frozen, paraffin or celloidin sections; nearly always after an aldehyde-containing fixative) with a solution containing silver ions, whose concentration may be anything from about 10^{-5} to 1.0 M.
2. Subsequent treatment of the specimen with a reducing agent capable of effecting the reaction: $Ag^+ + \varepsilon^- \rightarrow Ag\downarrow$.
3. The deposition of a darkly coloured material (consisting mainly or entirely of metallic silver) in the axons, which are said to be argyrophilic.

The chemical reactions involved in selective axonal staining with silver are fairly well understood as the result of the investigations of Holmes (1943), Samuel (1953a), and Peters (1955a, b). These studies were concerned principally with some of the technically simpler procedures applied to paraffin sections.

In the first step of such methods, sections are exposed for several hours to a mildly alkaline solution in which the concentration of free silver ions is low (10^{-5} to 10^{-3} M) but in which there is an adequate excess of the silver. The latter condition is attained by having a large volume of dilute aqueous $AgNO_3$ or by arranging for most of the metal to be present as a complex such as $[Ag(NH_3)_2]^+$, or a silver–protein compound, or by using a saturated solution of a sparingly soluble salt such as Ag_2O, Ag_2CO_3, or AgOCN in equilibrium with the undissolved solid. Methods in which concentrated silver nitrate solutions are used have not been studied, but it is probable that the same chemical reactions occur. Silver is taken up by the section in two ways. The larger quantity is bound chemically by protein throughout the tissue. This chemically bound silver is not specifically related to axons and can easily be removed by treating the tissue with complexing agents such as ammonium hydroxide, citric acid, or sodium sulphite. A much smaller amount of silver is reduced at sites in the axons and precipitated as tiny "nuclei" of the metal. These nuclei are too small to be resolved with the electron microscope (Peters, 1955c). Similar nuclei, formed by the action of light upon crystals of silver halides, form the latent image in an exposed but undeveloped photographic emulsion. A latent image nucleus consists of 2–6 atoms of silver (see James, 1966). The nuclei formed in axons may be similar in size. The reduced silver cannot be extracted by NH_4OH or Na_2SO_3.

In the second stage of the method, the sections are transferred to a solution similar to a photographic developer. A commonly used one contains sodium sulphite and hydroquinone. This mixture is alkaline. The sulphite removes the chemically bound silver, introducing $[Ag(SO_3)_2]^{3-}$ ions into the solution. Hydroquinone reduces this complex ion to silver (metal) on the surfaces of the previously formed nuclei of metallic silver present in the axons. The nuclei are thereby enlarged until they are (in light microscopy) coalescent and the axons appear as black or brown linear structures. In some techniques the chemically bound silver is removed by treatment with Na_2SO_3 alone and a developer containing a known concentration of silver is used. Such a mixture (known as a physical developer) must contain stabilizing agents to delay the spontaneous reduction of Ag^+ in the solution. Stability is obtained by including in the physical developer a substance which forms a soluble complex with silver, such as citric acid or sodium sulphite, so that the concentration of free silver ions is low. Addition of a macromolecular solute, commonly gelatine or gum acacia, provides additional stabilization.

If the processes of development in histological staining with silver are closely similar to the ones used in photography, it is likely that the silver precipitated as submicroscopic nuclei during the first stage of the procedure serves as a catalyst for reactions of the type

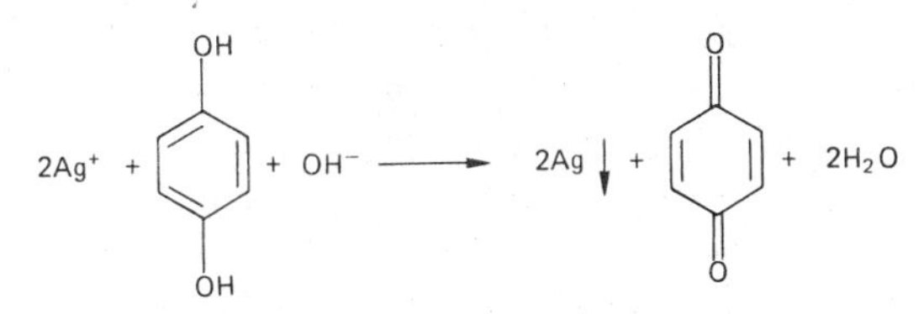

Metallic silver is insoluble, so it accumulates around the particles of the catalyst. Since many substances other than silver can catalyse the above reaction, physical developers have been used to produce visible end-products in a wide variety of histological and histochemical techniques (see Gallyas, 1971). The developing agent used in silver-staining methods should be one that is "weak" or "slow" by photographic standards. Hydroquinone, pyrogallol, and *p*-hydroxyphenylglycine are suitable, but the more powerful developers such as *N*-methyl-*p*-aminophenol (metol, elon) cause excessive deposition of silver which obscures the histological detail.

18.4.2. Gold toning

The coloration of axons impregnated with silver as described above does not always provide adequate contrast under the microscope. The contrast can be improved by adding a third stage of gold toning. To do this, the silver-stained sections are immersed in a solution of gold chloride. Sometimes this manoeuvre produces an adequate increase in contrast, but it is commonly necessary to add a further stage of reduction in oxalic acid. Finally, the sections are immersed in aqueous sodium thiosulphate to remove residual silver salts—a process analogous to the "fixing" of a developed photograph. The chemistry of gold toning was rather simply explained by Samuel (1953b) on the basis of the reaction

$$3Ag + AuCl_3 \rightarrow Au + 3AgCl$$

However, this explanation is unsatisfactory since it does not conform with the known chemistry of gold compounds and it assumes that one atom of gold would be optically more conspicuous than three of silver.

Histologists use "gold chloride" in two forms, a yellow solid, which is sodium tetrachloroaurate, $NaAuCl_4 . H_2O$, and a brown solid, which is chloroauric acid, $HAuCl_4$. Both dissolve readily in water to give solutions containing chloroaurate anions, $[AuCl_4]^-$. Auric chloride, $AuCl_3$, which dissolves to form $H^+ [AuCl_3OH]^-$, is not used in histology and neither is the sparingly soluble aurous chloride, AuCl.

The complex ion $[AuCl_4]^-$ is reduced by silver, with the formation of gold and silver chloride:

$$\underset{\text{(Solid)}}{3Ag} + \underset{\text{(Dissolved)}}{[AuCl_4]^-} \longrightarrow \underset{\text{(Solid)}}{Au} + \underset{\text{(Insoluble)}}{3AgCl} + \underset{\text{(in solution)}}{Cl^-}$$

This reaction is essentially similar to the one proposed by Samuel (1953b). Silver chloride can be removed by treatment with sodium thiosulphate:

$$\underset{\text{(Solid)}}{AgCl} + \underset{(Na_2S_2O_3\text{ in solution})}{2S_2O_3^{2-}} \longrightarrow \underset{\text{(Soluble)}}{[Ag(S_2O_3)_2]^{3-}} + \underset{\text{(in solution)}}{Cl^-}$$

In those techniques in which the toning procedure consists only of treatment with "gold chloride" followed by sodium thiosulphate, the increased contrast must be due to a reduction in the unwanted argyrophilia of the background rather than to an intensification of the metallic deposits in the axons.

The intensification of gold toning by treatment with oxalic acid is not so easily explained. Oxalic acid is not a photographic developer and does not reduce AgCl to Ag (which might have accounted for increased metallic deposition), but it does reduce chloroaurate anions to metallic gold. Now the particles of silver in stained axons are very small (Peters, 1955c; Sechrist, 1969), and they may well form, collectively, a colloidal suspension of the metal. After treatment with $HAuCl_4$, the silver is replaced by a mixture of Au and AgCl, presumably also in particles with colloidal dimensions. Since many colloids are able to adsorb ions and other molecules, it is not unreasonable to suppose that $[AuCl_4]^-$ ions are bound by physical forces at the sites of axonal staining. The bound chloroaurate ions would then be reduced on treatment with oxalic acid, possibly in the reaction

$$2[AuCl_4]^- + 3H_2C_2O_4 \rightarrow 2Au\downarrow + 6CO_2 + 6H^+ + 8Cl^-$$

Hence, substantial deposits of gold would be precipitated around each original particle of silver. The above speculations are supported by the observations of Zagon *et al.* (1970), who found by electron probe microanalysis that the metallic particles in silver-stained and gold-toned protozoa contained considerably more gold than silver.

18.4.3. Choice of silver methods

Axons in most parts of the CNS are most easily demonstrated in paraffin sections by methods in which the primary impregnation is with a solution containing a very low concentration of free silver ions, applied for 24–48 h at 37°C. The simplest technique is that of Holmes (1943). Another widely used technique is Bodian's (1936) protargol method. Protargol is a silver–protein complex, manufactured for use as an antiseptic, which slowly releases Ag^+ ions.† These procedures are also suitable for paraffin sections of peripheral nerves, but they often fail to demonstrate the innervation of tissues such as skin and viscera.

†Two types of silver proteinate, designated "strong" and "weak", have been used as antiseptics. Only the "strong" type, which is now obsolete as a pharmaceutical product, is suitable for Bodian's technique. Many unsatisfactory products are supplied by distributors of microscopical stains. The name "protargol-S" has been adopted by the Biological Stains Commission for silver proteinates tested and certified as satisfactory for the histological demonstration of axons.

To stain peripheral nerve endings with silver, frozen sections of material thoroughly fixed in formaldehyde should be used. Primary impregnation is accomplished with a concentrated (10–20%) solution of silver nitrate. A simple and versatile technique is Mitchell's modification of the Hirano–Zimmermann procedure. This method can be applied to frozen, paraffin, and celloidin sections, and although it is rather capricious it is usually possible to obtain satisfactory results by varying the times of exposure of sections to the reagents. Cutaneous innervation is notoriously difficult to demonstrate by silver methods, but a simple technique devised by Winkelmann & Schmit (1957) is often satisfactory for this purpose. If it fails, the rather more troublesome Gros–Schultze method can be tried. The latter procedure is derived from earlier ones introduced by Bielschowsky and makes use of an ammoniacal silver nitrate solution containing the silver diammine cation, $[Ag(NH_3)_2]^+$. The reduction of this complex ion by formaldehyde (see Chapter 10, p. 130) is probably catalysed by silver nuclei produced in axons during the primary impregnation. Thus, the ammoniacal solution appears to serve the same purpose as a physical developer. Indeed, much of the untrustworthiness of Bielschowsky-type methods disappears when a controlled physical development replaces the alternation of treatments with formalin and silver diammine solutions (see Section 18.9.8).

There are many methods for staining with silver in the block before embedding and sectioning (see Jones, 1950). Excellent results can sometimes be obtained with these methods in both CNS and PNS, but the effects of uneven penetration of the reagents are usually apparent in the sectioned material.

18.5. THE GOLGI METHODS

Camillo Golgi (1843–1926) discovered his famous method by chance when, in 1873, he observed the blackening of occasional whole neurons in dichromate-fixed blocks of tissue which had been subsequently immersed in a solution of silver nitrate. The serendipity of this Italian histologist led to the elucidation of the morphology of the neuron and the development of the neuron theory. The theory was championed by a Spaniard, Santiago Ramon y Cajal (1852–1934), who is generally held to be the greatest of all neurohistologists. Although they disagreed bitterly about the structure of the nervous system, the two men shared the Nobel Prize for Physiology and Medicine in 1906.

There are three major variants of the Golgi procedure, known respectively as the Golgi, rapid Golgi, and Golgi–Cox methods. With all of these methods, whole neurons (dendrites, perikarya, and axons) and glial cells are coloured in shades of deep reddish-brown and black. Only about 1% of the cells in the tissues are stained, however, so it is possible to see the shapes of the individual cells very clearly. Examination with the electron microscope has revealed that the deposits are located mainly within the cytoplasm, and that within individual stained cells the impregnation is sometimes incomplete. The myelinated parts of the axon are rarely darkened by Golgi procedures, and the methods are used principally in the study of dendritic morphology and of the relationships between axonal terminals and the cells with which they synapse.

In the Golgi methods most closely related to the original one, pieces of central nervous tissue are treated with silver nitrate after prolonged fixation in potassium dichromate, while in the rapid Golgi methods a shorter fixation in a mixture containing potassium dichromate and osmium tetroxide precedes the immersion in silver nitrate. In the Golgi–Cox and related methods, tissues are fixed for several weeks in a solution containing $K_2Cr_2O_7$, K_2CrO_4 and a mercuric salt. Either whole blocks or sections of nitrocellulose-embedded tissue are then treated with an alkali, which precipitates black material in the randomly impregnated cells. Related techniques are available for tissue fixed in formaldehyde.

No satisfactory explanations are yet available for the peculiar specificity of staining achieved with any of the Golgi techniques. Even the chemical natures of the final deposits are uncertain: recent opinions are that with the rapid Golgi method the end-product is silver chromate, Ag_2CrO_4 (Chan-Palay, 1973), and that with the Golgi–Cox technique it is either mercuric oxide chromate, $Hg_3O_2CrO_4$ (Blackstad *et al.*, 1973), or mercuric sulphide, HgS (Stean, 1974). Braitenberg *et al.* (1967) noticed that with a Golgi method of the classical type, axonal staining was associated with the appearance of filamentous crystals on the surface of the tissue. They suggested that similar linear crystalline growth might occur within the sporadically impregnated cells.

Before using a Golgi method for research purposes, the student is advised to consult the reviews by Ramon-Moliner (1970) and Kiernan & Berry (1975) in which the methods are discussed at some length and in which the advantages and disadvantages of the different techniques are evaluated in relation to particular neurohistological applications. A Golgi–Cox technique described below is suitable for demonstrating the dendritic architecture of most types of neurons in the CNS.

18.6. VITAL METHYLENE BLUE

Methylene blue has already been encountered as an ordinary cationic dye (see Chapters 5, 6, and 7), but its value in neurohistology depends on properties other than basicity. When living or freshly removed tissue is treated with methylene blue under suitable conditions, axons and their terminal branches are selectively stained. The biochemical mechanisms underlying this coloration are poorly understood despite several studies of the phenomenon (see Schabadasch, 1930; Richardson, 1969; Kiernan, 1974). The ability to stain nerve fibres is shared by some other N-methylated derivatives of thionine, but methylene blue itself is generally considered to be the best dye for the purpose.

For staining peripheral innervation, the most satisfactory results are obtained by immersing freshly removed tissue in a dilute oxygenated solution of methylene blue buffered to pH 5.0–7.0, for about 30 min at 30°C. It is also possible to inject the dye locally or systemically into the living animal. Since the deposits of dye in the axons are soluble in water and in alcohol, the stained tissues are fixed in ammonium molybdate, which forms an insoluble salt with methylene blue. Ammonium picrate has also been used for the same purpose, but the picrate of methylene blue is soluble in alcohol though not in water. Thin specimens are examined as whole mounts, but it is also possible to prepare frozen or paraffin sections of the vitally stained specimens.

Ideally, the axons stand out as blue lines on a colourless background, but commonly there is also some blue staining of non-neuronal elements. Both somatic and autonomic fibres take up vital methylene blue, the former at pH 5.5–7.0, the latter at pH 5.0–5.5. Above pH 7.0, only the coarser, myelinated sensory axons are stained (FitzGerald & Fitter, 1971).

Vital methylene blue can also be used for the CNS. Usually an animal is perfused with a solution of the dye (commonly a higher concentration than is used than for the immersion methods). Reducing agents in the anoxic tissues decolorize the methylene blue, forming a leuco-compound, which is thought to enter living cells more easily than the ionized dye (see Baker, 1958). Thin slices of the nervous tissue are removed and exposed to moist air to re-oxidize the dye, which is then fixed in ammonium molybdate. In a simpler method, methylene blue powder is sprinkled onto moist, freshly removed fragments of brain or spinal cord (see Gray, 1954). Vital staining with methylene blue is not a very reliable technique for the CNS, but when it works well, occasional whole neurons are coloured, as with the Golgi methods. Neuroglial cells are not stained. Cajal (1911) used the method to elucidate the morphological characteristics of many types of central neuron.

18.7. NEUROGLIA

One technique for normal neuroglial cells, Cajal's gold–sublimate method for astrocytes, will now be briefly discussed. It is somewhat unreliable and no satisfactory explanation is available for the moderately specific staining which can be obtained.

Frozen sections of tissue fixed in formal–ammonium–bromide are placed for up to 8 h in a solution of gold chloride and mercuric chloride. Cook (1974) prefers yellow gold chloride ($NaAuCl_4.2H_2O$), but most other authorities favour the "brown" variety ($HAuCl_4$). If material has been fixed in a simple formaldehyde-containing mixture, the sections can be pre-conditioned by Globus's treatment with ammonia and hydrobromic acid before staining. A modified technique for paraffin sections has been described by Naoumenko & Feigin (1961).

Astrocytes and their processes are coloured dark purple by the gold–sublimate method, and their end-feet on capillaries are discernible in well-stained sections. This technique, like the metal–impregnation methods for other types of neuroglia, is used by neuropathologists, but its applications in research are limited. Neuroglial cells can be identified on the basis of their nuclear morphology in Nissl preparations, but the full extent of their cytoplasmic processes is revealed only by electron microscopy.

18.8. MISCELLANEOUS METHODS

18.8.1. Demonstration of motor end-plates

Neuromuscular innervation can be stained by several empirical techniques (fully discussed by Zacks, 1973), notably Ranvier's lemon juice–gold chloride method.

Since the motor end-plate comprises an axon ending upon a specialized region of the sarcolemma (the subneural apparatus), it is reasonable to combine different techniques for the two components. Nerve fibres in muscle are fairly reliably stained by silver methods and the subneural apparatus is rich in acetylcholinesterase. Silver methods have been combined with thiocholine techniques for AChE (Gwyn & Heardman, 1965; Namba *et al.*, 1967; Cunningham & FitzGerald, 1972), but when this is done both components of the end-plate are coloured black. By using an admittedly less specific indigogenic method for the enzyme of the subneural apparatus, as in the procedure described below (Section 18.9.12), the pre- and post-synaptic parts can be displayed in different colours.

18.8.2. Neurosecretory material

In mammals the neurons of the supraoptic and paraventricular nuclei of the hypothalamus produce hormones (antidiuretic hormone and oxytocin) that are elaborated in granules with characteristic staining properties. The granules are present in the perikarya, the axons, and—in highest concentration—in the perivascular axonal terminals of the neurohypophysis. Neurosecretory material (NSM) with similar tinctorial properties occurs in the homologous hypothalamo-hypophysial systems of submammalian vertebrates and in various parts of the nervous systems of almost all invertebrate animals (see Gabe, 1966, for an extensive review).

The first method to be used specifically for staining NSM (Bargmann, 1949) was Gomori's chrome alum–haematoxylin, originally developed for the cell types of the pancreatic islets. It is not always possible, however, to obtain satisfactory results with this technique. Gabe (1953) showed that it was possible to stain the material with aldehyde–fuchsine after preliminary treatment of sections with acidified potassium permanganate. A slight modification of the original procedure is described below. It is quite trustworthy. Gabe (1976) pointed out that NSM is stainable by many cationic dyes at low pH as well as by aldehyde–fuchsine or a chromium–haematein complex after oxidation by aqueous potassium permanganate. He considered it likely that cystine in the NSM is oxidized to cysteic acid (see Chapter 10), which binds the colorant molecules. Since NSM contains much cystine, the performic acid–alcian blue method (see Chapter 10) can also be employed.

Aqueous fixatives must be used for material in which NSM is to be stained. Formaldehyde and Bouin are suitable but brighter coloration is seen after Susa or Helly's fluid. Although alcoholic fixatives render NSM extractable from tissues by water, the usual methods of dehydration and clearing produce no adverse effects when they follow fixation in an aqueous mixture. If an alcoholic fixative must be used, the NSM can be immobilized by floating out the paraffin sections on warm Bouin's fluid instead of water.

18.8.3. Iodide–osmium methods

In some tissues the sensory and autonomic innervation are shown with striking clarity by using one of the iodide–osmium methods. Specimens are fixed in a solution containing iodide ions and osmium tetroxide and are subsequently either examined as whole mounts or embedded and sectioned. The cation associated with the iodide was sodium or potassium in the earlier methods of this type (e.g. Champy *et al.*, 1946), but more recently zinc iodide has been preferred (Maillet, 1963).

Although several ideas have been suggested and disproved (see Maillet, 1968), there is still no satisfactory explanation for the mechanism of axonal staining by iodide–osmium mixtures. The chemistry of the reagent has been partially elucidated by Gilloteaux & Naud (1979). When solutions of zinc iodide and osmium tetroxide are mixed, the latter compound is reduced by iodide ions to the osmate(VI) ion:

The solution rapidly becomes orange as iodine is liberated. After several hours it turns black; a black precipitate settles out during the next 1 to 2 weeks. This precipitate is zinc osmate, $ZnOsO_4$. It has been shown to be identical to the black, electron-dense substance formed within tissues fixed in zinc iodide–osmium tetroxide mixtures. Reduction occurs more rapidly in tissue than in the solution and may be catalysed at the sites of organic reducing groups such as —SH and —CH=CH—. Gilloteaux and Naud have also obtained evidence indicating that deposition of zinc osmate can occur at intracellular sites which, in life, were occupied by calcium ions.

Examination of zinc iodide–osmium–fixed central nervous tissues with the electron microscope has revealed electron-dense deposits within synaptic vesicles (Akert & Sandri, 1968). The reaction in these organelles has been shown by Reinecke & Walther (1978) to depend upon the presence of free sulphydryl groups in the tissue. It is not known whether similarly produced deposits account for the optical blackness of stained peripheral axons.

Iodide–osmium methods have been found to be most valuable for the examination of nerve-endings in viscera, such as the gall bladder (Sutherland, 1963) and oesophagus (Rodrigo *et al.*, 1975). Both autonomic and sensory innervations are demonstrated. Central nervous tissue, which has a high content of osmiophilic membrane-bound lipids, is homogeneously blackened by fixation in iodide–osmium solutions and the preparations are uninformative when examined by light microscopy.

18.9. INDIVIDUAL METHODS

18.9.1. Chromoxane cyanine R method for myelin

Use paraffin sections of material fixed in neutral, buffered formaldehyde. Zenker, Helly, and Susa are also suitable fixatives, but when they are used it is not usually possible to eliminate nuclear staining by differentiation. Following alcoholic fixatives (e.g. Bodian's, see p. 22) the myelin sheaths are damaged but staining is still possible. Precipitates formed by the action of $HgCl_2$ must be removed as appropriate. This is the method of Page (1965).

Solutions required

A. Staining solution

Dissolve 0.5 g of chromoxane cyanine R (C.I. 43820) in 1.25 ml of concentrated sulphuric acid by stirring with a glass rod in a 300 ml flask. Dissolve 1.0 g of iron alum, $NH_4Fe(SO_4)_2 . 12H_2O$, in water to make 250 ml of solution. Add the iron-alum solution to the dissolved dye, mix, and filter. This solution is stable for about 2 years and may be used repeatedly. Alternatively, the sulphuric acid may be added after dissolving the dye and the iron alum in water. Equimolar quantities of ferric chloride (0.56 g of $FeCl_3 . 6H_2O$) or ferric nitrate (0.73 g of $Fe(NO_3)_3 . 6H_2O$) may be substituted for iron alum.

B. Iron-alum solution

Iron alum ($NH_4Fe(SO_4)_2 . 12H_2O$):	20 g
Water:	to 200 ml

Keeps indefinitely but can only be used once. Alternatively, use an equimolar solution of ferric chloride (5.6% $FeCl_3 . 6H_2O$) or ferric nitrate (7.3% $Fe(NO_3) . 6H_2O$).

C. Counterstain

0.5% aqueous neutral red or safranine is suitable.

Procedure

1. De-wax and hydrate paraffin sections.
2. Immerse in staining solution (A) for 15–20 min.
3. Wash in tap water (running, or several changes) until excess dye is removed (about 2 min).
4. Differentiate in 10% iron alum (solution B) until only the myelin (white matter of CNS) retains the stain. This usually takes 5–10 min. It is sometimes impossible to decolorize the nuclei completely without losing some intensity in myelin.
5. Wash in tap water (running, or three or four changes) for about 5 min.
6. Apply counterstain if desired. See following method (18.9.2) for details.
7. Wash, dehydrate through graded alcohols, clear in xylene, and cover, using a resinous mounting medium.

Result

Myelin and erythrocytes—deep blue. Nuclei and Nissl substance (with the counterstain suggested above)—red. Nuclei may be blue if perfect differentiation cannot be obtained.

18.9.2. Luxol fast blue for myelin

This is a slightly simplified form of the original method of Klüver & Barrera (1953). Fixation is as for the previous technique. Frozen sections of formaldehyde-fixed tissue are also satisfactorily stained by this procedure.

Solutions required

A. Staining solution

Luxol fast blue MBS:	0.25 g	Keeps indefinitely
95% ethanol:	250 ml	

B. Differentiating solution (0.05% Li_2CO_3)

Lithium carbonate (Li_2CO_3):	0.25 g
Water:	500 ml

This keeps indefinitely and can be used repeatedly, but should be discarded when it is more than faintly blue.

C. Counterstain

0.5% aqueous neutral red. The addition of a few drops of glacial acetic acid improves the staining properties of this solution, which keeps indefinitely and may be used many times.

Procedure

1. De-wax paraffin sections and take to absolute ethanol. Frozen sections should be dried onto slides (from water) and then equilibrated with absolute ethanol.
2. Stain in solution A in a screw-capped staining jar at 56–60°C (i.e. in wax oven) for 18–24 h.
3. Rinse in 70% ethanol and take to water.
4. Immerse in the differentiating solution (B) until grey and white matter can be distinguished. This commonly takes 20–30 s for thin paraffin sections, but thick frozen sections may require up to 30 min.
5. Transfer slides to 70% ethanol, two changes, each 1 min. More dye leaves the sections at this stage.
6. Rinse in water.
7. Counterstain for 1–5 min in 0.5% neutral red (solution C).
8. Wash in water and blot dry.
9. If necessary, differentiate the counterstain in 70% ethanol until only the nuclei and Nissl substance are red. The colour of the stained myelin is deepened by counterstaining.
10. Blot dry and dehydrate in two changes (each 3–5 min) of *n*-butanol. Clear in xylene and cover, using a resinous medium.

Result

Myelin—blue; nuclei and Nissl substance—red.

Notes

1. Some batches of luxol fast blue do not work very well. For other techniques, see Cook (1974). Suitable alternative dyes can be made in the laboratory by combining acid dyes with diphenylguanidine; for instructions, see Clasen *et al.* (1973).
2. An alternative counterstain is the PAS procedure (see Chapter 11). This provides a pink background without cellular staining.

18.9.3. A Marchi method for degenerating myelin

This is the method of Swank & Davenport (1935) and is described as applied to rats with experimental lesions in the CNS. For optimum results, the animals should be killed 2 weeks after placement of the lesion, and no departures should be made from the procedure given below.

Solutions required

A. Preliminary fixative

Magnesium sulphate ($MgSO_4.7H_2O$):	15 g	Keeps indefinitely
Potassium dichromate ($K_2Cr_2O_7$):	5 g	
Water: dissolve and make up to	250 ml	

B. Secondary fixative (10% formaldehyde)

Formalin (37–40% HCHO):	100 ml	Keeps for several weeks
Water:	900 ml	

C. Staining solution

Potassium chlorate ($KClO_3$):	1.5 g	Keeps for a few months at 4°C in a tightly stoppered bottle
Water:	200 ml	
Osmium tetroxide:	0.5 g	
Formalin (37–40% HCHO):	30 ml	
Glacial acetic acid:	2.5 ml	

Procedure

1. Inject the animal with a lethal intraperitoneal dose of a barbiturate. Do **not** use ether as an anaesthetic.
2. When the animal is deeply unconscious, open the thorax and insert a cannula into the left ventricle. Open the right atrium to drain out most of the blood and then perfuse about 100 ml of the preliminary fixative (solution A). See also *Note 3* below.
3. Immediately remove the brain and put it into 10% formaldehyde (solution B) for 48 h.
4. Trim the specimen into pieces no more than 3 mm thick and place the pieces (without washing) into the staining solution (C). The volume of this solution should be approximately fifteen times that of the tissue. Leave in a tightly capped specimen jar for 7–10 days, with daily agitation to expose all surfaces of the tissue evenly to the reagent.

5. Wash the specimens in running tap water for 24 h.
6. Dehydrate in graded alcohols and infiltrate with nitrocellulose. Either embed in nitrocellulose or double-embed in paraffin wax. Cut serial sections 20 μm thick and mount onto slides.
7. De-wax and clear the sections by passing through three changes of xylene. Cover, using a resinous mounting medium. (See *Note 2* below.)

Result

Degenerating myelin black. Normal myelin unstained. Fat in adipose tissue is also blackened. Occasional normal myelinated axons are commonly stained, but they can easily be recognized by their integrity. The degenerating fibres are fragmented and appear as rows of black dots.

Notes

1. For distinguishing between early and late products of degeneration, see the discussion of this method (Section 18.3.2 above).
2. A counterstain may be applied to some sections in order to facilitate topographical orientation. Alum–haematoxylin is a suitable nuclear stain, but cationic dyes do not work very well after fixation in OsO_4.
3. The perfusion with the preliminary fixative is not an essential part of the method. If it is to be omitted, start at stage 3.

18.9.4. Holmes's silver method for axons

The best results are obtained with tissue from the CNS fixed for 24 h in neutral, buffered formaldehyde. Paraffin sections are used. The technique described below differs from the original (Holmes, 1943) only in the addition of a little pyridine to the silver solution. The function of the pyridine is unknown, but if it is omitted there is a reduction in the contrast between the stained axons and their background. Addition of pyridine does not change the pH of the staining solution but it may reduce the concentration of free silver ions by forming complexes such as $[Ag(C_5H_5N)_2]^+$. A preliminary treatment with 20% $AgNO_3$ is often recommended (Luna, 1968; Culling, 1974) and is erroneously attributed to the original author. This appears to be an unnecessary elaboration of the method.

Solutions required

A. 1% silver nitrate ($AgNO_3$) (stock solution)

Keep in a dark place.

B. 1% (v/v) pyridine in water (stock solution)

Keeps for several months in a tightly stoppered bottle.

C. Staining solution

Borate buffer, pH 8.4:	100 ml	Mix just before using
1% $AgNO_3$ (solution A):	1.0 ml	
1% pyridine (solution B):	5.0 ml	
Water:	to 500 ml	

D. Developer

Hydroquinone (= quinol):	2.0 g	Mix on the day it is to be used
Sodium sulphite (Na_2SO_3):	10 g	
(**or** 20 g of $Na_2SO_3.7H_2O$)		
Water: Dissolve and make up to 200 ml		

E. 0.2% gold chloride (stock solution)

May be $NaAuCl_4.2H_2O$ (yellow) or $HAuCl_4$ (brown). This solution can be used repeatedly but should be discarded (and recycled; see Kiernan, 1977b) when it has a greyish tinge or when there is an appreciable quantity of precipitated material in the bottle.

F. 1% oxalic acid ($H_2C_2O_4.2H_2O$)

This stock solution keeps indefinitely, but should be used only once.

G. 5% sodium thiosulphate ($Na_2S_2O_3.5H_2O$)

The remarks for solution F also apply here.

Procedure

1. De-wax and hydrate paraffin sections. Wash the sections in three changes of water (which **must** be distilled or de-ionized) for a total of 10 min.

2. Place the slides, in a glass rack, in staining solution C for 24 h at 37°C. The staining tank must be covered to reduce evaporation and there must be at least 50 ml of the staining solution for each slide.
3. Remove the rack of slides from solution C, shake off excess liquid, but do not rinse in water. Transfer directly to the developer (solution D) for 3 min. (See *Note 3* below.)
4. Rinse in distilled or de-ionized water, then wash for 3 min in running tap water. Rinse again in distilled or de-ionized water.
5. Immerse in 0.2% gold chloride (solution E) for 3 min.
6. Wash in two changes of water.
7. Immerse in 1% oxalic acid (solution F) for 3–10 min until the sections are deep grey. Longer treatment than necessary produces a reddish colour and this is best avoided.
8. Wash in two changes (each 1 min) of tap water.
9. Immerse in sodium thiosulphate (solution G) for 5 min.
10. Wash in three changes of tap water, dehydrate, clear, and mount in a resinous medium.

Results

Axons black. Cell-bodies and nuclei are generally unstained and connective tissue fibres should be no darker than light grey. With over-treatment at stage 7, the grey and black tones are changed to pink and dark red.

Notes

1. The method as described is nearly always successful with formaldehyde-fixed CNS or peripheral nerve trunks. Peripheral nerve endings, however, are rarely demonstrable by this technique.
2. If poor results are obtained, variations in the composition of staining solution C should be tried. The volume of 1% silver nitrate may be halved or doubled; the pH may be varied between 7.0 and 9.0. All other aspects of the technique should be kept constant. The composition of solution C will usually need to be changed when fixatives other than neutral, buffered formaldehyde are used. Satisfactory results can be obtained with tissue fixed in most of the commonly used mixtures other than those containing potassium dichromate or osmium tetroxide.
3. The slides must not be washed in water after stage 2. It is necessary to carry some silver ions into the developer. The axonal staining is invisible, or at best very faint, after the development at stage 2. The first part of the toning procedure (stage 5) causes the axons to disappear almost completely, but they reappear and become intensely coloured during stage 7 of the method.

18.9.5. Mitchell's rapid silver method

This technique (R. Mitchell, personal communication) is described because it is rapid. When it fails, minor variations can be introduced for subsequent sections without wasting too much time. The method is similar to one for celloidin sections described by Hirano & Zimmermann (1962), which can also be used on paraffin sections. Silver nuclei in axons are probably deposited during the impregnation with 20% $AgNO_3$. In the later stages of the method it is likely that a silver diammine complex is transiently formed and then immediately reduced to the metal by formaldehyde at the sites of the silver nuclei.

Material should be fixed in formaldehyde for at least 12 h and frozen sections cut (usually 15–25 μm thick). The technique is applicable to any tissue containing nerve fibres.

Solutions required

A. 20% aqueous silver nitrate

Keeps for several months in a dark place. It may be used repeatedly if not contaminated by other reagents.

B. Dilute ammonia

Ammonium hydroxide (= ammonia solution; S.G. 0.9, 28% NH_3): 20 drops
Tap water: 50 ml
} Freshly mixed

C. Formalin solution

Formalin (37–40% HCHO): 10 ml
Tap water: 40 ml
} Freshly mixed

Procedure

1. Rinse the frozen sections in three changes of water (which **must** be distilled or de-ionized). They can be left overnight in water if necessary. The sections must be handled with glass hooks. They are taken individually through the following steps of the method.
2. Place section in 20% $AgNO_3$ (solution A), for about 10 s.
3. Transfer (without washing) to the dilute ammonia (solution B), for 10 s.
4. Transfer (without washing) to formalin (solution C) for about 30 s.
5. Rinse section in water and either mount in glycerine or dehydrate, clear, and mount in a resinous medium.
6. Examine the stained section. If it is unsatisfactory, proceed as directed in the *Notes* below.

Result

Axons black. Background (cells, collagen, etc.) in shades of yellow, gold, and brown.

Notes

1. If staining is **too light, either** rinse the section in water and repeat stages 2–5, **or** leave a subsequent section longer (up to 30 min) in the 20% $AgNO_3$ at stage 2 and proceed as before.
2. If staining (especially the background) is **too dark**, try one of the following modifications on subsequent sections:
   (a) Use a stronger solution of ammonia at stage 3.
   (b) Leave for longer in the dilute ammonia at stage 3.
   (c) Rinse the section in water between stages 2 and 3.

18.9.6. Winkelmann and Schmit method for peripheral nerve endings

Material should be fixed in 4% formaldehyde for several days. Small pieces of tissue are then dehydrated in three changes of ethanol (each 1 h), cleared in xylene (for 1 h), rehydrated, and, finally, returned to the fixative until they are sectioned (Winkelmann & Schmit, 1957). (See *Note 1* below.)

This method was devised for skin, but axons in other tissues, such as muscle, are also stained by it.

Solutions required

A. Silver solution

20% aqueous silver nitrate. Keeps indefinitely; may be used repeatedly. Filter after use to remove particles derived from sections.

B. Developer

Hydroquinone:	0.2 g	Make up just before using
Sodium sulphite (Na_2SO_3):	1.0 g	
Water:	100 ml	

C. 0.2% gold chloride

(See under solution E of Holmes's method, p. 268.)

D. 5% aqueous sodium thiosulphate ($Na_2S_2O_3.5H_2O$)

Keeps indefinitely, but use only once.

Procedure

1. Cut frozen sections 50 μm thick and collect them into 4% aqueous formaldehyde. Take the free-floating sections through stages 2–10 of the method.
2. Rinse the sections in three changes of water (**must** be distilled or de-ionized) and leave for 30 min in the last change.
3. Place sections in 20% $AgNO_3$ (solution A) for 20 min. The sections should not be creased or folded.
4. Rinse sections in three changes of water, 3 s in each.
5. Place sections (avoiding folds and creases) in the developer (solution B) for 10 min.
6. Rinse in two changes of water.
7. Transfer sections to toning solution (C) for 2 min.
8. Wash in two changes of water.
9. Immerse sections in sodium thiosulphate (solution D) for 5 min.
10. Wash in two changes of water, mount onto slides, dehydrate, clear, and cover, using a resinous medium.

Result

Axons black. Other structures (cells, connective tissue, etc.) in shades of grey.

Notes

1. The preliminary treatment of the fixed blocks with alcohol and xylene is to remove soluble lipids which can interfere with the staining of axons. Other reagents (e.g. methanol, acetone, etc.) may be substituted.
2. Many beautiful photomicrographs of the results obtained with this method can be found in Winkelmann (1960). It is important that the sections be adequately thick, so that enough chemically bound silver will be carried through into stage 5 to allow solution B to function as a physical developer. Thick sections are also needed in order to reveal the full extent of axonal end-formations in the skin.

18.9.7. Gros–Schultze method for axons

This rapid and fairly reliable Bielschowsky-type method can be used with frozen or paraffin sections of formaldehyde-fixed material. The procedure described below (based on technical details given by Culling, 1974) more closely resembles the method of Gros than that of Schultze (see Jones, 1950). It is also known as the "Gros–Bielschowsky" method.

Solutions required

A. 20% aqueous silver nitrate

Keeps indefinitely; may be used repeatedly. Filter after use to remove debris derived from sections.

B. Formalin solution

Prepare four dishes or staining tanks, each containing a freshly prepared mixture of 20 volumes of formalin (37–40% HCHO) and 80 volumes of tap water.

C. Ammoniacal silver solution

(See *Note 1* below.)

To 30 ml of 20% aqueous silver nitrate in a 100 ml conical flask add ammonium hydroxide (strong ammonia solution; S.G. 0.88–0.91) until the precipitate of brown silver oxide just redissolves. The ammonia should be added drop by drop, with swirling for a few seconds between each addition. About 2.5 ml of strong ammonia solution are needed. When the precipitate has dissolved, add a further 18 drops of the ammonium hydroxide.

D. Dilute ammonia solution

Strong ammonium hydroxide (as above):	5 ml	Freshly mixed
Water:	95 ml	

E. Dilute acetic acid

Glacial acetic acid:	1.0 ml	Freshly mixed
Water:	99 ml	

F. Toning solution

0.02% gold chloride. Conveniently made by tenfold dilution of 0.2% gold chloride (see under Holmes's method, p. 268). This very dilute gold chloride should be used for only one or two batches of sections.

G. 5% aqueous sodium thiosulphate ($Na_2S_2O_3.5H_2O$)

Keeps indefinitely, but use only once.

Procedure

1. De-wax and hydrate paraffin sections. Collect frozen sections into water. Wash the sections (frozen or paraffin) in three changes of water (**must** be distilled or de-ionized), for total time of 10 min.
2. Immerse in 20% silver nitrate (solution A) for 5 min (frozen sections) or 30 min (paraffin sections). (See also *Note 2* below.)
3. Without washing, transfer the sections directly to the first bath of formalin solution (B). Agitate and transfer sections successively through the other three formalin baths, allowing about 2 min in each. The last change of formalin solution must be free of turbidity. If it is not, move the sections into one or two more baths of solution B.
4. Without washing, transfer sections (or slides) directly into ammoniacal silver solution (C) for about 20 s. (Slides bearing paraffin sections can be placed in a clean coplin jar and solution C poured in. Allow about 10 ml for each slide.) The solution becomes grey and turbid and silver mirrors form on slides and

on the inside of the staining vessel. The sections go dark brown. (See also *Notes 2* and *3* below.)
5. Transfer sections to dilute ammonia (solution D) for 1 min.
6. Transfer to dilute acetic acid (solution E) for 1 or 2 min.
7. Wash in water for about 30 s.
8. Immerse in toning solution F for 10 min.
9. Wash in water for about 30 s.
10. Immerse in sodium thiosulphate (solution G) for 3 to 5 min.
11. Wash in two changes of water, dehydrate, clear, and mount in a resinous medium.

Results

Axons black. In the CNS, fibrillary structures (neurofibrils) in the perikarya and dendrites of some large neurons are also coloured black. Other tissue components, grey.

Notes

1. The ammoniacal silver solution should be mixed just before using. The ammonium hydroxide should come from a recently opened stock bottle since it loses NH_3 on standing. Handle strong ammonia carefully in a fume hood. Surplus and used solution should be discarded by washing down the sink. All ammoniacal silver solutions are potentially dangerous because if they evaporate to dryness the deposit may contain fulminating silver, which is probably a mixture of silver amide ($AgNH_2$) and nitride (Ag_3N). Fulminating silver explodes violently when touched.
2. The length of time in 20% $AgNO_3$ can be varied if unsatisfactory results are obtained with the times suggested. If no staining at all is obtained, it is probable that solution C contained too large an excess of ammonia.
3. Dirty precipitates on the sections can be due to too long a time in the ammoniacal silver solution. They are presumably due to the reaction

$$2Ag(NH_3)_2^+ + 2OH^- + HCHO \rightarrow 2Ag\downarrow + 4NH_3 + H_2O + HCOOH$$

18.9.8. A physical developer method for axons

This previously unpublished technique may be applied to frozen or paraffin sections of any formaldehyde-fixed tissue. Bouin's fluid is also a suitable fixative. This method differs from earlier techniques employing physical developers in two respects: (a) the initial impregnation is with a concentrated silver nitrate solution, so only a short time is required for the formation of nuclei within axons; (b) the action of the developer, which is modified from one used in photographic research (Berg & Ford, 1949) is more easily controlled than that of other physical developers previously used for histological purposes.

Solutions required

A. 20% aqueous silver nitrate

Keeps for several months and may be used repeatedly.

B. Sodium sulphite solution

Sodium sulphite (Na_2SO_3): 15.0 g
Water: to 500 ml

Keeps indefinitely, but may be used only once.

C. Stock solution for physical developer

Hydroquinone (= quinol):	4.0 g	Stable for 4–6 weeks
Citric acid ($C_6H_8O_7 . H_2O$):	4.0 g	
Gelatine (use a high quality bacteriological grade):	4.0 g	
Water:	400 ml	
Glacial acetic acid:	20 ml	
Dissolve all the above ingredients (30 min with occasional shaking) and add:		
Water:	to make 500 ml	

D. Physical developer (working solution)

Solution C: 100 ml
Solution A: 20 ml

This must be mixed immediately before use (see stage 6 of the procedure below).

E. Gold chloride solution

This is a 0.2% aqueous solution of either sodium

chloroaurate ($NaAuCl_4.2H_2O$) or chloroauric acid ($HAuCl_4$). It may be re-used many times.

F. 1% aqueous oxalic acid ($H_2C_2O_4.2H_2O$)

Keeps indefinitely, but should be used only once.

G. 5% aqueous sodium thiosulphate ($Na_2S_2O_3.5H_2O$)

Keeps indefinitely but should be used only once.

Procedure

1. De-wax and hydrate paraffin sections. Attach frozen sections to slides. Either Mayer's albumen or chrome–gelatine may be used as adhesive. Wash in three changes of water.
2. Immerse slides in 20% $AgNO_3$ (solution A) for 15 min (time is not critical and may range from 10 min to 2 h).
3. Wash in three changes of water.
4. Immerse in 3% Na_2SO_3 (solution B) for 5 min. (The slides may remain in this solution for 1 h without ill effect.)
5. Rinse in three changes of water. Drain.
6. Mix the working solution of the physical developer (D) and pour it into the vessel containing the slides. The time for development depends on the ambient temperature: usually 4 min at 25°C or 6 min at 20°C. If the developer becomes grey and cloudy, proceed immediately to stage 7, whatever time has elapsed.
7. Pour off the developer and rinse slides with three changes of water. Examine under a microscope. If axonal staining is adequate, counterstain, dehydrate, clear, and mount. If more intense staining is required, proceed with the remaining stages of the method.
8. Tone in gold chloride (solution E) for 5 min. Rinse in three changes of water. If staining is now satisfactory, proceed to stage 11. If not, proceed to stage 9.
9. Immerse in oxalic acid (solution F) for 5–10 min. The time is not critical.
10. Rinse in three changes of water.
11. Immerse in sodium thiosulphate (solution G) for 5 min.
12. Wash in running tap water for about 2 min.
13. Apply a counterstain if desired. Neutral red (see Chapter 6, p. 84) is suitable for the CNS; the van Gieson method (see Chapter 8, p. 99), for peripheral tissues.
14. Wash, dehydrate, clear, and mount in a resinous medium.

Result

Axons black. Other structures are usually completely colourless if the procedure is stopped at stage 7 or stage 8. Treatment with oxalic acid (stage 9) produces a grey background and sometimes causes blackening of connective tissue fibres. If neutral red is used as a counterstain, nuclei, Nissl substance (RNA), and proteoglycans are red. The van Gieson counterstain gives yellow cytoplasm and red collagen.

For some parts of the CNS, including the cerebral and cerebellar cortex, Holmes's method is superior. This method is superior to Holmes's technique for peripheral innervation and for the retina and optic nerve.

18.9.9. Golgi–Cox method

Although it is a slow method, this is possibly the most reliable of the Golgi techniques.

Solutions required

A. Golgi–Cox fixative

This is prepared from three stable stock solutions:

(i) 5% aqueous mercuric chloride ($HgCl_2$).
(ii) 5% aqueous potassium dichromate ($K_2Cr_2O_7$).
(iii) 5% aqueous potassium chromate (K_2CrO_4).

Working solution. Mix in order:

(i) 20 ml	Mix before using
(ii) 20 ml	
Water: 40 ml	
(iii) 16 ml	

B. Alkaline developer

Strong ammonia solution (17% NH_3):	5 ml	Mix just before using
Water:	95 ml	

Procedure

1. Fix pieces of nervous tissue (not more than 10 mm thick) in the working solution of

Golgi–Cox fixative for 6–8 weeks in a tightly capped container at 37°C. Pour off fixative and replace with fresh solution after the first day.
2. Wash in water, many changes, for 6–8 h, then dehydrate and embed in nitrocellulose. Cut sections 100 μm thick.
3. Treat the sections with the alkaline developer (solution B) for 2 or 3 min.
4. Wash in water, dehydrate, clear, and mount in a resinous medium. (See *Note* below.)

Result

Some neurons and neuroglial cells, including their cytoplasmic processes—black. Background—pale yellow.

Note

A mixture of chloroform (33 ml), xylene (33 ml), and absolute ethanol (33 ml) is preferred to absolute alcohol for the last stage of dehydration (following 95% alcohol). This can be followed by clearing for 10–15 min in creosote (beechwood) or in cedarwood oil, followed by a rinse in xylene.

The black colour fades with time if a coverslip is applied, so it is usual to mount the sections in **thick** Canada balsam, and allow to set at 40–45°C without a coverslip. Such preparations are stable for many years. If a synthetic mounting medium such as DPX is used, coverslips may be applied, but fading occurs after about 6 months.

18.9.10. Vital methylene blue

This is a method for freshly removed tissue, intended for the demonstration of peripheral innervation (Richardson, 1969; Kiernan, 1974).

Solutions required

A. Buffered diluent. (Store at 4°C)

Sodium succinate, $(CH_2COONa)_2 . 6H_2O$:	14.0 g
Sodium chloride (NaCl):	6.0 g
Glucose (dextrose):	2.0 g
Water:	800 ml
Concentrated hydrochloric acid:	Add carefully until the pH is 5.5
Water:	to make final volume 1000 ml

B. Methylene blue stock. (Keep at room temperature)

Methylene blue (C.I. 52015):	100 mg
Water:	100 ml

(It is advisable to use a grade of dye designated "for vital staining".)

C. Staining solution

(Usually mixed just before use, but can be kept for a few weeks at 4°C.)

Solution A:	100 ml
Solution B:	1.5 ml

Allow to warm to 37°C before using.

D. Ammonium molybdate fixative

Buffer (pH 5.5):

Sodium phosphate, dibasic (Na_2HPO_4):	2.0 g	Keeps for several months at 4°C
Citric acid ($C_6H_8O_7 . H_2O$):	1.2 g	
Water:	500 ml	

Working fixative solution:

Buffer (above):	50 ml	Prepare before using
Ammonium molybdate, $(NH_4)_6Mo_7O_{24} . 4H_2O$:	3 g	

It takes about 15 min for the ammonium molybdate to dissolve (magnetic stirrer). Make up the fixative while the specimens are incubating in the stain. Cool to 0–4°C for use.

Procedure

1. Kill a small animal. Remove pieces of tissue, no more than 1.0 mm thick. Collect into saline (0.9% NaCl) but do not leave them in it for more than 5 min.
2. Put the pieces in the warm staining solution (C) with continuous aeration at about 30°C. (The apparatus shown in Fig. 18.1, contained in a 37°C incubator, produces these conditions.) Leave for 30–40 min.
3. Remove specimens from the staining solution, rinse quickly (5–10 s) in water and place in the working fixative solution (D) for 12–18 h at 4°C.
4. Wash in water (4°C) for 10 min.
5. Transfer to absolute methanol at 4°C for 30 min.

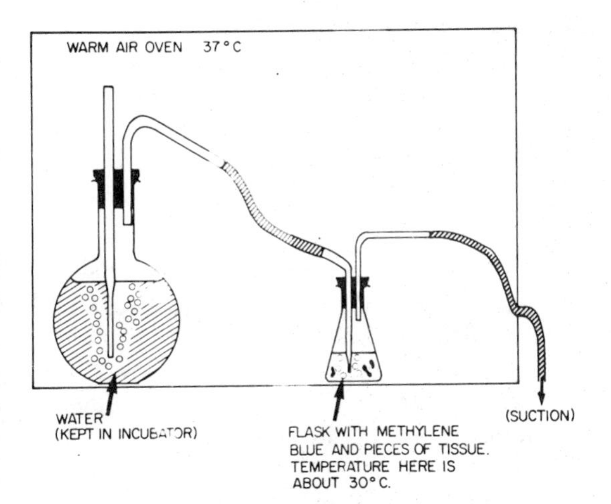

FIG. 18.1. Apparatus for bubbling warm moist air through a solution of methylene blue for vital staining. The air is saturated with water vapour before bubbling through the dye solution to avoid excessive cooling of the latter by evaporation.

6. Complete the dehydration in *n*-butanol at 4°C for 30 min.
7. Clear in benzene, 15 min (at room temperature).
8. Prepare as whole mounts in a resinous medium. (See also *Note 2* below.)

Result

Axons blue. The background should be largely unstained.

Notes

1. If the method fails, try varying the pH between 5 and 7.5. In general, thicker axons are stained at higher pH.
2. As an alternative to whole mounts, the specimens may be (a) sectioned on a freezing microtome and collected into ice-cold water after stage 4, and then mounted onto slides, dehydrated in **cold** alcohols, cleared, and covered, **or** (b) embedded in wax after stage 7 and the paraffin sections mounted onto slides, cleared, and covered. Sections should be thick: 50–100 μm.

18.9.11. Cajal's gold–sublimate method for astrocytes

The ideal fixative is formal–ammonium–bromide (see Chapter 2, p. 22). If material fixed in ordinary formaldehyde must be used, proceed as in *Note* below. Penfield and Cone (1950) state that the method works better with human, cat, and dog than with rabbit CNS. The rat's brain gives reasonably clear results.

Solutions required

A. Globus's reagents (see *Note*, p. 276)

Prepare as required.

(i) *Ammonia–water:* Ammonium hydroxide (S.G. 0.9; 27% NH_3): 5.0 ml
Water: 45 ml

(ii) *Globus's hydrobromic acid:* Concentrated hydrobromic acid (47% HBr): 5.0 ml
Water: 45 ml

B. Gold–sublimate mixture

A stock solution of gold chloride (1% aqueous brown gold chloride, $HAuCl_4$) is needed. (This should not be a solution which has been previously used for other purposes.)

Working solution

Mercuric chloride ($HgCl_2$; of the highest available purity):	0.5 g
Water:	60 ml

Dissolve the $HgCl_2$ (warm gently and stir with a glass rod). To this solution, add 10 ml of 1% brown gold chloride. This mixture should be made just before using.

C. 5% aqueous sodium thiosulphate ($Na_2S_2O_3 . 5H_2O$)

Procedure

1. Fix pieces of tissue up to 5 mm thick in formal–ammonium–bromide for 24–48 h. Cut frozen sections (about 20 μm) and collect into water containing a few drops of formalin.
2. Place about twenty sections in the gold–sublimate mixture (solution B) for 4–8 h in a petri dish in the dark. The sections must not be creased or folded. They should be manipulated with a glass hook. At hourly intervals remove one or two sections, take them

through stages 3, 4 and 5, and examine under a microscope. Grossly, the sections in the gold–sublimate solution become purple. This colour deepens with time. The hourly checking of sections need not begin until the purple coloration appears.

3. When staining of astrocytes is adequate, pass all the remaining sections through two changes of water.
4. Immerse in 5% $Na_2S_2O_3.5H_2O$ (solution C) for 5 min.
5. Wash in two changes of water, mount onto slides, dehydrate, clear, and cover, using a resinous mounting medium.

Result

Astrocytes dark red, dark purple, and black. The cytoplasmic processes and pericapillary end-feet stain more intensely than the perinuclear parts of the cells. Other cells, notably neurons, are coloured in lighter shades of purple.

Note

For material fixed in formaldehyde (without NH_4Br), cut frozen sections and wash them in three changes of water. Place the sections in ammonia water (solution A(i)) overnight in a tightly closed container, then rinse them quickly in two changes of water and put them into Globus's hydrobromic acid (solution A(ii)) for 1 h at 37°C. Rinse quickly in two changes of water and place the sections into solution B as stage 2 of the procedure given above.

18.9.12. Bromoindigo–silver method for motor end-plates

A combination of an indigogenic histochemical method for the AChE of the subneural apparatus with a silver method for axons was described by McIsaac and Kiernan (1974). The following technique differs only in the replacement of the original silver method by the one described in Section 18.9.8 of this chapter. The new silver method is more reliable than the one originally recommended and gives hardly any non-specific staining of blood-vessels and muscle cells. Consequently, it is possible to counterstain the preparations.

Muscles are fixed for 24 h **at 4°C** in neutral, buffered formaldehyde. Frozen sections 60 μm thick are collected into water.

Solutions required

A. Reagents for the indigogenic method for carboxylic esterases (Chapter 15, pp. 215–216).
B. Reagents for the physical developer method for axons (this chapter, p. 272).
C. van Gieson's stain (Chapter 8, p. 99).

Procedure

1. Wash frozen sections in water, then incubate according to the indigogenic method for esterases (Section 15.4.4, p. 216). Thirty minutes' incubation is usually sufficient. Check a wet section under a microscope to ensure that the subneural apparatuses are stained. They appear as blue dots, each about the size of a nucleus, and are usually confined to a band across the widest part of the muscle.
2. Rinse sections in two changes of water and mount onto slides. Mayer's albumen or chrome–gelatine may be used as adhesive. Allow to dry for about 20 min on a hotplate (45–50°C).
3. Wash slides in water and stain axons by the physical developer method (Section 18.9.8, p. 273). Gold toning is usually desirable, but treatment with oxalic acid is rarely needed.
4. If desired, counterstain the sections with van Gieson's picric acid–acid fuchsine (Section 8.5.1, p. 99) for about 1 min. (See *Note 1* below.)
5. Wash in tap water, dehydrate in 95% and three changes of 100% ethanol, clear in xylene, and cover, using a resinous mounting medium.

Result

Axons, including their terminal branches within motor end-plates—black; subneural apparatus—blue. If the counterstain (stage 4) is applied, muscle cytoplasm is yellow and collagen red.

Notes

1. The counterstaining should be much lighter than when it is used primarily for connective tissue. It is often advantageous to dilute the van Gieson mixture 4–5 times with water.
2. The resolution of detail within the subneural apparatus by this method is inferior to that obtainable with thiocholine methods for AChE (see Chapter 15). The latter are therefore preferred when demonstration of the innervating axons is not required.

18.9.13. Aldehyde–fuchsine for neurosecretory material

This method, which is a slight modification of that of Gabe (1953), differs from the aldehyde–fuchsine procedure for elastic fibres only in the addition of an oxidation before staining.

Suitable fixatives are discussed in Section 18.8.2, p. 265.

Solutions required

A. Acidified permanganate

Potassium permanganate ($KMnO_4$):	1.0 g	Mix just before using
Water:	200 ml	
Stir thoroughly and add, slowly with stirring:		
Concentrated sulphuric acid:	2.0 ml	

B. 1% aqueous oxalic acid

Keeps indefinitely, but use only once.

C. Aldehyde–fuchsine (working solution)

See Chapter 8, p. 102.

D. 1% v/v glacial acetic acid in 70% alcohol (freshly prepared)

E. A counterstain

For example: alum–haematoxylin for nuclei, fast green FCF for cytoplasm and collagen; or a blue cationic dye such as toluidine blue O.

Procedure

1. De-wax and hydrate paraffin sections. Remove mercurial deposits if necessary.
2. Immerse slides in acidified permanganate (solution A) for 5 min.
3. Wash in tap water to remove excess of the purple $KMnO_4$. The sections should be brown.
4. Immerse in oxalic acid (solution B) until the sections are white (about 2 min).
5. Wash in running tap water for 3–5 min.
6. Stain with aldehyde–fuchsine (solution C) for 15 min.
7. Rinse in 70% alcohol containing 1% v/v acetic acid (solution D) for 1 min (time not critical).
8. Rinse in two or three changes of 70% ethanol until no more colour comes out of the sections.
9. Wash in water and apply a counterstain if desired.
10. Dehydrate, clear, and mount in a resinous medium.

Result

Neurosecretory material purple.

Note

Elastin, mast cell granules, and cartilage matrix are also purple, even when the oxidation (stage 2) is omitted. The complete method can also be used with the pancreas as a selective stain for the β-cells of the islets of Langerhans.

18.9.14. Iodide–osmium method

This procedure is based on the method described by Rodrigo *et al.* (1970).

Pieces of tissue must be taken from a freshly killed animal and should be no more than 4.0 mm thick, preferably only about 1.0 mm. Hollow viscera are opened to make flat specimens, which can then be fixed to an improvised glass frame to facilitate penetration of the fixative.

Solutions required

A. 2% osmium tetroxide

1.0 g of OsO_4 is dissolved in 50 ml of water by breaking the cleaned ampoule under water in a clean, stoppered bottle. Do this in a fume hood. The solution keeps for several weeks at 4°C.

B. Zinc iodide

Put 6.0 g of zinc powder in a 500 ml conical flask. Add 10 g of iodine (resublimed) and then 200 ml of water. Swirl continuously. The elements combine with evolution of heat and the colour of the solution changes from reddish to grey. Allow the remaining solid material to settle and then filter. This solution should be prepared shortly before use and the excess discarded.

C. Working fixative

Add 25 ml of solution A to 75 ml of solution B (or smaller quantities in the same proportions). The mixture is at first yellow, but soon darkens. Mix immediately before using.

D. 2% potassium permanganate

Potassium permanganate ($KMnO_4$):	2.0 g
Water:	100 ml

Keeps for several months. Filter before using

Alternatively, dilute a stock solution containing 6% $KMnO_4$.

E. 2% oxalic acid

Oxalic acid ($H_2C_2O_4 . 2H_2O$):	5.0 g	Keeps indefinitely
Water:	250 ml	

Procedure

1. Immerse the specimens in at least twenty times their own volumes of Solution C, for 24 h.
2. Wash in running tap water for 4–6 h (if frozen sections are to be cut) or for 24 h if the specimens are to be embedded in wax.
3. **Either** cut frozen sections **or** dehydrate, clear, embed in wax, and cut paraffin sections. Mount the sections onto slides and take to water. (See *Note* below.)
4. Examine the sections. If staining is satisfactory, dehydrate, clear, and mount. If the colour is too dark and nerve fibres cannot be distinguished from background, proceed as follows.
5. To differentiate the stain, pass the slides individually through (a) 2% $KMnO_4$ (solution D), (b) water, (c) 2% oxalic acid (solution E), (d) water, for about 5 s in each. Re-examine the sections. Repeat the differentiating procedure until the desired appearance is obtained.
6. Wash in water, dehydrate through graded alcohols, clear in xylene, and mount in a resinous medium.

Result

Unmyelinated terminal branches of axons black. Myelin sheaths are also blackened. Other components of the tissue appear in shades of grey.

Note

Very thin specimens can be prepared as whole mounts. It is important to wash out all the osmium tetroxide before dehydrating, since alcohols reduce OsO_4 (see Chapter 2). Sections need to be fairly thick (20 to 100 μm).

18.10. EXERCISES

Theoretical

1. Which of the staining methods discussed in this chapter would be suitable for:

(a) Study of the sizes and shapes of neurons in the six layers of the cerebral cortex?
(b) Study of the extents of the dendritic trees of neurons?
(c) Examination of cell groups and major tracts in the brain stem, at low magnification?
(d) Comparison of the lengths and thicknesses of astrocytic processes in normal and pathological brains?

2. Why would Sudan IV (see Chapter 12) be unsatisfactory as a stain for myelinated nerve fibres in frozen sections?

3. Why does the developer used in Holmes's method for axons contain sodium sulphite?

4. If, in processing an exposed photographic film, you accidentally fixed it (in sodium thiosulphate) when you should have developed it, what could you do to obtain printable negatives? (If the fixer used were a simple aqueous solution of $Na_2S_2O_3$ it would be possible to do this. An acid-fixing-hardening solution, which is normally used in photography, would dissolve the latent image and the negative would be irretrievable.)

5. Many insoluble sulphides can be converted to silver sulphide by treatment with aqueous silver nitrate solution on account of the extremely low solubility product of Ag_2S. Particles of silver sulphide are able to serve as nuclei for physical development. These reactions form the basis of a histochemical method (Timm's sulphide–silver technique) of low specificity but high sensitivity for Cd, Cu, Fe, Hg, Pb, Zn, and several other metals. Devise a histochemical procedure, based on the above reactions, for the detection of heavy metals in tissues. How would you control for non-specific staining?

6. What methods are available for the demonstration of unmyelinated autonomic axons in tissues such as the skin or viscera? Evaluate the advantages and disadvantages of each technique.

Practical

7. Using frozen sections of formaldehyde-fixed skin or tongue and brain, demonstrate myelinated nerve fibres by means of (a) one of the methods described in this chapter, and (b) a suitable histochemical method from Chapter 12. Why does the histochemical technique give better results with peripheral than with central nervous tissue?

8. Demonstrate axons by silver methods in frozen and paraffin sections of brain and of a peripheral tissue such as skin or muscle, after fixation in at least two different mixtures (e.g. neutral buffered formaldehyde, Bouin, Heidenhain's Susa). It will be necessary to vary the technique to suit the fixative.

9. Demonstrate astrocytes in sections of brain.

10. Apply methods for peripheral nerve-endings (e.g. physical developer method; vital methylene blue; iodide–osmium; histochemical methods for choline esterases or monoamines) to a selection of tissues taken from a freshly killed rat or mouse. Suitable tissues include thin skin of the external ear, mesentery, urinary bladder, diaphragm, muscles from front of neck, salivary glands, and the right atrium of the heart. The thinner specimens, especially from the mouse, are easily prepared as whole mounts.

11. Show that the neurosecretory axons in the posterior lobe of the pituitary gland are unmyelinated.

19

Immunohistochemistry

19.1. Antigens and antibodies 280
19.2. Antibody molecules 281
19.3. Antigen–antibody complexes 282
19.4. Some definitions 283
19.5. Direct fluorescent antibody methods 283
19.5.1. The technique 283
19.5.2. Controls 285
19.5.3. Shortcomings of the direct method 285
19.6. Indirect fluorescence techniques 286
19.6.1. Method for detection of an antigen 286
19.6.2. Controls 287
19.6.3. Critique of the method 287
19.7. Enzyme-labelled antibody methods 288
19.8. The unlabelled antibody–enzyme method 289
19.8.1. Principle of the method 289
19.8.2. Procedure for detection of an antigen 289
19.8.3. Controls 290
19.8.4. Critique of the method 290
19.9. Other immunohistochemical methods 292
19.9.1. Sandwich technique for antibody in tissue 292
19.9.2. Detection of antibody in serum 292
19.9.3. Methods using staphylococcal protein A 292
19.10. Non-immunological affinity techniques 293
19.11. Exercises 293

In the preceding chapters attempts have been made to show how the principles of chemistry and biochemistry are applied in techniques for the localization of substances within tissues. In this chapter we shall consider some methods based on the precepts of immunology. Students totally unfamiliar with this science are advised to read an introductory text such as that by Cunningham (1978) or the relevant parts of a textbook of pathology. Fortunately, the immunological principles applicable to microscopical techniques are fairly easy to grasp. Valuable works of reference in the field of immunohistochemistry are the books by Goldman (1968), Nairn (1976), and Sternberger (1979).

Immunohistochemical methods are employed in many fields of biological research and a few are regularly used in diagnostic pathology. Every problem to which these techniques are to be applied must be considered individually and it is not possible to give detailed practical instructions applicable to all the uses of any individual method. This chapter therefore serves only as an introduction to the subject. Before attempting a method that has been used before, it is important to read thoroughly the original literature describing the experience of other workers. In developing a procedure that is not an exact repetition of someone else's work, attention must be paid to the principles outlined in this chapter and to the more comprehensive works mentioned above.

19.1. ANTIGENS AND ANTIBODIES

Vertebrate animals are able to defend themselves against the potentially harmful effects of macromolecules derived from other organisms. Any such foreign material that may enter an animal's body is called an **antigen**. When an antigen is introduced into the *milieu intérieur* of an animal, some of its molecules are carried, by lymph or blood-vessels, to lymph nodes. Here the antigen molecules come into contact with small lymphocytes of the B-type (bone-marrow derived), which react to the encounter by transforming themselves into plasma cells. The plasma cells synthesize and secrete **antibodies.** These are proteins of the γ-globulin class capable of combining specifically with the antigens that evoked their production. It will be noticed that the words "antigen" and "antibody" are difficult to define. An antigen is a substance produced in response to the presence of an antigen. The antibodies secreted by plasma cells in lymphoid tissues circulate in the blood plasma and so gain access to all parts of the body including the original sites of introduction of antigens. The combination of an antibody with its antigen commonly results in neutralization of the toxicity or pathogenicity of the latter.

When a molecule of antigen bumps into a molecule of its antibody, the two combine to form an **antigen–antibody complex.** This happens because part of the antibody molecule has been specially tailored to accommodate part of the antigen molecule.

The reaction is reminiscent of the "lock-and-key" mechanism by which an enzyme combines with its substrate. The two components of the complex are held together by non-covalent forces such as ionic attraction, hydrogen bonding, and hydrophobic interaction.

The part of an antigen molecule that joins it to the antibody is known as an **antigenic determinant**. A large protein molecule, or an object such as a bacterium that consists of many different macromolecules, will have many antigenic determinants and will therefore evoke the synthesis of many different species of antibody molecule, all of which will be capable of combining with the same antigen. It is, however, a property of all antibodies that they form complexes only with the antigens that stimulated their production.† This specificity is fundamental to the techniques of immunohistochemistry: no purely chemical reactions can be used to identify individual macromolecular substances.

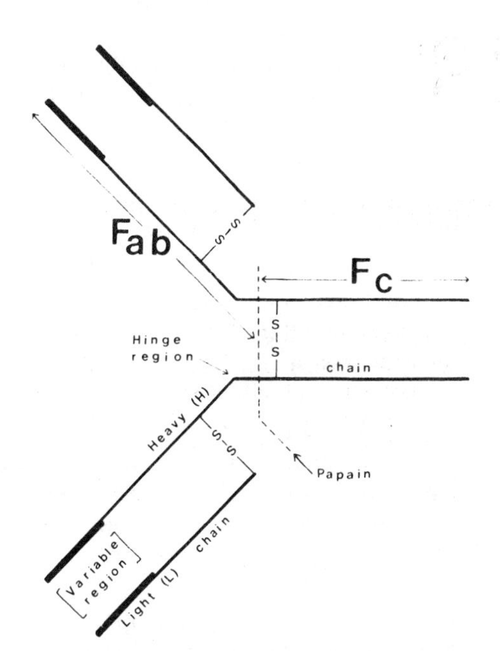

Fig. 19.1. Diagram showing the structure of one molecule of immunoglobulin-G (IgG). The antibody-combining sites are the variable regions of the two F_{ab} segments. When IgG serves as an antigen in immunohistochemical techniques it is the F_c segment of the molecule that is the antigenic determinant. Each IgG molecule is potentially capable of combining with two specific antigenic determinant sites. The distance between these two sites may vary since there is some flexibility at the hinge regions of the H chains. Note that papain releases one F_c and two F_{ab} fragments from each IgG molecule. Cleavage of the disulphide bridges yields two H chains and two L chains. Oligosaccharide chains are attached to the H chains in the F_c segment.

19.2. ANTIBODY MOLECULES

The antibodies circulating in the blood belong to the γ-globulin class of plasma proteins, and are known as **immunoglobulins**. The most abundant type is immunoglobulin-G (IgG). This is not the only kind of immunoglobulin. The other types, however, are relatively unimportant to the immunohistochemist. The IgG molecule consists of two identical subunits joined by a disulphide (cystine) bridge. Each subunit comprises two polypeptide chains: a long one, the heavy (H) chain, and a short one, the light (L) chain. The H chain is joined to the L chain by a disulphide bridge. The structure of IgG is shown diagrammatically in Fig. 19.1. It can be seen that the molecule is Y-shaped. The stem of the Y and the proximal parts of the arms have the same amino-acid composition in all the IgG molecules of a given species of animal. Specificity for antigens resides in the distal (variable) part of the H and L chains constituting the two limbs of the Y. Each limb is capable of combining with an antigenic determinant and is therefore called an antibody fragment or F_{ab}. The stem of the Y, consisting of parts of both H chains, is called the constant fragment or F_c. The F_{ab} and F_c fragments can be isolated by collecting the products of digestion of IgG by papain, a proteolytic enzyme. Other fragments have been isolated from the products of attack by other enzymes and by disulphide-splitting reagents.

†This statement is not strictly true. Two closely similar proteins may react with an antibody raised by only one of them. Such cross-reactivity occurs when the antigenic determinant site is a sequence of amino acids common to both the proteins. Cross-reactivity is sometimes a source of confusion in the interpretation of immunohistochemically stained preparations because it cannot be detected by the usual controls for specificity. The subject is discussed by Swaab *et al.* (1977), Hutson *et al.* (1979), and Vandesande (1979), but will not be further considered in this introductory account.

From the point of view of the immunohistochemist, the IgG molecule has three important features:

(i) There are two sites, the ends of the F_{ab} segments, each capable of binding to an antigenic determinant. Thus, the antibody molecule is **bivalent**.
(ii) Part of the IgG molecule, the F_c fragment, is common to all antibodies of the animal species concerned and is not involved in combining with antigens.
(iii) Immunoglobulins are themselves macromolecules and as such can behave as antigens when injected into different species of animals. Because the F_c fragment has a constant chemical structure, it is possible to raise in one species antibodies against all the possible immunoglobulins that might be produced by another species of animal. Such anti-antibodies (anti-γ-globulin antibodies) are important immunohistochemical reagents. As an antigen, the IgG molecule may have several antigenic sites and be able to bind more than one molecule of anti-antibody.

Antibody molecules are large: the M.W. of IgG is 150,000. For this reason it is possible to conjugate some of their amino-acid side-chains with other compounds such as fluorochromes or enzymes. This process, known as "labelling", usually involves the formation of covalent linkages with the ε-amino groups of lysine. The combining properties of an antibody will remain intact provided that the labelling molecule does not sterically obstruct the specific "keyholes" formed by the distal parts of the F_{ab} segments. Since, in solution, the molecules of a labelling agent react randomly with different parts of the molecules of IgG, some blocking of the combining sites is inevitable. Consequently the potency of an antibody-containing serum is always reduced by combination with a fluorescent or enzymatic label.

19.3. ANTIGEN–ANTIBODY COMPLEXES

When a solution containing an antigen is mixed with a solution containing appropriate antibodies, the antigen–antibody complex often precipitates. Insoluble aggregates are formed because each antibody molecule is bivalent and each antigen molecule has multiple determinant sites. If either component of the mixture is present in excess, the aggregates will be smaller and will remain soluble.

If an antigen forms part of a solid structure, such as a cell in a section of a tissue, it is able to bind antibody molecules from an applied solution:

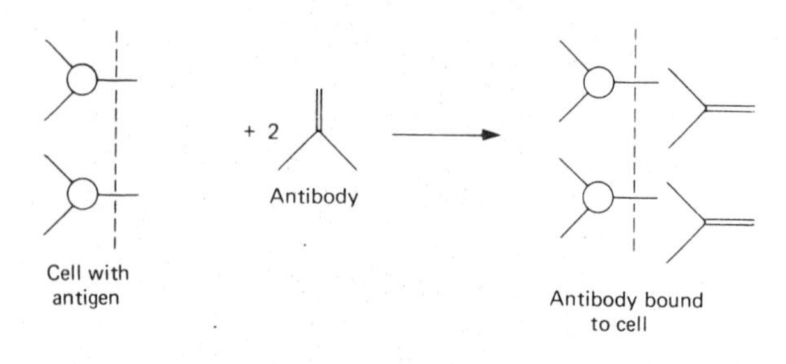

The antibody is quite firmly attached and is not removed when the excess solution is washed off. Similar attachment occurs when a dissolved antigen is brought into contact with an object such as a plasma cell which contains appropriate antibodies:

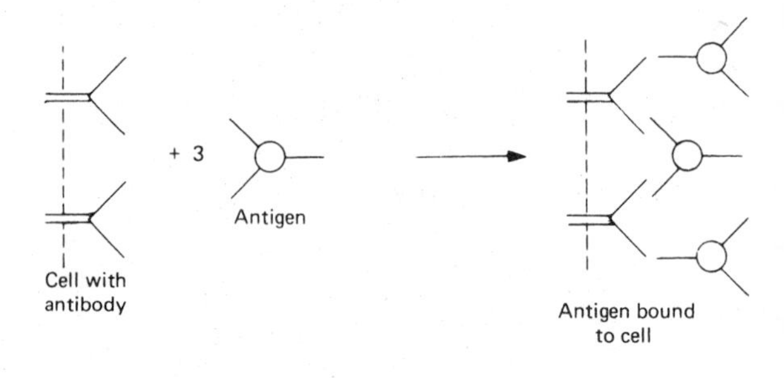

The number of molecules bound from an applied solution depends on the abundance of combining sites in the specimen. Steric factors are also important: if in the above example the two antibody molecules were further apart, they would be able to unite with four antigen molecules.

It is possible to label antibodies by conjugating them with fluorochromes or with histochemically demonstrable enzymes. Consequently, if a solution of a labelled antibody is applied to a section of tissue containing the appropriate antigen, the label will be detected at those sites at which antigen–antibody complexes have been formed. Since antigens and antibodies combine specifically with one another, it ought to be possible to localize individual macromolecular substances under the microscope. The techniques of immunohistochemistry are based on the use of labelled antibodies, but the methodology is, unfortunately, not quite as simple as might at first be thought. The more important techniques will be discussed in Sections 19.5–19.9.

19.4. SOME DEFINITIONS

The understanding of immunohistochemical methods will be easier if certain widely used terms are first defined.

Immunocytochemistry. Widely used as a synonym for immunohistochemistry but strictly applicable only when cells or intracellular structures are the objects of interest.

Immunofluorescence. Secondary fluorescence introduced into tissues by the application of immunohistochemical methods in which fluorochromes are used as labels.

Immunization. Administration of antigen to an animal to evoke the production of antibodies. Originally the word was applied to the induction of immunity to infectious diseases. It now enjoys wider usage. The act of injecting antigenic material is often called **inoculation.**

Serum. Blood plasma from which the fibrinogen has been removed. Mammalian sera contain about 8% w/v protein, consisting of approximately equal proportions of albumin and globulin.

Antiserum. Serum containing antibodies to an antigen. Loosely used also for diluted sera and diluted solutions of the globulin fraction.

Globulin. The proteins remaining in serum after removal of the albumin by chromatographic or other separation methods. The γ-globulin fraction, comprising the immunoglobulins, constitutes approximately one-third of the total globulin. The concentration of γ-globulin in whole serum is 7–15 mg per ml.

Label. A molecule artificially attached to a protein. The usual labels are either fluorescent substances or enzymes for which simple, reliable histochemical methods are available. Horseradish peroxidase (HRP) is the most popular label of the latter type. Ordinary dyes and radioactive isotopes are also used to label proteins, though mainly for applications other than immunohistochemistry.

Saline. Water is unphysiological and alcohol denatures proteins, so all immunological materials are dissolved in 0.9% sodium chloride, which is usually dissolved in 0.03 M phosphate buffer, pH 7.4.

Absorption. (i) Neutralization of specific antibody in an antiserum by adding an excess of the appropriate antigen. (ii) Removal of unreacted fluorochrome from a solution containing conjugated proteins by treatment with activated charcoal. (iii) Treatment of a labelled antiserum with powdered acetone-fixed tissue, such as liver or kidney, in order to remove unwanted labelled proteins that bind to tissues by non-immune mechanisms. The tissue powder must not, of course, contain the antigen to which the antiserum was raised.

19.5. DIRECT FLUORESCENT ANTIBODY METHODS

These are the simplest and oldest immunohistochemical methods. A known antigen in a section or smear of tissue is localized by virtue of its combination with fluorescently labelled molecules of its antibody. The technique has several shortcomings and is not often used, but it will be described in some detail in order to introduce principles that also apply to the more complicated modern methods.

19.5.1. The technique

It is first necessary to isolate the antigen and purify it as much as possible. An animal (of a species other than that from which the antigen was taken, if the antigen is of animal origin) is then immunized by injecting it, usually on several occasions, with the purified antigen. When the animal has a high circulating level of antibody, blood is collected and the serum, an antiserum, is separated. If possible, the globulin or, preferably, the γ-globulin, is isolated from the antiserum. The techniques for purification of antigen, immunization, and isolation of serum proteins require some expertise in practical biochemistry and immunology. Apparatus not usually available in a general histology laboratory is also needed. Very detailed descriptions of most of the practical procedures used in immunology are given by Garvey *et al.* (1977), but that book includes only a brief account of immunohistochemical methodology. Many pure antigenic substances and antisera are now available commercially.

The antiserum must next be labelled by conjugating its proteins with a fluorescent dye. The most widely used label is fluorescein isothiocyanate (FITC). In alkaline solution (pH 9–10), FITC combines covalently with proteins, reacting principally with the ε-amino group of lysine:

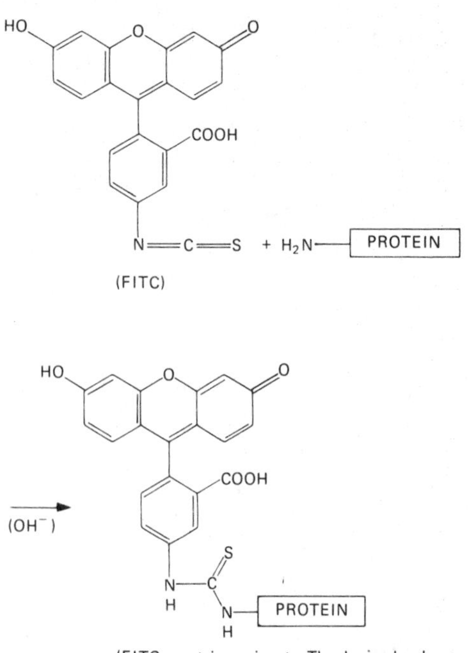

(FITC–protein conjugate. The dye molecule is covalently bound through a thiourea linkage)

Other fluorochromes are also used. The reactive group is usually isothiocyanate, sulphonyl chloride ($—SO_2Cl$), or dichlorotriazinyl (see p. 40). The most popular alternatives to FITC are l-dimethylaminonaphthalene-5-sulphonyl chloride (dansyl chloride) and various reactive derivatives of rhodamine B and related dyes.

The fluorochrome will, of course, conjugate with all the types of protein present in the serum, not just with the antibodies. Unbound fluorescent material will also be present in the conjugated serum because some hydrolysis of FITC occurs in alkaline solutions:

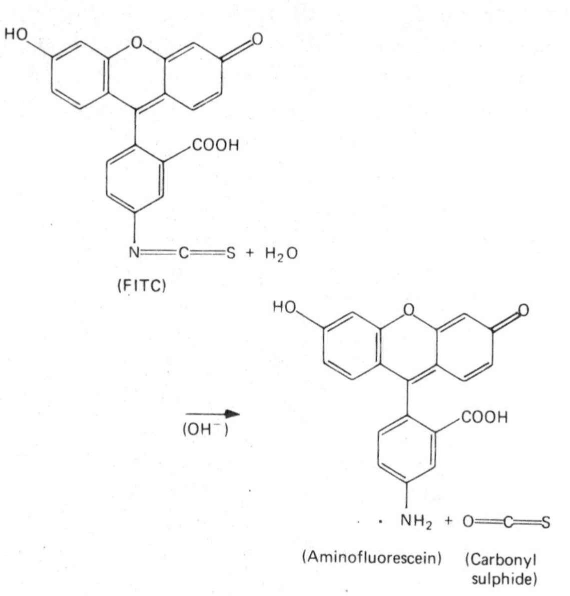

The unconjugated fluorescent material must be removed from the serum since it would be capable of staining tissues in its own right. The serum is therefore dialysed to remove small molecules (M.W. < about 5000) and to replace the alkaline solvent with a saline solution of physiological pH. It may also be passed through a gel-filtration column which retains small molecules or it may be treated with activated charcoal, which does much the same thing. Sometimes all three methods are used. Finally, the serum is concentrated by dialysis against an inert synthetic polymer such as polyvinylpyrollidone. The labelled antiserum is now ready for use. Practical instructions for preparing labelled proteins are given by Pearse (1968) and Nairn (1976).

Sections of the antigen-containing specimen are cut and mounted onto slides or coverslips. The tissue must not have been fixed or otherwise altered by processing to an extent sufficient to destroy or distort the antibody-binding sites of the antigen. Some degree of fixation is desirable, however, to prevent disintegration of the sections during the staining procedure. Commonly fresh tissue is cut on a cryostat and the sections are fixed for a few minutes in alcohol, acetone, or neutral, buffered formaldehyde. Cross-linking fixatives such as formaldehyde and glutaraldehyde are more likely to destroy antigenicity than are simple coagulants. Salts of heavy metals are usually avoided because fluorescence can be quenched in the presence of elements with high atomic numbers. Some antigens survive embedding in wax but others do not. The optimum treatment for every tissue and antigen has to be found by trial and error.

Some soluble antigens, including serum proteins, may diffuse from their normal loci in the tissue during the course of freezing or cutting of unfixed material, giving rise to artifactual false-positive localizations in the stained preparations (Sparrow, 1980). Errors from this cause should be suspected when an antigen is detected in an unexpected part of a tissue, especially when one of the more sensitive immunohistochemical techniques, such as the unlabelled antibody–enzyme method, is used.

The sections are now rinsed with saline and a drop of labelled antiserum is placed on each. Various twofold dilutions of the serum in saline are tried. Dilutions weaker than 1 in 8 are unlikely to be needed for direct fluorescent antibody methods. The serum is left in contact with the sections, in a humid atmosphere, for approximately 1 h. The serum is then washed off with saline and a coverslip is applied. Immunofluorescent preparations are usually mounted in a mixture of glycerol and phosphate buffer, pH 7.4–7.6. Permanent mounting media are likely to weaken the intensity of the fluorescence.

The stained, mounted sections are examined by fluorescence microscopy. For FITC conjugates the optimum wavelength for excitation is 490 nm (blue). The maximum emission occurs at 550 nm (green–yellow). Exciting light of 320 nm (in the ultraviolet) may also be used, but the emission will be less intense. However, with ultraviolet excitation and a colourless barrier filter it is easier to distinguish between the specific emission of FITC and the blue autofluorescence of the specimens. When blue exciting light is used, a yellow or orange barrier filter is necessary and the autofluorescence appears to be green. This colour is difficult to distinguish from the specific emission of FITC. Rhodamine B derivatives are optimally excited by green light (546 nm) and have orange–red emissions (580–600 nm). Dansylated proteins absorb in the ultraviolet and emit blue–green light.

19.5.2. Controls

The following control procedures must be carried out alongside the definitive staining technique:

1. Incubate in saline alone. The only fluorescence seen will be autofluorescence.
2. Incubate in a fluorescently labelled non-immune serum derived from the same species as that in which the antiserum was raised. Any fluorescence seen in this control but not in No. 1 will be due to non-specific attachment of fluorescent proteins to the section. Absorption of the labelled antiserum with a tissue powder (see p. 283) may help to eliminate this artifact.
3. Incubate with labelled antiserum an identically processed section known not to contain the antigen. This is only possible when the antigen is one foreign to the tissue (e.g. bacteria, foreign protein). The specimens for this control should be of the same tissue as that used for the definitive procedure.
4. Mix some labelled antiserum with a solution in saline of the purified antigen. A large excess of the latter should be present in the mixture. The added antigen combines with the antibody and little or no specific fluorescence should be seen in a section stained with this absorbed antiserum.

19.5.3. Shortcomings of the direct method

With the direct fluorescent antibody method it is not possible to localize antigens present at low concentrations in a tissue. There are two reasons for this lack of sensitivity:

1. The process of conjugation with FITC or any other reactive fluorochrome will cause blocking of the combining sites of some of the antibody molecules, reducing the potency of the antiserum. The likelihood of attachment of label to the combining site will be reduced if the concentration of the labelling agent is low, but then the number of fluorescent groups usefully attached to antibody molecules will also be less. An ideally labelled antiserum has about ten fluorochrome molecules attached to each molecule of 1gG.

2. The immunological reaction in the direct method is

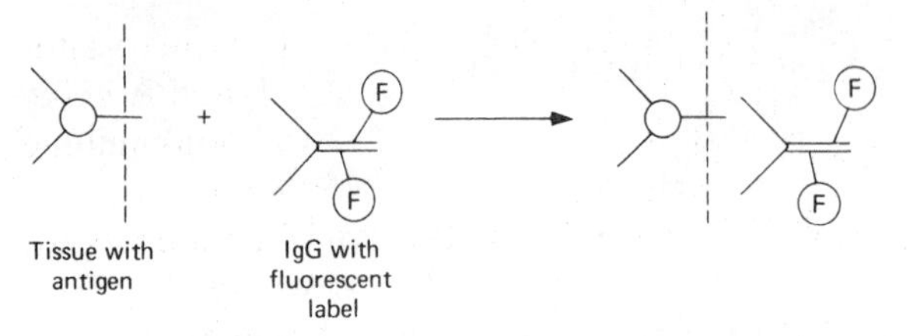

Thus only one molecule of antibody, with its attached fluorochrome, combines with each antigenic determinant site in the tissue. If the concentration of antigen is low the density of bound fluorochrome molecules may not be high enough to permit their detection under the microscope. In other methods it is possible to accumulate greater numbers of molecules of visible substances at each antigenic site.

Another objection to this technique arises from the necessity to label a specific antiserum. Antisera are troublesome to prepare and often available only in small quantities. The conjugation also requires time and effort. An investigation of the localizations of several antigens would demand the preparation of as many labelled antisera, and this would entail an inordinate amount of work. In the methods to be described next, a single labelled antiserum will render visible the sites of localization of any number of different antigens.

19.6. INDIRECT FLUORESCENCE TECHNIQUES

With indirect methods it is possible to identify either antigens or antibodies (see Section 19.9.1) in a tissue. The reagent common to the procedures of this type is a fluorescently labelled antiserum to immunoglobulin: an anti-antibody. The method for demonstration of a tissue antigen will illustrate the principle of the method.

19.6.1. Method for detection of an antigen

A section containing the antigen is first incubated for 30–60 min with a suitable dilution (commonly 1 in 20 or 1 in 50) of an appropriate antiserum. This antiserum, which is **not** labelled, is called the **primary antiserum**:

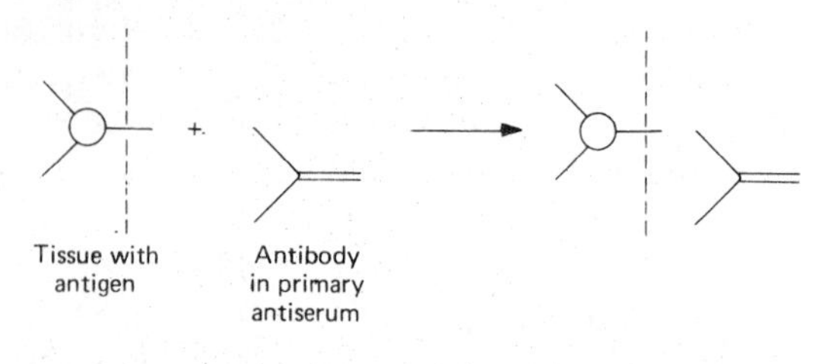

The antibody molecules which attach to the antigenic sites in the tissue are immunoglobulins. The F_c segments of their molecules are the same as those of all other immunoglobulins of the same species and can themselves serve as antigens. It is possible to raise an anti-immunoglobulin antiserum by injecting into another species of animal some γ-globulin from the blood of the species that was the source of the primary antiserum. This **secondary antiserum** may be labelled by conjugating it with a fluorescent dye.

In the next stage of the staining method, the section is treated with fluorescently labelled secondary antiserum. The anti-IgG molecules in it will bind to the F_c segments of the already attached immunoglobulin molecules derived from the primary antiserum:

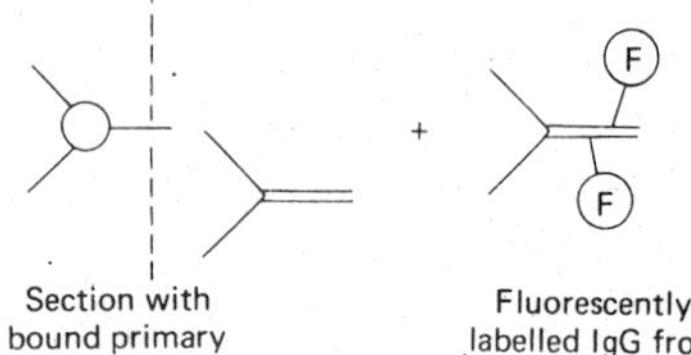

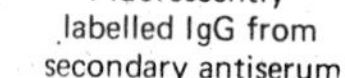

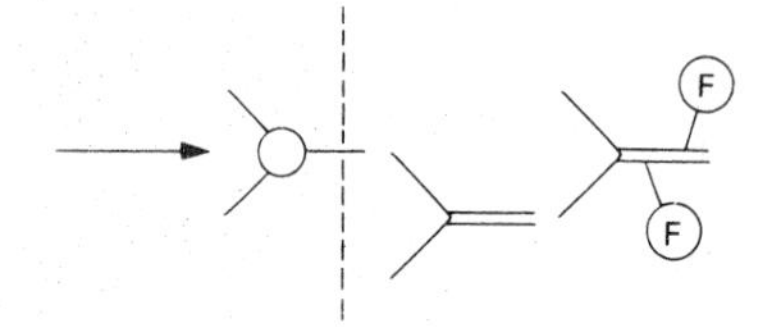

Thus the fluorochrome will be seen at the sites occupied by the antigen.

A specific example will further clarify the technique. Let us suppose that a guinea-pig has been given an intravenous injection of egg albumen. Several hours later the foreign protein will be present in the renal tubules. We have available an antiserum to egg albumen raised in a rabbit. This is **rab-**

bit antiserum to egg albumen (or "rabbit anti-egg albumen"). We also have a labelled antiserum to rabbit immunoglobulin. The latter antiserum was raised in a goat, so it is **goat antiserum to rabbit immunoglobulin** (often also known as "goat anti-rabbit γ-globulin"). Cryostat sections of one of the guinea-pig's kidneys are briefly fixed in alcohol and then treated as follows:

1. Rinse in saline.
2. Apply rabbit antiserum to egg albumen. Use serial twofold dilutions in saline, from 1 in 10 to 1 in 160. Leave in contact with the sections for 30–60 min at room temperature in a humid atmosphere.
3. Rinse sections with three changes of saline.
4. Apply labelled goat antiserum to rabbit immunoglobulin (try 1 in 10 and 1 in 40 dilutions) for 30–60 min.
5. Rinse sections with three changes of saline.
6. Mount in buffered glycerol and examine by fluorescence microscopy.

The sera are diluted in order to minimize non-specific adherence of protein molecules to the sections. This could result in fluorescence at sites other than those containing the antigen. Generally the dilution of the primary antiserum is more critical than that of the labelled secondary antiserum. Diluted sera containing 1.0–5.0 mg of protein per ml are often the most satisfactory for indirect immunofluorescence methods.

19.6.2. Controls

As with the direct fluorescent antibody method, control procedures are necessary. These are:

1. Omit treatment with both sera to reveal the autofluorescence of the tissue.
2. Omit treatment with the primary antiserum. Incubate with saline instead. Any fluorescence seen cannot be due to the antigen and is probably autofluorescence or a result of non-specific binding of fluorescently labelled proteins from the secondary antiserum.
3. In place of the primary antiserum, use non-immune serum from the same species as that in which the primary antiserum was raised. If fluorescence is seen that does not appear in controls 1 and 2, it is possible that the primary antiserum used in the definitive method is being bound to the tissue by a non-immune mechanism.
4. Mix the primary antiserum with an excess of the purified antigen. Absent or greatly diminished immunofluorescence should be obtained after application of this absorbed antiserum.

19.6.3. Critique of the method

This technique is superior to the direct method for two reasons:

1. It is not necessary to label and thereby reduce the potency of the valuable primary antiserum. Labelled secondary antisera can be obtained commercially and are potentially useful for the detection of an infinite number of antigens.
2. The determinant sites of antigens are multiple (see p. 281). This is true of the antigen in the tissue and of the F_c segments of the antibody molecules derived from the primary antiserum. Consequently, more than one molecule of fluorescent antibody from the secondary antiserum will be able to attach to each molecule of the primary antibody. For example, two suitably spaced antigenic determinant sites in a section will bind two antibody molecules from a primary antiserum:

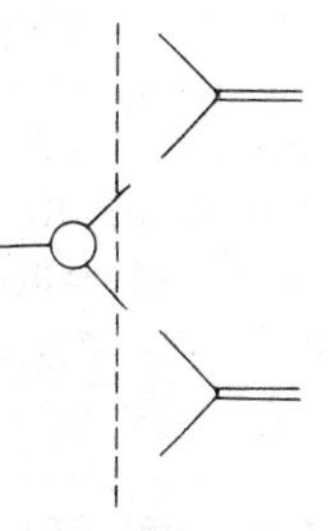

If each of these bound antibody molecules also has two antigenic determinant sites on its F_c segment, four molecules of fluorescent anti-antibody will be bound:

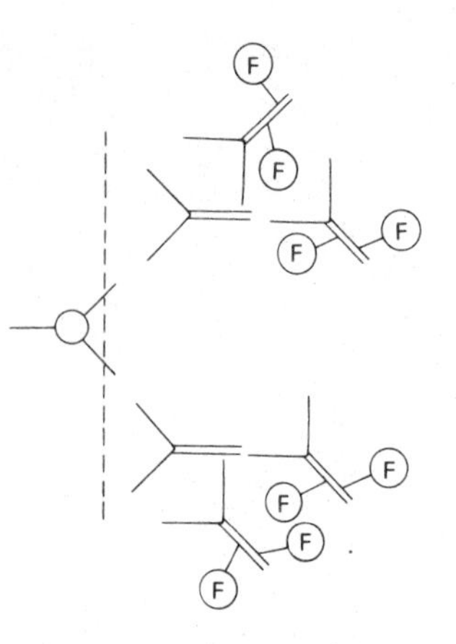

In this example, each antigen molecule has accreted twice as many fluorescent molecules as would have been possible had a direct immunofluorescent method been used. In reality the amplification factor may be much greater than 2, so indirect immunohistochemical techniques are always more sensitive than direct ones.

Even so, some antigens occur in tissues in quantities too small to be detected by the indirect fluorescent antibody method. Greater sensitivity is achieved when the label is a histochemically demonstrable enzyme rather than a fluorochrome.

19.7. ENZYME-LABELLED ANTIBODY METHODS

In these techniques antibodies are labelled by conjugation with enzymes. The enzyme currently most in favour is horseradish peroxidase (HRP). Simple and reliable histochemical methods are available for the localization of peroxidases and some of them yield products which can be detected by both light and electron microscopy (see Chapter 16). Other enzymes, including acetylcholinesterase and cytochrome oxidase, have occasionally been used for the same purpose.

Various methods of conjugation are feasible. A simple one consists of mixing solutions containing the antibody and HRP and then adding a little glutaraldehyde. Cross-linking of protein molecules occurs by the same chemical reactions as those involved in the fixation of tissues by glutaraldehyde (see Chapter 2). After reaction the solution can be expected to contain the following:

Unreacted glutaraldehyde
Unreacted HRP
Unreacted antibody
Unreacted other proteins
(Antibody)—$(CH_2)_5$—(Antibody)
(HRP)—$(CH_2)_5$—(HRP)
(Antibody)—$(CH_2)_5$—(HRP)
(Other proteins)—$(CH_2)_5$—(Other proteins)
(Other proteins)—$(CH_2)_5$—(HRP)

Larger aggregates derived from three or more protein molecules form a precipitate, which is removed by centrifugation. Any excess of glutaraldehyde is removed by dialysis or gel-filtration. Some of the double molecules: (Antibody)—$(CH_2)_5$—(HRP) have the structure

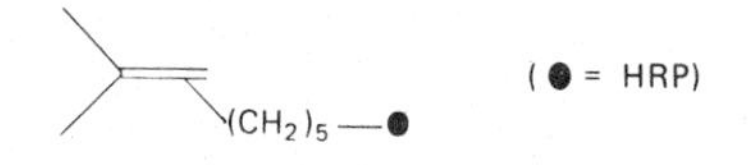

with the F_c segment labelled and the F_{ab} segments unhindered. This is the only useful component of the solution, but further purification is not usually necessary.

The solution containing HRP-labelled antibody is used in exactly the same way as a fluorochrome-labelled antiserum. The sites of attachment of HRP to the tissue are detected histochemically, usually by means of the DAB–hydrogen peroxide reaction (see Chapter 16). Enzyme-labelled antibodies may be used in direct or indirect techniques. The latter are generally preferred, for the reasons set out in Section 19.6.3 of this chapter. The bound HRP is capable of catalysing the oxidation of many molecules of DAB by hydrogen peroxide with consequent accumulation of substantial amounts of the insoluble brown product. The apparent size of each antigenic site in the tissue is therefore enlarged until such time as the enzyme is inhibited by the accumulated products of its activity. The "amplification factor" is greater than can be obtained in fluorescent antibody techniques.

As with the immunofluorescence methods, control procedures must accompany the use of enzyme-labelled antibodies. Indeed, false-positive results are more likely to be seen with the latter than with the former techniques, since many tissues contain endogenous peroxidases. With the indirect enzyme-labelled antibody methods the following controls are required:

1. Omit all immunological reagents. Carry out only the histochemical method for the enzymatic label. This will reveal sites of endogenous enzyme.
2. Omit treatment with the primary antiserum. This will reveal any non-specific attachment of enzyme-labelled proteins or of unconjugated enzyme. HRP has a rather strong tendency to bind to tissues in this way.
3. Substitute non-immune serum for the primary antiserum to detect non-immunological binding of γ-globulin.
4. Treat with primary antiserum that has been absorbed with an excess of its antigen (see p. 283). This should result in absent or greatly diminished intensity of staining.

The advantage of this technique over the indirect fluorescent antibody method lies in its higher sensitivity. The higher probability of obtaining false-positive results is a relative disadvantage. Both methods involve the use of covalently labelled antisera, whose potencies are reduced as a result of hindrance of the combining sites of many of their specific antibody molecules.

When antigenic sites are few and far between, the indirect fluorescent antibody technique is to be preferred. It is easier to see isolated fluorescent spots on a dark background than to find occasional small accumulations of enzymatic reaction product against a bright background. Histochemical methods yielding fluorescent end-products are available for a few enzymes, including peroxidases (Papadimitriou *et al.*, 1976) but have not yet been used in conjunction with immunohistochemical methods. Enzyme-labelled antibodies are valuable for ultrastructural immunohistochemistry because the end-products of many histochemical reactions for enzymes, including the DAB–H_2O_2 method, can be detected by electron microscopy.

The next method to be described is much more sensitive than the indirect enzyme-labelled antibody procedure. Chemically labelled proteins are not required, and the likelihood of false-positive results is no higher than with other techniques.

19.8. THE UNLABELLED ANTIBODY–ENZYME METHOD

19.8.1. Principle of the method

Horseradish peroxidase (HRP) can serve as an antigen when it is injected into an animal (usually a rabbit). By careful mixing of the resultant immunoglobulin with HRP it has been possible to isolate a stable antigen–antibody complex. This is known as peroxidase–antiperoxidase (PAP) and has the structure

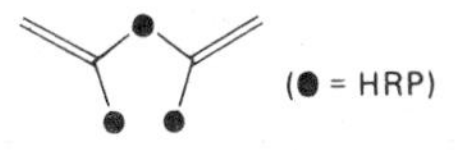

Two molecules of specific IgG are associated with three of HRP. The complex is soluble and is available commercially. Combination with antibody does not inhibit the activity of the enzyme. Each complex molecule of PAP has two F_c segments, which are available for attachment to suitable anti-IgG molecules. For example, the F_c segments of rabbit PAP will bind to the F_{ab} segments of the specific IgG in a goat antiserum to rabbit γ-globulin. The PAP complex is thus a valuable and versatile immunohistochemical reagent for the detection of the sites of binding of anti-antibodies.

19.8.2. Procedure for detection of an antigen

Theoretical and practical aspects of the unlabelled antibody–enzyme method will be illustrated by considering the technique for localization of an antigen, X, in sections of a tissue. An antiserum to X is raised in a rabbit. The other reagents needed are goat antiserum to rabbit immunoglobulin (ideally an antiserum specific for the F_c segment of IgG), rabbit PAP, and the chemicals required for histochemical demonstration of peroxidase. Both the secondary antiserum and the PAP are available from several commercial suppliers. The procedure is as follows:

1. Apply rabbit antiserum to X (the primary antiserum) to the sections for 1 h at room temperature. Several dilutions (in phosphate-buffered saline) are used, from 1 in 50 to 1 in

2000. A longer exposure (12–24 h, at 4°C) is sometimes advantageous when an exceedingly dilute primary antiserum is necessary.

2. Rinse in three or four changes of phosphate-buffered saline.
3. Apply goat antiserum to rabbit immunoglobulin for 30 min at room temperature. A 1 in 10 or 1 in 20 dilution of this serum is used. The concentration is not critical but it must be much higher than that of the primary antiserum or the PAP.
4. Rinse in three or four changes of phosphate-buffered saline.
5. Apply rabbit peroxidase–antiperoxidase (rabbit PAP) for 30 min at room temperature. The stock solution (containing 1.0 mg protein per ml) of the commercially available product is diluted 1 in 50 with phosphate-buffered saline.
6. Rinse in three or four changes of phosphate-buffered saline.
7. Incubate for demonstration of peroxidase activity (see Chapter 16). The DAB–H_2O_2 method (p. 242) is usually used with the incubation medium buffered to pH 7.6.
8. Wash in water, dehydrate, clear, and mount in a resinous medium.

A three-layered "sandwich" is built up by the three reagents employed:

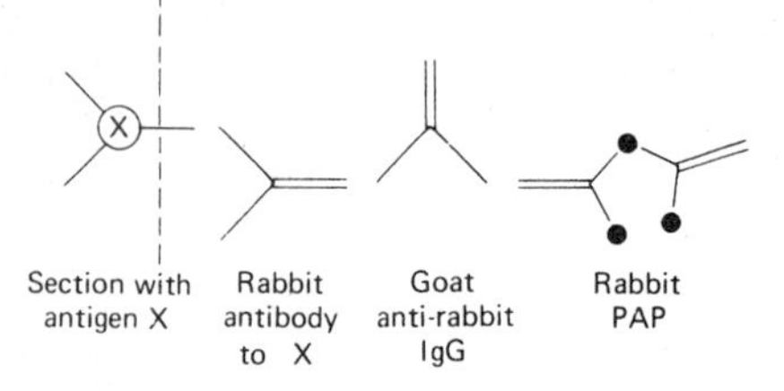

Because the goat antibody to rabbit IgG is bivalent, its molecules attach to the F_c segments of the primary rabbit anti-X antibodies **and** to those of the rabbit antiperoxidase antibodies in the PAP complex. The products of the histochemical reaction for HRP therefore accumulate at the sites of the antigen X.

The intensity of the colour of the end-product is influenced by the dilution of the primary antiserum. When the antigen is present at high concentration in the tissue, the density of the final product is higher after the application of more dilute solutions of the primary antiserum. The probable reason for this apparently paradoxical effect is given in Fig. 19.2.

19.8.3. Controls

The following control procedures should accompany any investigation in which the PAP method is used for the localization of an antigen in a tissue:

1. Omit treatment with both the antisera and the PAP. Incubate for peroxidase to demonstrate the endogenous enzymatic activity of the tissue.
2. Omit treatment with the primary antiserum. Any staining then observed cannot be due to the antigen.
3. Substitute a non-immune serum for the primary antiserum (e.g. normal rabbit serum, in the method for antigen X described above). This control will reveal any non-immunological attachment of γ-globulins to the tissue.
4. Omit treatment with both the primary and the secondary antisera. Non-specific binding of PAP, should it occur, will then be demonstrated. If non-specific binding is a function of the HRP moiety of PAP, it can be prevented by treating the sections (before stage 1 of the above procedure) with a solution of HRP, followed by 1% H_2O_2 in 30% methanol, which irreversibly inhibits the enzymatic activity of the bound HRP (Minard & Cawley, 1978).
5. Mix the primary antiserum with an excess of the purified antigen and apply this mixture to the sections instead of the primary antiserum. Greatly diminished staining should be observed. It is rarely possible to inhibit staining by the PAP method completely by the use of absorbed antisera.

19.8.4. Critique of the method

The main advantage of the unlabelled antibody–enzyme method is its sensitivity, which is considerably greater than that of any other immunohistochemical technique. Positive results can often be

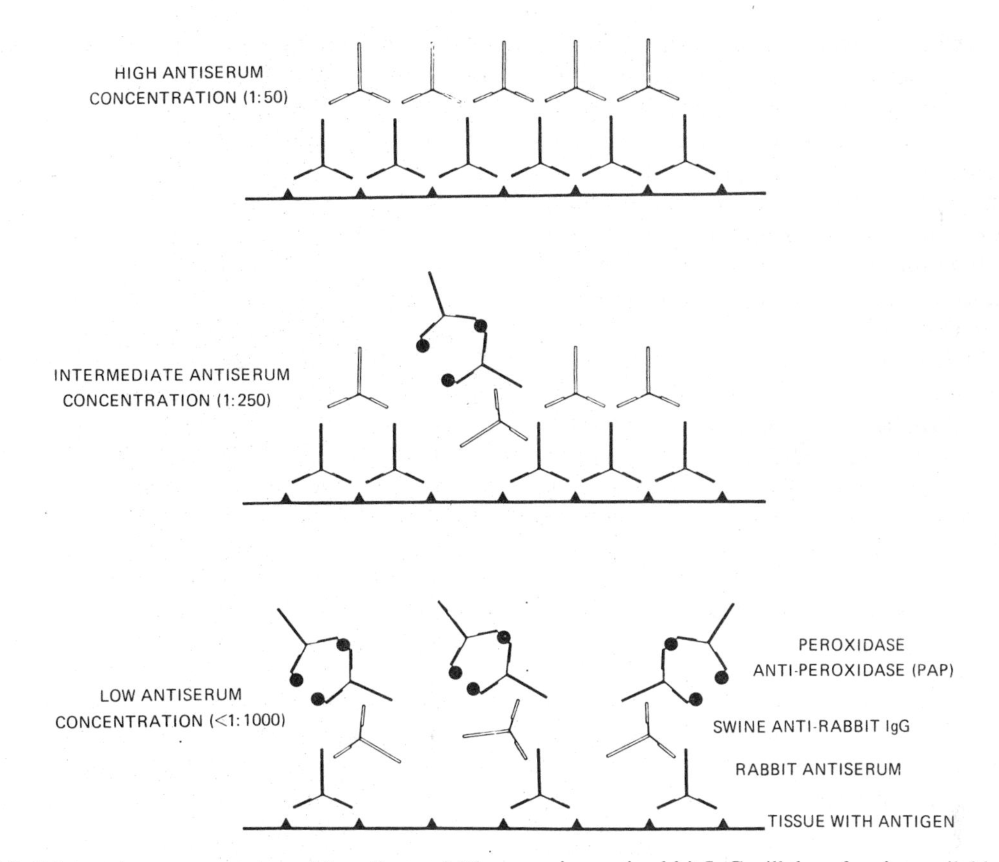

FIG. 19.2. Schematic representation of the effects of different primary antiserum concentrations on the staining of a tissue with high density of antigenic sites (unlabelled antibody–enzyme method). High concentration of primary antiserum (top diagram) results in binding of the secondary antibody (swine anti-rabbit IgG) by both F_{ab} segments. With low concentration of primary antiserum, the F_c segments of the primary antibody molecules will be further apart, so that each will bind only one F_{ab} segment of the secondary antibody: the free F_{ab} segments of the swine anti-rabbit IgG will therefore be available for combination with F_c segments of the PAP complex. Thus, when antigenic determinant sites in the tissue are close together, maximum staining is obtained with a dilute primary antiserum, but no staining occurs after application of a concentrated primary antiserum. (From Bigbee *et al.*, 1977. Reproduced by permission of the authors and the Williams & Wilkins Co. Copyright 1977. The Histochemical Society Inc.)

obtained with sections of tissues fixed and processed by methods that denature most of the antigen present. It has even been possible to use old H. & E. stained slides from which the coverslips, mounting media, and dyes have been removed. The high sensitivity is due to the attachment of three molecules of peroxidase at each site of binding of the PAP complex. Amplification due to multiplicity of antigenic sites on the F_c segment of the primary antibodies may also occur, as with indirect immunofluorescence techniques, though the large size of the PAP complex molecule would be expected to limit the amount of PAP binding to the secondary antibody. Although control procedures are necessary in order to exclude non-specific staining, this is a "cleaner" method than most others, probably because the PAP reagent is a pure antigen–antibody complex, unlikely to associate itself with proteins other than its own specific antibodies.

It has been pointed out that the intensity of staining is commonly increased when the primary antiserum is diluted. Consequently it is not possible when using the PAP method to draw even approximate conclusions concerning the quantity of antigen in a specimen or even to make meaningful comparisons between strongly and weakly stained regions within a single section. It can only be concluded that an antigen is either present or undetectable. With other immunohistochemical methods, especially those involving fluorescently labelled antibodies, it is widely assumed that the intensity of the observed staining has some quantitative significance, though there is little evidence to justify such an assertion.

19.9. OTHER IMMUNOHISTOCHEMICAL METHODS

The techniques described above are those most widely used for the demonstration of antigens in tissues. Other ingenious techniques also exist for the localization of both antigens and antibodies. Three of these will be described briefly to give the reader some idea of the scope and versatility of immunohistochemical methodology.

19.9.1. Sandwich technique for antibody in tissue

Antibodies in a tissue are demonstrated by applying the purified antigen followed by a fluorescently labelled antiserum to the antigen. The method is often used with impression smears of cells from lymph nodes and with cryostat sections. For example, a section of human tissue containing antibodies to an antigen Y could be treated with a solution of Y, following by a fluorescently labelled rabbit antiserum to Y.

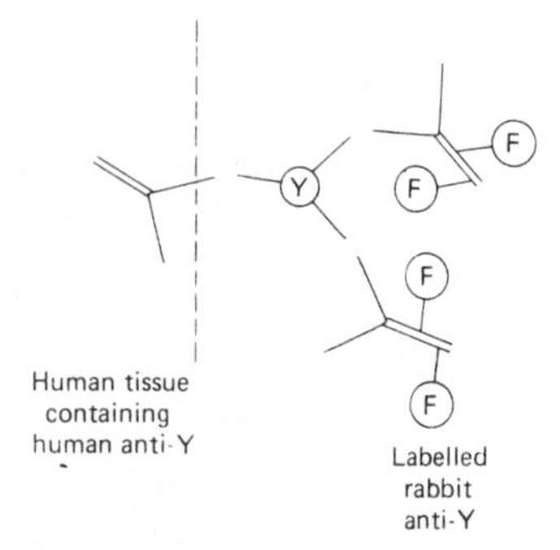

This is a method for demonstrating specific antibodies. Antibodies in general (i.e. immunoglobulins) are detected by considering them to be antigens. Human immunoglobulin would be localized in sections by means of a rabbit antiserum to human immunoglobulin and a labelled goat antiserum to rabbit immunoglobulin.

19.9.2. Detection of antibody in serum

In autoimmune diseases the blood contains circulating antibodies to certain components of a patient's or animal's own tissues. The existence and specificity of such auto-antibodies can be detected immunohistochemically by applying the suspect serum (usually as a 1 in 10 dilution in saline) to unfixed cryostat sections of tissues known to contain appropriate auto-antigens. Different tissues serve as controls. The sites of binding of antibodies to the sections are made visible by the application of a fluorescently labelled antiserum to the immunoglobulins of the species from which the serum under test was obtained.

For example, to test human serum for antibodies against thyroglobulin one would require some sections of normal human thyroid gland and a fluorescently labelled rabbit antiserum to human γ-globulin. The stained preparation would have the structure:

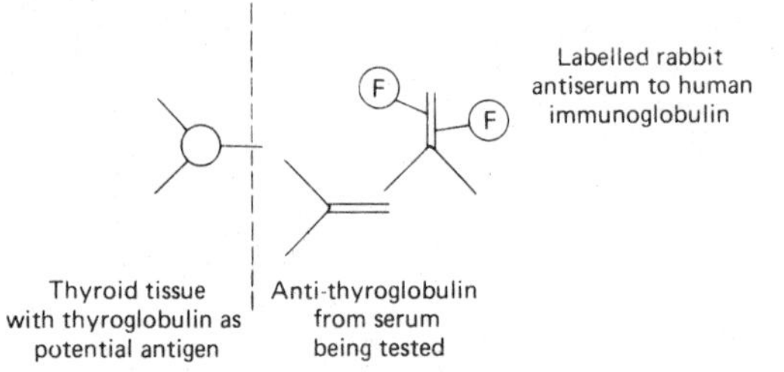

A positive result would appear as fluorescence in the follicles of the thyroid tissue. A section of another human organ, such as the liver or a kidney, serves as a control for specificity. Often the sections used are composite blocks containing several tissues, so that it is possible to seek the presence of several auto-antigens in a single drop of serum.

19.9.3. Methods using staphylococcal protein A

The cell-wall of *Staphylococcus aureus*, a common pathogen, contains an isolable component known as protein A which is able to bind to the F_c segments of mammalian IgG molecules. The binding is non-immune but results from an affinity similar to that existing between antigens and antibodies. The molecule of protein A is bivalent and can therefore be made to serve as a bridge between the F_c portions of IgG molecules of different kinds. It is also possible to label protein A with fluorochromes or with HRP (Dubois-Dalcq *et al.*, 1977). Thus protein A may be used for the same purposes as an antiserum to IgG. Like the commoner anti-immunoglobulin sera, protein A is commercially available.

A conjugate of protein A with HRP is employed in a technique closely similar to an enzyme-labelled antibody method. The specific primary antiserum is applied to a section of tissue and followed by HRP–protein A conjugate:

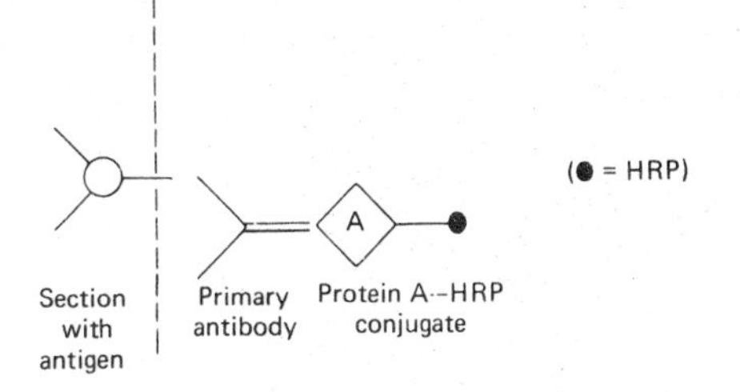

The sites of bound HRP are demonstrated histochemically in the usual way. This method may provide "cleaner" staining, with less non-immunological background coloration, than the equivalent enzyme-labelled antibody technique.

Protein A has also been employed instead of a secondary antiserum in the unlabelled antibody–enzyme method. Application of a specific primary antiserum is followed by protein A and then by PAP:

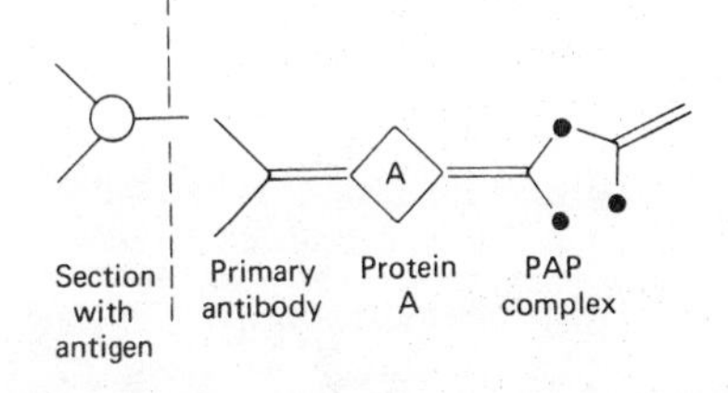

This technique, however, is somewhat less sensitive than those in which secondary anti-immunoglobulin sera are used (Celio *et al.*, 1979). Some other immunohistochemical applications of protein A are described by Notani *et al.* (1979).

The main advantage of protein A is that it is able to combine with IgG of all mammals, so that it has a greater number of potential uses than an antiserum to the immunoglobulin of a single species. Obviously protein A is not suitable for use with any specimen that already contains IgG.

19.10. NON-IMMUNOLOGICAL AFFINITY TECHNIQUES

Specific attachments between molecules are by no means confined to the reactions of antigens with antibodies. The affinities of biologically significant substances for one another have been exploited in a variety of histochemical methods, and the number of such applications will certainly increase in the future. The following types of technique are worthy of mention:

1. The use of lectins in carbohydrate histochemistry (see Chapter 11).
2. The use of labelled inhibitors for the localization of enzymes (see Chapter 14, p. 202).
3. The use of labelled drugs to bind to physiologically or pharmacologically defined receptor sites in tissues. For example, fluorescently labelled α-bungarotoxin (from a snake venom) has been used to demonstrate receptors for acetylcholine, including those at neuromuscular junctions (Anderson & Cohen, 1974). With some fluorescent derivatives of drugs, however, the microscopically observed binding has been shown not to coincide with the sites of pharmacological action (Corréa *et al.*, 1980).
4. The filamentous protein actin, which is important in the contraction and locomotion of cells, selectively binds heavy meromyosin, which is isolated from the actomyosin of striated muscle. The binding results in thickening of the actin filaments, as observed in the electron microscope. The histochemically demonstrable adenosine triphosphatase activity of heavy meromyosin may also be exploited.

19.11. EXERCISES

1. Why are antisera labelled by conjugating them with fluorochromes rather than with ordinary dyes?

2. Suggest reasons why alcohol and picric acid are more suitable fixatives for antigens than formaldehyde and glutaraldehyde.

3. Devise an immunohistochemical method suitable for localizing the large amounts of HRP present in the root of the horseradish plant.

4. It has occasionally been possible to localize antibodies in tissues by a direct method using fluorescently labelled antigens. Why is this simple technique usually unsuccessful?

5. What are the advantages of the indirect over the direct fluorescent antibody techniques?

6. Devise an immunohistochemical method for the localization of an antigen A making use of the following:

(a) purified A;
(b) antiserum to A;
(c) covalent conjugate of A with an enzyme such as HRP.

Discuss the value and limitations of the method (described by Mason & Sammons, 1979).

7. When a tissue has been fixed in glutaraldehyde, widely distributed non-specific staining by the unlabelled antibody–enzyme (PAP) method is observed in addition to the specific staining of antigenic sites in the sections. This non-specific staining is suppressed if the sections are incubated with non-immune goat serum before applying the primary antiserum. Explain.

8. In the unlabelled antibody–enzyme method it is necessary to use a high concentration of the secondary antiserum but a low concentration of primary antiserum. Why is a high concentration of secondary antiserum necessary if adequate quantities of PAP are to be bound in the third stage of the technique?

9. In an attempt to stain a hormone in sections of an endocrine gland, the following reagents were applied sequentially to sections:

(a) dog antiserum to the hormone;
(b) goat antiserum to dog γ-globulin;
(c) rabbit peroxidase–antiperoxidase (PAP);
(d) histochemical detection of peroxidase.

Why did this procedure fail to demonstrate the hormone?

Using the same primary antiserum, what could be done to obtain a successful immunohistochemical result?

10. It is possible to attach molecules of staphylococcal protein A to the particles of gold in a colloidal gold suspension (Roth *et al.*, 1978). Suggest some ways in which the protein A–gold complex might be used as an immunohistochemical reagent.

20

Miscellaneous Data

20.1. Buffer solutions 295
20.1.1. pH 0.7–5.2: acetate–hydrochloric acid buffer 295
20.1.2. pH 3.6–5.6: acetate–acetic acid buffer 296
20.1.3. pH 2.7–7.7: phosphate–citrate buffer 296
20.1.4. pH 5.3–8.0: phosphate buffer 297
20.1.5. pH 7.2–9.0: TRIS buffer 297
20.1.6. pH 5.0–7.4: cacodylate–hydrochloric acid buffer 298
20.1.7. pH 7.4–9.0: borate buffer 298
20.1.8. pH 7.0–9.6: barbitone buffer 298
20.1.9. pH 8.5–12.8: glycine–sodium hydroxide buffer 299
20.1.10. pH 9.2–10.6: carbonate–bicarbonate buffer 299
20.2. Dilution of acids and alkalis 299
20.3. Atomic weights 301
20.4. Suitable tissues for histochemical techniques 301

20.1. BUFFER SOLUTIONS

A buffer is chosen for (a) its efficacy in the required pH range, and (b) the absence of components which would react undesirably with other substances in the solution which is to be buffered. Prescriptions for the commonly used buffers follow. The pH values are given to the nearest 0.1 pH units, since greater accuracy is rarely required in histochemistry. For more extensive tables of buffers, see Pearse (1968), Dean (1973), and Lillie & Fullmer (1976). Some useful practical information concerning the preparation of buffers is given by Kalimo & Pelliniemi (1977). Variations in temperature affect the pH of any solution, but most of the buffers given below should be accurate to ± 0.1 pH unit over the range 15–25°C.

The pH of a mixture containing a buffer should be checked with a pH meter. It is usual to standardize the meter against buffer solutions obtained from a chemical supply house. However, the following standard solutions are easily made in the laboratory:

0.05M *potassium hydrogen phthalate*

$KHC_8H_4O_4$: 10.21 g; water to 1000 ml.
The pH is 4.0 from 0°C to 40°C.

0.01M *borax*

$Na_2B_4O_7.10H_2O$: 3.81 g; water to 1000 ml.
The pH is 9.3 from 10°C to 15°C; 9.2 from 20°C to 25°C; 9.1 from 30°C to 35°C.

20.1.1. pH 0.7–5.2: acetate–hydrochloric acid buffer

This buffer is compatible with all reagents other than those metals whose chlorides are insoluble (e.g. Ag, Pb).

Stock solutions

A. 1.0M sodium acetate

Either CH_3COONa: 82.04 g; water to 1000 ml;
or $CH_3COONa.3H_2O$: 136.09 g; water to 1000 ml.

B. 1.0N hydrochloric acid (see Section 20.2, p. 300)

pH	A ml 1.0M sodium acetate	B ml 1.0N HCl	Water: to make total volume (ml)
0.7	50	95	250
0.9	50	80	250
1.1	50	70	250
1.2	50	65	250
1.4	50	60	250
1.7	50	55	250
1.9	50	53	250
2.3	50	51	250
3.2	50	48	250
3.6	50	45	250
3.8	50	42.5	250
4.0	50	40	250
4.2	50	35	250
4.4	50	30	250
4.6	50	25	250
4.8	50	18	250
4.9	50	15	250
5.2	50	10	250

The final volume of 250 ml must include all other ingredients of a buffered mixture. It is important to check the final pH with a meter. The buffering capacity is poor between pH 1.8 and 3.5.

20.1.2. pH 3.6–5.6: acetate–acetic acid buffer

This buffer is compatible with all commonly used reagents.

The instructions here are for preparation of a 0.1 M buffer. Some techniques require 0.2 M acetate buffer, so it is convenient to keep 0.2 M stock solutions, with double the strengths of those described below. (See also *Note* following the table.)

Stock solutions

A. 0.1 M sodium acetate

Either CH_3COONa: 8.20 g; water to 1000 ml;
or $CH_3COONa.3H_2O$: 13.61 g; water to 1000 ml.

B. 0.1 M acetic acid (see Section 20.2, p. 300)

pH	A ml 0.1 M sodium acetate	B ml 0.1 M acetic acid
3.6	15	185
3.8	24	176
4.0	36	164
4.2	53	147
4.4	74	126
4.6	98	102
4.8	120	80
5.0	141	59
5.2	158	42
5.4	171	29
5.6	181	19

Note

This buffer may also be made up with 0.2 M, 0.05 M, or 0.01 M reagents, but the pH values for the proportions given in the table are not accurate for concentrations other than 0.1 M. Check with a meter and adjust the pH when making acetate buffer at strengths other than 0.1 M.

20.1.3. pH 2.7–7.7: phosphate–citrate buffer

This buffer is useful because it is effective over a wide range. It should not be used in the presence of any metal ions that form insoluble phosphates or complexes with citric acid (e.g. Ag, Al, Ba, Ca, Co, Cu, Fe, Mg, Mn, Ni, Pb, Zn).

Stock solutions

A. 0.2 M disodium hydrogen phosphate (= sodium phosphate, dibasic)

Either Na_2HPO_4: 28.39 g; water to 1000 ml;
or $Na_2HPO_4.2H_2O$: 35.60 g; water to 1000 ml;
or $Na_2HPO_4.7H_2O$: 53.61 g; water to 1000 ml.

(The dihydrate is the most stable form, since it is neither hygroscopic nor efflorescent.)

B. 0.1 M citric acid

Either $C_6H_8O_7$ (anhydrous): 19.21 g; water to 1000 ml;
or $C_6H_8O_7.H_2O$: 21.01 g; water to 1000 ml.

(The monohydrate is the form of citric acid most commonly used.)

pH	A ml 0.2 M Na_2HPO_4	B ml 0.1 M citric acid
2.6	22	178
2.8	32	168
3.0	41	159
3.2	49	151
3.4	57	143
3.6	64	136
3.8	71	129
4.0	77	123
4.2	83	117
4.4	88	112
4.6	93.5	106.5
4.8	99	101
5.0	103	97
5.2	107	93
5.4	111	89
5.6	116	84
5.8	121	79
6.0	126	74
6.2	132	68
6.4	138.5	61.5
6.6	145.5	54.5
6.8	154.5	45.5
7.0	165	35
7.2	174	26
7.4	182	18
7.6	187	13
7.8	191.5	8.5

20.1.4. pH 5.3–8.0: phosphate buffer

The table gives quantities for preparing 0.1 M sodium phosphate buffer solutions. These may be diluted with water to obtain weaker solutions (e.g. 0.06 M, 0.05 M, etc.). Twofold dilution increases the pH by approximately 0.05. Phosphate buffers cannot be used in the presence of those metal ions which would be precipitated as insoluble phosphates (i.e. all common cations other than Na^+, K^+, and NH_4^+). Equimolar quantities of potassium salts may be substituted for the sodium phosphates prescribed here.

pH	A ml 0.1 M NaH_2PO_4	B ml 0.1 M Na_2HPO_4
5.3	192	8
5.5	188	12
5.7	184	16
5.8	180	20
5.9	174	26
6.0	168	32
6.1	162	38
6.2	154	46
6.3	146	54
6.4	136	64
6.5	128	72
6.6	112	88
6.7	104	96
6.8	96	104
6.9	82	118
7.0	68	132
7.1	56	144
7.2	48	152
7.3	40	160
7.4	32	168
7.5	28	172
7.6	23	177
7.7	17	183
7.8	12	188
7.9	8	192

Stock solutions

A. 0.1 M sodium dihydrogen phosphate (= sodium phosphate, monobasic; sodium acid phosphate)

Either NaH_2PO_4: 12.40 g; water to 1000 ml;
or $NaH_2PO_4.H_2O$: 13.80 g; water to 1000 ml;
or $NaH_2PO_4.2H_2O$: 15.60 g; water to 1000 ml.

(The dihydrate is preferred, since it is neither hygroscopic nor efflorescent.)

B. 0.1 M disodium hydrogen phosphate (= sodium phosphate, dibasic)

Either Na_2HPO_4: 14.20 g; water to 1000 ml;
or $Na_2HPO_4.2H_2O$: 17.80 g; water to 1000 ml;
or $Na_2HPO_4.7H_2O$: 26.81 g; water to 1000 ml.

(The dihydrate is preferred, since it is neither hygroscopic nor efflorescent.)

20.1.5. pH 7.2–9.0: TRIS buffer

This is a 0.05 M TRIS buffer. It is compatible with salts of all heavy metals other than those with insoluble chlorides (e.g. Ag, Pb). The effective concentrations of some metal ions may be reduced by complex formulation.

Stock solutions

A. 0.2 M TRIS

Tris (hydroxymethyl) aminomethane: 24.2 g
Water: to 1000 ml

B. 0.1 N hydrochloric acid (see p. 300)

pH (see *Note*, p. 298)	A ml 0.2 M TRIS	B ml 0.1 N HCl	Water; to make total volume (ml)
7.2	50	89.5	200
7.4	50	84	200
7.6	50	77	200
7.8	50	69	200
8.0	50	58.5	200
8.2	50	46	200
8.4	50	35	200
8.6	50	25	200
8.8	50	17	200
9.0	50	11.5	200

Note

TRIS buffers are more strongly affected by temperature than most others. The pH values in this table are correct at 25°C. For each degree C below 25°C the pH will be higher by 0.03. Thus the mixture in the eighth line of the table (pH 8.6 at 25°C) will have pH 8.75 at 20°C and pH 8.39 at 32°C.

20.1.6. pH 5.0–7.4: cacodylate–hydrochloric acid buffer

This buffer is much used in fixatives for electron microscopy, but it is doubtful whether it has any advantage over cheaper, less toxic buffers. Sodium cacodylate is poisonous. Its solution must not be pipetted by mouth.

Stock solutions

A. 0.2 M sodium cacodylate

$Na(CH_3)_2AsO_2.3H_2O$: 42.8 g; water to 1000 ml.

This reagent is expensive, so use only as required (e.g. 2.16 g for 50 ml solution).

B. 0.2 N hydrochloric acid

See Section 20.2, p. 300.

pH	A ml 0.2 M sodium cacodylate	B ml 0.2 M HCl	Water: to make total volume (ml)
5.0	50	47	200
5.2	50	45	200
5.4	50	43	200
5.6	50	39	200
5.8	50	35	200
6.0	50	29.5	200
6.2	50	24	200
6.4	50	18.5	200
6.6	50	13.5	200
6.8	50	9.5	200
7.0	50	6.5	200
7.2	50	4.2	200
7.4	50	2.5	200

20.1.7. pH 7.4–9.0: borate buffer

This buffer (Holmes, 1943), which is compatible with low concentrations of silver nitrate, should not be confused with other borax-containing buffers.

Stock solutions

A. 0.2 M boric acid

H_3BO_3: 12.4 g; water to 1000 ml.

Takes 1–2 h to dissolve with vigorous magnetic stirring.

B. 0.05 M borax

$Na_2B_4O_7.10H_2O$: 19.0 g; water to 1000 ml.

Dissolves easily.

pH	A ml 0.2 M boric acid	B ml 0.5 M borax
7.4	180	20
7.6	170	30
7.8	160	40
8.0	140	60
8.2	130	70
8.4	110	90
8.7	80	120
9.0	40	160

20.1.8. pH 7.0–9.6: barbitone buffer

This is compatible with most commonly used reagents. Barbitone sodium (also known as barbital sodium, sodium diethylbarbiturate, veronal, and medinal) is subject to dangerous drug control legislation in most countries. It is a toxic substance and should not be pipetted by mouth. It is probably rarely necessary to use barbitone buffer, since several other buffer systems cover the same pH range.

Stock solutions

A. 0.1 M barbitone sodium

Barbitone sodium ($C_6H_{11}O_3N_2Na$): 20.62 g; water to 1000 ml.

B. 0.1 N hydrochloric acid (see Section 20.2, p. 300)

pH	A ml 0.1 M barbitone sodium	B ml 0.1 N HCl
7.0	107	93
7.2	111	89
7.4	116	84
7.6	123	77
7.8	132.5	67.5
8.0	143	57
8.2	154	46
8.4	164.5	35.5
8.6	174	26
8.8	181.5	18.5
9.0	187	13
9.2	190.5	9.5
9.4	195	5
9.6	197	3

20.1.9. pH 8.5–12.8: glycine–sodium hydroxide buffer

Stock solutions

A. 0.1 M glycine with 0.1 M sodium choride

Glycine (aminoacetic acid; H_2NCH_2COOH):	7.51 g
Sodium chloride (NaCl):	5.84 g
Water:	to make 1000 ml

B. 0.1 M sodium hydroxide (see Section 20.2, p. 300)

pH	A ml 0.1 M glycine + NaCl	B ml 0.1 M NaOH
8.5	190	10
8.8	180	20
9.2	160	40
9.6	140	60
10.0	120	80
10.3	110	90
10.9	102	98
11.1	100	100
11.4	98	102
11.9	90	110
12.2	80	120
12.5	60	140
12.7	40	160
12.8	20	180

20.1.10. pH 9.2–10.6: carbonate–bicarbonate buffer

Stock solutions

A. 0.2 M sodium carbonate

Sodium carbonate (Na_2CO_3): 21.2 g; water to 1000 ml.

B. 0.2 M sodium bicarbonate

Sodium bicarbonate ($NaHCO_3$): 16.8 g; water to 1000 ml.

pH	A ml 0.2 M Na_2CO_3	B ml 0.2 M $NaHCO_3$	Water: to make total volume (ml)
9.2	4	46	200
9.4	9.5	40.5	200
9.6	16	34	200
9.8	22	28	200
10.0	27.5	22.5	200
10.2	33	17	200
10.4	38.5	11.5	200
10.6	42.5	7.5	200

20.2. DILUTION OF ACIDS AND ALKALIS

Concentrations of acids and bases are commonly expressed as **normalities** rather than as molarities. A normal solution of an acid or alkali contains the equivalent weight of the substance in a volume of 1 litre.

$$\text{Equivalent weight} = \frac{\text{Molecular weight}}{\text{Number of available } H^+ \text{ or } OH^- \text{ ions per molecule}}$$

Thus the equivalent weight of NaOH (M.W. 40) is $40 \div 1 = 40$, so that a 1.0 N solution contains 40 g of NaOH per litre. For sulphuric acid (H_2SO_4, M.W. 98.1) the equivalent weight is $98.1 \div 2 = 49.05$, so a 1.0 N solution will contain 49.05 g of 100% H_2SO_4 per litre.

The common laboratory acids are liquids, so it is convenient to dispense them by volume rather than by weight. Furthermore, these reagents are rarely 100% pure. The bottle in which a concentrated acid is supplied bears a label on which will be found the

TABLE 20.1. *Data for preparation of 1 l of 1.0 N or 1.0 M solutions of some acids and alkalis*. Add the "quantity required" to 800 ml water, mix well, then add water to 1000 ml

Name	Concentrated product			Quantity required	Strength of dilute solution
	Assay (w/w) (%)	S.G.	Normality or molarity		
Hydrochloric acid	36	1.18	12 N (= 12 M)	83 ml	1.0 N (= 1.0 M)
Hydrobromic acid	40	1.38	6.8 N (= 6.8 M)	147 ml	1.0 N (= 1.0 M)
Nitric acid	71	1.42	16 N (= 16 M)	63 ml	1.0 N (= 1.0 M)
Perchloric acid	60	1.54	9.5 N (= 9.2 M)	109 ml	1.0 N (= 1.0 M)
Sulphuric acid	96	1.84	36 N (= 18 M)	28 ml	1.0 N (= 0.5 M)
Acetic acid	99.5	1.05	17.4 M (= 17.4 N)	57	1.0 M (= 1.0 N)
Formic acid	90	1.20	23.4 M (= 23.4 N)	42.5 ml	1.0 M (= 1.0 N)
Sodium hydroxide	100 (solid)	2.13	—	40 g	1.0 N (= 1.0 M)
Potassium hydroxide	100 (solid)	2.04	—	56 g	1.0 N (= 1.0 M)
Ammonium hydroxide	27% NH_3	0.901	14.3 M	70 ml	1.0 M
(= ammonia water)	35% NH_3	0.880	18.2 M	55 ml	1.0 M

molecular weight, the specific gravity, and the w/w percentage assay (g of 100% acid per 100 g of the liquid in the bottle). From these data, a solution of any normality may be prepared by calculating the volume of the acid needed to make 1 litre of the solution.

$$V = \frac{100\,MN}{BPD}$$

where V is the required volume of concentrated acid (**in ml**) to make 1 l; N is the desired normality of the solution; M is the molecular weight; P is the percentage assay of concentrated acid (**w/w**); D is the specific gravity of concentrated acid; B is the basicity (i.e. number of available H^+ per molecule). The only common mineral acid whose basicity is greater than 1 is H_2SO_4 ($B = 2$). For all organic acids and for weak inorganic acids the concentration is best expressed as molarity, not as normality.

Table 20.1 is useful when preparing 1.0 N or 1.0 M† solutions of common acids and alkalis. When the specific gravity and percentage assay stated on the label differ appreciably from the data in the table, it will be necessary to use the formula explained above.

Notes

1. **Caution.** Always add the concentrated acid slowly to the larger volume of water, stirring thoroughly to avoid overheating. Sodium or potassium hydroxide should also be added to water in the same way.

   Concentrated HCl, HBr, HNO_3, CH_3COOH, HCOOH, and NH_4OH have pungent vapours. They should be poured in a fume cupboard.

   All the concentrated acids and alkalis in the table are strongly caustic. Avoid contact with skin. If you get concentrated acid on your skin

† Do not confuse "molar" (moles per litre of solution, abbreviated to M) with "molal" (moles per kilogram of solution). In histochemical practice, concentrations are hardly ever expressed as molalities.

it must be washed off with copious running tap water within 5–15 s if burning is to be prevented.

Never pipette concentrated acids or alkalis by mouth.

2. **Accuracy.** Volatile substances (HCl, HBr, and, especially, NH_4OH) lose potency after the bottles have been opened. When a reagent from an old, two-thirds-empty bottle is used in the preparation of a buffer solution, it is important to check the pH with a meter. Be sure that the meter has been standardized against a reliable buffer solution.

20.3. ATOMIC WEIGHTS

Table 20.2 may be used in calculating molecular weights of compounds. In histology and histochemistry, molecular weights accurate to the nearest whole number are adequate. The values of atomic weights in this table are approximated to the first decimal place.

20.4. SUITABLE TISSUES FOR HISTOCHEMICAL TECHNIQUES

This list includes some mammalian tissues with which positive histochemical reactions may easily be obtained. For a longer list, including invertebrate material, see Gabe (1976). For botanical materials, see Jensen (1962) and Klein & Klein (1970).

NUCLEIC ACIDS

DNA	Any cellular tissue (nuclei)
RNA	Brain, spinal cord, ganglia (Nissl substance of neurons); glands (e.g. salivary glands, pancreas, stomach, intestine, pituitary); active lymph nodes (plasma cells)

TABLE 20.2. *Approximate atomic weights of the commoner elements*

Element Name	Symbol	Atomic weight
Aluminium	Al	27.0
Antimony	Sb	121.8
Arsenic	As	74.9
Barium	Ba	137.3
Beryllium	Be	9.0
Bismuth	Bi	209.0
Boron	B	10.8
Bromine	Br	79.9
Cadmium	Cd	112.4
Calcium	Ca	40.1
Carbon	C	12.0
Cerium	Ce	140.1
Chlorine	Cl	35.5
Chromium	Cr	52.0
Cobalt	Co	58.9
Copper	Cu	63.5
Fluorine	F	19.0
Gold	Au	197.0
Hydrogen	H	1.0
Iodine	I	126.9
Iron	Fe	55.8
Lanthanum	La	138.9
Lead	Pb	207.2
Lithium	Li	6.9
Magnesium	Mg	24.3
Manganese	Mn	54.9
Mercury	Hg	200.6
Molybdenum	Mo	95.9
Nickel	Ni	58.7
Nitrogen	N	14.0
Osmium	Os	190.2
Oxygen	O	16.0
Palladium	Pd	106.4
Phosphorus	P	31.0
Platinum	Pt	195.1
Potassium	K	39.1
Ruthenium	Ru	101.1
Selenium	Se	79.0
Silicon	Si	28.1
Silver	Ag	107.9
Sodium	Na	23.0
Strontium	Sr	87.6
Sulphur	S	32.1
Tellurium	Te	127.6
Thallium	Tl	204.4
Thorium	Th	232.0
Tin	Sn	118.7
Titanium	Ti	47.9
Tungsten	W	183.9
Uranium	U	238.0
Vanadium	V	50.9
Zinc	Zn	65.4
Zirconium	Zr	91.2

Substance	Tissue
PROTEINS AND FUNCTIONAL GROUPS	
Protein (general, amino and carboxyl groups, tyrosine)	Any tissue (cytoplasm, collagen)
Arginine	Intestine (Paneth cells); lymphoid tissue; any tissue with many nuclei
Tryptophan	Pancreas (exocrine cells, α-cells of islets); amyloid; fibrin
Cysteine	Skin (hair follicles)
Cystine	Skin (stratum corneum, hair shafts); pituitary (neurosecretory material); pancreas (β-cells of islets)
Aldehyde groups	Arterial elastic laminae in young rodents; any tissue fixed in glutaraldehyde (especially cytoplasm and collagen)
CARBOHYDRATES	
Glycogen	Liver (hepatocytes)
Proteoglycans:	
Hyaluronic acid	Umbilical cord (Wharton's jelly); eye (vitreous); joints (synovial fluid)
Chondroitin sulphates	Cartilage matrix
Dermatan sulphate	Skin (dermis); tendon; lung (connective tissue)
Keratan sulphates	Cornea
Heparin	Mast cells (e.g. in skin, tongue, mesentery); blood basophils
Neutral glycoproteins	Stomach (surface mucus); thyroid (follicular cells); salivary glands (serous cells); collagen, reticulin
Acid glycoproteins:	
Sulphated	Rat or mouse tongue (mucous glands); rat or mouse duodenum (Brunner's glands, goblet cells); colon (goblet cells)
With sialic acids, labile to neuraminidase	Rat or mouse rectum (goblet cells); mouse sublingual salivary gland (mucous cells)
With sialic acids, labile to neuraminidase only after saponification	Rat sublingual salivary gland (mucous cells)
LIPIDS	
Neutral fats	Adipose connective tissue
Phospholipids	Brain, peripheral nerve (myelin); heart, kidney (mitochondria); erythrocytes
Cholesterol esters	Degenerating myelin (2–4 weeks after transection of axons or destruction of neuronal somata in CNS); atherosclerotic lesions in human arteries
Steroids	Adrenal cortex; testis (Leydig cells)
Cholesterol	Brain (myelin)
INORGANIC IONS	
Calcium (phosphate and carbonate)	Sites of pathological or senile calcification (kidney, tendons, human pineal gland); incompletely decalcified bones or teeth
Calcium (soluble salts)	Kidney, muscle, nervous tissue
Iron	Liver, spleen, bone marrow (phagocytic cells); any tissue at site of old injury or haemorrhage
Zinc	Blood, haemopoietic tissue (granular leukocytes); prostate gland (cells of ducts); pan-

	creas (β-cells of islets); brain (neuropil of various regions, especially hippocampus)
ENZYMES	
Acid phosphatase	Kidney (proximal tubule epithelium); liver (hepatocytes, phagocytic cells); prostate; intestine (cytoplasm of epithelial cells). Nuclear staining is a common artifact
Alkaline phosphatase	Kidney (brush-border of proximal tubules); intestine (brush-border of epithelium); rat brain (endothelium)
Esterases ("non-specific")	Liver (hepatocytes); kidney (tubules); brain (neuroglia, pericytes)
Acetylcholinesterase	Muscle (motor end-plates); brain (some neurons, neuropil, and axons)
Cholinesterase ("pseudocholinesterase")	Brain (some neurons, capillary endothelium in rat)
Dehydrogenases	Liver, kidney, heart, intestine, etc. (cytoplasm, mitochondria)
Cytochrome oxidase	Liver, kidney, heart, intestine, etc. (mitochondria)
Catechol oxidase	Skin (melanocytes in epidermis and dermis); eye (retina, choroid, iris). Do not use albino animal
Peroxidase	Blood, haemopoietic tissue (granular leukocytes). Exogenous HRP in motor neurons 24–48 h after injection into muscle. Erythrocytes exhibit peroxidase-like activity due to haemoglobin
AMINES	
Serotonin	Intestine (large amounts in argentaffin cells of epithelium); rat or mouse mast cells (large amounts in granules). Brain (medulla), spinal cord (small amounts in some neurons and axons)
Noradrenaline	Adrenal medulla (large amounts in some chromaffin cells); brain (small amounts in some axons); ductus deferens (sympathetic axons)
Adrenaline	Adrenal medulla (large amounts in some chromaffin cells)
Dopamine	Mast cells in lungs of ruminants (large amounts); stomach of rat (small amounts in endocrine cells in mucosa of fundus)

Glossary

MANY terms are defined in the text. For these, consult the index. The following list includes a variety of chemical and histological terms with which some readers may be unfamiliar.

Acetal. Compound formed by condensation of one molecule of an aldehyde with two molecules of an alcohol to give the structure:

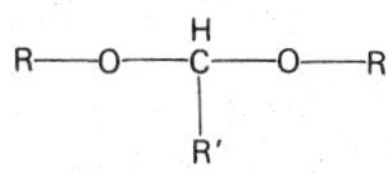

Adduct. Compound formed by combination of two others without the loss of any atoms. Often used when the precise structure of the addition compound is uncertain, or when one of the components is a reagent used for analytical purposes.

Albumen. The principal protein of egg-white. The penultimate *e* distinguishes it from the **albumins**, which are proteins soluble in water and precipitated by high concentrations of salts (e.g. saturation with $(NH_4)_2SO_4$), such as serum albumin.

Aprotic solvent. A polar solvent which does not contain an ionizable hydrogen atom. Its molecules cluster around (solvate) cations, but leave anions relatively unimpeded, so that the latter will be more reactive than when dissolved in an ordinary (protic) polar solvent. Examples are dimethylsulphoxide and *N*-dimethylformamide.

Astrocyte. A neuroglial cell with numerous cytoplasmic processes, some of which form end-feet on capillary blood vessels.

Axon. That cytoplasmic process of a neuron which is specialized for the conduction of trains of impulses, usually away from the soma.

Canonical forms. The different structures which may exist, at instants in time, of an organic compound in which resonance occurs. In writing canonical structures, only bonds and sites of electrical charge may be varied; the positions of the atoms may not be changed.

Caudal. Towards the tail of an animal. Used mainly to refer to relative positions along the axis of the central nervous system.

Chromatin. The material in the nucleus of a cell (excluding the nucleolus) which is stained by cationic dyes and by some dye–metal complexes such as aluminium–haematein. Consists of the DNA and nucleoprotein of the chromosomes.

Common ion effect. The tendency of a salt to become less soluble when the concentration of one of its ions in a solution greatly exceeds that of the other ion.

Condensation. Combination of two molecules with elimination of a compound of low molecular weight such as water.

Delocalized π-electrons. Electrons, associated with double or triple bonds between atoms or with resonant structures such as aromatic rings, which are shared by more than two atoms and cannot therefore be said to form part of any individual covalent bond.

Dendrites. Processes of a neuron specialized for receiving synaptic connections and conducting impulses towards the soma. Dendrites are usually multiple and shorter than the axon. They also differ ultrastructurally and electrophysiologically from axons.

Dialysis. Passage of small, but not large molecules through a semi-permeable membrane. A technique for the purification or concentration of solutions of proteins or other macromolecular substances.

Dimer. A molecule formed by the union of two molecules of the same compound.

Enantiomers. Isomers whose three-dimensional structures are mirror-images of one another.

Furanose. A sugar whose ring structure consists of four carbon atoms and one oxygen atom, so that it could be thought of as a derivative of **furan:**

Gel. A colloidal solution with a semi-solid consistency due to extensive hydrogen bonding between the suspended macromolecules and the "solvent", which is usually water.

Gel filtration. A technique whereby molecules of different size are separated by virtue of their entry or non-entry into the pores contained in beads of a suitably designed polymer. Usually, the polymer is packed in a column and a solution containing the substances to be separated is applied. When the column is eluted with a suitable solvent, the larger molecules are released first and the smaller molecules later.

Glycocalyx. The carbohydrate-containing material present on the outer surfaces of all plasmalemmae.

Haem. The non-protein portion of the haemoglobin molecule. Often used more generally for iron–porphyrin prosthetic groups of proteins, such as occur in many enzymes and cytochromes.

Haematin. A pigment formed when haemoglobin is degraded in acid conditions. Also known as acid haematin. Do not confuse with haematein.

Haemopoietic tissue. Tissue such as red bone marrow in which the cells of the blood are produced.

Hemiacetal. Compound formed by condensation of one molecule of an aldehyde with one molecule of an alcohol to give the structure:

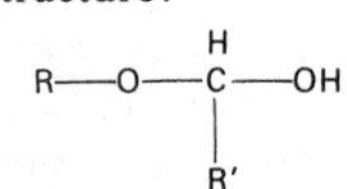

Hydrophilic. Describes substances which attract water: water molecules are able to come into intimate contact with a hydrophilic compound because the latter contains oxygen or nitrogen atoms with which hydrogen bonds can be formed.

Hydrophobic. Describes substances which repel water: a hydrophobic compound has few or no atoms capable of forming hydrogen bonds.

Hydroxyalkylation. Addition of an aldehyde or ketone to an aromatic ring.

Hypertonic. Having a higher osmotic pressure than blood or extracellular fluid.

Hypotonic. Having a lower osmotic pressure than blood or extracellular fluid.

Imide. A compound in which the two bonds of the >NH radical are joined to acyl groups, to give the structure:

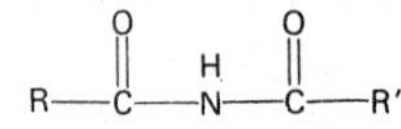

Imine. A compound containing the configuration —C(H)=N—. Such compounds are also known as azomethines, anils, or Schiff's bases. The term "imino" is sometimes applied (though not in this book) to the >NH radical of secondary amines.

Isotonic. Having the same osmotic pressure as blood or extracellular fluid.

Ketal. A compound formed by condensation of a ketone with an alcohol, to give the structure:

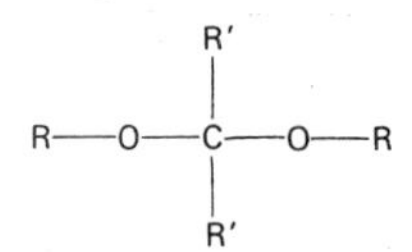

Le Chatelier's principle. When a constraint is applied to any system in equilibrium, the system will always react in a direction which will tend to counteract the applied constraint. For chemical equilibria, the "constraint" may be a change in concentration of a reactant, or a change of temperature, etc. The common ion effect is a consequence of this principle.

Lipofuscin. A pigment containing lipids and proteins, found as granules within the cytoplasm of cells, especially in old animals. Thought to be the indigestible remains of phagocytosed material.

Metabolite. Any substance participating in a chemical reaction in a living organism.

Milieu intérieur. The "internal environment" of the cells of an organism. In higher animals, this is the extracellular fluid.

Monoamine oxidase. An enzyme which catalyses the oxidative degradation of biogenic monoamines such as noradrenaline and serotonin.

Myelin. The sheath surrounding many of the axons of the central and peripheral nervous systems of vertebrate animals. It contains numerous proteins and lipids and is formed from the plasmalemmae of the non-neuronal cells that ensheath the axon. The myelin is trophically dependent upon the axon: its disintegrates and is phagocytosed if the axon dies.

Neuroglia. The cells other than neurons that occur in the central nervous system. Often the term is used more generally to include also the non-neuronal cells of peripheral nerves (Schwann cells) and ganglia (satellite cells) and the non-neuronal cells present in the nervous systems of invertebrate animals.

Neurohypophysis. The portion of the pituitary gland derived from the central nervous system. Comprises the median eminence of the ventral surface of the brain, the stalk of the gland (exclusive of the surrounding, glandular, pars tuberalis), and the infundibular process. The last-named structure is the posterior lobe of the gland, exclusive of the pars intermedia.

Neuropil. Tissue within the nervous system consisting of axons and dendrites, with numerous synapses, but without neuronal somata or tracts of myelinated axons.

Neurosecretion. The secretion of a substance by a neuron for release into the blood. The only neurosecretory cells found in mammals are located in the hypothalamus; their axons end on capillaries in the neurohypophysis. The word neurosecretion is sometimes also applied to neurons whose axons terminate upon endocrine cells. Neurons of the latter type do not usually contain products stainable by the traditional methods for neurosecretory material.

Non-polar solvent. A hydrophobic liquid, not miscible with water, such as benzene or carbon tetrachloride. A molecule is non-polar because its electrons are symmetrically distributed.

Oligodendrocyte. A neuroglial cell with few cytoplasmic processes; responsible for formation of myelin sheaths in the central nervous system.

Periodontal membrane. The connective tissue which anchors a tooth into its bony socket.

Plasmalemma. The membrane forming the outside surface of a cell. Also called the cell membrane. Not to be confused with the cell-wall in plants, which is external to the plasmalemma.

Polar solvent. A liquid miscible with water and capable of dissolving ionized substances. A molecule is polar because its electrons are unevenly distributed, so that one end is relatively electropositive and the other end relatively electronegative. Such a molecule is known as a dipole.

Pyranose. A sugar whose ring structure consists of five carbon atoms and one oxygen atom, so that it can be thought of as a derivative of the hypothetical substance **pyran:**

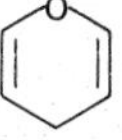

Quinaldine. An aromatic heterocyclic compound, also known as 2-methylquinoline:

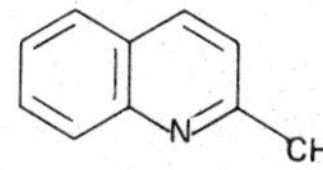

Quinhydrone. The darkly coloured substance formed when hydroquinone is partially oxidized to quinone. Formed by hydrogen bonding of hydroquinone to *p*-quinone:

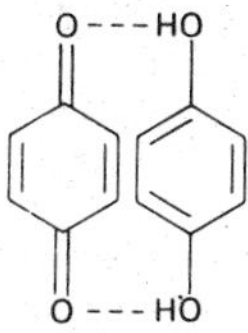

Reserpine. A drug which causes biogenic monoamines to be released from the cells in which they are stored. The cells are thereby depleted of amines.

Rostral. Towards the beak or nose of an animal. Mainly applied to levels on the axis of the central nervous system.

Salting out. Precipitation of an ionic compound due to addition of an excess of one of its ions to the solution. Also applied to precipitation of protein by addition of an inorganic salt to its solution.

Sol. A colloidal dispersion of an inorganic substance, such as gold, ferric hydroxide or sulphur, in a "solvent", which is usually water. Unlike a gel, a sol is a mobile liquid. The particles suspended in a sol are charged; the balancing opposite charge is carried by the solvent molecules surrounding each particle.

Soma (plural: **somata**). A body. Applied to the cell-body of a neuron, in which the nucleus is contained.

Sulphoamino. The radical $-\overset{H}{N}-SO_3H$.

Tunicates. A subphylum of the Chordata, also known as *Urochorda*, including the ascidians or sea-squirts. Only the larval form has a notochord. The adult animal is tubular and is covered externally by a "test" composed of cellulose.

Vital staining. The application of dyes to living cells. With intravital staining, the dye is administered to the whole animal or plant; with supravital (or supervital) staining, freshly excised tissue is treated with a dye solution. Effects due to vital staining cannot be obtained when the tissue is dead.

Wallerian degeneration. The fragmentation and eventual disappearance of axons and their myelin sheaths following severance from or destruction of neuronal somata.

Bibliography

ABADIA-FENOLL, F. (1968) Staining pericellular boutons of glutaraldehyde-fixed nervous tissue: a chromatin-silvering technique for frozen sections. *Stain Technology* **43**, 190–195.

ABRAHART, E. N. (1968) *Dyes and their Intermediates*. Oxford: Pergamon Press.

ADAMS, C. W. M. (1965) Histochemistry of lipids. Chapter 2 in *Neurohistochemistry* (ed. C. W. M. Adams, pp. 6–66). Amsterdam, London, and New York: Elsevier.

ADAMS, C. W. M., ABDULLAH, Y. H., and BAYLISS, O. B. (1967) Osmium tetroxide as a histochemical and histological reagent. *Histochemie* **9**, 68–77.

ADAMS, C. W. M. and BAYLISS, O. B. (1962) The release of protein, lipid and polysaccharide components of the arterial elastica by proteolytic enzymes and lipid solvents. *Journal of Histochemistry and Cytochemistry* **10**, 222–226.

ADAMS, C. W. M. and BAYLISS, O. B. (1968) Reappraisal of osmium tetroxide and OTAN histochemical reactions. *Histochemie* **16**, 162–166.

ADAMS, C. W. M. and SLOPER, J. C. (1955) Technique for demonstrating neurosecretory material in the human hypothalamus. *Lancet* 1955-**I**, 651–652.

ADAMS, C. W. M. and TUQAN, N. A. (1961) The histochemical demonstration of protease by a gelatin–silver film substrate. *Journal of Histochemistry and Cytochemistry* **9**, 469–472.

AKERT, K. and SANDRI, C. (1968) An electron microscope study of zinc–iodide–osmium impregnation of neurons. I. Staining of synaptic vesicles and cholinergic junctions. *Brain Research* **7**, 286–295.

ALLEN, A. K., NEUBERGER, A., and SHARON, N. (1973) The purification, composition and specificity of wheat-germ agglutinin. *Biochemical Journal* **131**, 155–162.

ALLEN, R. L. M. (1971) *Colour Chemistry*. London: Nelson.

ALROY, J., TEMAURA, K., DAVIDSOHN, I., and WEINSTEIN, R. S. (1978) A method for demonstrating blood group isoantigens in permanent tissue sections. *Stain Technology* **53**, 53–56.

ANDERSON, M. J. and COHEN, M. W. (1974) Fluorescent staining of acetylcholine receptors in vertebrate skeletal muscle. *Journal of Physiology* **237**, 385–400.

ARMSTRONG, J., RICHARDSON, K. C., and YOUNG, J. Z. (1956) Staining neural end-feet and mitochondria after postchroming and Carbowax embedding. *Stain Technology* **31**, 263–270.

ARMSTRONG, J. and STEPHENS, P. R. (1960) A modified chrome–silver paraffin wax technique for staining neural end-feet. *Stain Technology* **35**, 71–75.

AXELSSON, S., BJÖRKLUND, A., FALCK, B., LINDVALL, O., and SVENSSON, L. A. (1973) Glyoxylic acid condensation: a new fluorescence method for the histochemical demonstration of biogenic monoamines. *Acta Physiologica Scandinavica* **87**, 57–62.

BACKSTROM, G., HALLEN, A., HOOK, M., JANSSON, L. and LINDAHL, U. (1975) Biosynthesis of heparin. In *Heparin* (ed. R. A. Bradshaw and S. Wessler), (*Advances in Experimental Medicine and Biology*, vol. **52**), pp. 61–72. New York and London: Plenum Press.

BAHR, G. F. (1954) Osmium tetroxide and ruthenium tetroxide and their reactions with biologically important substances. *Experimental Cell Research* **7**, 457–479.

BAKER, J. R. (1946) The histochemical recognition of lipine. *Quarterly Journal of Microscopical Science* **87**, 441–470.

BAKER, J. R. (1947) The histochemical recognition of certain guanidine derivatives. *Quarterly Journal of Microscopical Science* **88**, 115–121.

BAKER, J. R. (1956) The histochemical recognition of phenols, especially tyrosine. *Quarterly Journal of Microscopical Science* **97**, 161–164.

BAKER, J. R. (1958) *Principles of Biological Microtechnique* (reprinted 1970 with revisions). London: Methuen.

BAKER, J. R. (1960) Experiments on the action of mordants. 1. "Single-bath" mordant dyeing. *Quarterly Journal of Microscopical Science* **101**, 255–272.

BAKER, J. R. (1962) Experiments on the action of mordants. 2. Aluminium–haematein. *Quarterly Journal of Microscopical Science* **103**, 493–517.

BAKER, J. R. (1966) *Cytological Technique*, 5th edn. London: Methuen.

BAKER, J. R. and WILLIAMS, E. G. M. (1965) The use of methyl green as a histochemical reagent. *Quarterly Journal of Microscopical Science* **106**, 3–13.

BANCROFT, J. D. (1967) *An Introduction to Histochemical Techniques*. London: Butterworths.

BANCROFT, J. D. and STEVENS, A. (eds.) (1977) *Theory and Practice of Histological Techniques*. Edinburgh, London, and New York: Churchill–Livingstone.

BANGLE, R. (1954) Gomori's paraldehyde–fuchsin stain. I. Physico-chemical and staining properties of the dye. *Journal of Histochemistry and Cytochemistry* **2**, 291–299.

BANGLE, R. and ALFORD, W. C. (1954) The chemical basis of the periodic acid–Schiff reaction of collagen fibers with reference to periodate consumption by collagen and reticulin. *Journal of Histochemistry and Cytochemistry* **2**, 62–76.

BARGMANN, W. (1949) Über die neurosekretorische Verknupfung von Hypothalamus und Neurohypophyse. *Zeitschrift für Zellforschung* **34**, 610–634.

BARKA, T. and ANDERSON, P. J. (1963) *Histochemistry. Theory, Practice and Bibliography*. New York: Harper & Row.

BARRETT, A. J. (1971) The biochemistry and function of mucosubstances. *Histochemical Journal* **3**, 213–321.

BARRNETT, R. J. and SELIGMAN, A. M. (1958) Histochemical demonstration of protein-bound alpha-acylamido carboxyl groups. *Journal of Biophysical and Biochemical Cytology* **4**, 169–176.

BAYLISS, O. B. and ADAMS, C. W. M. (1972) Bromine–Sudan black: a general stain for lipids including free cholesterol. *Histochemical Journal* **4**, 505–515.

BAYLISS, O. B. and ADAMS, C. W. M. (1979) The pH dependence of borohydride as an aldehyde reductant. *Histochemical Journal* **11**, 111–116.

BAYLISS HIGH, O. (1977) Lipids. Chapter 11 in *Theory and Practice of Histological Techniques* (ed. J. D. Bancroft and A. Stevens), pp. 168–185. Edinburgh, London and New York: Churchill–Livingstone.

BELL, C. (1966) Use of the direct-coloring thiocholine technique for demonstration of intracellular neuronal cholinesterases. *Journal of Histochemistry and Cytochemistry* **14**, 567–570.

BENSON, R. C., MEYER, R. A., ZARUBA, M. E., and MCKHANN, G. M. (1979) Cellular autofluorescence—is it due to flavins? *Journal of Histochemistry and Cytochemistry* **27**, 44–48.

BERG, W. F. and FORD, D. G. (1949) Latent image distribution as shown by physical development. *Abridged Scientific Publications from the Kodak Research Laboratories* **31**, 194–199. Rochester, NY: Eastman-Kodak.

BERGERON, J. A. and SINGER, M. (1958) Metachromasy: an experimental and theoretical evaluation. *Journal of Biophysical and Biochemical Cytology* **4**, 433–457.

BERMAN, L. and ANDREWS, M. (1970) The modulation of human lymphocytes in response to phytohemagglutinin stimulation. I. Observations on the localization of PHA in leukocyte cultures utilizing fluorescein isothiocyanate-labeled PHA. *Journal of the Reticuloendothelial Society* **8**, 74–90.

BERNHARD, W. and AVRAMEAS, S. (1971) Ultrastructural visualization of cellular carbohydrate components by means of concanavalin A. *Experimental Cell Research* **64**, 232–236.

BERTALANFFY, L. VON and BICKIS, I. (1956) Identification of cytoplasmic basophilia (ribonucleic acid) by fluorescence microscopy. *Journal of Histochemistry and Cytochemistry* **4**, 481–493.

BIGBEE, J. W., KOSEK, J. C., and ENG, L. F. (1977) Effects of primary antiserum dilution on staining of "antigen-rich" tissues with the peroxidase–antiperoxidase technique. *Journal of Histochemistry and Cytochemistry* **25**, 443–447.

BILLMAN, J. H. and DIESING, A. C. (1957) Reduction of Schiff bases with sodium borohydride. *Journal of Organic Chemistry* **22**, 1068–1070.

BITTIGER, H. and SCHNEBLI, H. P. (1976) *Concanavalin A as a Tool*. London, New York, Sydney and Toronto: Wiley.

BJÖRKLUND, A., FALCK, B., and LINDVALL, O. (1975) Microspectro-fluorometric analysis of cellular monoamines after formaldehyde or glyoxylic acid condensation. Chapter 5 in *Methods in Brain Research* (ed. P. B. Bradley), pp. 249–294. London, New York, Sydney and Toronto: Wiley.

BLACKSTAD, T. W., FREGERSLEV, S., LAURBERG, S., and ROKKEDAL, K. (1973) Golgi impregnation with potassium dichromate and mercurous or mercuric nitrate: identification of the precipitate by X-ray and electron diffraction methods. *Histochemie* **36**, 247–268.

BLOOM, F. E. and BATTENBERG, E. L. F. (1976) A rapid, simple and sensitive method for the demonstration of central catecholamine-containing axons. II. A detailed description of methodology. *Journal of Histochemistry and Cytochemistry* **24**, 561–571.

BODIAN, D. (1936) A new method for staining nerve fibers and nerve endings in mounted paraffin sections. *Anatomical Record* **65**, 85–97.

BODIAN, D. (1937) The staining of paraffin sections of nervous tissue with activated protargol. The role of fixatives. *Anatomical Record* **69**, 153–162.

BOTTCHER, C. J. F. and BOELSMA-VAN HOUTE, E. (1964) Method for the histochemical identification of choline-containing compounds. *Journal of Atherosclerosis Research* **4**, 109–112.

BOYD, W. C. (1970) Lectins. *Annals of the New York Academy of Science* **169**, 168–190.

BRAAK, H. and JACOB, K. (1973) Dye staining of neural end-feet for light microscopy after glutaraldehyde fixation. *Stain Technology* **48**, 181–183.

BRACHET, J. (1953) The use of basic dyes and ribonuclease for the cytochemical detection of ribonucleic acid. *Quarterly Journal of Microscopical Science* **94**, 1–10.

BRADBURY, P. and GORDON, K. C. (1977) Tissue processing. Chapter 3 in *Theory and Practice of Histological Techniques* (ed. J. D. Bancroft and A. Stevens), pp. 29–45. Edinburgh, London, and New York: Churchill–Livingstone.

BRADBURY, S. (1973) *Peacock's Elementary Microtechnique*, 4th edn. London: Arnold.

BRAITENBERG, V., GUGLIELMOTTI, V., and SADA, E. (1967) Correlation of crystal growth with the staining of axons by the Golgi procedure. *Stain Technology* **42**, 277–283.

BRIDGES, J. W. (1968) Fluorescence of organic compounds. Chapter 6 in *Luminescence in Chemistry* (ed. E. J. Bowen), pp. 77–115. London, Princeton, Toronto, and Melbourne: van Nostrand.

BRIMACOMBE, J. S. and WEBBER, J. M. (1964) *Mucopolysaccharides*. Amsterdam, New York, and London: Elsevier.

BRODERSON, S. H., WESTRUM, L. E., and SUTTON, A. E. (1974) Studies of the direct coloring method for localizing cholinesterase activity. *Histochemistry* **40**, 13–23.

BROOKER, L. G. S. (1966) Sensitizing and desensitizing dyes. Chapter 11 in *The Theory of the Photographic Process*, 3rd edn. (ed. T. H. James), pp. 198–232. New York and London: Macmillan.

BUEHNER, T. S., NETTLETON, G. S., and LONGLEY, J. B. (1979) Staining properties of aldehyde fuchsin analogues. *Journal of Histochemistry and Cytochemistry* **27**, 782–787.

BULMER, D. (1962) Observations on histological methods involving the use of phosphotungstic and phosphomolybdic acids, with particular reference to staining with phosphotungstic acid/haematoxylin. *Quarterly Journal of Microscopical Science* **103**, 311–323.

BURGER, M. M. and GOLDBERG, A. R. (1967) Identification of a tumor-specific determinant on neoplastic cell surfaces. *Proceedings of the National Academy of Science* **57**, 359–366.

BURSTONE, M. S. (1961) Modifications of histochemical techniques for the demonstration of cytochrome oxidase. *Journal of Histochemistry and Cytochemistry* **9**, 59–65.

BURSTONE, M. S. (1962) *Enzyme Histochemistry*. New York and London: Academic Press.

CAJAL, S. RAMON Y (1911) *Histologie du Système Nerveux de l'Homme et des Vertébrés*, 2 vols. (transl. L. Azoulay). Paris: Maloine.

CANNON, H. G. (1937) A new biological stain for general purposes. *Nature* **139**, 549.

CARTER, H. E., GLICK, F. J., NORRIS, W. P., and PHILLIPS, G. E. (1947) Biochemistry of the sphingolipides. III. The structure of sphingosine. *Journal of Biological Chemistry* **170**, 285–294.

CASON, J. E. (1950) A rapid one-step Mallory–Heidenhain stain for connective tissue. *Stain Technology* **25**, 225–226.

CAWOOD, A. H., POTTER, U., and DICKINSON, H. G. (1978) An evaluation of coomassie brilliant blue as a stain for quantitative microdensitometry of protein in section. *Journal of Histochemistry and Cytochemistry* **26**, 645–650.

CELIO, M. R., LUTZ, H., BINZ, H., and FEY, H. (1979) Protein A in immunoperoxidase techniques. *Journal of Histochemistry and Cytochemistry* **27**, 691–698.

CHABEREK, S. and MARTELL, A. E. (1959) *Organic Sequestering Agents*. New York: Wiley.

CHAMPY, C., COUJARD, R., and COUJARD-CHAMPY, C. (1946) L'innervation sympathétique des glandes. *Acta Anatomica* **1**, 233–283.

CHAN-PALAY, V. (1973) A brief note on the chemical nature of the precipitate within nerve fibers after the rapid Golgi reaction: selected area diffraction in high voltage electron microscopy. *Zeitschrift für Anatomie und Entwicklungs-Geschichte* **139**, 115–117.

CHAYEN, J., BITENSKY, L., and BUTCHER, R. G. (1973) *Practical Histochemistry*. London, New York, Sydney, and Toronto: Wiley.

CLARK, G. (ed.) (1973) *Staining Procedures Used by the Biological Stains Commission*. 3rd edn. Baltimore: Williams & Wilkins.

CLARK, P. (1954) A comparison of decalcifying methods. *American Journal of Clinical Pathology* **24**, 1113.

CLARK, W. M. (1972) *Oxidation–Reduction Potentials of Organic Systems*. Huntington, NY: Krieger (Williams & Wilkins).

CLASEN, R. A., SIMON, G., SCOTT, R. V., PANDOLFI, S., and LESAK, R. (1973) The staining of the myelin sheath by luxol dye techniques. *Journal of Neuropathology and Experimental Neurology* **32**, 271–283.

CLELAND, W. W. (1964) Dithiothreitol, a new protective reagent for SH groups. *Biochemistry* **3**, 480–482.

COMBS, J. W., LAGUNOFF, D., and BENDITT, E. P. (1965) Differentiation and proliferation of embryonic mast cells of the rat. *Journal of Cell Biology* **25**, 577–592.

CONTESTABILE, A. and ANDERSEN, H. (1978) Methodological aspects of the histochemical localization and activity of some cerebellar dehydrogenases. *Histochemistry* **56**, 117–132.

COOK, G. M. W. and STODDART, R. W. (1973) *Surface Carbohydrates of the Eukaryotic Cell*. London and New York: Academic Press.

COOK, H. C. (1974) *Manual of Histological Demonstration Methods*. London: Butterworths.

CORBETT, J. F. (1971) Hair dyes. Ch. 7 in *The Chemistry of Synthetic Dyes* (ed. K. Venkataraman), pp. 475–534. New York and London: Academic Press.

CORRÉA, F. M. A., INNIS, R. B., ROUOT, B., PASTERNAK, G. W., and SNYDER, S. H. (1980) Fluorescent probes of β-adrenergic and opiate receptors: biochemical and histochemical evaluation. *Neuroscience Letters* **16**, 47–53.

CORRODI, H. and JONSSON, G. (1967) The formaldehyde fluorescence method for the histochemical demonstration of biogenic monoamines. *Journal of Histochemistry and Cytochemistry* **15**, 65–78.

COTMAN, C. W. and TAYLOR, T. (1974) Localization and characterization of concanavalin A receptors in the synaptic cleft. *Journal of Cell Biology* **62**, 236–242.

COTTON, F. A. and WILKINSON, G. (1972) *Advanced Inorganic Chemistry*, 3rd edn. New York, London, Sydney, and Toronto: Wiley.

COUPLAND, R. E., KOBAYASHI, S., and CROWE, J. (1976) On the fixation of catecholamines including adrenaline in tissue sections. *Journal of Anatomy* **122**, 403–413.

COUPLAND, R. E., PYPER, A. S., and HOPWOOD, D. (1964) A method for differentiating between noradrenaline- and adrenaline-storing cells in the light and electron microscope. *Nature* **201**, 1240–1242.

COWAN, W. M., GOTTLIEB, D. I., HENDRICKSON, A., PRICE, J. L., and WOOLSEY, T. A. (1972) The autoradiographic demonstration of axonal connections in the central nervous system. *Brain Research* **37**, 21–51.

CULLING, C. F. A. (1974) *Handbook of Histopathological and Histochemical Techniques*, 3rd edn. London: Butterworths.

COWDEN, R. R. and CURTIS, S. K. (1970) Demonstration of protein-bound sulfhydryl and disulfide groups with fluorescent mercurials. *Histochemie* **22**, 247–255.

CULLING, C. F. A., REID, P. E., and DUNN, W. L. (1976) A new histochemical method for the identification and visualization of both side chain acylated and nonacylated sialic acids. *Journal of Histochemistry and Cytochemistry* **24**, 1225–1230.

CUNNINGHAM, A. J. (1978) *Understanding Immunology*. New York, San Francisco, and London: Academic Press.

CUNNINGHAM, F. O. and FITZGERALD, M. J. T. (1972) Encapsulated nerve endings in hairy skin. *Journal of Anatomy* **112**, 93–97.

CUNNINGHAM, L. (1967) Histochemical observations of the enzymatic hydrolysis of gelatin films. *Journal of Histochemistry and Cytochemistry* **15**, 292–298.

CURRAN, R. C. (1964) The histochemistry of mucopolysaccharides. *International Review of Cytology* **17**, 149–212.

DABELSTEEN, E., FEJERSKOV, O., NOREN, O., and MACKENZIE, I. C. (1978) Concanavalin A and *Ricinus communis* receptor sites in normal human oral mucosa. *Journal of Investigative Dermatology* **70**, 11–15.

DAHLSTROM, A. (1971) Regional distribution of brain catecholamines and serotonin. *Neurosciences Research Program Bulletin* **9**, 197–205.

DANIELLI, J. F. (1947) A study of techniques for the cytochemical demonstration of nucleic acids and some components of protein. *Symposia of the Society for Experimental Biology* **1**, 101–113.

DAVIS, R. P. and JANIS, R. (1966) Free aldehydic groups in collagen and other tissue components. *Nature* **210**, 318–319.

DAVIS, R. and KOELLE, G. B. (1967) Electron microscopic localization of acetylcholinesterase and nonspecific cholinesterase at the neuromuscular junction by the gold–thiocholine and gold–thiolacetic acid methods. *Journal of Cell Biology* **34**, 157–171.

DEAN, J. A. (ed.) (1973) *Lange's Handbook of Chemistry*, 11th edn. New York: McGraw-Hill.

DEIERKAUF, F. A. and HESLINGA, F. J. M. (1962) The action of formaldehyde of rat brain lipids. *Journal of Histochemistry and Cytochemistry* **10**, 79–82.

DEMALSY, P. and CALLEBAUT, M. (1967) Plain water as a rinsing agent preferable to sulfurous acid after the Feulgen nucleal reaction. *Stain Technology* **42**, 133–136.

DEMÊMES, D., FUENTES, C., and MARTY, R. (1974) Cinétique des processus de dégénérescence axonique dans le système nerveux central: étude expérimentale à court terme dans le corps calleux chez le rat. *Acta Neuropathologica* **29**, 311–323.

DESCLIN, J. C. (1973) A simplified silver impregnation of neural end-feet in paraffin sections. *Stain Technology* **48**, 327–331.

DIXON, M. and WEBB, E. C. (1964) *Enzymes*, 2nd edn. London: Longmans.

DRURY, P. (1973) Methode zur histologischen Untersuchung von Knorpelgewebe. *Das Medizinische Laboratorium* **26**, 21–22.

DRYSDALE, R. G., HERRICK, P. R., and FRANKS, D. (1968) The specificity of the haemagglutinin of the castor bean, *Ricinus communis*. *Vox Sanguinis* **15**, 194–202.

DRZENIEK, R. (1973) Substrate specificity of neuraminidases. *Histochemical Journal* **5**, 271–290.

DUBOIS-DALCQ, M., MCFARLAND, H., and MCFARLIN, D. (1977) Protein A–peroxidase: a valuable tool for the localization of antigens. *Journal of Histochemistry and Cytochemistry* **25**, 1201–1206.

DUJINDAM, W. A. L. and VAN DUIJN, P. (1975) The interaction of apurinic aldehyde groups with pararosaniline in the Feulgen–Schiff and related staining procedures. *Histochemistry* **44**, 67–85.

DWYER, F. R. and MELLOR, D. P. (ed.) (1964) *Chelating Agents and Metal Chelates*. New York and London: Academic Press.

EGGERT, F. M. and GERMAIN, J. P. (1979) Rapid demineralization in acidic buffers. *Histochemistry* **59**, 215–224.

EINARSON, L. (1951) On the theory of gallocyanin–chromalum staining and its application for quantitative estimation of basophilia. *Acta Pathologica et Microbiologica Scandinavica* **28**, 82–102.

ELLEDER, M. and LOJDA, Z. (1970) Studies in lipid histochemistry. III. Reaction of Schiff's reagent with plasmalogens. *Histochemie* **24**, 328–335.

ELLEDER, M. and LOJDA, Z. (1971) Studies in lipid histochemistry. VI. Problems of extraction with acetone in lipid histochemistry. *Histochemie* **28**, 68–87.

ELLEDER, M. and LOJDA, Z. (1972) Studies in lipid histochemistry. IX. The specificity of Holczinger's reaction for fatty acids. *Histochemie* **32**, 301–305.

ELLIOTT, K. A. and GREIG, M. E. (1938) The distribution of the succinic oxidase system in animal tissues. *Biochemical Journal* **32**, 1407–1423.

ENERBACK, L. (1969) Detection of histamine in mast cells by *o*-phthalaldehyde reaction after liquid fixation. *Journal of Histochemistry and Cytochemistry* **17**, 757–759.

ERANKO, O., KOELLE, G. B., and RAISANEN, L. (1967) A thiocholine–lead ferricyanide method for acetylcholinesterase. *Journal of Histochemistry and Cytochemistry* **15**, 674–679.

ESSNER, E., SCHREIBER, J., and GRIEWSKI, R. A. (1978) Localization of carbohydrate components in rat colon with fluoresceinated lectins. *Journal of Histochemistry and Cytochemistry* **26**, 452–458.

ETZLER, M. E. and BRANSTRATOR, M. L. (1974) Differential localization of cell surface and secretory components in rat intestinal epithelium by use of lectins. *Journal of Cell Biology* **63**, 329–343.

ETZLER, M. E. and KABAT, E. A. (1970) Purification and characterization of a lectin (plant hemagglutinin) with blood group A specificity from *Dolichos biflorus*. *Biochemistry* **9**, 869–877.

EVERETT, M. M. and MILLER, W. A. (1974) The role of phosphotungstic and phosphomolybdic acids in connective tissue staining. I. Histochemical studies. *Histochemical Journal* **6**, 25–34.

FEIGL, F. (1960) *Spot Tests in Organic Analysis*, 6th edn. (transl. R. E. Oesper). Amsterdam, London, New York, and Princeton: Elsevier.

FEIGL, F. and ANGER, V. (1972) *Spot Tests in Inorganic Analysis*, 6th English edn. (transl. R. E. Oesper). Amsterdam, London, and New York: Elsevier.

FINK, R. P. and HEIMER, L. (1967) Two methods for selective silver impregnation of degenerating axons and their synaptic endings in the central nervous system. *Brain Research* **4**, 369–374.

FITZGERALD, M. J. T. and FITTER, W. F. (1971) Significance of pH in staining nerves with methylene blue. *Laboratory Practice* **20**, 793, 800.

FRASER, F. J. (1972) Degenerating myelin: comparative histochemical studies using classical myelin stains and an improved Marchi technique minimizing artifacts. *Stain Technology* **47**, 147–154.

FREDERIKS, W. M. (1977) Some aspects of the value of Sudan black B in lipid histochemistry. *Histochemistry* **54**, 27–37.

FRIED, B., GILBERG, J. J., and FEESE, R. C. (1976) Mercuric bromophenol blue to reveal gelatin substrates for protease. *Stain Technology* **51**, 140–141.

FULLER, P. M. and PRIOR, D. J. (1975) Cobalt iontophoresis techniques for tracing afferent and efferent connections in the vertebrate CNS. *Brain Research* **88**, 211–220.

FURNESS, J. B. and COSTA, M. (1975) The use of glyoxylic acid for the fluorescence histochemical demonstration of peripheral stores of noradrenaline and 5-hydroxytryptamine in whole mounts. *Histochemistry* **41**, 335–352.

FURNESS, J. B., COSTA, M., and WILSON, A. J. (1977) Water-stable fluorophores, produced by reaction with aldehyde solutions, for the histochemical localization of catechol- and indolethylamines. *Histochemistry* **52**, 159–170.

FURNESS, J. B., HEATH, J. W., and COSTA, M. (1978) Aqueous aldehyde (Faglu) methods for the fluorescence histochemical localization of catecholamines and for ultrastructural studies of central nervous tissue. *Histochemistry* **57**, 285–295.

GABE, M. (1953) Sur quelques applications de la coloration par la fuchsine-paraldéhyde. *Bulletin de Microscopie Appliqué* (2nd series) **3**, 153–162.

GABE, M. (1966) *Neurosecretion* (transl. R. Crawford). Oxford: Pergamon Press.

GABE, M. (1976) *Histological Techniques*, English edn. (transl. E. Blackith and A. Kavoor). Paris, New York, Barcelona, and Milan: Masson.

GALLYAS, F. (1971) A principle for silver staining of tissue elements by physical development. *Acta Morphologica Hungariae* **19**, 57–71.

GARVEY, J. S., CREMER, N. E., and SUSSDORF, D. H. (1977) *Methods in Immunology*, 3rd edn. Reading, Mass.: Benjamin.

GATENBY, J. B. and BEAMS, H. W. (1950) *The Microtomist's Vade-Mecum (Bolles Lee)*, 11th edn. London: Churchill.

GILL, J. E. and JOTZ, M. M. (1976) Further observations on the chemistry of pararosaniline–Feulgen staining. *Histochemistry* **46**, 147–160.

GILLOTEAUX, J. and NAUD, J. (1979) The zinc iodide–osmium tetroxide staining-fixative of Maillet. Nature of the precipitate studied by X-ray microanalysis and detection of Ca^{2+}-affinity subcellular sites in a tonic smooth muscle. *Histochemistry* **63**, 227–243.

GLAZER, A. N. (1976) The chemical modification of proteins by group-specific and site-specific reagents. In *The Proteins*, 3rd edn. (ed. H. Neurath and R. L. Hill), vol **2**, pp. 1–103. New York: Academic Press.

GLEGG, R. E., CLERMONT, Y., and LEBLOND, C. P. (1952) The use of lead tetraacetate, benzidine, *o*-anisi-
dine, and a "film test" in investigating the periodic acid–Schiff technique. *Stain Technology* **27**, 277–

GLEGG, R. E., CLERMONT, Y., and LEBLOND, C. P. (1952) The use of lead tetraacetate, benzidine, *o*-anisidine, and a "film test" in investigating the periodic acid–Schiff technique. *Stain Technology* **27**, 277–305.

GLENNER, G. G. (1957) The histochemical demonstration of indole derivatives by the rosindole reaction of E. Fischer. *Journal of Histochemistry and Cytochemistry* **5**, 297–304.

GOLAND, P., GRAND, N. G., and KATELE, K. V. (1967) Cyanuric chloride and *N*-methylmorpholine in methanol as a fixative for polysaccharides. *Stain Technology* **42**, 41–51.

GOLDMAN, M. (1968) *Fluorescent Antibody Methods*. London and New York: Academic Press.

GOLDSTEIN, D. J. (1962) Ionic and non-ionic bonds in staining, with special reference to the action of urea and sodium chloride on the staining of elastic fibres and glycogen. *Quarterly Journal of Microscopical Science* **103**, 477–492.

GOLDSTEIN, M. J., HOLLERMAN, C. E., and SMITH, E. E. (1965) Protein–carbohydrate interaction. II. Inhibition studies on the interaction of concanavalin A with polysaccharides. *Biochemistry* **4**, 876–883.

GORDON, H. and SWEETS, H. H. (1936) A simple method for the silver impregnation of reticulum. *American Journal of Pathology* **12**, 545–552.

GOTTSCHALK, A. (1972) *Glycoproteins*, 2nd edn., 2 vols.: A and B. Amsterdam, New York, and London: Elsevier.

GOYER, R. A. and CHERIAN, M. G. (1977) Tissue and cellular toxicology of metals. In *Clinical Chemistry and Chemical Toxicology of Metals* (ed. S. S.

Brown), pp. 89–103. Amsterdam, New York, and Oxford: Elsevier–North Holland.

GRAHAM, R. C. and KARNOVSKY, M. J. (1966) The early stages of absorption of injected horseradish peroxidase in the proximal tubules of mouse kidney: ultrastructural cytochemistry by a new technique. *Journal of Histochemistry and Cytochemistry* **14**, 291–302.

GRAY, P. (1954) *The Microtomist's Formulary and Guide*. London: Constable.

GREEN, M. R. and PASTEWKA, J. V. (1974a) Simultaneous differential staining by a cationic carbocyanine dye of nucleic acids, proteins and conjugated proteins. I. Phosphoproteins. *Journal of Histochemistry and Cytochemistry* **22**, 767–773.

GREEN, M. R. and PASTEWKA, J. V. (1974b) Simultaneous differential staining by a cationic carbocyanine dye of nucleic acids, proteins and conjugated proteins. II. Carbohydrate and sulphated carbohydrate-containing proteins. *Journal of Histochemistry and Cytochemistry* **22**, 774–781.

GREEN, M. R. and PASTEWKA, J. V. (1979) The cationic carbocyanine dyes stains-all, DBTC and ethyl stains-all, DBTC-3,3′,9 triethyl. *Journal of Histochemistry and Cytochemistry* **27**, 797–799.

GREENHALGH, C. W. (1976) Aspects of anthraquinone dyestuff chemistry. *Endeavour* **35**, 134–140.

GROS, D., OBRENOVITCH, A., CHALLICE, C. E., MONSIGNY, M., and SCHREVEL, J. (1977) Ultrastructural visualization of cellular carbohydrate components by means of lectins on ultrathin glycol methacrylate sections. *Journal of Histochemistry and Cytochemistry* **25**, 104–114.

GURR, E. (1971) *Synthetic Dyes in Biology, Medicine and Chemistry*. London: Academic Press.

GUSTAVSON, K. H. (1956) *The Chemistry of Tanning Processes*. New York: Academic Press.

GWYN, D. G. and HEARDMAN, V. (1965) A cholinesterase–Bielschowsky staining method for mammalian motor and plates. *Stain Technology* **40**, 15–18.

HAASE, P. (1975) The development of nephrocalcinosis in the rat following injections of neutral sodium phosphate. *Journal of Anatomy* **119**, 19–37.

HADLER, W. A., LUCCA, O. DE, ZITI, L. M., and PATELLI, A. S. (1969) An analysis of the effect of some fixatives on the histochemical detection of nonhaem ferric iron in spleen sections. *Revista Brasileira de Pesquisas Medicas e Biologicas* **2**, 378–383.

HADLER, W. A. and SILVEIRA, S. R. (1978) Histochemical technique to detect choline-containing lipids. *Acta Histochemica* **63**, 265–270.

HAHN VON DORSCHE, H., KRAUSE, R., FEHRMANN, P., and SULZMANN, R. (1975) Histochemische Nachweismethoden für biogene Aminė. *Acta Histochemica* **52**, 281–302.

HAKANSON, R., OWMAN, C., and SUNDLER, E. (1972) *o*-Phthalaldehyde (OPT). A sensitive detection reagent for glucagon, secretin and vasoactive intestinal peptide. *Journal of Histochemistry and Cytochemistry* **20**, 138–140.

HANKER, J. S., THORNBURG, L. P., YATES, P. E., and MOORE, H. G. (1973) The demonstration of cholinesterase by the formation of osmium blacks at the sites of Hatchett's brown. *Histochemie* **37**, 233–242.

HANKER, J. S., YATES, P. E., METZ, C. B., and RUSTIONI, A. (1977) A new specific, sensitive and non-carcinogenic reagent for the demonstration of horseradish peroxidase. *Histochemical Journal* **9**, 789–792.

HARRIS, C. M. and LIVINGSTONE, S. E. (1964) Bidentate chelates. Chapter 3 in *Chelating Agents and Metal Chelates* (ed. F. P. Dwyer and D. P. Mellor), pp. 95–141. New York and London: Academic Press.

HASEGAWA, J. and HASEGAWA, J. (1977) Substrate limitations of the color film technique (Fratello) for the localization of proteases. *Journal of Histochemistry and Cytochemistry* **25**, 234.

HAYAT, M. A. (1975) *Positive Staining for Electron Microscopy*. New York: van Nostrand–Reinhold.

HENDRICKSON, J. B., CRAM, D. J., and HAMMOND, G. S. (1970) *Organic Chemistry*, 3rd edn. New York: McGraw-Hill.

HESLINGA, F. J. M. and DEIERKAUF, F. A. (1961) The action of histological fixatives on tissue lipids. Comparison of the action of several fixatives using paper chromatography. *Journal of Histochemistry and Cytochemistry* **9**, 572–577.

HESS, A. (1978) A simple procedure for distinguishing dopamine from noradrenaline in peripheral nervous structures in the fluorescence microscope. *Journal of Histochemistry and Cytochemistry* **26**, 141–144.

HIRANO, A. and ZIMMERMANN, H. M. (1962) Silver impregnation of nerve cells and fibers in celloidin sections. *Archives of Neurology* **6**, 114–122.

HOGG, R. M. and SIMPSON, R. (1975) An evaluation of solochrome cyanine RS as a nuclear stain similar to haematoxylin. *Medical Laboratory Technology* **32**, 301–306.

HOLMES, W. (1943) Silver staining of nerve axons in paraffin sections. *Anatomical Record* **86**, 157–187.

HOLT, S. J. and WITHERS, R. F. J. (1952) Cytochemical localization of esterases using indoxyl derivatives. *Nature* **170**, 1012–1014.

HOOGHWINKEL, G. J. M. and SMITS, G. (1957) The specificity of the periodic acid–Schiff technique studied by a quantitative test-tube method. *Journal of Histochemistry and Cytochemistry* **5**, 120–126.

HOPWOOD, D. (1969) Fixatives and fixation: a review. *Histochemical Journal* **1**, 323–360.

HOPWOOD, D. (1977) Fixation and fixatives. Chapter 2 in *Theory and Practice of Histological Techniques* (ed. J. D. Bancroft and A. Stevens), pp. 16–28. Edinburgh, London and New York: Churchill–Livingstone.

HOROBIN, R. W. (1977) Theory of staining. Chapter 6 in *Theory and Practice of Histological Techniques* (ed. J. D. Bancroft and A. Stevens), pp. 75–84. Edinburgh, London and New York: Churchill–Livingstone.

HOROBIN, R. W. and GOLDSTEIN, D. J. (1974) The influence of salt on the staining of tissue sections with basic dyes: an investigation into general applicability of the critical electrolyte concentration theory. *Histochemical Journal* **6**, 599–609.

HOROBIN, R. W. and JAMES, N. T. (1970) The staining of elastic fibres with Direct blue 152. A general hypothesis for the staining of elastic fibres. *Histochemie* **22**, 324–336.

HOROBIN, R. W. and KEVILL-DAVIES, I. M. (1971a) Basic fuchsin in acid alcohol: a simplified alternative to Schiff reagent. *Stain Technology* **46**, 53–58.

HOROBIN, R. W. and KEVILL-DAVIES, I. M. (1971b) A mechanistic study of the histochemical reaction between aldehydes and basic fuchsin in acid alcohol used as a simplified Schiff's reagent. *Histochemical Journal* **3**, 371–378.

HUGHES, R. C. (1976) *Membrane Glycoproteins*. London and Boston: Butterworths.

HUMASON, G. (1972) *Animal Tissue Techniques*, 3rd edn. San Francisco: Freeman.

HUTSON, J. C., CHILDS, G. V., and GARDNER, P. J. (1979) Considerations for establishing the validity of immunocytochemical studies. *Journal of Histochemistry and Cytochemistry* **27**, 1201–1202.

INTERNATIONAL UNION OF BIOCHEMISTRY (1961) *Report of the Commission on Enzymes*. Oxford and New York: Pergamon Press.

IRONS, R. D., SCHENK, E. A., and LEE, C. K. (1977) Cytochemical methods for copper. *Archives of Pathology and Laboratory Medicine* **101**, 298–301.

IVERSON, L. L. (1967) *The Uptake and Storage of Noradrenaline in Sympathetic Nerves*. Cambridge: Cambridge University Press.

JAKOVLEVA, I. V., DANILOVA, O. A., DONEV, C., and POLENOV, A. L. (1968) Some remarks to the Sterba's fluorescent method in application to the neurosecretory cells. *Histochemie* **13**, 305–311.

JÁMBOR, B. (1954) Reduction of tetrazolium salt. *Nature* **173**, 774–775.

JAMES, T. H. (ed.) (1966) *The Theory of the Photographic Process,* 3rd edn. New York and London: Macmillan.

JANSSON, L., OGREN, S., and LINDAHL, U. (1975) Macromolecular properties and end-group analysis of heparin isolated from bovine liver capsule. *Biochemical Journal* **145**, 53–62.

JAQUES, L. B. (1978) The nature of mucopolysaccharides. *Medical Hypotheses* **4**, 123–135.

JAQUES, L. B., MAHADOO, J., and RILEY, J. F. (1977) The mast cell/heparin paradox. *Lancet* 1977-**I**, 411–413.

JASMIN, G. and BOIS, P. (1961) Coloration différentielle des mastocytes chez le rat. *Revue Canadienne de Biologie* **20**, 773–774.

JEANLOZ, R. W. (1975) The chemistry of heparin. In *Heparin: Structure, Function and Clinical Implications* (ed. R. A. Bradshaw and S. Wessler) (*Advances in Experimental Medicine and Biology*, vol. **52**), pp. 3–15. New York and London: Plenum Press.

JENSEN, W. A. (1962) *Botanical Histochemistry*. San Francisco and London: Freeman.

JONES, D. (1972) Reactions of aldehydes with unsaturated fatty acids during histological fixation. *Histochemical Journal* **4**, 421–465.

JONES, R. MCCLUNG (ed.) (1950) *McClung's Handbook of Microscopical Technique*, 3rd edn. New York: Hoebner.

JURAND, A. and GOEL, S. C. (1976) The use of methyl green–pyronin staining after glutaraldehyde fixation and paraffin or Araldite embedding. *Tissue and Cell* **8**, 389–394.

KALIMO, H. and PELLINIEMI, L. J. (1977) Pitfalls in the preparation of buffers for electron microscopy. *Histochemical Journal* **9**, 241–246.

KARNOVSKY, M. J. and FASMAN, G. D. (1960) A histochemical method for distinguishing between side-chain and terminal (α-acylamido) carboxyl groups of proteins. *Journal of Biophysical and Biochemical Cytology,* **8**, 319–325.

KARNOVSKY, M. J. and MANN, M. S. (1961) The significance of the histochemical reaction for carboxyl groups of proteins in cartilage matrix. *Histochemie* **2**, 234–243.

KARNOVSKY, M. J. and ROOTS, L. (1964) A "direct coloring" thiocholine method for cholinesterases. *Journal of Histochemistry and Cytochemistry* **12**, 219–221.

KASHIWA, H. K. and ATKINSON, W. B. (1963) The applicability of a new Schiff base, glyoxal *bis* (2-hydroxyanil), for the cytochemical localization of ionic calcium. *Journal of Histochemistry and Cytochemistry* **11**, 258–264.

KASHIWA, H. K. and HOUSE, C. M. (1964) The glyoxal *bis* (2-hydroxyanil) method modified for localizing insoluble calcium salts. *Stain Technology* **39**, 359–367.

KASTEN, F. H. and LALA, R. (1975) The Feulgen reaction after glutaraldehyde fixation. *Stain Technology* **50**, 197–201.

KATER, S. B. and NICHOLSON, C. (ed.) (1973) *Intracellular Staining in Neurobiology*. New York, Heidelberg, and Berlin: Springer.

KATSUYAMA, T. and SPICER, S. S. (1978) Histochemical differentiation of complex carbohydrates with variants of the concanavalin A–horseradish peroxidase method. *Journal of Histochemistry and Cytochemistry* **26**, 232–250.

KEILIN, D. and HARTREE, E. F. (1938) Cytochrome oxidase. *Proceedings of the Royal Society B* **125**: 171–186.

KENT, S. P. (1964) The demonstration and distribution of water soluble blood group O(H) antigen in tissue sections using a fluorescein labeled extract of *Ulex europeus* seeds. *Journal of Histochemistry and Cytochemistry* **12**, 591–599.

KIERNAN, J. A. (1964) Carboxylic esterases of the hypothalamus and neurohypophysis of the hedgehog. *Journal of the Royal Microscopical Society* **83**, 297–306.

KIERNAN, J. A. (1970) Silver staining of axons in subcellular fractions of nervous tissue. *Experientia* **26**, 1352.

KIERNAN, J. A. (1974) Effects of metabolic inhibitors on vital staining with methylene blue. *Histochemistry* **40**, 51–57.

KIERNAN, J. A. (1975) Localization of α-D-glucosyl and α-D-mannosyl groups of mucosubstances with concanavalin A and horseradish peroxidase. *Histochemistry* **44**, 39–45.

KIERNAN, J. A. (1977a) Histochemical demonstration of unsaturated hydrophilic lipids with palladium chloride. *Journal of Histochemistry and Cytochemistry* **25**, 200–205.

KIERNAN, J. A. (1977b) Recycling procedure for gold chloride used in neurohistology. *Stain Technology* **52**, 245–248.

KIERNAN, J. A. (1978) Recovery of osmium tetroxide from used fixative solutions. *Journal of Microscopy* **113**, 77–82.

KIERNAN, J. A. and BERRY, M. (1975) Neuroanatomical methods. Chapter 1 in *Methods in Brain Research* (ed. P. B. Bradley), pp. 1–77. London, New York, Sydney, and Toronto: Wiley.

KIERNAN, J. A. and STODDART, R. W. (1973) Fluorescent-labelled aprotinin: a new reagent for the histochemical detection of acid mucosubstances. *Histochemie* **34**, 77–84.

KJELLSTRAND, P. T. T. (1977) Temperature and acid concentration in the search for optimum Feulgen hydrolysis conditions. *Journal of Histochemistry and Cytochemistry* **25**, 129–134.

KLEIN, R. M. and KLEIN, D. T. (1970) *Research Methods in Plant Science*. Garden City, NY: American Museum of Natural History.

KLESSEN, C. (1974) Histochemical demonstration of thyrotropic and gonadotropic cells in the pituitary gland of the rat by the use of a lead tetraacetate–sodium bisulphite technique. *Histochemical Journal* **6**, 311–318.

KLÜVER, H. and BARRERA, E. (1953) A method for the combined staining of cells and fibres in the central nervous system. *Journal of Neuropathology and Experimental Neurology* **12**, 400–403.

KOELLE, G. B. and FRIEDENWALD, J. S. (1949) A histochemical method for localizing cholinesterase activity. *Proceedings of the Society for Experimental Biology and Medicine* **70**, 617–622.

KOELLE, G. B. and GROMADZKI, C. G. (1966) Comparison of the gold–thiocholine and gold–thiolacetic methods for the histochemical localization of acetylcholinesterase and cholinesterase. *Journal of Histochemistry and Cytochemistry* **14**, 443–454.

KORN, E. G. (1967) A chromatographic and spectrophotometric study of the products of the reaction of osmium tetroxide with unsaturated lipids. *Journal of Cell Biology* **34**, 627–638.

KRAJIAN, A. A. and GRADWOHL, R. B. M. (1952) *Histopathological Technic*, 2nd edn. St. Louis: Mosby.

KRAMER, H. and WINDRUM, G. M. (1955) The metachromatic staining reaction. *Journal of Histochemistry and Cytochemistry* **3**, 226–237.

KUGLER, P. and WROBEL, K. H. (1978) Meldola blue: a new electron carrier for the histochemical demonstration of dehydrogenases (SDH, LDH, G-6-PDH). *Histochemistry* **59**, 97–109.

KURNICK, N. B. (1955) Pyronin Y in the methyl-green–pyronin histological stain. *Stain Technology* **30**, 213–230.

KUYPERS, H. G. J. M., CATSMAN-BERREVOETS, C. E., and PADT, R. E. (1977) Retrograde axonal transport of fluorescent substances in the rat's forebrain. *Neuroscience Letters* **6**, 127–135.

LAI, M., LAMPERT, I. A., and LEWIS, P. D. (1975) The influence of fixation on staining of glycosaminoglycans in glial cells. *Histochemistry* **41**, 275–279.

LATIES, A. M., LUND, R., and JACOBOWITZ, D. (1967) A simplified method for the histochemical localization of cardiac catecholamine-containing nerve fibres. *Journal of Histochemistry and Cytochemistry* **15**, 535–541.

LAVAIL, J. H., WINSTON, K. R., and TISH, A. (1973) A method based on retrograde intraaxonal transport of protein for identification of cell bodies of origin of axons terminating within the CNS. *Brain Research* **58**, 470–477.

LEBLOND, C. P., GLEGG, R. E., and EIDINGER, D. (1957) Presence of carbohydrates with free 1,2-glycol groups in sites stained by the periodic acid–Schiff technique. *Journal of Histochemistry and Cytochemistry* **5**, 445–458.

LEVER, J. D., SANTER, R. M., LU, K. S., and PRESLEY, R. (1977) Electron probe X-ray microanalysis of small granulated cells in rat sympathetic ganglia after sequential aldehyde and dichromate treatment. *Journal of Histochemistry and Cytochemistry* **35**, 275–279.

LEVINSON, J. W., RETZEL, S., and MCCORMICK, J. J. (1977) An improved acriflavin–Feulgen method. *Journal of Histochemistry and Cytochemistry* **25**, 355–358.

LHOTKA, J. F. (1952) Histochemical use of sodium bismuthate. *Stain Technology* **27**, 259–262.

LHOTKA, J. F. (1956) On tissue argyrophilia. *Stain Technology* **31**, 185–188.

LILLIE, R. D. (1952) Ethylenic reaction of ceroid with performic acid and Schiff reagent. *Stain Technology* **27**, 37–45.

LILLIE, R. D. (1962) The histochemical reaction of aryl amines with tissue aldehydes produced by periodic and chromic acids. *Journal of Histochemistry and Cytochemistry* **19**, 303–314.

LILLIE, R. D. (1964) Histochemical acylation of hydroxyl and amino groups. Effect on the periodic acid Schiff reaction, anionic and cationic dye and van Gieson collagen stains. *Journal of Histochemistry and Cytochemistry* **12**, 821–841.

LILLIE, R. D. (1969) Mechanisms of chromatin hematoxylin stains. *Histochemie* **20**, 338–354.

LILLIE, R. D. (1977) *H. J. Conn's Biological Stains*, 9th edn. Baltimore: Williams & Wilkins.

LILLIE, R. D. and BURTNER, H. J. (1953) The ferric ferricyanide reduction test in histochemistry. *Journal of Histochemistry and Cytochemistry* **1**, 87–92.

LILLIE, R. D. and DONALDSON, P. T. (1974) The mechanism of the ferric-ferricyanide reduction reaction. *Histochemical Journal* **6**, 679–684.

LILLIE, R. D. and FULLMER, H. M. (1976) *Histopathologic Technic and Practical Histochemistry*, 4th edn. New York: McGraw-Hill.

LILLIE, R. D., HENDERSON, R., and GUTIERREZ, A. (1968) The diazosafranin method: control of nitrate concentration and refinements in specificity. *Stain Technology* **43**, 311–313.

LILLIE, R. D., PIZZOLATO, P., DESSAUER, H. C., and DONALDSON, P. T. (1971) Histochemical reactions at tissue arginine sites with alkaline solutions of *β*-naphthoquinone-4-sodium sulfonate and other *o*-quinones and oxidized *o*-diphenols. *Journal of Histochemistry and Cytochemistry* **19**, 487–497.

LILLIE, R. D., PIZZOLATO, P., and DONALDSON, P. T. (1976a) Nuclear stains with soluble metachrome metal mordant lake dyes. The effect of chemical endgroup blocking reactions and the artificial introduction of acid groups into tissues. *Histochemistry* **49**, 23–35.

LILLIE, R. D., PIZZOLATO, P., and DONALDSON, P. T. (1976b) Hematoxylin substitutes: a survey of mordant dyes tested and consideration of the relation of their structure to performance as nuclear stains. *Stain Technology* **51**, 25–41.

LINDVALL, O. and BJÖRKLUND, A. (1974) The glyoxylic acid fluorescence histochemical method: a detailed account of the methodology for the visualization of central catecholamine neurons. *Histochemistry* **39**, 97–127.

LIS, H., SELA, B. A., SACHS, L., and SHARON, N. (1970) Specific inhibition by *N*-acetyl-D-galactosamine of the interaction between soybean agglutinin and animal cell surfaces. *Biochemica et Biophysica Acta* **211**, 582–585.

LLEWELLYN, B. D. (1974) Mordant blue 3: a readily available substitute for hematoxylin in the routine hematoxylin and eosin stain. *Stain Technology* **49**, 347–349.

LLEWELLYN, B. D. (1978) Improved nuclear staining with Mordant blue 3 as a hematoxylin substitute. *Stain Technology* **53**, 73–77.

LOACH, P. A. (1976) Oxidation–reduction potentials, absorbance bands and molar absorbance of compounds used in biochemical studies. In *Handbook of Biochemistry and Molecular Biology*, 3rd edn., *Physical and Chemical Data* (ed. G. D. Fasman), vol. **1**, pp. 122–130. Cleveland: CRC Press.

LOREN, I., BJÖRKLUND, A., FALCK, B., and LINDVALL, O. (1976) An improved histofluorescence procedure for freeze-dried paraffin-embedded tissue based on combined formaldehyde–glyoxylic acid perfusion with high magnesium content and acid pH. *Histochemistry* **49**, 177–192.

LOREN, I., BJÖRKLUND, A., FALCK, B., and LINDVALL, O. (1980) The aluminium–formaldehyde (ALFA) method for improved visualization of catecholamines and indoleamines. 1. A detailed account of the methodology for central nervous tissue using paraffin, cryostat or Vibratome sections. *Journal of Neuroscience Methods* **2**, 277–300.

LUNA, L. G. (1968) *Manual of Histologic Staining Methods of the Armed Forces Institute of Pathology*, 3rd edn. New York: McGraw-Hill.

LYCETTE, R. M., DANFORTH, W. F., KOPPEL, J. L., and OLWIN, J. H. (1970) The binding of Luxol fast blue ARN by various biological lipids. *Stain Technology* **45**, 155–160.

LYNCH, G., SMITH, R. L., MENSAH, P., and COTMAN, C. (1973) Tracing the dentate gyrus mossy fiber system with horseradish peroxidase histochemistry. *Experimental Neurology* **40**, 516–524.

MACCALLUM, D. K. (1973) Positive Schiff reactivity of aortic elastin without prior HIO_4 oxidation: influence of maturity and a suggested source of the aldehyde. *Stain Technology* **48**, 117–122.

MCISAAC, G. and KIERNAN, J. A. (1974) Complete staining of neuromuscular innervation with bromoindigo and silver. *Stain Technology* **49**, 211–214.

MCKAY, R. B. (1962) An investigation of the anomalous staining of chromatin by the acid dyes, methyl blue and aniline blue. *Quarterly Journal of Microscopical Science* **103**, 519–530.

MAILLET, M. (1963) Le réactif au tetraoxyde d'osmium–iodure du zinc. *Zeitschrift für mikroskopisch-anatomische Forschung* **76**, 397–425.

MAILLET, M. (1968) Étude critique des fixations au tetraoxyde d'osmium–iodure. *Comptes Rendus de l'Association des Anatomistes* **53**, 231–394.

MÅKELÅ, O. (1957) Studies in haemagglutinins of Leguminosae seeds. *Acta Medicinae Experimentalis et Biologicae Fenniae* **35**, Suppl. **11**, 1–113.

MALININ, G. I. (1977) Stable sudanophilia of "bound" lipids in tissue culture cells is a staining artifact. *Journal of Histochemistry and Cytochemistry* **25**, 155–156.

MALININ, G. I. (1980) The *in situ* determination of melting–solidification points of lipid inclusions in fixed cultured cells. *Journal of Histochemistry and Cytochemistry* **28**, 708–709.

MALM, M. (1962) *p*-toluenesulphonic acid as a fixative. *Quarterly Journal of Microscopical Science* **103**, 163–171.

MALMGREN, H. and SYLVEN, B. (1955) On the chemistry of the thiocholine method of Koelle. *Journal of Histochemistry and Cytochemistry* **3**, 441–448.

MANN, G. (1902) *Physiological Histology. Methods and Theory*. Oxford: Clarendon Press.

MARCH, J. (1977) *Advanced Organic Chemistry*, 2nd edn. New York: McGraw-Hill.

MARCHALONIS, J. J. and EDELMAN, G. M. (1968) Isolation and characterization of a hemagglutinin from *Limulus polyphemus*. *Journal of Molecular Biology* **32**, 453–465.

MARCHI, V. (1892) Sur l'origine et la cours des pédoncules cérébelleux et sur leurs rapports avec les autres centres nerveux. *Archives Italiennes de Biologie* **17**, 190–201.

MAREEL, M., DRAGONETTI, C., and VAN PETECHEM, M. C. (1976) Cytochemistry of colloidal iron binding to the surface of HeLa cells and human erythrocytes. *Histochemistry* **48**, 71–80.

MARSHALL, P. N. (1977) Thin layer chromatography of Sudan dyes. *Journal of Chromatography* **136**, 353–357.

MARSHALL, P. N. (1978) Romanowsky-type stains in haematology. *Histochemical Journal* **10**, 1–29.

MARSHALL, P. N. and HOROBIN, R. W. (1972) The oxidation products of haematoxylin and their role in biological staining. *Histochemical Journal* **4**, 493–503.

MARSHALL, P. N. and HOROBIN, R. W. (1973) The mechanisms of action of "mordant" dyes—a study using preformed metal complexes. *Histochemie* **35**, 361–371.

MARSHALL, P. N. and HOROBIN, R. W. (1974) A simple assay procedure for mixtures of hematoxylin and hematein. *Stain Technology* **49**, 137–142.

MARSLAND, T. A., GLEES, P., and ERIKSON, L. B. (1954) Modification of the Glees silver impregnation method for paraffin sections. *Journal of Neuropathology and Experimental Neurology* **13**, 587–591.

MASON, D. Y. and SAMMONS, R. E. (1979) The labeled antigen method of immunoenzymatic staining. *Journal of Histochemistry and Cytochemistry* **27**, 832–840.

MATSUMOTO, I. and OSAWA, T. (1969) Purification and characterization of an anti-H (O) phytohaemagglutinin of *Ulex europeus*. *Biochemica et Biophysica Acta* **194**, 180–189.

MAZURKIEWICZ, J. E., HOSSLER, F. E., and BARRNETT, R. J. (1978) Cytochemical demonstration of sodium, potassium–adenosine triphosphatase by a hemepeptide derivative of ouabain. *Journal of Histochemistry and Cytochemistry* **26**, 1042–1052.

MAZZUCA, M., ROCHE, A. C., LHERMITTE, M., and ROUSSEL, P. (1977) *Limulus polyphemus* lectin sites in human bronchial mucosa. *Journal of Histochemistry and Cytochemistry* **25**, 470–473.

MELOAN, S. N., VALENTINE, L. S., and PUCHTLER, H. (1971) On the structure of carminic acid and carmine. *Histochemie* **27**, 87–95.

MESULAM, M. M. (1978) Tetramethyl benzidine for horseradish peroxidase neurohistochemistry: a non--carcinogenic blue reaction-product with superior sensitivity for visualizing neural afferents and efferents. *Journal of Histochemistry and Cytochemistry* **26**, 106–117.

MESULAM, M. M. and ROSENE, D. L. (1979) Sensitivity in horseradish peroxidase neurohistochemistry: a comparative and quantitative study of nine methods. *Journal of Histochemistry and Cytochemistry* **27**, 763–773.

MINARD, B. J. and CAWLEY, L. P. (1978) Use of horseradish peroxidas to block nonspecific enzyme uptake in immunoperoxidase microscopy. *Journal of Histochemistry and Cytochemistry* **26**, 685–687.

MOLNAR, J. (1952) The use of rhodizonate in enzymatic histochemistry. *Stain Technology* **27**, 221–222.

MORRIS, S. M., STONE, P. J., ROSENKRANS, W. A., CALORE, J. D., ALBRIGHT, J. T., and FRANZENBLAU, C. (1978) Palladium chloride as a stain for elastin at the ultrastructural level. *Journal of Histochemistry and Cytochemistry* **26**, 635–644.

MORRISON, R. T. and BOYD, R. N. (1973) *Organic Chemistry*, 3rd edn. Boston: Allyn & Bacon.

MORTON, D. (1978) A comparison of iron histochemical methods for use on glycol methacrylate embedded tissues. *Stain Technology* **53**, 217–223.

MOWRY, R. W. (1978) Aldehyde fuchsin staining, direct or after oxidation: problems and remedies, with special references to human pancreatic B cells, pituitaries and elastic fibers. *Stain Technology* **53**, 141–154.

MOWRY, R. W. and EMMEL, V. M. (1966) The coloration of carbohydrate polyanions by National Fast Blue compared with that obtained with Alcian Blue 8GX. *Journal of Histochemistry and Cytochemistry* **14**, 799–800.

MOWRY, R. W. and EMMEL, V. M. (1977) The production of aldehyde fuchsin depends on the pararosaniline (C.I. No. 42500) content of basic fuchsins which is sometimes negligible and is sometimes mislabelled. *Journal of Histochemistry and Cytochemistry* **25**, 239.

NAIRN, R. C. (1976) *Fluorescent Protein Tracing*, 4th edn. Edinburgh and London: Churchill–Livingstone.

NAKAO, K. and ANGRIST, A. A. (1968) A histochemical demonstration of aldehyde in elastin. *American Journal of Clinical Pathology* **49**, 65–67.

NAMBA, T., NAKAMURA, T., and GROB, D. (1967) Staining for nerve fiber and cholinesterase activity in fresh frozen sections. *American Journal of Clinical Pathology* **47**, 74–77.

NAOUMENKO, J. and FEIGIN, I. (1961) A modification for paraffin sections of the Cajal gold–sublimate stain for astrocytes. *Journal of Neuropathology and Experimental Neurology* **20**, 602–604.

NAUTA, W. J. H. (1957) Silver impregnation of degenerating axons. In *New Research Techniques of Neuroanatomy* (ed. W. F. Windle), pp. 17–26. Springfield, Ill.: Thomas.

NAUTA, W. J. H. and EBBESSON, S. O. E. (ed.) (1970)

Contemporary Research Methods in Neuroanatomy. Berlin, Heidelberg, and New York: Springer.

Nauta, W. J. H. and Gygax, P. A. (1951) Silver impregnation of degenerating axon terminals in the central nervous system: (1) Technic. (2) Chemical notes. *Stain Technology* **26**, 5–11.

Nauta, W. J. H. and Gygax, P. A. (1954) Silver impregnation of degenerating axons in the central nervous system: a modified technic. *Stain Technology* **29**, 91–93.

Nedzel, G. A. (1951) Intranuclear birefringent inclusions, an artifact occurring in paraffin sections. *Quarterly Journal of Microscopical Science* **92**, 343–346.

Nettleton, G. S. and Carpenter, A. M. (1977) Studies on the mechanism of the periodic acid–Schiff histochemical reaction for glycogen using infrared spectroscopy and model chemical compounds. *Stain Technology* **52**, 63–77.

Nicolet, B. H. and Shinn, L. A. (1939) The action of periodic acid on α-amino alcohols. *Journal of the American Chemical Society* **61**, 1615.

Nicolson, G. L. (1974) The interactions of lectins with animal cell surfaces. *International Review of Cytology* **39**, 89–190.

Nicolson, G. L. and Singer, S. J. (1971) Ferritin-conjugated plant agglutinins as specific saccharide stains for electron microscopy: application to saccharides bound to cell membranes. *Proceedings of the National Academy of Science* **68**, 942–945.

Nieland, M. L. (1973) Epidermal intercellular staining with fluorescein-conjugated phytohemagglutinins. *Journal of Investigative Dermatology* **60**, 61–66.

Nielson, A. J. and Griffith, W. P. (1978) Tissue fixation and staining with osmium tetroxide: the role of phenolic compounds. *Journal of Histochemistry and Cytochemistry* **26**, 138–140.

Nielson, A. J. and Griffith, W. P. (1979) Tissue fixation by osmium tetroxide. A possible role for proteins. *Journal of Histochemistry and Cytochemistry* **27**, 997–999.

Noller, C. R. (1965) *Chemistry of Organic Compounds*, 3rd edn. Philadelphia and London: Saunders.

Norton, W. T., Korey, S. R., and Brotz, M. (1962) Histochemical demonstration of unsaturated lipids by a bromine–silver method. *Journal of Histochemistry and Cytochemistry* **10**, 83–88.

Notani, G. W., Parsons, J. A., and Erlandsen, S. L. (1979) Versatility of *Staphylococcus aureus* protein A in immunocytochemistry. Use in unlabelled antibody enzyme system and fluorescent methods. *Journal of Histochemistry and Cytochemistry* **27**, 1438–1444.

Ohnishi, T., Yamamoto, K., and Terayama, H. (1973) Polysaccharides associated with chromosomes and their behaviour in the cell cycle. *Histochemie* **35**, 1–10.

Ornstein, L., Mautner, W., Davis, B. J., and Tamura, R. (1957) New horizons in fluorescence microscopy. *Journal of the Mount Sinai Hospital* **24**, 1066–1078.

Osterberg, R. (1974) Metal ion–protein interactions in solution. Chapter 2 in *Metal Ions in Biological Systems* (ed. H. Sigel), vol. **3**, pp. 45–88. New York: Marcel Dekker.

Overend, W. G. and Stacey, M. (1949) Mechanism of the Feulgen nucleal reaction. *Nature* **163**, 538–540.

Page, K. M. (1965) A stain for myelin using solochrome cyanin. *Journal of Medical Laboratory Technology* **22**, 224–225.

Page, K. M. (1977) Bone and the preparation of bone sections. Chapter 15 in *Theory and Practice of Histological Techniques* (ed. J. D. Bancroft and A. Stevens), pp. 223–248. Edinburgh, London, and New York: Churchill–Livingstone.

Paladino, G. (1890) D'un nouveau procédé pour les recherches microscopiques du système nerveux central. *Archives Italiennes de Biologie* **13**, 484–486.

Paljarvi, L., Garcia, J. H., and Kalimo, H. (1979) The efficiency of aldehyde fixation for electron microscopy: stabilization of rat brain tissue to withstand osmotic stress. *Histochemical Journal* **11**, 267–276.

Papadimitriou, J. M., van Duijn, P., Brederoo, P., and Streefkerk, J. G. (1976) A new method for the cytochemical demonstration of peroxidase for light, fluorescence and electron microscopy. *Journal of Histochemistry and Cytochemistry* **24**, 82–90.

Parmley, R. T., Martin, B. J., and Spicer, S. S. (1973) Staining of blood cell surfaces with a lectin–horseradish peroxidase method. *Journal of Histochemistry and Cytochemistry* **21**, 912–922.

Parmley, R. T., Spicer, S. S., and Alvarez, C. J. (1978) Ultrastructural localization of nonheme cellular iron with ferrocyanide. *Journal of Histochemistry and Cytochemistry* **26**, 729–741.

Pasteels, J. L. and Herlant, M. (1962) Notions nouvelles sur la cytologie de l'antéhypophyse chez le rat. *Zeitschrift für Zellforschung* **56**, 20–39.

Patterson, A. M., Capell, L. T., and Walker, D. F. (1960) *The Ring Index*, 2nd edn. Washington: American Chemical Society.

Pauling, L. (1947) *General Chemistry*. San Francisco: Freeman.

Paulova, M., Ticha, M., Entlicher, G., Kostir, J., and Kokourek, J. (1970) Relationship between red blood cell agglutination and polysaccharide precipitation by phytohemagglutinin of *Pisum sativum* L. *FEBS Letters* **9**, 345–347.

Pearse, A. G. E. (1968) Common cytochemical and ultrastructural characteristics of cells producing polypeptide hormones (the APUD series) and their relevance to thyroid and ultimobranchial C-cells and calcitonin. *Proceedings of the Royal Society B* **170**, 71–80.

Pearse, A. G. E. (1968 and 1972) *Histochemistry, Theoretical and Applied*, 3rd edn., 2 vols. London and Edinburgh: Churchill–Livingstone.

Pearse, A. G. E. and Polak, J. H. (1975) Bifunctional reagents as vapour- and liquid-phase fixatives for immunohistochemistry. *Histochemical Journal* **7**, 179–186.

Pearson, C. K. (1963) A formalin–sucrose ammonia fixative for cholinesterases. *Journal of Histochemistry and Cytochemistry* **11**, 665–666.

Penfield, W. and Cone, W. V. (1950) Neuroglia and microglia (the metallic methods). In *McClung's Handbook of Microscopical Technique*, 3rd edn. (ed. R. McClung Jones), pp. 399–431. New York: Hoeber.

Pepler, W. J. and Pearse, A. G. E. (1957) The histochemistry of the esterases of rat brain with special reference to those of the hypothalamic nuclei. *Journal of Neurochemistry* **1**, 193–202.

Peters, A. (1955a) Experiments on the mechanism of silver staining. Part I. Impregnation. *Quarterly Journal of Microscopical Science* **96**, 84–102.

Peters, A. (1955b) Experiments on the mechanism of silver staining. Part II. Development. *Quarterly Journal of Microscopical Science* **96**, 103–115.

Peters, A. (1955c) Experiments on the mechanism of silver staining. Part III. Electron microscope studies. *Quarterly Journal of Microscopical Science* **96**, 317–322.

Püller, Y., Franz, H., and Preiss, A. (1977) Sudan black B: chemical structure and histochemistry of the blue main components. *Histochemistry* **54**, 237–250.

Pochhammer, C., Dietsch, P., and Sigmund, P. R. (1979) Histochemical detection of carbonic anhydrase with dimethylaminonaphthalene-5-sulfonamide. *Journal of Histochemistry and Cytochemistry* **27**, 1103–1107.

Pouradier, J. and Burness, D. M. (1966) The hardening of gelatins and emulsions. Chapter 3, Part II, in *The Theory of the Photographic Process*, 3rd edn. (ed. T. H. James), pp. 54–60. New York and London: Macmillan.

Prentø, P. (1978) Rapid dehydration-clearing with 2,2-dimethoxypropane for paraffin embedding. *Journal of Histochemistry and Cytochemistry* **26**, 865–867.

Puchtler, H. and Isler, H. (1958) The effect of phosphomolybdic acid on the stainability of connective tissues by various dyes. *Journal of Histochemistry and Cytochemistry* **6**, 265–270.

Puchtler, H. and Sweat, F. (1964a) Histochemical specificity of staining methods for connective tissue fibers: resorcin-fuchsin and van Gieson's picro-fuchsin. *Histochemie* **4**, 24–34.

Puchtler, H. and Sweat, F. (1964b) Effect of phosphomolybdic acid on the binding of Sudan black B. *Histochemie* **4**, 20–23.

Puchtler, H., Sweat, F., and Kuhns, J. G. (1964) On the binding of direct cotton dyes by amyloid. *Journal of Histochemistry and Cytochemistry* **12**, 900–907.

Puchtler, H. and Waldrop, F. S. (1978) Silver impregnation methods for reticulum fibers and reticulin: a reinvestigation of their origins and specificity. *Histochemistry* **57**, 177–187.

Quintarelli, G., Scott, J. E., and Dellovo, M. C. (1964a) The chemical and histochemical properties of alcian blue. II. Dye binding by tissue polyanions. *Histochemie* **4**, 86–98.

Quintarelli, G., Scott, J. E., and Dellovo, M. C. (1964b) The chemical and histochemical properties of alcian blue. III. Chemical blocking and unblocking. *Histochemie* **4**, 99–112.

Ralis, H. M., Beesley, R. A., and Ralis, Z. A. (1973) *Techniques in Neurohistology*. London: Butterworths.

Ramon-Moliner, E. (1970) The Golgi–Cox technique. In *Contemporary Research Methods in Neuroanatomy* (ed. W. J. H. Nauta and S. O. E. Ebbesson), pp. 32–55. Berlin, Heidelberg, and New York: Springer.

Rasmussen, G. L. (1957) Selective silver impregnation of synaptic endings. In *New Research Techniques of Neuroanatomy* (ed. W. F. Windle), pp. 27–39. Springfield, Ill.: Thomas.

Reid, L. and Clamp, J. R. (1978) The biochemical and histochemical nomenclature of mucus. *British Medical Bulletin* **34**, 5–8.

Reid, P. E., Culling, C. F. A., Dunn, W. L., Clay, M. G., and Ramey, C. W. (1978) A correlative chemical and histochemical study of the *O*-acetylated sialic acids of human colonic epithelial glycoproteins in formalin fixed paraffin embedded tissues. *Journal of Histochemistry and Cytochemistry* **26**, 1033–1041.

Reinecke, M. and Walther, C. (1978) Aspects of turnover and biogenesis of synaptic vesicles at locust neuromuscular junctions as revealed by zinc iodide–osmium tetroxide (ZIO) reacting with intravesicular SH-groups. *Journal of Cell Biology* **78**, 839–855.

Reiner, A. and Gamlin, P. (1980) On noncarcinogenic chromogens for horseradish peroxidase histochemistry. *Journal of Histochemistry and Cytochemistry* **28**, 187–189.

Richardson, K. C. (1969) The fine structure of autonomic nerves after vital staining with methylene blue. *Anatomical Record* **164**, 359–378.

Rieck, G. D. (1967) *Tungsten and its Compounds*. Oxford: Pergamon Press.

Roberts, G. P. (1977) Histochemical detection of sialic acid residues using periodate oxidation. *Histochemical Journal* **9**, 97–102.

Rodrigo, J., Hernandez, C. J., Vidal, M. A., and Pedrosa, J. A. (1975) Vegetative innervation of the oesophagus. II. Intraganglionic laminar endings. *Acta Anatomica* **92**, 79–100.

Rodrigo, J., Nava, B. E., and Pedrosa, J. (1970) Study of vegetative innervation in the oesophagus. I. Perivascular endings. *Trabajos del Instituto Cajal de Investigaciones Biologicas* **62**, 39–65.

Rogers, A. W. (1975) Autoradiography and the study of the central nervous system. Chapter 2 in *Methods in Brain Research* (ed. P. B. Bradley), pp. 79–112. London, New York, Sydney, and Toronto: Wiley.

ROMMANYI, G., DEAK, G., and FISCHER, J. (1975) Aldehyde–bisulphite–toluidine blue (ABT) staining as a topo-optical reaction for demonstration of linear order of vicinal OH groups in biological structures. *Histochemistry* **43**, 333–348.

ROSENE, D. L. and MESULAM, M. M. (1978) Fixation variables in horseradish peroxidase neurohistochemistry. I. The effects of fixation time and perfusion procedures upon enzyme activity. *Journal of Histochemistry and Cytochemistry* **26**, 28–39.

ROTH, J., BENDAYAN, M., and ORCI, L. (1978) Ultrastructural localization of intracellular antigens by the use of protein A–gold complex. *Journal of Histochemistry and Cytochemistry* **26**, 1074–1081.

ROTH, J., BINDER, M., and GERHARD, U. J. (1978) Conjugation of lectins with fluorochromes: an approach to histochemical double labeling of carbohydrate components. *Histochemistry* **56**, 265–273.

SALTHOUSE, T. N. (1962) Luxol fast blue ARN: a new solvent azo dye with improved staining qualities for myelin and phospholipids. *Stain Technology* **37**, 313–316.

SALTHOUSE, T. N. (1963) Reversal of solubility characteristics of "Luxol" dye–phospholipid complexes. *Nature* **199**, 821.

SAMUEL, E. P. (1953a) The mechanism of silver staining. *Journal of Anatomy* **87**, 278–287.

SAMUEL, E. P. (1953b) Gold toning. *Stain Technology* **28**, 225–229.

SANTINI, M. (ed.) (1975) *Golgi Centennial Symposium: Perspectives in Neurobiology*. New York: Raven Press.

SANWICKI, E., HAUSER, T. R., STANLEY, T. W., and ELBERT, W. (1961) The 3-methyl-2-benzothiazolone hydrazone test. *Analytical Chemistry* **33**, 93–96.

SCARSELLI, V. (1961) Histochemical demonstration of aldehydes by *p*-phenylenediamine. *Nature* **190**, 1206–1207.

SCHABADASCH, A. (1930) Untersuchungen zur Methodik der Methylenblaufärbung des vegetativen Nervensystems. *Zeitschrift für Zellforschung* **10**, 221–243.

SCHUBERT, M. and HAMERMAN, D. (1956) Metachromasia: chemical theory and histochemical use. *Journal of Histochemistry and Cytochemistry* **4**, 159–189.

SCOTT, J. E. (1967) On the mechanism of the methyl green–pyronin stain for nucleic acids. *Histochemie* **9**, 30–47.

SCOTT, J. E. (1972a) Histochemistry of Alcian blue. II. The structure of Alcian blue 8GX. *Histochemie* **30**, 215–234.

SCOTT, J. E. (1972b) Histochemistry of Alcian blue. III. The molecular biological basis of staining by Alcian blue 8GX and analogous phthalocyanins. *Histochemie* **32**, 191–212.

SCOTT, J. E. and DORLING, J. (1969) Periodate oxidation of acid polysaccharides. III. A PAS method for chondroitin sulphates and other glycosamino-glycuronans. *Histochemie* **19**, 295–301.

SCOTT, J. E. and HARBINSON, R. J. (1969) Periodate oxidation of acid polysaccharides. II. Rates of oxidation of uronic acids in polyuronides and acid mucopolysaccharides. *Histochemie* **19**, 155–161.

SCOTT, J. E., QUINTARELLI, G., and DELLOVO, M. C. (1964) The chemical and histochemical properties of alcian blue. I. The mechanism of alcian blue staining. *Histochemie* **4**, 73–85.

SECHRIST, J. W. (1969) Neurocytogenesis. I. Neurofibrils, neurofilaments and the terminal mitotic cycle. *American Journal of Anatomy* **124**, 117–134.

SHACKLEFORD, J. M. (1963) Histochemical comparison of mucous secretions in rodent, carnivore, ungulate and primate major salivary glands. *Annals of the New York Academy of Sciences* **106**, 572–582.

SHARON, N. and LIS, H. (1972) Lectins: cell-agglutinating and sugar-specific proteins. *Science* **177**, 949–959.

SHORE, P. A., BURKHALTER, A., and COHN, V. H. (1959) A method for the fluorometric assay of histamine in tissues. *Journal of Pharmacology and Experimental Therapeutics* **127**, 182–186.

SILBERT, J. E., KLEINMAN, H. K., and SILBERT, C. K. (1975) Heparin and heparin-like substances of cells. In *Heparin* (ed. R. A. Bradshaw and S. Wessler), *Advances in Experimental Medicine and Biology*, vol. **52**, pp. 51–60. New York and London: Plenum Press.

SILVEIRA, S. R. and HADLER, W. A. (1978) Catalases and peroxidases histochemical detection; techniques suitable to discriminate these enzymes. *Acta Histochemica* **63**, 1–10.

SMITH, G. L., JENKINS, R. A., and GOUGH, J. F. (1969) A fluorescent method for the detection and localization of zinc in human granulocytes. *Journal of Histochemistry and Cytochemistry* **17**, 749–750.

SOLCIA, E., VASSALLO, G., and CAPELLA, C. (1968) Selective staining of endocrine cells by basic dyes after acid hydrolysis. *Stain Technology* **43**, 257–263.

SOLLENBERGER, P. Y. and MARTIN, R. B. (1968) Carbon–nitrogen and nitrogen–nitrogen double bond condensation reactions. Chapter 7 in *The Chemistry of the Amino Group* (ed. S. Patai), pp. 349–406. London, New York, and Sydney: Wiley-Interscience.

SORVARI, T. E. and LAURÉN, R. A. (1973) The effect of various fixation procedures on the digestibility of sialomucins with neuraminidase. *Histochemical Journal* **5**, 405–412.

SORVARI, T. E. and STOWARD, P. J. (1970) Some investigations of the mechanism of the so-called "methylation" reactions used in mucosubstance histochemistry. *Histochemie* **24**, 106–119.

SPARROW, J. R. (1980) Immunohistochemical study of the blood–brain barrier. Production of an artifact. *Journal of Histochemistry and Cytochemistry* **26**, 570–572.

SPICER, S. S. (1960) Siderosis associated with increased lipofuscins and mast cells in aging mice. *American Journal of Pathology* **37**, 457–475.

STEAN, J. P. B. (1974) Some evidence of the nature of the

Golgi–Cox deposit and its biochemical origin. *Histochemistry* **40**, 377–383.

STEEDMAN, H. F. (1960) *Section Cutting in Microscopy*. Oxford: Blackwell.

STERNBERGER, L. A. (1979) *Immunocytochemistry*, 2nd edn. Englewood Cliffs, NJ: Prentice-Hall.

STEVENS, A. (1977) The haematoxylins. Chapter 7 in *Theory and Practice of Histological Techniques* (ed. J. D. Bancroft and A. Stevens), pp. 85–94. Edinburgh, London, and New York: Churchill–Livingstone.

STODDART, R. W., COLLINS, R. D., and JACOBSON, W. (1974) The microanalysis of saccharide structures of normal and neoplastic tissues. *Biochemical Society Transactions* **2**, 481–483.

STODDART, R. W. and KIERNAN, J. A. (1973a) Histochemical detection of the α-D-arabinopyranoside configuration using fluorescent-labelled concanavalin A. *Histamine* **33**, 87–94.

STODDART, R. W. and KIERNAN, J. A. (1973b) Aprotinin, a carbohydrate-binding protein. *Histochemie* **34**, 275–280.

STOWARD, P. J. (1968a) Studies in fluorescence histochemistry. V. The influence of trace metals on the fluorescence of periodate-oxidized mucosubstance salicylhydrazones. *Journal of the Royal Microscopical Society* **88**, 571–585.

STOWARD, P. J. (1968b) Fluorescence microscopy and histochemistry. Chapter 11 in *Luminescence in Chemistry* (ed. E. J. Bowen), pp. 222–249. London: van Nostrand.

STOWARD, P. J. and BURNS, J. (1971) Studies in fluorescence histochemistry. VII. The mechanism of the complex reactions that may take place between protein carboxyl groups and hot mixtures of acetic anhydride and pyridine in the acetic anhydride–salicylhydrazide–zinc (or fluorescent ketone) method for localizing protein C-terminal carboxyl groups. *Histochemical Journal* **3**, 127–141.

STRAUS, W. (1964) Factors affecting the cytochemical reaction of peroxidase with benzidine and the stability of the blue reaction product. *Journal of Histochemistry and Cytochemistry* **12**, 462–469.

STREEFKERK, J. G. and VAN DER PLOEG, M. (1974) The effect of methanol on granulocyte and horseradish peroxidase quantitatively studied in a film model system. *Histochemistry* **40**, 105–111.

STREIT, P. and REUBI, J. C. (1977) A new and sensitive staining method for axonally transported horseradish peroxidase (HRP) in the pigeon visual system. *Brain Research* **126**, 530–537.

STRICH, S. J. (1968) Notes on the Marchi method of staining degenerating myelin in the peripheral and central nervous system. *Journal of Neurology, Neurosurgery and Psychiatry* **31**, 110–114.

SUMNER, B. E. H. (1965) A histochemical study of aldehyde–fuchsin staining. *Journal of the Royal Microscopical Society* **84**, 329–338.

SUTHERLAND, S. D. (1963) The use of zinc iodide in the Champy osmic acid technique. *Journal of Anatomy* **97**, 624–625.

SWAAB, D. G., POOL, C. W., and VAN LEEUWEN, F. W. (1977) Can specificity ever be proved in immunocytochemical staining? *Journal of Histochemistry and Cytochemistry* **25**, 388–390.

SWANK, R. L. and DAVENPORT, H. A. (1935) Chlorate–osmic–formalin method for staining degenerating myelin. *Stain Technology* **10**, 87–90.

SWASH, M. and FOX, K. P. (1972) Techniques for the demonstration of human muscle spindle innervation in neuromuscular disease. *Journal of the Neurological Sciences* **15**, 291–302.

SWIFT, H. H. (1950) The desoxyribose nucleic acid content of animal nuclei. *Physiological Zoology* **23**, 169–200.

TAS, J. (1977) The alcian blue and combined alcian blue-safranin O staining of glycosaminoglycans studied in a model system and in mast cells. *Histochemical Journal* **9**, 205–230.

TAYLOR, K. B. (1961) The influence of molecular structure of thiazine and oxazine dyes on their metachromatic properties. *Stain Technology* **36**, 73–83.

TERNER, J. Y. and HAYES, E. R. (1961) Histochemistry of plasmalogens. *Stain Technology* **36**, 265–278.

THOMPSON, S. W. (1965) *Selected Histochemical and Histopathological Methods*. Springfield, Ill.: Thomas.

THORSTENSEN, T. C. (1969) *Practical Leather Technology*. New York: van Nostrand Reinhold.

THOSS, K. and ROTH, J. (1976) Histochemical lectin affinity technique by means of FITC-labeled serum protein fractions. *Histochemistry* **49**, 67–72.

TILAK, B. D. (1971) Naphthoquinonoid dyes and pigments. Chapter 1 in *The Chemistry of Synthetic Dyes* (ed. K. Venkataraman), vol. **5**, pp. 1–55. New York and London: Academic Press.

TOMS, G. C. and WESTERN, A. (1971) Phytohaemagglutinins. Chapter 10 in *Chemotaxonomy of the Leguminosae* (ed. J. B. Harborne, D. Boulton, and B. L. Turner), pp. 367–462. London and New York: Academic Press.

TOYOSHIMA, S., OSAWA, T., and TONOMURA, A. (1970) Some properties of purified phytohemagglutinin from *Lens culinaris* seeds. *Biochemica et Biophysica Acta* **221**, 514–521.

TRAMEZZANI, J. H., CHIOCCHIO, S., and WASSERMANN, G. F. (1964) A technique for light and electron microscopic identification of adrenalin- and noradrenalin-storing cells. *Journal of Histochemistry and Cytochemistry* **12**, 890–899.

TRANZER, J. P. and RICHARDS, J. G. (1976) Ultrastructural cytochemistry of biogenic amines in nervous tissue: methodologic improvements. *Journal of Histochemistry and Cytochemistry* **24**, 1178–1193.

TSUJI, S. (1974) On the chemical basis of thiocholine methods for demonstration of acetylcholinesterase. *Histochemistry* **42**, 99–110.

VANDESANDE, F. (1979) A critical review of immunocytochemical methods for light microscopy. *Journal of Neuroscience Methods* **1**, 3–23.

VELICAN, C. and VELICAN, D. (1970) Structural heterogeneity of basement membranes and reticular fibres. *Acta Anatomica* **77**, 540–559.

VELICAN, C. and VELICAN, D. (1972) Silver impregnation techniques for the histochemical analysis of basement membranes and reticular fiber networks. In *Techniques of Biochemical and Biophysical Morphology* (ed. D. Glick and R. M. Rosenblum), vol. **1**, pp. 143–190. New York, London, Sydney, and Toronto: Wiley.

VERMEER, B. J., VAN GENT, C. M., DE BRUIJN, W. C., and BOONDERS, T. (1978) The effect of digitonin-containing fixatives on the retention of free cholesterol and cholesterol esters. *Histochemical Journal* **10**, 287–298.

VIDAL, B. DE C. (1978) The use of the fluorescent probe 8-anilinonaphthalene sulfate (ANS) for collagen and elastin histochemistry. *Journal of Histochemistry and Cytochemistry* **26**, 196–201.

VOLLMANN, H. (1971) Phthalogen dyestuffs. Chapter 5 in *The Chemistry of Synthetic Dyes* (ed. K. Venkataraman), vol. **5**, pp. 283–311. New York and London: Academic Press.

WALKER, J. F. (1964) *Formaldehyde*, 3rd edn. New York: Reinhold; also London: Chapman & Hall.

WATERS, S. E. and BUTCHER, R. G. (1980) Studies on the Gomori acid phosphatase reaction: the preparation of the incubation medium. *Histochemical Journal* **12**, 191–200.

WATSON, J. D. and CRICK, F. H. C. (1953) Molecular structure of nucleic acids. A structure for deoxyribose nucleic acid. *Nature* **171**, 737–738.

WATSON, S. J. and BARCHAS, J. D. (1977) Catecholamine histofluorescence using cryostat sectioning and glyoxylic acid in unperfused frozen brain: a detailed description of the technique. *Histochemical Journal* **9**, 183–195.

WATSON, S. J. and ELLISON, J. P. (1976) Cryostat technique for central nervous system histofluorescence. *Histochemistry* **50**, 119–127.

WATTENBERG, L. W. and LEONG, J. L. (1960) Effects of coenzyme Q_{10} and menadione on succinic dehydrogenase activity as measured by tetrazolium salt reduction. *Journal of Histochemistry and Cytochemistry* **8**, 296–303.

WEBER, P., HARRISON, F. W., and HOF, L. (1975) The histochemical application of dansylhydrazine as a fluorescent labeling reagent for sialic acids in glycoconjugates. *Histochemistry* **45**, 271–277.

WEIR, E. E., PRETLOW, T. G., PITTS, A., and WILLIAMS, E. E. (1974) Destruction of endogenous peroxidase activity in order to locate antigens by peroxidase-labeled antibodies. *Journal of Histochemistry and Cytochemistry* **22**, 51–54.

WEISS, L. P., TSOU, K, C., and SELIGMAN, A. M. (1954) Histochemical demonstration of protein-bound amino groups. *Journal of Histochemistry and Cytochemistry* **2**, 29–49.

WEISSBERGER, A. (1966) Principles and chemistry of color photography. Chapter 17 in *The Theory of the Photographic Process*, 3rd edn. (ed. T. James), pp. 382–396. New York and London: Macmillan.

WEST, E. S. and TODD, W. R. (1956) *Textbook of Biochemistry*, 2nd edn. New York: Macmillan.

WHITE, A., HANDLER, P., and SMITH, E. L. (1973) *Principles of Biochemistry*, 5th edn. New York: McGraw-Hill.

WHITE, E. H. and WOODCOCK, D. J. (1968) Cleavage of the carbon–nitrogen bond. Chapter 8 in *The Chemistry of the Amino Group* (ed. S. Patai), pp. 407–497. London, New York, and Sydney: Wiley-Interscience.

WHYTE, A., LOKE, Y. W., and STODDART, R. W. (1978) Saccharide distribution in human trophoblast demonstrated using fluorescein-labelled lectins. *Histochemical Journal* **10**, 417–423.

WIGGLESWORTH, V. B. (1957) The use of osmium tetroxide in the fixation and staining of tissue. *Proceedings of the Royal Society B* **147**, 185–199.

WILLIAMS, G. and JACKSON, D. S. (1956) Two organic fixatives for acid mucopolysaccharides. *Stain Technology* **31**, 189–191.

WINKELMANN, R. K. (1960) *Nerve Endings in Normal and Pathologic Skin*. Springfield, Ill.: Thomas.

WINKELMANN, R. K. and SCHMIT, R. W. (1957) A simple silver method for nerve axoplasm. *Proceedings of Staff Meetings of the Mayo Clinic* **32**, 217–222.

WOLLIN, A. and JAQUES, L. B. (1973) Metachromasia: an explanation of the colour change produced in dyes by heparin and other substances. *Thrombosis Research* **2**, 377–382.

WOLTERS, G. J. H., PASMA, A., KONIJNENDIJK, W., and BOUMAN, P. R. (1979) Evaluation of the glyoxal-bis-(2-hydroxyanil) method for staining of calcium in model gelatin films and pancreatic islets. *Histochemistry* **62**, 137–151.

YAMADA, K. and SHIMIZU, S. (1976) Concanavalin A-peroxidase-diaminobenzidine (Con A-PO-DAB)-alcian blue (AB). A reliable method for dual staining of complex carbohydrates. *Histochemistry* **47**, 159–169.

YAMADA, K. and SHIMIZU, S. (1977) The histochemistry of galactose residues of complex carbohydrates as studied by peroxidase-labeled *Ricinus communis* agglutinin. *Histochemistry* **53**, 143–156.

YAMADA, K. and SHIMIZU, S. (1979) The use of peroxidase-labelled *Limulus polyphemus* agglutinin for the histochemistry of sialic acid-containing glycoproteins in light microscopy. *Histochemical Journal* **11**, 457–471.

YARIV, J., KALB, A., and KATCHALSKI, E. (1967) Isolation of an L-fucose binding protein from *Lotus tetragonobolus* seed. *Nature* **215**, 890–891.

YOUNG, N. M., LEON, M. A., and TAKAHASHI, T. (1971) Studies on a phytohemagglutinin from the lentil. III. Reaction of *Lens culinaris* hemagglutinin with polysaccharides, glycoproteins and lymphocytes. *Journal of Biological Chemistry* **246**, 1596–1601.

ZACKS, S. I. (1973) *The Motor Endplate*, 2nd edn. Huntington, NY: Krieger.

ZAGON, I. S., VAVRA, J., and STEELE, I. (1970) Microprobe analysis of protargol stain deposition in two protozoa. *Journal of Histochemistry and Cytochemistry* **18**, 559–564.

ZIEGLÄNSBERGER, W. and REITER, C. (1974) Interneuronal movement of procion yellow in cat spinal neurones. *Experimental Brain Research* **20**, 527–530.

ZOLLINGER, H. (1961) *Azo and Diazo Chemistry. Aliphatic and Aromatic Compounds* (transl. H. E. Nursten). London and New York: Interscience Publishers.

Index

See also Abbreviations (pp. x–xii) and Glossary (pp. 305–307)

Absorption 283
Acetal lipids *see* Plasmalogens
Acetic acid
  extraction of zinc 197
  fixative 10–11, 12
  hydrolysis of lipoproteins 188
  physical properties 33, 300
Acetic anhydride
  for acetylation 117, 120, 140
  for carboxyl groups 118, 133
Acetone
  anhydrous: preparation 188
  fixative 10–11, 12, 228, 236
  lipid extraction 176, 188
  properties 32
Acetylation
  methods 140, 167
  of amino groups 120
  of hydroxyl groups 117, 124
  of mucosubstances 161, 167
  of phenols (tyrosine) 124
O-acetyl-5-bromoindoxyl 210, 216
O-acetyl-4-chloro-5-bromoindoxyl 216
Acetylcholine receptors 293
Acetylcholinesterase
  histochemical methods 210–212, 216–217
  inhibitors 212, 214–215
  motor end-plate 264
  synonyms 208
Acetylesterase
  histochemical detection 210–211, 215–216
  inhibitors 209–210, 214–215
  synonyms 208
N-acetylgalactosamine
  in lipids 170, 174
  in mucosubstances 148–149
  lectin affinities 158
  structure 146
N-acetylglucosamine
  in lipids 170
  in mucosubstances 148–149
  lectin affinities 158
  structure 146
N-acetylneuraminic acid
  acetyl derivatives 155–156
  in glycoproteins 149
  in lipids 170, 174
  lectin affinities 158
  periodate oxidation 154–155
  structure 147
  see also Sialic acids
Acetylthiocholine 211–212, 216
Acid anhydride method for —COOH 133–134
Acid dyes *see* Anionic dyes
Acid fuchsine (C.I. 42685) 65
  in trichrome methods 94–97
  in van Gieson's method 93–94
Acid–haematein test 182, 187–188
Acid phosphatase *see* Phosphatases
Acidophilia
  effect of heteropolyacids on 95–96
  reversal by OsO_4 19
Acridine dyes 98
Acridine orange (C.I. 46005) 68
  nucleic acid staining 109
Acriflavine (C.I. 46000) 68
  alcoholic solution 130, 140
  amine–aldehyde condensation 130
  method for aldehydes 140
  pseudo-Schiff reagent from 129
Acrolein 18, 19
Actin 239
Acylation *see* Acetylation, Benzoylation
Adenosine triphosphatase 202
Adhesives for sections 34–35
Adrenaline (ADR)
  chromaffin reaction 247, 250
  occurrence 246
  structure 245
Albumen, adhesive 34–35
Alcian blue (C.I. 74240)
  chemistry 74
  for acid mucosubstances 150–151
  for sulphation-induced basophilia 133
  in method for aldehydes 131
  in method for cystine 126, 137–138
  nucleic acids not stained 151
  with safranine 151
Alcian green 75
Alcian yellow 75
Alcohol dehydrogenase 237–240
Alcohols
  for dehydration 30–32
  long-chain 169
  terminology ix
  see also Glycols, Hydroxyl group
Aldehyde fuchsine
  chemistry 98–99
  in method for cystine 126
  in method for elastin 102
  in method for NSM 265, 277
  preparation of 102
  uses 99
Aldehydes
  bisulphite compounds 131

Aldehydes (*cont.*)
blocking reactions 131–132, 143
from DNA 106
from lipids 179–180, 181
from mucosubstances 153–156
from reticulin 97
in collagen 92
in elastin 93
methods for 128–131, 140
Alizarin (C.I. 58000) 73
Alizarin red S (C.I. 58005) 72–73
method for calcium 190–191, 194–195
Alkaline phosphatase *see* Phosphatases
Altmann's fixative 22
Alum–tetracycline method for nuclei 83
Aluminium–haematein 79, 81, 82
Aminals 130
Amines
aromatic 130
biogenic 245–255
reagents for aldehydes 130, 132, 143
secondary 130
Amino groups
acylation 120
anionic dye binding 42, 80, 119
azo coupling 127
blocking reactions 119–120
fixation in tissue 119
histochemical reactions 119–120, 127
hydroxynaphthaldehyde method 119
in lipids 170
methylation 120
removal (deamination) 19, 120
sulphation 120
p-aminodiphenylamine 232, 241
Aminoketone dyes 60–62
8-amino-1-naphthol-3,6-disulphonic acid *see* H-acid
m-aminophenol 130, 143
Ammonia 300
Amyl acetate 32
Amylase, for glycogen 160, 166
Amyloid 149–150
Aniline blue WS (C.I. 42755)
chemistry 65–66
collagen staining 94
nuclear staining 94
trichrome methods 94–97
8-anilino-1-naphthalene sulphonate 98
Anils *see* Imines
Anionic dyes 42, 48, 50
counterstains 80
for collagen and cytoplasm 93–94
in trichrome methods 94–97
molecular size and staining 78, 93–94, 96
nuclear staining 78–79
protein staining 42, 80, 119, 120, 126–127
Anthraquinone dyes 72–74
Antibody
definition 280–281
detection in serum 292
detection in tissue 292
–enzyme method, unlabelled 289–292
to immunoglobulin 286
Antigen
–antibody complexes 280, 282, 289
definition 280–281
detection in tissue 283–293
determinant site 281
immunoglobulin as 282, 286
Antiserum
definition 283
fluorescent labelling 284
peroxidase-labelled 288–289
primary 286
secondary 286
to immunoglobulin 286
Apathy's mounting medium 37
Apoenzyme 222
Aprotinin 158, 167
Argentaffin
cells 246, 250
reaction 246
Arginine 120–122
blocking reaction 121–122, 142
methods for 134–135
Aromatic character 39
Arylesterase
histochemical detection 210–211, 215–216
inhibitors 209–210, 214–215
synonyms 208
Arylguanidinium salts of dyes 75, 258
Arylmethane dyes 62–66
Ascorbic acid 254
Astrocytes 264, 275–276
Atomic weights 301
Auramine O (C.I. 41000) 62–63
Autofluorescence 2
Autoimmune diseases 292
Auxochromes 39–40
Axons
amines in 248–250
degenerating, methods for 256–257
iodide–osmium methods 265, 277–278
myelinated 258–259, 266–267
silver methods for 260–262, 268–273
tracer methods 257
vital methylene blue staining 263–264, 274–275
AZAN staining method 69
Azide
catechol oxidase inhibition 233, 242
cytochrome oxidase inhibition 229, 231, 241
electron transport inhibition 229, 237–238
Azine dyes 68–69
Azo dyes 52–58
synthesis 52–53, 76
types of 53–56
Azo group
chemistry 52–53

Azo group (*cont.*)
  chromophore 39
Azocarmine B (C.I. 50090) 69
Azocarmine G (C.I. 50085) 69
Azoic dyes 50, 56–58
  coupling components 58
  diazo components 57–58
Azomethines *see* Imines
Azure I 70
Azure II 90
Azure A (C.I. 52005) 70
  eosinate 81, 86
Azure B (C.I. 52010)
  in blood stains 88
  structure 70
Azure C (C.I. 52002) 70
Azure–eosin staining 81, 86–87, 90–91

Baker's acid–haematein test 182, 187–188
Baker's haematal-16 82
Baker's views on dye diffusion 93–94, 96
Bandrowski's base 131
Basic dyes *see* Cationic dyes
Basic fuchsine
  amine–aldehyde condensation with 130
  in acid–alcohol 111, 130, 140
  in aldehyde–fuchsine 102
  in Schiff's reagent 110–111
Basophilia
  definition 78
  effects of fixation on 10–11, 78
  effects of pH on 78
  induced by heteropolyacids 95–96
  induced by OsO_4 19
  induced by sulphation 132, 167
  see also Cationic dyes
Benzene
  properties 32
  resonance in x, 39
Benzidine
  peroxidase method 234
  tetra-azotized 127
Benzil 122
  arginine blockade 142
Benzosalicylanilide 58
Benzothiazole 60
Benzoyl chloride 117, 120, 140
Benzoylation
  method 140
  of amino groups 120
  of hydroxyl groups 117, 124
  of phenols (tyrosine) 124
3,4-benzpyrene 75
Benzyl benzoate 32
Biebrich scarlet (C.I. 26905) 53
Bismarck brown Y(G) (C.I. 21000) 55
Bismuthate, sodium 154
Bisulphite method for aldehydes 131
Blocking reactions
  for functional groups 140–143
  table of 132
Blood cells
  agglutination by lectins 157
  in films 89–90
  in sections 90–91
  staining 88–91
Blood group antigens
  chemistry 149
  lectin binding 157
Bodian's fixative 22
Bonds, chemical
  charge-transfer 43
  coordinate 44
  covalent 42
  dative 44
  double *see* Unsaturation
  electrovalent 42
  hydrogen 43, 94, 99, 150
  hydrophobic interaction 44, 98
  ionic 42
  semi-polar 44–46
  van der Waals forces 44, 46, 79, 98
Borohydride, sodium 143
  in method for sialic acids 155
  reduction of aldehydes and ketones 132, 143
Borrel's methylene blue 85
Bouin, alcoholic 21
Bouin's fluid 21
  weak 188
Brazalum, Mayer's 83
Brazilein 62
  aluminium– 80, 83
  nuclear stain 80, 83
Brazilin (C.I. 75280) 61–62
  see also Brazilein
Brilliant blue R (C.I. 42660) 125
Brilliant indocyanine 6B (C.I. 42660) 126
  method for protein 126–127, 138–139
Brilliant sulphoflavine FF (C.I. 56205) 62
Bromination
  blockade of olefins 179
  effect on sudanophilia 177
  procedure 188
  –silver method 179
  –Sudan black B method 183–184
Bromoindigo
  in esterase methods 210
  motor end-plate method 264, 276
5-bromoindoxyl acetate 210, 216
Buffers
  for azure–eosin staining 86
  for blood stains 89–91
  for fixatives 20
  tables of 295–299
α-bungarotoxin 293
2,3-butanedione 122

n-butanol
  dehydration of sections 7
  properties 32
t-butanol 33
Butyl alcohols 7, 32, 33
Butyrylthiocholine 212, 216
B.W. 284C51 212, 215

Cadmium salts in fixation 19
Cajal, S. Ramon y 263
  gold–sublimate method 264, 275–276
Calcium
  histochemical detection 190–192, 194–196
  in fixation of lipids 175
  in hard tissues 25
Calcium red (C.I. 60760) 73
Canada balsam 2, 36
Carbocyanine DBTC 59
Carbodiimides as fixatives 19
Carbohydrates
  histochemical methods 145–168
  in collagen and reticulin 92, 149
  in lipids 174, 179–180
  in nuclear chromatin 151
  lectin binding 157
  see also Glycoproteins, Monosaccharides, Mucosubstances, Polysaccharides, Proteoglycans
Carbon tetrachloride 32, 184
Carbonate
  detection in tissue 192
  von Kossa technique 196
Carbonic anhydrase
  affinity labelling 203
  zinc in 193
Carbonyl groups *see* Aldehydes, Ketones
Carboxyl groups
  acid anhydride method 118, 133–134
  alcian blue staining 150–151, 162
  blocking 118–119, 167
  cationic dye binding 78, 118, 150
  esterification 118–119, 167
  histochemical reactions 118–119
  of lipids 199
  of mucosubstances 150–152
  of proteins, detection 118–119, 133–134
  see also Fatty acids, Sialic acids, Uronic acids
Carboxylesterase
  histochemical detection 210–211, 215–216
  inhibitors 209–210, 214–215
  synonyms 208
Carboxylic esterases 208–212, 214–217
Cardiolipin 173–174
Carmalum, Mayer's 83, 183
Carmine (C.I. 74570) 73–74
Carminic acid (C.I. 75470) 73
  aluminium– 80, 83
  nuclear stain 80, 83

Carnoy's fluid 21
Cartilage, softening of 27
Cason's trichrome method 100
Catalase 236, 241, 243
Catechol oxidase
  biochemistry 232–233
  histochemical technique 241–242
  inhibition 233, 242
Catecholamines
  chemistry 245
  chromaffin reaction 247, 250
  fluorescence methods 248–250, 252–254
  glutaraldehyde–osmium method 247, 251
Cationic dyes 42, 48, 50
  aldehyde bisulphite compounds, staining 131
  carbohydrate staining 78, 150–152
  nucleic acid staining 77–78, 84–85, 105–106
  oversight stains 80, 85
  pH of 78, 118
  protein staining 78
  see also Basophilia
Cedarwood oil
  method of use 30–31
  properties 32
Celestine blue B (C.I. 51050) 70
Celloidin *see* Nitrocellulose
Cellosolve 33
Cellulose 148
Cephalins *see* Phosphatidyl ethanolamines
Ceramides
  chemistry 174
  resistance to saponification 176
Cerebrosides 174, 182
Cetyl alcohol 169
Cetylpyridinium chloride 24, 148
Chelation
  decalcification 25–27
  definition 25–26
  dyes 44–46, 54
  stabilization of colour 232
Chitin
  chemistry 148
  histochemistry 168
  softening of 27
Chloral hydrate
  fixative 19, 26
  in de Castro's fluid 26–27
  in Mayer's haemalum 81
Chlorazol black E (C.I. 30235) 56
  staining method 81, 85–86
Chloride, histochemical detection 199
4-chloro-5-bromoindoxyl acetate 216
Chloroform
  –methanol lipid extraction 176, 188
  properties 32
p-chloromercuribenzoate 210, 215, 240
p-chloromercuriphenylazo-2-naphthol 125
Chloroplatinic acid, fixative 19
Chlorpromazine 75

Cholamine *see* Ethanolamine
Cholesterol
  digitonin adduct 182, 187
  lysochrome staining 177, 183–184
  method for 181–182, 186–187
  physical properties 172, 177, 182
  structure 170
Cholesterol esters
  bromine–Sudan black B method 177, 183–184
  method for 181–182, 186–187
  properties 172, 177, 182
Choline
  -containing lipids 173–174, 182, 187–188
  structure 170
Choline esterases
  histochemical methods 210–212, 216–217
  inhibitors 209–210, 212, 214–215
  nomenclature 208
Cholinesterase
  histochemical methods 210–212, 216–217
  inhibitors 212, 214–215
  synonyms 208
Chondroitin sulphates
  chemistry 148–149
  enzymatic extraction 160, 166
Chromaffin
  cells 246
  reaction 247
  technique 250
Chromate
  in Golgi techniques 263
  method for lead 194
  oxidation of catecholamines 247
Chromatin
  definition 304
  fluorochroming 83–84
  staining 77–80, 81–85
Chromation of lipids 175
Chrome–gelatine 35
Chrome orange GR (C.I. 26520) 54
Chromic acid, fixative 10–11, 14–15
Chromic ions
  action on gelatine 14
  fixation by 19, 27
  in Perenyi's fluid 27
Chromium compounds *see* Chromate, Chromation, Chromic acid, Chromic ions, Dichromate
Chromogen 39
Chromophores
  chemistry 39–40, 60, 72
  definition 38
Chromosomes 77
  mucosubstances in 151
Chromotrope 2R (C.I. 16570) 53
Chromotropic substances 152
Chromoxane cyanine R (C.I. 43820) 66
  iron– 80, 82–83
  myelin staining 258, 266
  nuclear staining 80, 82–83
  oversight stain 81
Clearing blocks 1, 30, 34
Clearing sections 1, 6
Clearing, solvents for 32–33
Cleland's reagent *see* Dithiothreitol
Cobalt ions
  fixation by 19
  in cytochrome oxidase method 232, 241
  in TPPase method 218
  neuronal tracer 257
Cochineal 73
Coenzyme
  definition 222
  function 223–224
  in incubation media 228–229
Colamine *see* Ethanolamine
Collagen
  anionic dye binding 93–94
  chemistry 92, 149
  PAS staining 156
  staining techniques 99–100
  trichrome staining 94–97
Collodion *see* Nitrocellulose
Colloidal ferric hydroxide
  method 163
  properties 152
Colour Index 46
Colour Index names of dyes
  Acid blue 22 65
  Acid blue 74 72
  Acid blue 93 65
  Acid green 5 65
  Acid orange 10 53
  Acid orange 52 52
  Acid red 29 53
  Acid red 51 67
  Acid red 66 53
  Acid red 87 67
  Acid red 91 67
  Acid red 92 67
  Acid red 94 67
  Acid red 95 67
  Acid red 98 67
  Acid red 101 69
  Acid red 103 69
  Acid violet 19 65
  Acid yellow 7 62
  Acid yellow 24 52
  Acid yellow 73 67
  Azoic diazo 4 57
  Azoic diazo 5 57
  Azoic diazo 24 57
  Azoic diazo 38 57–58
  Azoic diazo 48 57
  Basic blue 6 230
  Basic blue 8 65
  Basic blue 9 70
  Basic blue 11 65
  Basic blue 12 70, 189

Colour Index names of dyes (*cont.*)
Basic blue 17 71
Basic blue 20 64
Basic blue 24 71
Basic blue 25 71
Basic blue 26 65
Basic brown 1 55
Basic green 4 105
Basic green 5 71
Basic orange 14 68
Basic red 2 69
Basic red 5 68
Basic red 9 63
Basic violet 2 64
Basic violet 3 64
Basic violet 5 71
Basic violet 10 66
Basic violet 14 64
Basic yellow 1 60
Basic yellow 2 62
Direct black 38 56
Direct blue 86 74
Direct red 28 56
Direct yellow 59 60
Fluorescent brightening agent 32 59–60
Food green 3 65
Ingrain blue 1 74
Mordant black 1 54
Mordant black 37 62
Mordant blue 3 66
Mordant blue 10 69
Mordant blue 14 70
Mordant green 4 51
Mordant orange 6 54
Mordant red 3 73, 191
Mordant red 11 73
Natural black 1 60
Natural red 4 73
Natural red 24 61
Natural yellow 11 191
Reactive red 1 55, 218
Solvent black 3 177
Solvent blue 3 66
Solvent blue 22 58
Solvent blue 37 75
Solvent blue 38 75
Solvent red 23 176
Solvent red 24 55, 176
Solvent red 27 55, 176
Solvent red 45 68
Solvent yellow 2 52
Vat blue 1 71–72
Vat yellow 5 62
Colour Index number of dyes
10005 51
10315 52
10316 52
11020 52
13025 52
15710 54
16230 53
16570 53
18158 55, 218
21000 55
22120 56
26100 176
26105 55, 176
26126 55, 176
26150 177
26520 54
26905 53
30235 56
37125 57
37155 57
37190 57–58
37210 57
37235 57
40620 59–60
41000 62
42000 105
42053 65
42095 65
42500 63
42510 64
42520 64
42555 64
42563 65
42585 64
42685 65
42755 65
42775 66
42780 65
43820 66
44040 65
44045 65
45005 66
45010 66
45170 66
45350 67
45380 67
45386 68
45400 67
45405 67
45410 67
45425 67
45430 67
45440 67
46000 68
46005 68
49000 60
49005 60
49705 58
50040 68
50085 69
50090 69
50205 71
50240 69
51030 69

Colour Index number of dyes (*cont.*)
51050 70
51175 230
51180 70, 189
52000 70
52010 70
52015 70
52020 71
52025 71
52030 71
52040 71
52041 71
56005 62
56205 62
57010 62
58000 73
58005 73, 191
60760 73
73001 71–72
73015 72
74180 74
74240 74
75280 61
75290 60
75470 73
75660 191
Complexes, metal 44–45
Concanavalin A
control procedures with 166
method 165–166
specificity 158, 159
Condense dyes 49
Congo red (C.I. 22120) 56
Coomassie brilliant blue R-250 (C.I. 42660) 125
Coordination number 45
Copper
acetate method for fatty acids 180
dithiooxamide method for 198–199
ferrocyanide 217
fixation by salts of 19
histochemical detection 193–194
in oxidases 231–232
in thiocholine methods 211
in tissues 193, 194
phthalocyanine 74
unmasking treatment 199
Corrosive sublimate *see* Mercuric chloride
Coupled tetrazonium reaction
chemistry 127–128
technique 139
p-cresol 235, 243
Cresyl fast violet 69
Cresyl violet acetate 69
Nissl staining 258
Cresylecht violet 69
Critical electrolyte concentration 150
Cryostat 1, 204
Crystal violet (C.I. 42555) 64, 65
Cyanide
catechol oxidase inhibition 233, 242
cytochrome oxidase inhibition 231, 241
electron transport inhibition 229
reduction of cystine 124
safety precautions 195
Cyanine dyes 58–59
Cyanuric chloride
fixative 19, 148
reactive dyes 40, 43
Cysteic acid 124
method for cystine 125–126, 137–138
Cysteine
alkaline phosphatase inhibition 207, 218
blocking reactions 126, 142
chemistry 124–125
effect on catechol oxidase 233
fixatives for 124
histochemical reactions 125, 136–137
Cystine
histochemical reactions 125–126, 137–138
oxidation 124, 137–138
reduction 124, 142
Cytochrome oxidase
action 223
confusion with peroxidase 235
histochemical methods for 231–232, 241
inhibition 229, 231
Cytochromes 223–224

Dansyl chloride 284, 285
Dansylhydrazine 129
De Castro's fluid 26
Deamination 120, 141
Decalcification
chemistry of 25–26
reasons for 4
techniques 26–28
Dehydration
blocks 1, 30–31
chemical 32–33
hardening due to 9
sections: method 6–7
solvents for 32–33
Dehydrogenases
definition 223
histochemical methods 224–230, 236–240
Deoxyribonuclease
histochemical demonstration 218–219
reagent 108, 112
Deoxyribonucleic acid
chemistry 104–105
extraction by DNase 108, 112
extraction by TCA 108
Feulgen reaction 106–107, 108–110
methyl green–pyronine method 105–106, 109–110
staining by dyes 105–106, 108–110
see also Nucleic acids
Dermatan sulphate 149

Desmosine 93
Desulphation
  methods 141, 167
  of proteoglycans 150, 161
Detergents, fixative effects 19
Developers 260–261
Dewaxing of sections 5
Diacetyl trimer 122
3,3′-diaminobenzidine
  in immunohistochemistry 288, 290
  in peroxidase methods 234–235, 242–243
  oxidation of 235
  structure 234
o-dianisidine, tetra-azotized *see* Fast blue B salt
Diaphorases
  definition 222
  histochemical detection 227–228, 240
  physiological functions 227–228
Diazonium salts
  in dye syntheses 52–53
  in enzyme histochemistry 207
  in method for serotonin 246, 250
  reaction with phenols 127, 246
  reactions with proteins 127–128
  stabilized 57–58
Dichromate
  effect of pH 14
  fixation by 10–11, 14–15
  in acid–haematein method 182
  in chromaffin reaction 247
  in Golgi techniques 263
  lipids, effects on 14, 175, 182
  reduction by alcohol 14, 27
  washing out 14
Diethylene oxide 33
Diethyl-*p*-nitrophenyl phosphate (E600)
  actions 209–210
  method of use 214
Diethylpyrocarbonate 19
Differentiation
  alum–haematein 82
  cationic dyes 84–85
  definition 3
  eosin 82
  myelin stains 266–267
  van Gieson staining 99–100
Digitonin 182, 187
Diisopropylfluorophosphate (DFP)
  enzyme inhibition 209–210, 212
  method of use 214–215
  reaction with serine 117, 209
Dilution
  of acids, etc. 299–301
  of antisera 285, 286, 291
2,2-dimethoxypropane (DMP) 31, 253
p-dimethylaminobenzaldehyde (DMAB) 122
  method for tryptophan 135–136
p-dimethylaminobenzylidenerhodanine 194
N-dimethyl-*p*-phenylenediamine 231–232
2,4-dinitrofluorobenzene 120
Dioxane 33, 184
N-diphenylbenzidine 75
Diphenylguanidine 75
Diphenylmethane dyes 62–63
Diphenylthiocarbazone 193
Diphosphatidyl glycerols 173
Dipicrylamine 75
Direct dyes 48, 55–56
  amyloid staining 150
  collagen staining 94
  elastin staining 98
Disperse dyes 50
Disulphide group *see* Cystine
Dithionite
  effect on catechol oxidase 233
  extraction of iron 192, 197
Dithiooxamide
  chelation by 193–194
  in Holczinger's method 180, 185–186
  method for copper 198–199
Dithiothreitol 124–125, 142
Dithizone
  metal chelation 193, 198
  method for zinc 197–198
Donors of electrons
  in ligands 44
  in peroxidase oxidations 233–235
  reducing agents 220
DOPA oxidase 232
Dopamine
  chromaffin reaction 247
  fluorescence methods 248–250, 252–254
  occurrence 246
  structure 245
Double bonds *see* Unsaturation
DPX mounting medium 2, 36
Dyes
  chemistry 38–76
  classification 48–75
  colour of 39–40
  combination with substrates 40–44
  diffusion in tissues 78, 93, 96
  fixative effects 19
  fluorescence of 40, 50
  nomenclature 46–47
  purity 47–48
  textile applications 48–50
  see also names of groups of dyes, individual dyes and methods

Elastin
  aldehyde–fuchsine method 98–99, 102
  chemistry 92–93
  dye affinities 98
  palladium binding 179
Electron carriers
  artificial 225

Electron carriers (*cont.*)
definition 222
in cells 223
inhibition of 229
intermediate 229–230
Electron microscopy
ammoniacal silver staining 131
fixation for 11, 17–18
PTA staining 95
Embedding 1–2, 29–34
agar 29
double 34
gelatine 29
nitrocellulose 31
paraffin wax 30–31
Enterochromaffin cells 246, 250
Enzyme Commission numbers
1.1.1.1. 237
1.1.1.8. 237
1.1.1.22. 237
1.1.1.27. 237
1.1.1.49. 237
1.1.2.1. 239
1.3.99.1. 236
1.4.1.2. 237
1.4.1.3. 237
1.6.4.3. 227
1.6.99.1. 228
1.9.3.1. 231
1.10.3.1. 232
1.11.1.6. 233, 236
2.7.7.16. 108
3.1.1. 208
3.1.1.1. 208
3.1.1.2. 208
3.1.1.3. 201, 208
3.1.1.5. 208
3.1.1.6. 208
3.1.1.7. 208
3.1.1.8. 208
3.1.3. 206
3.1.3.1. 207
3.1.3.2. 206
3.1.4.5. 108
3.1.4.6. 108
3.2.1.1. 160
3.2.1.2. 160
3.2.1.3. 160
3.2.1.18. 160
3.4.4. 212
3.6.1.4. 202
4.2.2.1. 203
4.2.99.1. 160
Enzymes
affinity labelling 202–203
classification 204
fixation 204–205
histochemistry: generalities 203–204
immunocytochemistry 202
in biological oxidations 222–224
nomenclature 201–202
Enzymes as reagents
carbohydrate histochemistry 160–161, 166
DNase 107–108, 112
general considerations 107
immunohistochemistry 288–293
labels for lectins 157, 159
RNase 107–108, 112
Eosin (C.I. 45380) 67
alcohol-soluble 68
counterstain 80, 81
ethyl 68, 81
Eosin B (C.I. 45400) 67
–azure A (Lillie's method) 86
Eosin–methyl blue stain 78, 81, 85
Eosinates of thiazine dyes 81, 86, 88–90
Epinephrine *see* Adrenaline
Eriochrome cyanine R (C.I. 43820) 66
Erythrocytes
agglutination by lectins 157
anionic dye staining 93–94
blood stains 89–90
cationic dye staining 78
trichrome staining 94–97
Erythrosin (Y) (C.I. 45425) 67
Erythrosin B (C.I. 45430) 67
Eserine 209, 214
Esterases 208–212, 214–217
Esterification *see* Methylation
Ethanol
fixative 10–11, 12
properties 2–3, 32
Ethanolamine
in lipids 173
structure 170
Ether–alcohol 31–32
Ether, diethyl 33
Ether phosphatides 173
Ethopropazine 212, 215
Ethoxyethanol 33
Ethyl eosin (C.I. 45386) 68
Ethylene glycol
physical properties 33
solvent for haematein 82
Ethylene glycol monoethyl ether 33
Ethylenediamine tetraacetic acid (EDTA)
calcium chelate 26
chemistry 26
decalcification method 27
dehydrogenase inhibition 240
Ethylenic bonds *see* Unsaturation
N-ethylmaleimide 126, 142, 240
Euparal mounting medium 2–3, 37
Extraction
of lipids 176, 178
of nucleic acids 108

Farrant's mounting medium 3
Fast black K salt (C.I. 37190) 57
  in method for aldehydes 143
  in method for proteins 128, 139
Fast blue B salt (C.I. 37235) 57
  in acid anhydride method 133
  in amine–aldehyde condensation 143
  in coupled tetrazonium reaction 127–128, 139
  in hydroxynaphthaldehyde method 134
Fast blue RR salt (C.I. 37155) 57
  in alkaline phosphatase method 207
Fast garnet GBC salt (C.I. 37210) 57
Fast green FCF (C.I. 42053) 65
  counterstain 80, 82, 195
Fast red B salt (C.I. 37125) 57
Fat *see* Lipids, Neutral fats
Fat blue Z (C.I. 49705) 58
Fatty acids
  free 171, 180
  histochemical detection 180, 185–186
  in lipids 169
Ferric ferricyanide reaction 125, 136–137
Ferric hydroxide *see* Colloidal ferric hydroxide
Ferritin 192
Feulgen method 106–107, 110–112
Fixation 8–24
  chemical 9–20
  duration of 9
  for enzymes 204, 228
  for exogenous HRP 235
  for functional group methods 115–116
  for immunohistochemistry 284–285
  for lipids 18, 22, 175
  freeze-drying 8
  freeze substitution 8
  heat 8
  reasons for 4
  washing after 20
Fixatives
  additive 9–11
  coagulant 10–11
  hardening by 4, 10
  immersion and perfusion 23–24
  individual agents described 10–19
  mixtures 21–23
  non-additive 9–11
  non-coagulant 10–11
  penetration 9
  properties 9–12, 20
  salts added to 20
  shrinkage in 9
  sucrose in 20
  swelling in 9
  unusual 19
Flavianic acid 52
Flavin nucleotides 223, 228
Flavoproteins
  autofluorescence of 2
  oxidoreductases 223, 238
Fluorescein
  isothiocyanate 67, 284
  labelled lectins 157–159
  mercuric acetate 125
  sodium (C.I. 45350) 67
Fluorescence
  antibody methods 283–288
  auto- 2
  definition 2, 39
  efficiency 41
  labelling of lectins 157–159
  method for zinc 197
  methods for amines 248–250, 252–254
  microscope 2
  of dyes 39, 40–41, 50
  secondary 2
Fluorescent brighteners 50, 59
Fluorescent brightening agent 32 (C.I. 40620) 59–60
Fluorochromes
  definition 39
  for immunofluorescence 284
  general chemistry 40–41
  nuclear staining 83–84
  nucleic acid staining 109
Food colours 50
Formal–acetic–alcohol 22
Formal–ammonium bromide 22
Formal–calcium 22
Formal–sucrose–ammonia 216
Formaldehyde
  chemistry 15
  deterioration 15
  fixation by 10–11, 16–17, 148
  –glutaraldehyde method for amines 249–250, 253
  -induced fluorescence of amines 248–249, 252
  lipids, effects on 17
  minimal fixation 228, 236
  neutral buffered 17, 22
  polymers 15
  proteins, reactions with 16–17
  solution (formalin) ix, 15, 16
Formazans 225, 226, 230
Formic acid
  decalcification 26
  formalin contaminant 15
Freeze-drying 8, 252
Freeze substitution 8, 195
Freezing microtome 1
Fuchsines *see* Acid fuchsine, Basic fuchsine, Magenta, Pararosaniline, Rosaniline
Fucose
  in glycoproteins 149
  lectin affinities 158
  periodate oxidation 154
  structure 146
Functional groups 114–143
  blocking reactions (table) 132
  in tissues (table) 115–116

Galactose
in glycolipids 170, 174
in glycoproteins 149
in keratan sulphates 149
lectin affinities 158
periodate oxidation 154
structure 146
Gallocyanine (C.I. 51030) 69
chromium complex 46, 105
Gangliosides 174, 180
Gegen-ion 39, 75, 258
Gelatine
adhesive for sections 35
embedding 29
film method for proteinases 212–213, 217–218
Gendre's fluid 21
Giemsa's stain 90–91
Globulin 283
see also Immunoglobulins
Globus' reagents 275
Glucose
concanavalin A method for 159, 165–166
in carminic acid 73
in glycoproteins 92, 149
in polysaccharides 148
lectin affinities 158
periodate oxidation 154
structure 146
Glucose-6-phosphate dehydrogenase 237–240
Glucuronic acid
in proteoglycans 148–149
structure 146
see also Uronic acids
Glutamate dehydrogenases 237–240
Glutaraldehyde
artifacts, histochemical 10–11, 17–18
fixative 10–11, 17–18
–formaldehyde method for amines 250, 253–254
–osmium method for NA 247, 251
solution for fixation 22
Glycans *see* Polysaccharides
Glycerol
in blood stains 90
in lipids 170
physical properties 33
Glycerol jelly 36
Glycerolphosphate dehydrogenases 237–240
Glycocalyx 149
Glycogen
amylase extraction 160, 166
chemistry 148
fixation of 12, 21
PAS staining 153–156
Glycolipids
chemistry 170, 174
classification 171
distinction from glycoproteins 150
extraction of 174, 176
histochemical identification 179–180, 189
PAS reaction 156, 180
resistance to saponification 176
solubility 174
Glycols
in lipids 179–180
in mucosubstances 146–150, 153–156
periodate oxidation 153–156
Glycoproteins
alcian blue staining 150–151, 162
colloidal ferric hydroxide staining 152
composition 149
definition 147
PAS staining 153–156, 163–164
Glycosaminoglycans *see* Proteoglycans
Glycosaminoglycuronans *see* Proteoglycans
Glycoside linkage
enzymatic hydrolysis 160–161
in carbohydrates 146
in carminic acid 73
Glycosphingosides 174, 176
Glyoxal-*bis*-(2-hydroxyanil) (GBHA)
chemistry 191–192
method for calcium 195–196
Glyoxylic acid
method 253
reaction with amines 249–250
Gold
chloride method 257
in method for astrocytes 264
–protein A complex 294
stabilization of Bandrowski's base 131
–sublimate method 275–276
toning 261–262
Golgi
apparatus 218
Camillo 262–263
–Cox method 263, 273–274
methods, nervous system 262–263
Gros–Schultze method for axons 271–272
Ground substance 92
Grübler, G. 47–48
Guanidine 75
Guanidino group
anionic dye binding 42, 80
blocking reactions 121–122, 142
histochemical reactions 120–122, 134–135
structure 121
Gum acacia (arabic) 37, 215
Gum dammar 36
Gum sandarac 36
Gum sucrose 215

H-acid 58, 128, 139
Haemalum, Mayer's 81
Haematal-16, Baker's 82
Haematein
acid–, for phospholipids 182
aluminium– 79

Haematein (*cont.*)
chemistry 61
deterioration 61
iron– 79–80, 83, 99
mordants 61
nuclear staining 79–80, 81–82
Haematoxylin (C.I. 75290)
and eosin method 81–82
chemistry 60–61
Heidenhain's 61, 75, 81
Mayer's haemalum 81
mordants for 61
oxidation of 61
Weigert's 75, 80, 99
see also Haematein
Haemoglobin
eosin staining 88
iron in 190, 192
peroxidase activity 234
Haemosiderin 192
Heidenhain's AZAN stain 69
Heidenhain's haematoxylin method 61, 75, 81
Heidenhain's "Susa" 21
Helindon yellow CG (C.I. 56005) 62
Helly's fluid 15, 21
Heparan sulphate 149
Heparin 149, 168
Heparin monosulphuric acid 149
Heparitin sulphate 149
Heteropolyacids 95–97
Hexazonium
pararosaniline 58
salts 57
Histamine
occurrence 246
OPT method for 247–248, 251–252
structure 245
Histidine 127
Histochemistry
amines 245–255
carbohydrates 145–168
definition 2
enzymes 201–244
immuno- 280–294
lipids 169–189
nucleic acids 104–113
organic functional groups 114–144
proteins 126–128, 138–139
suitable tissues for 301–303
Histone *see* Nucleoprotein
Holczinger's method for fatty acids 180, 185–186
Holmes' method for axons 268–269
Horseradish peroxidase *see* Peroxidase
Hyaluronic acid
chemistry 148
enzymatic extraction 160, 166
Hyaluronidases 160, 166
Hydration of sections 4
Hydrazides 129
detection of carbonyl groups 118, 129–130, 140
Hydrazines 129
blocking of carbonyl groups 132, 143
detection of carbonyl groups 129
in azo dye syntheses 53
Hydrazones 132
Hydrogen acceptors
artificial 225
definition 222
in cells 223
intermediate 229–230
Hydrogen bonding
chemistry 43
dye binding by 94, 99, 150
Hydrophilic lipids
classification 171
extraction by solvents 176, 188
lysochrome affinities 178
palladium chloride method 179, 184–185
staining methods 183–185
Hydrophobic interactions 44, 98
Hydrophobic lipids
classification 171
extraction by solvents 176–188
staining 178, 183, 184
Hydroquinone 220–221, 261
Hydroxyadipaldehyde 19
Hydroxyapatite 25
Hydroxyketone dyes 60–62
Hydroxyl group
histochemical reactions 117
in lipids 169, 170
in tissues 115, 117
see also Glycols
Hydroxylysine 92, 117, 156
Hydroxynaphthaldehyde method 119, 134
2-hydroxy-3-naphthoic acid hydrazide (HNAH)
detection of carbonyl groups 130, 140
in acid anhydride method 118, 133
8-hydroxy-1,4-naphthoquinone 232, 241
Hydroxyproline 117
8-hydroxyquinoline 193, 197
5-hydroxytryptamine (5HT) *see* Serotonin

Iduronic acid
in proteoglycans 148–149
structure 146
see also Uronic acids
Imines
aldehyde–fuchsine 98–99
from amine–aldehyde reactions 130
reduction of 129, 132
Immersion fixation 23
Immunization 283
Immunocytochemistry, definition 283
Immunofluorescence
definition 283
direct method 284–286

Immunofluorescence (*cont.*)
indirect methods 286–288
Immunoglobulins
properties 282
protein A binding 293
structure 281
Immunohistochemical methods 280–293
Indamine
chromophore 39
dyes 58, 231–232
Indigo (C.I. 73001) 71–72
Indigocarmine (C.I. 73015) 72
Indigogenic methods 210–211, 215–216
Indigoid
chromophore 72
dyes 71–72
histochemical reaction products 210
Indole 122
Indophenol
blue 231–232
dyes 58
Indoxyls 71, 210
Ingrain dyes 49
Inhibition
acid phosphatase 213
alkaline phosphatase 214, 218
carboxylic esterases 209–210, 214–215
choline esterases 209–210, 214–215
Ink blue (C.I. 42780) 65–66
Inoculation 283
Inositol
glycol groups in 180–181
in lipids 173
structure 170
Iodate
method for NA 247
oxidation of haematoxylin 61
Iodide–osmium methods 265, 277–278
Iodination 124, 142
Iodine–thiosulphate treatment 36
Iodoacetic acid 126, 142
Iron
extraction from tissue 192–193, 197
histochemical detection 192
Perls' method for 196–197
Iron, colloidal *see* Colloidal ferric hydroxide
Iron–haematoxylins 61, 75, 80–82, 99
Isodesmosine 93
Isopentane 33
Isoprene 170–171
Isopropanol 32

Keratan sulphates 149
Kernechtrot (C.I. 60760) 73, 80
Ketones
blocking reactions 131–132, 143
histochemical detection 128, 130, 140
in lipids 172

Label 283
Lactate dehydrogenase
action 223–224, 237
histochemical method 238
Lead
fixation by salts of 19
histochemical detection 194, 199
in tissues 194
method for acid phosphatase 206–207, 213
tetraacetate 154
Leather brown (C.I. 21000) 55
Lecithins *see* Phosphatidyl cholines
Lectins
concanavalin A technique 165–166
histochemical reagents 157–160
labelled 157–159
properties 157
specificity of binding 157, 158
table of 158–159
Leishman's stain 89–90
Leukocytes, tinctorial properties 89
Ligand 44
Light green SF (C.I. 42095) 65
Ligroin 33
Lillie's azure–eosin method 86–87
Limulin (*Limulus* agglutinin) 158
Lipase 201, 208, 209
Lipids
chemistry 169–174
definition 199
extraction by solvents 176, 188
fixation 18, 22, 175, 182
histochemical methods 174–188
hydrophilic and hydrophobic 171
in degenerating myelin 259
masked 188
storage diseases 180
Lipoamide dehydrogenase 227
Lipoproteins 174, 176, 188
Luxol fast blue
ARN 75, 259
G 259
MBS 75, 259
myelin staining 258–259, 266–267
LVN *see* Nitrocellulose
Lysochromes 49, 55
lipid staining 176–178, 183–184

Magenta II 64
Magenta III (C.I. 42520) 64
Magnesium
cofactor for dehydrogenases 240
protection of mitochondria 229
Malachite green (C.I. 42000) 105
Malaria parasites 89
Mannose
concanavalin A method for 159, 165–166
in glycoproteins 149
lectin affinities 158

Mannose (*cont.*)
  periodate oxidation 154
  structure 146
Mann's eosin–methyl blue 78, 81, 85
Marchi's method 259–260, 267–268
Martius yellow (C.I. 10315) 52
Mast cells
  dopamine in (ruminant lungs) 246
  fixation 21
  heparin in 149, 168
  histamine in 246, 248, 251
  serotonin in (rodents) 246
Mayer's albumen 34–35
Mayer's brazalum 83
Mayer's carmalum 83, 183
Mayer's haemalum 81
Melanin 233
Meldola's blue (C.I. 51175) 229–230, 239
Menadione 229–230
Menaphthone 75
Mepacrine 75
Merbromin 125, 137
Mercaptan *see* Sulphydryl groups
Mercuric chloride
  chemistry 13–14
  fixative 10–11, 13–14
  in Golgi–Cox method 263
  in method for astrocytes 264
  phenyl- 125
  precipitates in tissues 14, 36
  reaction with plasmalogens 181
  staining, effect on 14
  sulphydryl group reactions 13–14, 124–125
Mercurochrome 125
  method for cysteine 137
Mercury
  dithizone chelate 193
  orange 125
  organic compounds, —SH reagents 125
  removal of precipitates 36
Metachromasia 151–152
  acid mucosubstances 152
  nucleic acids 109
  staining method 162–163
Methanol
  –chloroform, lipid extraction 176, 188
  fixative 10–11, 12
  in blood stains 88, 89, 90
  in formalin 15
  physical properties 32
Methyl benzoate 32
3-methyl-2-benzothiazolone hydrazone (MBTH) 131
Methyl blue (C.I. 42780) 65–66
  –eosin, Mann's 78, 81, 85
  nuclear staining 78
Methyl green (C.I. 42585)
  nuclear staining 84–85
  purification of 109
  –pyronine method 105–106, 109–110
Methyl orange (C.I. 13025) 52
Methyl salicylate 32, 34
Methylation
  effect on sialic acids 161
  effect on sulphate esters 117, 161
  methods 141, 167
  of amino groups 120
  of carboxyl groups 118, 150
  of mucosubstances 150, 161, 167
  of sulphoamino compounds 120
Methylene azure (C.I. 52010) 70
Methylene blue (C.I. 52015) 70–71
  Borrel's 85
  eosinate 88–90
  in blood stains 88, 90
  new (C.I. 52030) 71
  polychrome, staining with 85
  polychroming of 70–71, 88
  structure 70
  vital, for nervous tissue 263–264, 274–275
Methylene green (C.I. 52020) 71
Methylene violet (Bernthsen) (C.I. 52041) 71, 88
Methylene violet RR (C.I. 50205) 71
Microtomes, types of 1, 2
Millon reaction 123, 136
Mitchell's rapid method for axons 269–270
Mitochondria
  acid–haematein method 182, 187–188
  cardiolipin in 173–174
  electron transport system 223
  fixation of 10–11, 23
  in presynaptic boutons 257
  protection of 229
Molality 300
Molarity 300
Molybdate, ammonium 76, 84, 263
Monosaccharides
  chemistry 145–146
  in lipids 170
  lectin affinities 158
  structural formulae 145–147
Mordant dyes 48
  myelin staining 258, 266
  nuclear staining 79–80, 81–83
Mordants
  chemistry of 44–46, 54
  definition 44
Morin (C.I. 75660) 190–191
Motor end-plates 257, 264, 276
Mounting media 2, 36–37
Mucopolysaccharides *see* Proteoglycans
Mucosubstances
  classification 147
  definition 145
  enzyme extractions 160–161, 166
  histochemistry 145–168
  terminology 147
  see also Glycoproteins, Polysaccharides,
    Proteoglycans

Myelin
acid–haematein method 182
chromoxane cyanine R method 258, 266
degenerating 259–260, 267–268
luxol fast blue method 258–259, 266–267
mordant dye methods 258
osmium tetroxide methods 178, 259–260
palladium chloride method 179
plasmal and pseudoplasmal reactions 189

NADI reactions 231–232, 235
Naphthazarin (C.I. 57010) 62
Naphthol green Y (C.I. 10005) 51
Naphthol yellows (C.I. 10315–16) 52
Naphthols
azoic couplers 58
in peroxidase method 235
naphthol-AS 58
oxidation of 231
Naphthoquinone dyes 72
1,2-naphthoquinone-4-sulphonic acid (NQS)
in method for cholesterol 181–182
reaction for arginine 121, 135
α-naphthyl phosphate 207, 213–214
National fast blue 8XM 74
Necolloidin *see* Nitrocellulose
Nedzel's paraffin artifact 189
Nervous tissue, methods for 256–278
Neuraminidase 160–161, 166
Neuroglia, methods 257, 264
Neurosecretory material (NSM)
fixation of 12, 265
methods for 264–265, 277
Neutral fats
chemistry 172
saponification 176
staining 176, 183
Neutral red (C.I. 50040) 68–69
Nissl staining 258, 266–267
nuclear staining 84
Neutral stains 88
New fuchsine (C.I. 42520) 64
New methylene blue (C.I. 52030) 71
Nicotinamide dinucleotides 223
Nile blue (C.I. 51180) 70, 189, 229
Nissl substance 258
Nitric acid
decalcifier 26
fixative 19, 27
precautions 26, 300
Nitro blue tetrazolium (nitro-BT)
methods using 236–240
properties 226
structure 225
Nitro group
chromophore 39
dyes 51–52
Nitrocellulose
coating slides with 35
double embedding 34
embedding method 31–34
properties and uses 2, 31–34
Nitrogen, liquid 33
Nitroso group
chromophore 39
dyes 51
in tyrosine histochemistry 123
Nitrous acid
deamination 120, 141
diazotization 52, 123
in rosindole reaction 122, 135
nitrosation 123
Non-specific esterases 208–211, 214–216
Noradrenaline (NA)
chromaffin reaction 247, 250
fluorescence methods 248–250, 252–254
glutaraldehyde–osmium method 247, 251
iodate reaction 247
occurrence 246
structure 245
Norepinephrine *see* Noradrenaline
Normality 299–300
Nuclear fast red (C.I. 60760) 73, 80
Nucleic acids
acetic acid fixation 10–11
chemistry 104–105
extraction by enzymes 107–108, 112
extraction by perchloric acid 108
extraction by TCA 12, 108
fixation of 10–11
methylation of 119, 132, 141
staining 77–78, 105–107
Nucleolus
RNA in 104
staining 85, 109–110
Nucleoprotein
chemistry 104
in nucleus 77
staining 79, 134
Nucleus
composition 77, 104
fixation of 10–11, 12
staining of 77–80, 81–85, 110–112

Oil red O (C.I. 26125) 55, 176
Oil yellow II (C.I. 11020) 52
Old yellow enzyme 228
Olefins *see* Unsaturation
Orange G (C.I. 16230) 53
Organophosphorus inhibitors
actions 209–210
methods of use 214–215
Osmium tetroxide
acidophilia, effect on 19
fixation by 10–11, 18–19
intensification of oxidized DAB 211–212, 217

Osmium tetroxide (*cont.*)
–iodide methods 265–266, 277–278
lipids, reaction with 18–19, 178
Marchi's method 259–260, 267–268
–α-naphthylamine (OTAN) method 178
penetration 19
phenols, reaction with 178, 247
properties 18, 184
proteins, effects on 18, 19
staining method 184, 259
vapour 19
Osmotic pressure of fixatives 20
Ouabain 202
Oxalate test for decalcification 27–28
Oxalic acid
extraction of iron 192, 197
reduction of gold toning 261–262
Oxazine dyes 69–70, 229
Oxidases
definition 224
histochemical detection 231–232, 241–242
Oxidation bases 50, 235
Oxidation number
carbon, in organic compounds 221
chromium 175
copper, in tissues 194
definition 45
iron, in tissues 192
osmium 18–19
Oxidation–reduction
chemistry 220–221
potentials 221–222, 223, 226–227
reactions in cells 222–224
Oxine 193
Oxyhaematein 61

Palladium chloride
method for hydrophilic lipids 179, 184–185
myelin staining 179, 184
protein binding 179
Paraffin *see* Waxes
Paraformaldehyde 15, 22
Pararosaniline (C.I. 42500)
chemistry 63–64
hexazonium derivative 58
in acid–alcohol 111
in aldehyde fuchsine 98–99, 102
in Schiff's reagent 110–111, 128–129
Parlodion *see* Nitrocellulose
Penetration
dyes 78, 93, 96
fixatives 10–11, 20
Peracetic acid
oxidation of cystine 124
preparation 138
Perchloric acid 135
extraction of RNA 108
in method for cholesterol 181
in rosindole reaction 122
–naphthoquinone reaction 186–187
Perenyi's fluid 27
Performic acid
oxidation of cystine 124
oxidation of olefins 178
preparation 137
–Schiff method 178–179
Perfusion fixation 23
Periodic acid
chemical reactions of 153–156
oxidation of glycols 153
oxidation of olefins 180
Periodic acid–Schiff method
artifacts with 156–157, 180
chemistry 152–157
techniques 163–165
Perls' method for iron 192, 196–197
Permanganate
oxidation of cystine 124, 138
oxidation of glycols 97, 100
Peroxidase
actions 233–234
complex with antibody (PAP) 289
confusion with catalase 236
confusion with cytochrome oxidase 235
distribution 234
histochemical methods 234–236, 242–243
horseradish 234, 235, 242
in lectin methods 159, 165–166
inhibition 236
-labelled antibody methods 288–289, 293
neuroanatomical tracing 257
Peroxidase–antiperoxidase (PAP)
structure 289
technique 289–292
use of protein A 293
Persulphate oxidation 123
tryptophan blockade 136
Petroleum ether 33
pH
azure–eosin mixtures 86
blood stains 89–91
effect on dyeing 42, 78
Phase contrast microscope 2
Phenazine methosulphate 229–230, 239
Phenol oxidase *see* Catechol oxidase
Phenolic benzene 34
Phenols
azo coupling 52, 127
histochemical reactions 123–124, 127, 246–247
occurrence in tissues 115, 246
reduction of osmium tetroxide 178, 247
p-phenylenediamine 131, 235
Phenylhydrazine 132, 143
Phenylhydrazine-4-sulphonic acid 129
Phenylmercuric chloride 125
N-phenyl-*p*-phenylenediamine 232, 241
Phloxine (C.I. 45405) 67

Phloxine B (C.I. 45410) 67
Phosphatases 206–208
  acid: technique 206–207, 213
  alkaline: technique 207–208, 213–214
  inhibitors 207
Phosphate
  detection in tissue 192
  von Kossa technique 196
Phosphatidyl
  cholines 173, 182
  ethanolamines 173, 182
  inositols 173
  serines 173
Phosphoglycerides 173–174
Phospholipase B 208, 209
Phospholipids
  chemistry 173–174
  classification 171
  fixation of 175
  histochemical methods 182–188
  solubility 173–174
Phosphomolybdic acid 95–97
Phosphoric acid derivatives
  cationic dye staining 78
  lipids 170, 173–174
  nucleic acids 104–105
Phosphotungstic acid 95–97
Photographic developers 260–261, 278
o-phthaldialdehyde 247–248, 251–252
Phthalocyanines 74–75
Physical development
  chemistry 260–261
  in method for axons 272–273
  in sulphide–silver method 278
Phytohaemagglutinins *see* Lectins
Picric acid
  chemistry 12–13, 51
  cytoplasmic staining 93–94
  dye (C.I. 10305) 51–52
  fixative 10–11, 12–13
  –indigocarmine, stain 72
  removal from tissues 13
Pigments 50
Pinacyanol 59
Plasmal reaction
  chemistry 180–181
  effect of fixation 17
  method 186
Plasmalogens
  chemistry 173, 181
  histochemical demonstration 180–181, 186
Platinic chloride, fixative 19
Polychrome, polychroming *see* Methylene blue
Polycondensation dyes 49
Polyphenol oxidase *see* Catechol oxidase
Polysaccharides 147–148
Polystyrene mounting medium 36
Polyvinyl alcohol (PVA)
  in incubation media 237
  protection of mitochondria 229
Polyvinylpyrollidone (PVP)
  in incubation media 237
  mounting medium 37
  protection of mitochondria 229
Precipitates, mercury 14, 36
Primuline (C.I. 49000) 60
Processing tissues 29–34
Procion brilliant red M2B (C.I. 18158) 55, 218
Procion yellow 4MR 55, 257
Propylene glycol 33
Prosthetic group
  definition 222
  flavin nucleotides 223, 228
Protargol 262
Protein
  A (staphylococcal) 292–293, 294
  coupled tetrazonium reaction 127–128, 139
  dye methods for 126–127, 138–139
  fixation of 9–20, 115–116
  histochemistry 114–144
  in lipids 171, 174
  in mucosubstances 146, 148–149
  Millon reaction 123, 136
  serum 149
Proteinases 212–213, 217–218
Proteoglycans
  alcian blue staining 150–151, 162
  classification 148–149
  colloidal ferric hydroxide for 152, 163
  definition 147
  fixation of 148
  metachromasia of 151–152
  toluidine blue staining 162–163
Proteolipids 174
Protozoa in blood 88, 89
Prussian blue
  chemistry 125
  method for iron 192, 196–197
Pseudocholinesterase *see* Cholinesterase
Pseudoisocyanines 59
Pseudoplasmal reaction 180, 181
Pseudo-Schiff reagents 129
Pyrazine ring 68
Pyridine
  lipid extraction 176, 182, 188
  palladium complexes 179
Pyronine B (C.I. 45010) 66
Pyronine Y (=G) (C.I. 45005) 66
  stain for RNA 105–106, 109–110

Quinoline dyes 60
8-quinolinol 193
Quinone–imine chromophore 40
Quinones
  fixatives 19
  in azo dye syntheses 53
  oxidation–reduction reactions 220–221

Quinonoid chromophore 39
Quinophthalone 60

Radiography in decalcification 28
Reactive auxochromes 40, 43, 55
Reactive dyes 49, 55, 62
Redox potential *see* Oxidation–reduction
Refractive index
  contrast due to 2
  mounting media 2, 36, 37
  solvents 32–33
Reinecke salt 255
Resins 2, 36
Resonance 39
Reticulin
  chemistry 92
  silver methods for 97–98
  staining technique 100–102
Rhodamine B (C.I. 45170) 66–67
  fluorescent label 284, 285
  isothiocyanate 67
Rhodizonic acid
  full name 62
  metal chelation 194
  method for lead 199
Ribonuclease 108, 112
Ribonucleic acid
  chemistry 104–105
  extraction by perchloric acid 108
  extraction by RNase 108, 112
  extraction by TCA 13, 108
  hydrolysis by acids 106–107
  methyl green–pyronine method for 105–106, 109–110
  staining by dyes 105–106, 108–110
Rinsing sections 6
Rosaniline (C.I. 42510) 64
Rose Bengal (C.I. 45440) 67
Rosindole reaction 122–123, 135–136
Rubeanic acid *see* Dithiooxamide

Safety: general precautions ix
Safranine O (=T) (C.I. 50240) 69
  diazonium salt 58
  nuclear staining 84
  with alcian blue 151
Sakaguchi reaction 121, 134–135
Salicin black EAG (C.I. 15710) 54
Salicylhydrazide 130
Saline 283
Sandwich technique 292
Saponification
  before neuraminidase treatment 160–161
  methods 141, 167, 188
  of acylated amines 120
  of acylated hydroxyls 117
  of lipids 176, 188
  of methyl esters 119, 150, 161
  of sialic acids 155–156
Schiff's reagent
  chemistry 128–129
  in Feulgen method 106
  in PAS method 156
  in plasmal reaction 181
  preparation 110–111
  substitute 111
  washing after 111, 164, 186
Sections
  for enzyme histochemistry 204–205
  handling of 5–7
  types of 1, 2
Serine 117, 170
  in lipids 173
  reaction with DFP and E600 209
Serotonin (5HT)
  azo-coupling method 246, 250
  fluorescence methods 248–250, 252, 253–254
  occurrence 246
  structure 245
Serum
  anti- 283
  definition 283
  proteins 149, 159
Sialic acids
  acetylated 154–156, 164–165
  aprotinin affinity 158
  chemistry 147, 155
  enzyme extraction 160–161, 166
  hydrolysis 162, 167
  in lipids 170, 174
  limulin affinity 158
  PAS method for 164–165
  periodate oxidation 154–156
Sialidase *see* Neuraminidase
Silver methods
  aldehyde detection 130–131
  ascorbic acid 254
  axons 260–262, 268–273
  chloride 199
  Golgi (dichromate–silver) 262–263
  neuroglia 257
  phosphates and carbonates 192, 196
  reticulin 97–98, 100–102
  Timm's sulphide–silver 278
  unsaturated lipids 179
Sirius light turquoise blue GL (C.I. 74180) 74
Softening
  bone and teeth 25
  hard tissues 4, 27
Solochrome black A (C.I. 15710) 54
Solochrome cyanine R (C.I. 43820) 66
Solochrome orange GRS (C.I. 26520) 54
Soluble blue (C.I. 42780) 65–66
Solvent dyes 49, 55
  lipid staining 176, 183–184

Solvents
  aprotic 125
  for lipids 176, 188
  properties of 32–33
Sphingomyelins
  chemistry 174
  histochemical detection 182
  resistance to saponification 176
  solubility 174
Sphingosines 170, 174
Spirit blue (C.I. 42775) 66
Squalene 171
Staining
  collagen and cytoplasm 93–97
  effects of fixatives on 10–11
  elastin 97–98
  general instructions 5–6
  nuclei 77–78, 81–85
  nucleic acids 105–107, 108–112
  reasons for 2
  reticulin 97–98
  vital 263–264
  see also Dyes
Stains-all 59
Staphylococcal protein A 292–293, 294
Starch 148
Steroids 172
Stilbene dyes 59
Sublimate, corrosive *see* Mercuric chloride
Substrate film methods
  for DNase 218
  for proteinases 212–213, 217–218
  general considerations 203
Succinate dehydrogenase 224, 228, 236–237
Sucrose
  in fixatives 20, 216
  in gum sucrose 215
  inhibition of con A 165
Sudan III (C.I. 26100) 176
Sudan IV (C.I. 26105) 55, 176, 183
Sudan black B (C.I. 26150) 176–177, 183–184
Sugar *see* Monosaccharides, Sucrose
Sulphate ester groups
  alcian blue staining 150–151, 162
  effect on periodate oxidation 154
  from hydroxyls 117
  in lipids 174
  of mucosubstances 150–152
  removal of 117, 141, 161–162, 167
  see also Proteoglycans
Sulphatides 174, 180, 189
Sulphation
  –acetylation method 140–141
  -induced basophilia 132
  of amino groups 120
  of carbohydrates 167
  of hydroxyl groups 117
Sulphide–silver method 278
Sulphoamino groups
  from amino 120
  in carbohydrates 147, 149
  removal of 117, 141, 167
Sulphur dyes 49
Sulphuric acid
  fixation by 19
  precautions 300
  sulphation with 132–133
Sulphydryl groups
  blocking reactions 126, 142
  chemistry 124–125
  histochemical detection 125, 136–137
  iodide–osmium reaction 265
  production from cystine 124, 142
Susa 21

Tannic acid as fixative 19
Terpenes 171
Terpineol
  method of use 30–31
  physical properties 32
Tetracycline 84
  aluminium chelate, stain 84
Tetrahydrofuran 33
Tetrahydrofurfuryl alcohol 33, 251
Tetramethylbenzidine (TMB) 235
Tetramethylene oxide 33
Tetranitromethane 124
Tetrazolium reductases *see* Diaphorases
Tetrazolium salts
  chemistry 225
  histochemical uses 227–228
  in incubation media 229
  table of 226
Tetrazonium
  reaction, coupled 127–128, 139
  salts 57, 127–128
Thiamine pyrophosphatase 218
Thiazine dyes 70–71
  eosinates of 81, 86, 88–90
  for sulphation-induced basophilia 133
  in method for cystine 125
  nucleic acid staining 108
Thiazole dyes 60
Thiocholine methods
  chemistry 211–212
  technique 216–217
Thioflavine TCN (C.I. 49005) 60
Thioglycollate 124–125, 142
Thioindigoid dyes 72
Thiol *see* Sulphydryl groups
Thionine (C.I. 52000) 70, 258
Thionine blue (C.I. 52025) 71
Thionolines 71
Threonine 117
Timm's sulphide–silver method 278
Toluene 32
p-toluenesulphonic acid as fixative 19

Toluidine blue O (C.I. 52040) 71
in method for aldehydes 131
metachromatic staining, method 162–163
Nissl staining 258
nuclear staining 84
Triarylmethane dyes 63–66
Triazenes 127, 128
Trichloroacetic acid (TCA)
extraction of nucleic acids 13, 108
fixative 13
Trichrome methods
Cason's technique 100
theory 94–97
Triolein 189
Triphenylmethane dyes 63–66
Tristearin 188
Trypaflavine (C.I. 46000) 68
Tryptophan
blocking reactions 123
histochemical reactions 122–123, 135–136
in tissues 122
Turnbull's blue
chemistry 125
method for iron 192
Tyrosinase *see* Catechol oxidase
Tyrosine
blocking reactions 124, 142
histochemical reactions 123–124, 136
in tissues 115, 123

Ubiquinone
in biological oxidations 223
in histochemical methods 228
UDPG dehydrogenase 237–240
Unsaturation
atmospheric oxidation 180
blocking reactions 179, 188
bromination 179
histochemical detection 177–179, 184–185
in fatty acids 169, 177–178
in lipids 171–174, 175
in sphingosine 170
osmium tetroxide methods 178
palladium chloride reduction 179
performic acid oxidation 180
pseudoplasmal reaction 180, 181
Uranin (C.I. 45350) 67
Uranyl salts as fixatives 19
Uronic acids
alcian blue staining 150–151, 162
aprotinin affinity 158
in proteoglycans 148–149
PAS method for 165
periodate oxidation 154
structure 146
see also Proteoglycans

van der Waals forces 44
in elastin staining 98
in nuclear staining 79
van Gieson's method 93–94, 99–100
Vat dyes 49, 71, 72
Vesuvin (C.I. 21000) 55
Vibrating microtome (Vibratome) 1
Victoria blue B (C.I. 44045) 65
Victoria blue R (C.I. 44040) 65
Victoria blue 4R (C.I. 42563) 65
von Kossa method 192, 196

Warburg's old yellow enzyme 228
Washing
fixed blocks 20
sections 6
Water
properties 2–3, 33
purity of x
Water blue I (C.I. 42755) 65–66
Waxes
embedding methods 30, 34
lipids 172
paraffin 1, 2, 189
polyester 1
Weigert's iron–haematoxylin 75, 80, 99
Winkelmann and Schmit silver method 270–271
Wright's stain 89–90

Xam, mounting medium 2
Xanthene dyes 66–68
X-rays in decalcification 28
Xylene 2–3, 33
Xylose
in mucosubstances 146, 148–149
structure 146

Zenker's fluid 22
Zinc
histochemical detection 193, 197–198
in tissues 193
iodide–osmium tetroxide 265–266, 277–278